FOURTH EDITION

Solutions to IRODOV's PROBLEMS IN GENERAL PHYSICS

VOLUME I

AF653311

FOURTH EDITION

Solutions to IRODOV's PROBLEMS IN GENERAL PHYSICS

VOLUME I

ABHAY KUMAR SINGH
Director
Abhay's IIT Physics Teaching Centre
Patna

WILEY

Solutions to

IRODOV's

PROBLEMS IN GENERAL

PHYSICS

VOLUME I

Copyright © 2014 by Wiley India Pvt. Ltd., 4436/7, Ansari Road, Daryaganj, New Delhi-110002.

All rights reserved. No part of this book may be reproduced, stored in a retrieval system, or transmitted in any form or by any means, electronic, mechanical, photocopying, recording or scanning without the written permission of the publisher. All rights remain with respective holders.

Limits of Liability: While the publisher and the author have used their best efforts in preparing this book, Wiley and the author make no representation or warranties with respect to the accuracy or completeness of the contents of this book, and specifically disclaim any implied warranties of merchantability or fitness for any particular purpose. There are no warranties which extend beyond the descriptions contained in this paragraph. No warranty may be created or extended by sales representatives or written sales materials. The accuracy and completeness of the information provided herein and the opinions stated herein are not guaranteed or warranted to produce any particular results, and the advice and strategies contained herein may not be suitable for every individual. Neither Wiley India nor the author shall be liable for any loss of profit or any other commercial damages, including but not limited to special, incidental, consequential, or other damages.

Disclaimer: The contents of this book have been checked for accuracy. Since deviations cannot be precluded entirely, Wiley or its author cannot guarantee full agreement. As the book is intended for educational purpose, Wiley or its author shall not be responsible for any errors, omissions or damages arising out of the use of the information contained in the book. This publication is designed to provide accurate and authoritative information with regard to the subject matter covered. It is sold on the understanding that the Publisher is not engaged in rendering professional services.

Trademarks: All brand names and product names used in this book are trademarks, registered trademarks, or trade names of their respective holders. Wiley is not associated with any product or vendor mentioned in this book.

Other Wiley Editorial Offices:

John Wiley & Sons, Inc. 111 River Street, Hoboken, NJ 07030, USA

Wiley-VCH Verlag GmbH, Pappellaee 3, D-69469 Weinheim, Germany

John Wiley & Sons Australia Ltd, 42 McDougall Street, Milton, Queensland 4064, Australia

John Wiley & Sons (Asia) Pte Ltd, 1 Fusionopolis Walk #07-01 Solaris, South Tower, Singapore 138628

John Wiley & Sons Canada Ltd, 22 Worcester Road, Etobicoke, Ontario, Canada, M9W 1L1

Third Edition: 2010

Fourth Edition: 2014

Reprint: 2024

ISBN: 978-81-265-5118-7

www.wileyindia.com

Printed at: Yash Printographics

Dedicated to

My teacher
Prof. (Dr.) J. Thakur
(Ex V.C. and HOD Department of Physics,
Patna University)

PREFACE TO THE FOURTH EDITION

It is my privilege to bring to you the fourth edition of SOLUTIONS TO IRODOV'S PROBLEMS IN GENERAL PHYSICS. This edition is an enhanced version of the third edition, with a few newly added alternate solutions, and about 11 modified solutions, better diagrams to illustrate the solutions, and crisper explanations. I have also changed some solutions altogether, with the view that each problem has many methods of solving, but only one, or maximum two of them are precise and fitting.

To ensure that all solutions in this edition are concise as well as most compact, I have tried to solve each problem in minimum possible steps. Besides this, I did not want to spoon-feed the students, but to make them think and apply their intelligence. So, I have tried to present the solution in a manner which will challenge their knowledge of concepts of physics. I chose this method because I recognize that the students of modern India need to be armed with a strong knowledge of concepts to be able to compete with the best physicists in the world. And they will only reach this standard if they are proficient at problem-solving and not afraid of challenges.

Similar to the previous editions, as you know, the first volume includes the first three parts, containing 1052 problems, and this second volume contains the remaining parts and 826 problems.

I conclude here with the belief that nothing succeeds like honest, hard work and conviction in one's capability. I extend my deep gratitude and heartfelt wishes to all those who have directly or indirectly contributed to not only the creation of this edition, but also those who have shown faith in the authenticity and usefulness of the solutions. I would like to mention especially the editorial team at Wiley India, who were dedicated throughout to making this edition the best.

Finally, my best wishes to all my students and students throughout the country who will be helped by this book.

You can email me at: **authorakssa@gmail.com**

Abhay Kumar Singh
Mata Khudi lane
Mahendru
Patna-6
Aug 2014

PREFACE TO THE FIRST EDITION

When you envisage to write a book of solutions to problems, one pertinent question crops up in the mind that—why solution! Is this to prove one's erudition? My only defence against this is that the solution is a challenge to save the scientific man hours by channelizing thoughts in a right direction.

The book entitled "Problems in General Physics" authored by I.E. Irodov (a noted Russian physicist and mathematician) contains 1878 intriguing problems divided into six chapters.

After the acceptance of my first book "Problems in Physics", published by Wiley Eastern Limited, I have got the courage to acknowledge the fact that good and honest ultimately win in the market place. This stimulation provided me insight to come up with my second attempt—"Solutions to I.E. Irodov's Problems in General Physics."

The first volume encompasses solutions of first three chapters containing 1052 problems. Although a large number of problems can be solved by different methods, I have adopted standard methods and in many of the problems with helping hints for other methods.

In the solutions of chapter three, the emf of cell is represented by (xi) in contrast to the notation used in figures and in the problem books, due to some printing difficulty.

I am thankful to my students Mr. Om Prakash, Miss Neera and Miss Punam for their valuable co-operation even in my hard days while authoring the present book. I am also thankful to my younger sister Prof. Ranju Singh, my younger brother Adv. Ratan Kumar Singh, other well wishers and friends for their emotional support. At last and above all I am grateful to my Ma and Pappaji for their blessings and encouragement.

Abhay Kumar Singh

ABOUT THE AUTHOR

Although, Mr. Abhay Kumar Singh, studied both Physics and Statistics upto the postgraduate level, he chose Physics as his area of expertise due to his interest and an aptitude for the subject. He started his career as a teacher in June, 1984, guiding the IIT-JEE aspirants. His teaching methods have always been innovative and intellectual, helping the students in understanding the concepts and developing a genuine liking for the subject in the long run.

His first book titled as "Problem in Physics" was published in March, 1994. Even though he was a novice then, the hard work and efforts of the young author fulfilled a long felt need of the students in Physics. One of the distinctive features of this book is that it presents to its readers, various alternative methods to solve a problem, rather than focussing on the traditional solution, thus clarifying the concepts by involving the use of the reader's grey cells in every problem. He had the option of releasing his book inside the lavish premises of Raj Bhawan by the then Governor of Bihar, yet he chose his alma-mater, the renowned and prestigious Science College, Patna as the venue for book release as a mark of tribute to all the knowledge the place had given him.

In June 1995 Mr. Singh come up with his second book "Solutions to Irodov's Problems in General Physics" in two volumes, which created history. His books helpful to students as well as faculty have been out in the market for two decades now. The response of his readers and the enthusiasm of his juniors to these books has initiated him to work on a theory book, to be published soon. To those new in this field, he might be just a name, but for most he is a paragon in the field of teaching physics.

FOREWORD

Science, in general, and physics, in particular, have evolved out of man's quest to know beyond unknowns. Matter, radiation and their mutual interactions are basically studied in physics. Essentially, this is an experimental science. By observing appropriate phenomena in nature one arrives at a set of rules which goes to establish some basic fundamental concepts. Entire physics rests on them. Mere knowledge of them is however not enough. Ability to apply them to real day-to-day problems is required. Prof. Irodov's book contains one such set of numerical exercises spread over a wide spectrum of physical disciplines. Some of the problems of the book long appeared to be notorious to pose serious challenges to students as well as to their teachers. This book by Prof. Singh on the solutions to problems of Irodov's book, at the outset, seems to remove the sense of awe which at one time prevailed. Traditionally a difficult exercise to solve continues to draw the attention of concerned persons over a sufficiently long time. Once a logical solution for it becomes available, the difficulties associated with its solutions are forgotten very soon. This statement is not only valid for the solutions of simple physical problems but also to various physical phenomena.

Nevertheless, Prof. Singh's attempt to write a book of this magnitude deserves an all out praise. His ways of solving problems are elegant, straight forward, simple and direct. By writing this book he has definitely contributed to the cause of physics education. A word of advice to its users is however necessary. The solution to a particular problem as given in this book is never to be consulted unless an all out effort in solving it independently has been already made. Only by such judicious uses of this book one would be able to reap better benefits out of it.

As a teacher who has taught physics and who has been in touch with physics curricula at IIT, Delhi for over thirty years, I earnestly feel that this book will certainly be of benefit to younger students in their formative years.

Dr. Dilip Kumar Roy
Professor of Physics
IIT, Delhi
New Delhi

CONTENTS

PHYSICAL FUNDAMENTALS OF MECHANICS

1

1.1 Kinematics

1.1 Let v_0 be the stream velocity and v' the velocity of motorboat with respect to water. The motorboat reached point B while going downstream with velocity $(v_0 + v')$ and then returned with velocity $(v' - v_0)$ and passed the raft at point C. Let t be the time for the raft (which flows with stream with velocity v_0) to move from point A to C, during which the motorboat moves from A to B and then from B to C.

Therefore,

$$t = \frac{l}{v_0} = \tau + \frac{(v_0 + v')\tau - l}{(v' - v_0)}$$

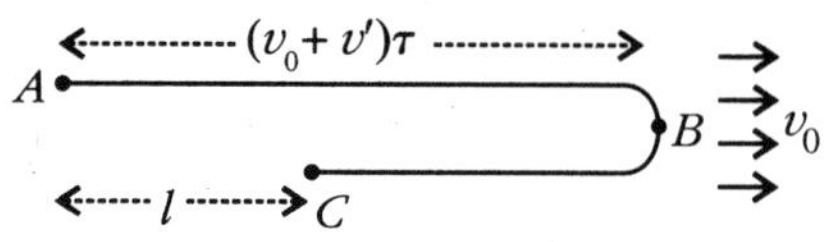

On solving, we get

$$v_0 = \frac{l}{2\tau} = 3 \text{ km/h} \quad \text{(on substituting values)}$$

Alternate:

In the frame of reference moving with the stream velocity the raft becomes stationary. The boat travels for time τ and turns and then travels back to reach the raft. As the raft is still at its original position, hence it would take exactly further time τ (the same time interval) for the boat to reach the raft. The boat thus takes time 2τ to reach the raft in the frame of stream. In Newtonian mechanics time interval between two events is frame independent, so 2τ is also the time interval in the frame of river bank. In the mean time the raft has travelled l. Hence the speed v_0 of the river is

$$v_0 = \frac{l}{2\tau} = 3 \text{ km/h}$$

1.2 Let s be the total distance traversed by the point and t_1 the time taken to cover half the distance. Further let $2t$ be the time to cover the rest half of the distance.

Therefore, $$\frac{s}{2} = v_0 t_1 \quad \text{or} \quad t_1 = \frac{s}{2v_0}$$

and $$\frac{s}{2} = (v_1 + v_2)t \quad \text{or} \quad 2t = \frac{s}{v_1 + v_2}$$

Hence the sought average velocity is given by

$$\langle v \rangle = \frac{s}{t_1 + 2t} = \frac{s}{[s/2v_0] + [s/(v_1 + v_2)]} = \frac{2v_0(v_1 + v_2)}{v_1 + v_2 + 2v_0}$$

1.3 As the car starts from rest and finally comes to a stop, and the rate of acceleration and deceleration are equal, the distances covered as well as the times taken are same in these phases of motion.

Let Δt be the time for which the car moves uniformly. Then the acceleration/deceleration time is $(\tau - \Delta t)/2$ each. So,

$$<v>\,\tau = 2\left\{\frac{1}{2}w\frac{(\tau-\Delta t)^2}{4}\right\} + w\frac{(\tau-\Delta t)}{2}\Delta t$$

or $$\Delta t^2 = \tau^2 - \frac{4<v>\tau}{w}$$

Hence, $$\Delta t = \tau\sqrt{1-\frac{4<v>}{w\tau}} = 15 \text{ s}$$

1.4 (a) Sought average velocity

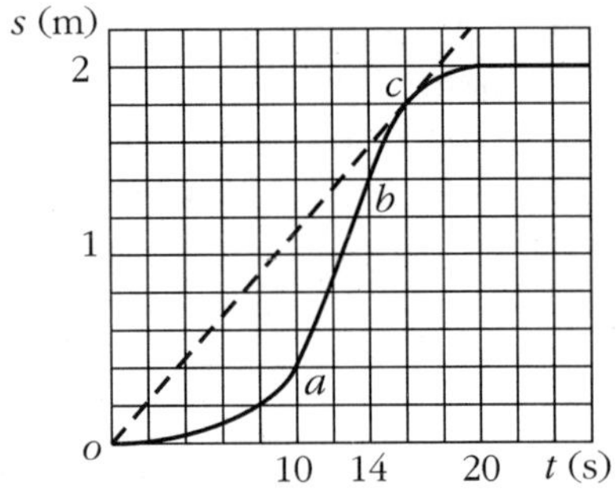

$$<v> = \frac{s}{t} = \frac{200 \text{ cm}}{20 \text{ s}} = 10 \text{ cm/s}$$

(b) For the maximum velocity, ds/dt should be maximum. From the figure ds/dt is maximum for all points on the line ab thus the sought maximum velocity becomes average velocity for the line ab and is

$$= \frac{100 \text{ cm}}{4 \text{ s}} = 25 \text{ cm/s}$$

(c) Time t_0 should be such that corresponding to it the slope ds/dt should coincide with the chord oc, to satisfy the relationship $ds/dt = s/t_0$. From figure the tangent at point c passes through the origin and thus corresponding time $t = t_0 = 16$ s.

1.5 Let the particles collide at the point A (see figure), whose position vector is $\mathbf{r}_3$ (say). If t be the time taken by each particle to reach point A, then from triangle law of vector addition:

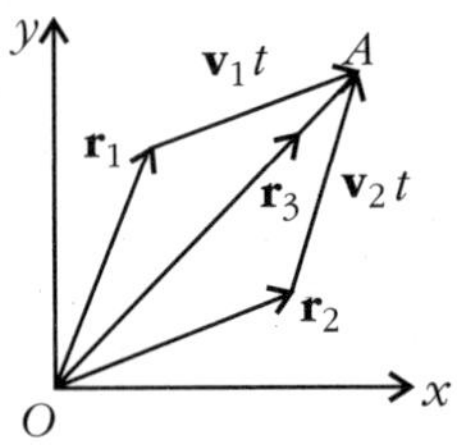

$$\mathbf{r}_3 = \mathbf{r}_1 + \mathbf{v}_1 t = \mathbf{r}_2 + \mathbf{v}_2 t$$

So, $$\mathbf{r}_1 - \mathbf{r}_2 = (\mathbf{v}_2 - \mathbf{v}_1)t \qquad (1)$$

Therefore, $$t = \frac{|\mathbf{r}_1 - \mathbf{r}_2|}{|\mathbf{v}_2 - \mathbf{v}_1|} \qquad (2)$$

From Eqs. (1) and (2)

$$\mathbf{r}_1 - \mathbf{r}_2 = (\mathbf{v}_2 - \mathbf{v}_1)t$$

Thus, $$\frac{\mathbf{r}_1 - \mathbf{r}_2}{|\mathbf{r}_1 - \mathbf{r}_2|} = \frac{\mathbf{v}_2 - \mathbf{v}_1}{|\mathbf{v}_2 - \mathbf{v}_1|},$$ which is the sought relationship.

1.6 We have

$$\mathbf{v}' = \mathbf{v} - \mathbf{v}_0$$

In the vector diagram for above relation, let us drop a dotted perpendicular from the tip of $\mathbf{v}$ onto the line of action of $\mathbf{v}_0$. Using the property of right angle triangle,

$$v' = \sqrt{(v\cos\varphi + v_0)^2 + (v\sin\varphi)^2} = \sqrt{v_0^2 + v^2 + 2v_0 v\cos\varphi}$$

$$= 39.7 \text{ km/h} \quad \text{(on substituting values)}$$

and $$\tan \varphi' = \frac{v \sin \varphi}{v \cos \varphi + v_0}$$

On substitution $$\varphi' = 19.1°$$

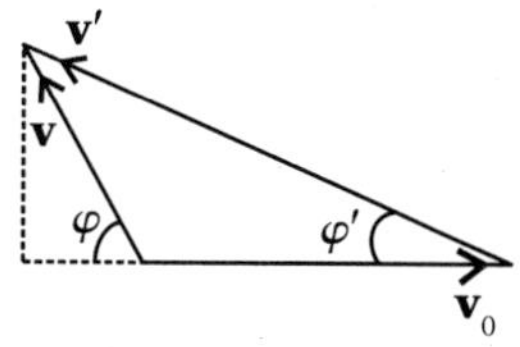

1.7 Let us suppose one of the swimmers (say 1) crosses the river along AB, which is obviously the shortest path. Time taken to cross the river by the swimmer 1

$$t_1 = \frac{d}{\sqrt{v'^2 - v_0^2}} \quad \text{(where } AB = d \text{ is the width of the river)} \tag{1}$$

For the other swimmer (say 2), who follows the quickest path, the time taken to cross the river,

$$t_2 = \frac{d}{v'} \tag{2}$$

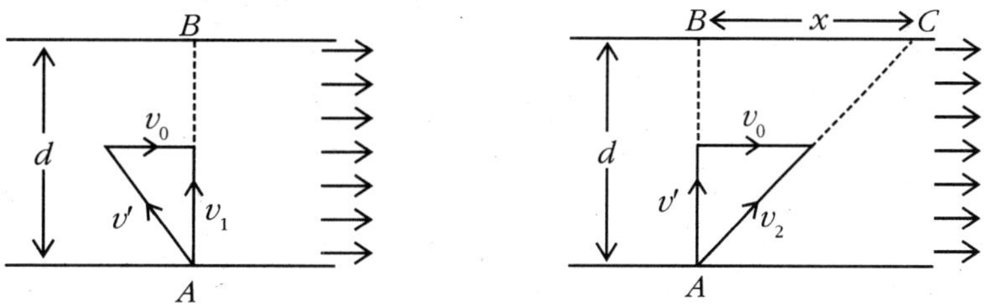

In the time t_2, drifting of the swimmer 2, becomes

$$x = v_0 t_2 = \frac{v_0}{v'} d \quad \text{(using Eq. 2)} \tag{3}$$

If t_3 be the time for swimmer 2 to walk the distance x to come C to B (see figure), then

$$t_3 = \frac{x}{u} = \frac{v_0 d}{v' u} \quad \text{(using Eq. 3)} \tag{4}$$

According to the problem $t_1 = t_2 + t_3$

or $$\frac{d}{\sqrt{v'^2 - v_0^2}} = \frac{d}{v'} + \frac{v_0 d}{v' u}$$

$\Rightarrow$ $$\frac{v_0}{u} = \frac{v'}{v'^2 - v_0^2} - 1$$

Thus, $$u = \frac{v_0}{(1 - v_0^2/v'^2)^{-1/2} - 1} = 3 \text{ km/h}$$

1.8 Let l be the distance covered by the boat A along the river as well as by the boat B across the river. Let v_0 be the stream velocity and v' the velocity of each boat with respect to water. Therefore time taken by the boat A in its journey

$$t_A = \frac{l}{v' + v_0} + \frac{l}{v' - v_0} = \frac{2lv'}{v'^2 - v_0^2}$$

and for the boat B

$$t_B = \frac{l}{\sqrt{v'^2 - v_0^2}} + \frac{l}{\sqrt{v'^2 - v_0^2}} = \frac{2l}{\sqrt{v'^2 - v_0^2}}$$

Hence,

$$\frac{t_A}{t_B} = \frac{v'}{\sqrt{v'^2 - v_0^2}} = \frac{\eta}{\sqrt{\eta^2 - 1}} \quad \left(\text{where } \eta = \frac{v'}{v_0}\right)$$

On substitution,

$$\frac{t_A}{t_B} = 1.8$$

1.9 Let v_0 be the stream velocity and v' the velocity of boat with respect to water. As $v_0/v' = n = 2 > 0$, some drifting of the boat is inevitable.

Let v' make an angle θ with flow direction (see figure), then the time taken to cross the river

$$t = \frac{d}{v' \sin\theta} \quad \text{(where } d \text{ is the width of the river)}$$

In this time interval, the drifting of the boat

$$x = (v' \cos\theta + v_0)t$$

$$= (v' \cos\theta + v_0)\frac{d}{v' \sin\theta} = (\cot\theta + \eta \operatorname{cosec}\theta)d$$

For x_{min} (minimum drifting)

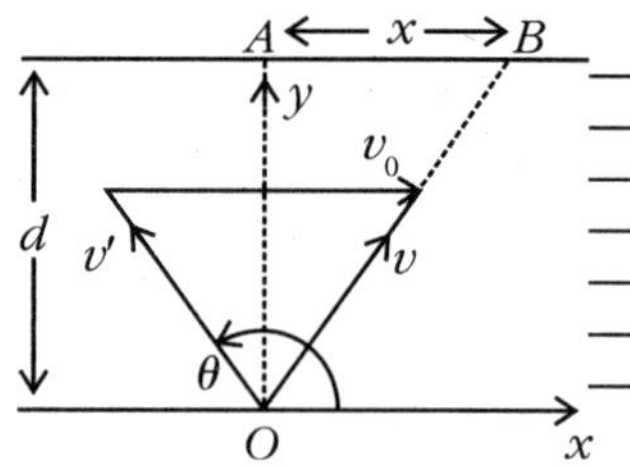

$$\frac{d}{d\theta}(\cot\theta + \eta \operatorname{cosec}\theta) = 0, \text{ which yields}$$

$$\cos\theta = -\frac{1}{n} = -\frac{1}{2}$$

Hence,

$$\theta = 120°$$

Alternate:

Let $\mathbf{v}_0$ be the stream velocity, $\mathbf{v}'$ the velocity of boat with respect to water and $\mathbf{v}$ be the resultant velocity of boat, i.e., $\mathbf{v} = \mathbf{v}_0 + \mathbf{v}'$.

The angle θ from the direction of stream at which boat can be rowed lies in the interval 0 to $\pi (0 \le \theta \le \pi)$. Now let us draw the vector diagram of velocity vectors. In the figure, a semicircle has been drawn whose radius is v'. The tip of vector $\mathbf{v}'$ lies on this semicircle. One can observe that for minimum drifting, angle α should be maximum and the line AB representing vector $\mathbf{v}$ becomes tangent to the semicircle so that line OB representing $\mathbf{v}'$ becomes perpendicular to it. The minimum drifting is CD.

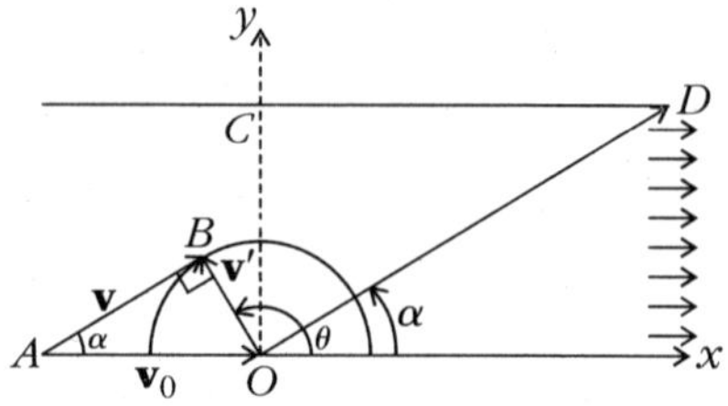

From the figure

$$\sin\alpha = \frac{OB}{AO} = \frac{v'}{v_0} = \frac{1}{n} \tag{1}$$

$$\theta = \frac{\pi}{2} + \alpha \quad \text{(see figure)}$$

Hence,

$$\theta = \sin^{-1}(1/n) + \pi/2 = 120° \quad \text{(on substituting values)}$$

In the triangle OCD

$$\sin\alpha = \frac{OC}{OD} \quad (2)$$

From Eqs. (1) and (2)

$$\frac{OC}{OD} = \frac{1}{n}$$

or $$OD = n(OC)$$

So, $$CD = \sqrt{(OD)^2 - (OC)^2} = OC\sqrt{n^2 - 1}$$

Hence, $$x_{\min} = d\sqrt{n^2 - 1}$$

1.10 Let us suppose one of the bodies, say 1, is thrown upwards and the other body (say 2) at an angle θ to the horizontal. Indicate y-axis vertically upward and x-axis along horizontal. Let after the time interval t, the particles 1 and 2 have coordinates $(0, y_1)$ and (x_2, y_2)

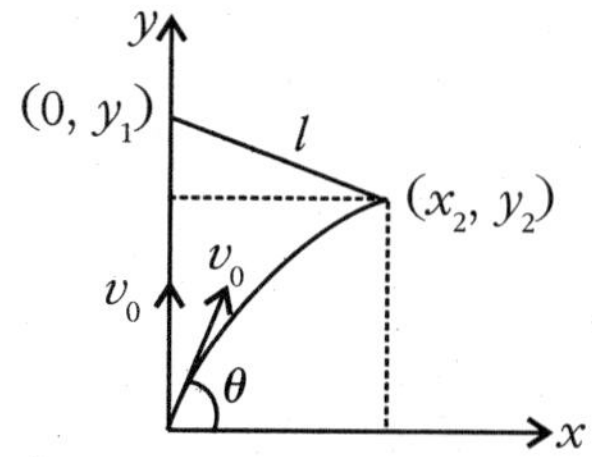

For body 1,

$$y_1 = v_0 t - \frac{1}{2}gt^2$$

and for body 2,

$$x_2 = v_0\cos\theta t \text{ and } y_2 = v_0\sin\theta t - \frac{1}{2}gt^2$$

Thus the sought separation between the particles 1 and 2 after the time interval t

$$l = \sqrt{x_2^2 + (y_1 - y_2)^2} = \sqrt{(v_0\cos\theta t)^2 + \{v_0 t(1 - \sin\theta)^2\}}$$

$$= v_0 t\sqrt{2(1 - \sin\theta)}$$

$= 22$ m, on putting the values of v_0, t and θ

Alternate:

The solution of this problem becomes interesting in the frame attached with one of the bodies. Let the body thrown straight up be 1 and the other body be 2, then for the body 1 in the frame of 2 from the kinematical equation for constant acceleration (since both are moving under constant acceleration) is

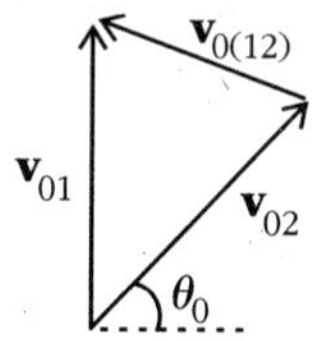

$$\mathbf{r}_{12} = \mathbf{r}_{0(12)} + \mathbf{v}_{0(12)}t + \frac{1}{2}\mathbf{w}_{12}t^2$$

So, $$\mathbf{r}_{12} = \mathbf{v}_{0(12)}t \quad (\text{because } \mathbf{w}_{12} = 0 \text{ and } \mathbf{r}_{0(12)} = 0)$$

or $$|\mathbf{r}_{12}| = |\mathbf{v}_{0(12)}|t$$

But, $$|\mathbf{v}_{01}| = |\mathbf{v}_{02}| = v_0$$

So, from properties of triangle

$$|\mathbf{v}_{0(12)}| = \sqrt{v_0^2 + v_0^2 - 2v_0v_0\cos(\pi/2 - \theta_0)}$$

Hence, the sought distance

$$|\mathbf{r}_{12}| = v_0 t\sqrt{2(1 - \sin\theta_0)} = 22 \text{ m}$$

1.11 Let the velocities of the particles (say $\mathbf{v}_1'$ and $\mathbf{v}_2'$) become mutually perpendicular after time t. Then their velocities become

$$\mathbf{v}_1' = \mathbf{v}_1 + \mathbf{g}t;\ \mathbf{v}_2' = \mathbf{v}_2 + \mathbf{g}t$$

As $$\mathbf{v}_1' \perp \mathbf{v}_2' \text{ so, } \mathbf{v}_1' \cdot \mathbf{v}_2' = 0$$

or $$(\mathbf{v}_1 + \mathbf{g}t) \cdot (\mathbf{v}_2 + \mathbf{g}t) = 0$$

or $$-v_1 v_2 + g^2 t^2 = 0$$

Hence, $$t = \frac{\sqrt{v_1 v_2}}{g}$$

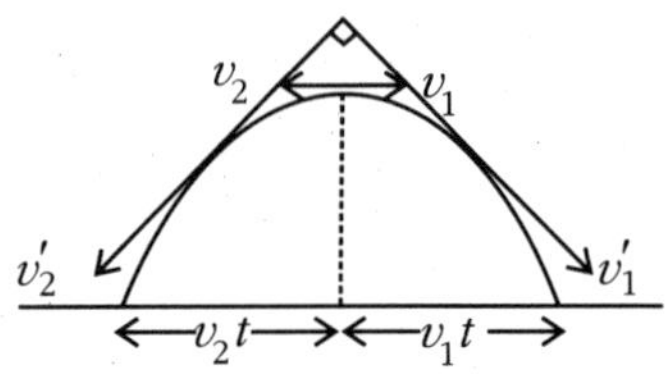

In the frame attached with 2 for the particle 1

$$\mathbf{r} = \mathbf{r}_0 + \mathbf{v}_0 t + \frac{1}{2}\mathbf{w}t^2$$

Both the particles are initially at the same position and have same acceleration $\mathbf{g}$, so $\mathbf{r}_0 = 0$ and $\mathbf{w} = 0$, and $\mathbf{v}_0 = |\mathbf{v}_1 - \mathbf{v}_2|$. Thus the sought distances

$$|\mathbf{r}| = |\mathbf{v}_0| t = (v_1 + v_2)t$$

$$= \frac{v_1 + v_2}{g}\sqrt{v_1 v_2} \quad \text{(using value of } t\text{)}$$

$$= 2.5 \text{ m, on putting the values of } v_1,\ v_2 \text{ and } g$$

1.12 From the symmetry of the problem all the three points are always located at the vertices of equilateral triangles of varying side length and finally meet at the centroid of the initial equilateral triangle. Symmetry of the problem tells us that the velocity with which point 1 approaches 2 is equal to the velocity by which 2 approaches 3 and 3 approaches 1.

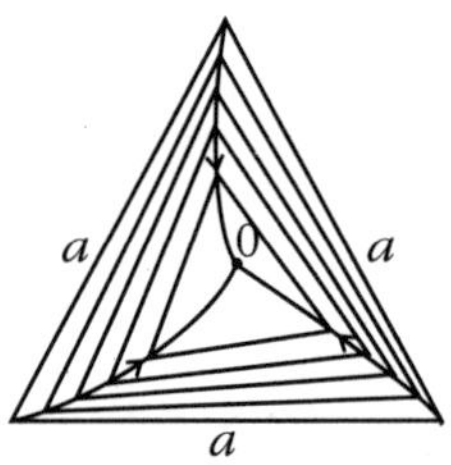

Method 1:

The rate at which 1 approaches 2 or the separation between 1 and 2 decreases with time is the projection of $\mathbf{v}_{12}$ towards $\mathbf{r}_{21}$. At an arbitrary instant when the equilateral triangle has edge length $l < a$, the velocity with which 1 approaches 2 becomes

$$\frac{-dl}{dt} = v - v\cos\left(\frac{2\pi}{3}\right)$$

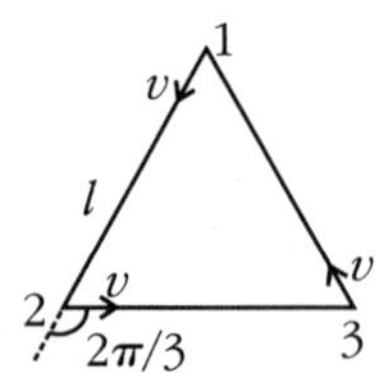

On integrating: $$\int_a^0 -dl = \frac{3v}{2}\int_0^t dt \quad \text{(where } t \text{ is the sought time)}$$

or $$a = \frac{3}{2}vt \text{ so, } t = \frac{2a}{3v}$$

Method 2:

If we concentrate on the motion of any one point say 3, one can observe that at all the time its velocity makes an angle 30° with 3O and is equal to $v\cos 30° = \sqrt{3}/2v$.

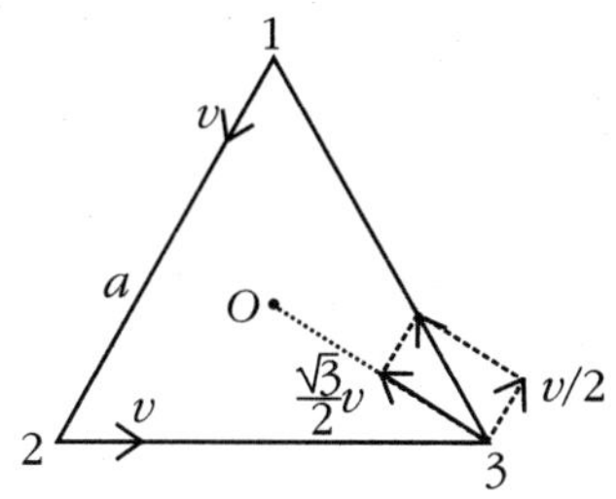

Initially 3O equals $a/\sqrt{3}$, so the sought time

$$t = \frac{a/\sqrt{3}}{(\sqrt{3}/2)v} = \frac{2a}{3v}$$

(In the figure, velocity of point 3 has been resolved into two rectangular components, one pointing towards the centre of the triangle and other perpendicular to it.)

1.13 Let us locate the points A and B at an arbitrary instant of time (see figure).

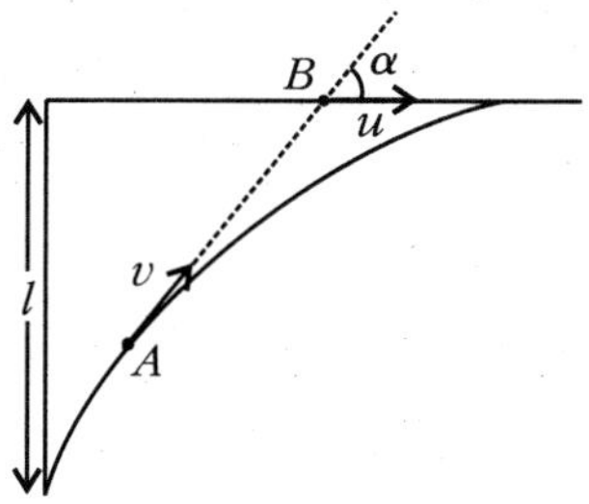

If A and B are separated by the distance s at this moment, then the points converge or point A approaches B with velocity $-ds/dt = v - u\cos\alpha$ where angle α varies with time. On integrating,

$$-\int_l^0 ds = \int_0^T (v - u\cos\alpha)\,dt \quad \text{(where } T \text{ is the sought time)}$$

or

$$l = \int_0^T (v - u\cos\alpha)\,dt$$

$$= vT - u\int_0^T \cos\alpha\,dt \qquad (1)$$

As both A and B cover the same distance in x-direction during the sought time interval, so the other condition which is required, can be obtained by the equation,

$$uT = \int_0^T v\cos\alpha\,dt \qquad (2)$$

Solving Eqs. (1) and (2), we get

$$T = \frac{ul}{v^2 - u^2}$$

One can see that if $u = v$, or $u < v$, point A cannot catch B.

1.14 In the reference frame fixed to the train, the distance between the two events is obviously equal to l. Suppose the train starts moving at time $t = 0$ in the positive direction and take the origin ($x = 0$) at the head-light of the train at $t = 0$. Then the coordinate of first event in the earth's frame is

$$x_1 = \frac{1}{2}wt^2$$

and similarly the coordinate of the second event is

$$x_2 = \frac{1}{2}w(t+\tau)^2 - l$$

The distance between the two events is obviously

$$x_1 - x_2 = l - w\tau(t + \tau/2) = 0.242 \text{ km}$$

in the reference frame fixed on the Earth.

For the two events to occur at the same point in the reference frame K, moving with constant velocity V relative to the earth, the distance traveled by the frame in the time interval τ must be equal to the above distance.

Thus, $$V\tau = l - w\tau(t + \tau/2)$$

So, $$V = \frac{l}{\tau} - w(t + \tau/2) = 4.03 \text{ m/s}$$

The frame K must clearly be moving in a direction opposite to the train so that if (for example) the origin of the frame coincides with the point x_1 on the earth at time t, it coincides with the point x_2 at time $t + \tau$.

1.15 (a) As time interval between two events is reference independent in Newtonian mechanics, so it is better to solve the problem in the elevator's frame having the observer at its floor (see figure). Let us denote separation between floor and ceiling by $h = 2.7$ m and the acceleration of the elevator by $w = 1.2 \text{ m/s}^2$.

From the kinematical formula

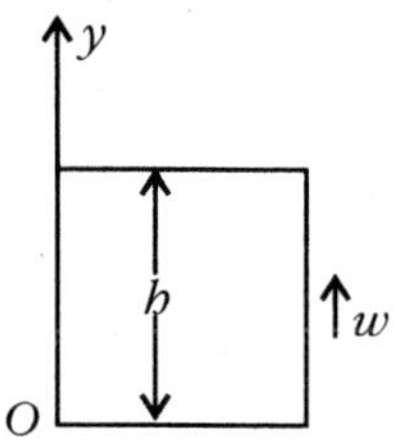

$$y = y_0 + v_{0y}t + \frac{1}{2}w_y t^2$$

Here, $$y = 0,\ y_0 = +h,\ v_{0y} = 0$$

and $$w_y = w_{\text{bolt}(y)} - w_{\text{ele}(y)}$$

$$= (-g) - (w) = -(g + w)$$

So, $$0 = h + \frac{1}{2}[-(g + w)]t^2$$

or $$t = \sqrt{\frac{2h}{g + w}} = 0.7 \text{ s} \quad \text{(on substituting values)}$$

(b) At the moment the bolt loses contact with the elevator, it has already acquired the velocity equal to elevator, given by;

$$v_0 = (1.2)(2) = 2.4 \text{ m/s}$$

In the reference frame attached with the elevator shaft (ground) and pointing the y-axis upward, we have for the displacement of the bolt,

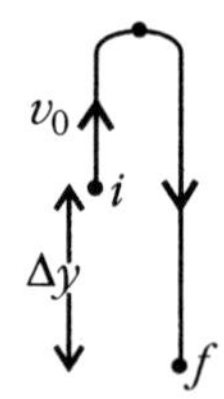

$$\Delta y = v_{0y}t + \frac{1}{2}w_y t^2$$

$$= v_0 t + \frac{1}{2}(-g)t^2$$

or $$\Delta y = (2.4)(0.7) + \frac{1}{2}(-9.8)(0.7)^2$$

$$= -0.7 \text{ m}$$

Hence the bolt comes down or displaces downward relative to the point, when it loses contact with the elevator by the amount 0.7 m (see figure).

Obviously the total distance covered by the bolt during its free fall time

$$s = |\Delta y| + 2\left(\frac{v_0^2}{2g}\right) = 0.7 \text{ m} + \frac{(2.4)^2}{(9.8)} \text{ m} = 1.3 \text{ m}$$

1.16 After time $t > 0$, the separation between the particles

$$l = \sqrt{(l_1 - v_1 t)^2 + (l_2 - v_2 t)^2}$$

For l and hence l^2 to be minimum

$$\frac{dl^2}{dt} = 0$$

or

$$\frac{d}{dt}\{(l_1 + v_1 t)^2 + (l_2 - v_2 t)^2\} = 0$$

which gives

$$t = \frac{(v_1 l_1 - v_2 l_2)}{\sqrt{v_1^2 + v_2^2}}$$

On using the value of t, in expression of l

$$l_{min} = \frac{|v_1 l_2 - v_2 l_1|}{\sqrt{v_1^2 + v_2^2}}.$$

Alternate:

Let the particles 1 and 2 be at points B and A respectively at $t = 0$, at the distances l_1 and l_2 from intersection point O.

Let us fix the inertial frame with the particle 2. Now the particle 1 moves relative to this reference frame with a relative velocity $\mathbf{v}_{12} = \mathbf{v}_1 - \mathbf{v}_2$, and its trajectory is the straight line BP. Obviously, the minimum distance between the particles is equal to the length of the perpendicular AP dropped from point A on to the straight line BP (Fig. (a)).

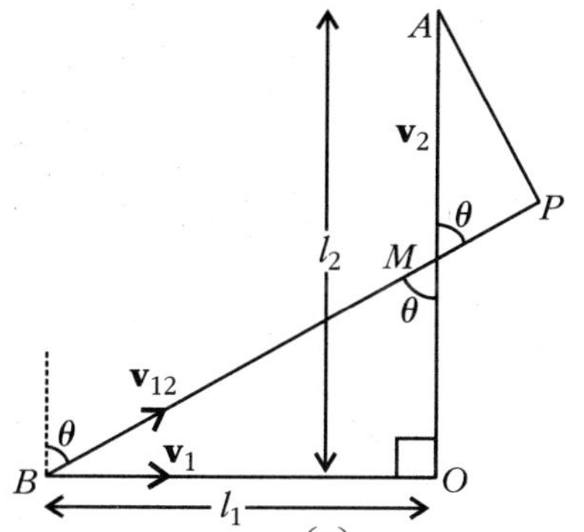

(a)

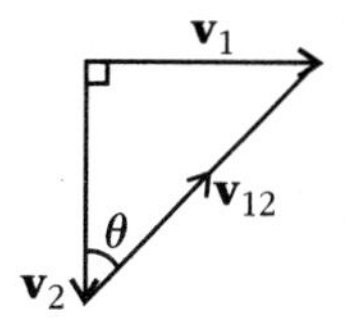

(b)

From Fig. (b), $v_{12} = \sqrt{v_1^2 + v_2^2}$, $\tan\theta = \frac{v_1}{v_2}$,

So, $$\cos\theta = \frac{v_2}{\sqrt{v_1^2 + v_2^2}} \quad \text{and} \quad \sin\theta = \frac{v_1}{\sqrt{v_1^2 + v_2^2}} \qquad (1)$$

From Fig. (a), the shortest distance

$$AP = AM \sin\theta = (OA - OM)\sin\theta = (l_2 - l_1 \cot\theta)\sin\theta$$

or

$$AP = \left(l_2 - l_1 \frac{v_2}{v_1}\right)\frac{v_1}{\sqrt{v_1^2 + v_2^2}} = \frac{|v_1 l_2 - v_2 l_1|}{\sqrt{v_1^2 + v_2^2}} \quad \text{(using Eq. 1)}$$

The sought time

$$t = \frac{BP}{|\mathbf{v}_{12}|} = \frac{BM + MP}{|\mathbf{v}_{12}|} = \frac{l_1 \operatorname{cosec}\theta + (l_2 - l_1 \cot\theta)\cos\theta}{\sqrt{v_1^2 + v_2^2}}$$

$$= \frac{l_2 \cos\theta + l_1 \sin\theta}{\sqrt{v_1^2 + v_2^2}} = \frac{l_2 v_2 + l_1 v_1}{v_1^2 + v_2^2} \quad \text{(using Eq. 1)}$$

1.17 Let the car turns off the highway at a distance x from the point D. So, $CD = x$, and if speed of car in the field is v, then the time taken by the car to cover the distance $AC = (AD - x)$ on the highway

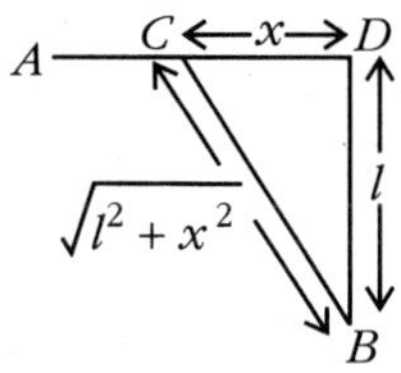

$$t_1 = \frac{AD - x}{\eta v} \tag{1}$$

and the time taken to travel the distance CB in the field

$$t_2 = \frac{\sqrt{l^2 + x^2}}{v} \tag{2}$$

So, the total time elapsed to move the car from point A to B

$$t = t_1 + t_2 = \frac{AD - x}{\eta v} + \frac{\sqrt{l^2 + x^2}}{v}$$

For t to be minimum

$$\frac{dt}{dx} = 0 \quad \text{or} \quad \frac{1}{v}\left[-\frac{1}{\eta} + \frac{x}{\sqrt{l^2 + x^2}}\right] = 0$$

or $$\eta^2 x^2 = l^2 + x^2 \quad \text{or} \quad x = \frac{l}{\sqrt{\eta^2 - 1}}$$

1.18 From question $v_x(0) = 0$, so the motion starts from origin, thus from the plot of $v_x(t)$ we have

$$v_x = \begin{cases} t; & 0 \le t \le 1 \\ 1; & 1 < t \le 3 \\ 4 - t; & 3 < t \le 6 \\ 2t - 14; & 6 < t \le 7 \end{cases}$$

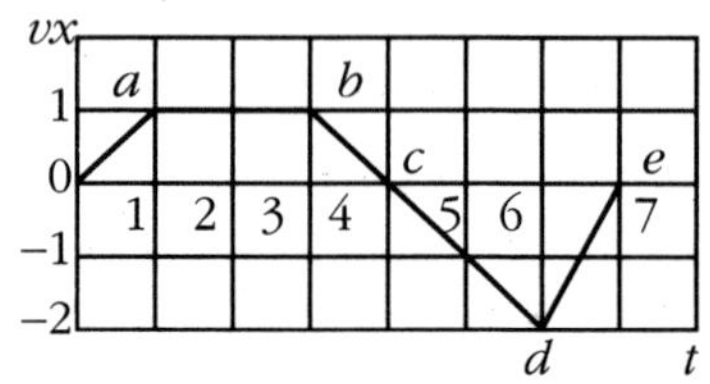

and $v_x = 0$ for $t > 7$.

As at $t = 0$, $x = 0$,

So, $$x(t) = \int_0^t v_x \, dt = \begin{cases} \dfrac{t^2}{2}; & 0 \le t \le 1 \\ t - \dfrac{1}{2}; & 1 < t \le 3 \\ 4t - \dfrac{t^2}{2} + 5; & 3 < t \le 6 \\ t^2 - 14t + 49; & 6 < t \le 7 \end{cases}$$

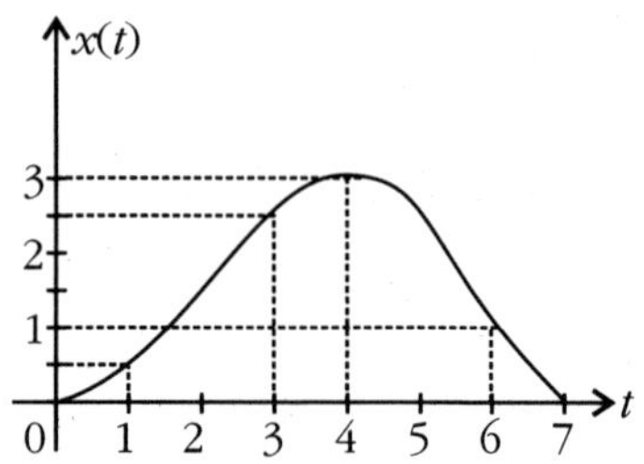

Now,
$$s(t)=\int_0^t |v_x|\,dt=\begin{cases}\dfrac{t^2}{2}; & 0\le t\le 1\\ t-\dfrac{1}{2}; & 1<t\le 3\\ 4t-\dfrac{t^2}{2}-5; & 3<t\le 4\\ \dfrac{t^2}{2}-4t+11; & 4<t\le 6\\ 14t-t^2-43; & 6<t\le 7\end{cases}$$

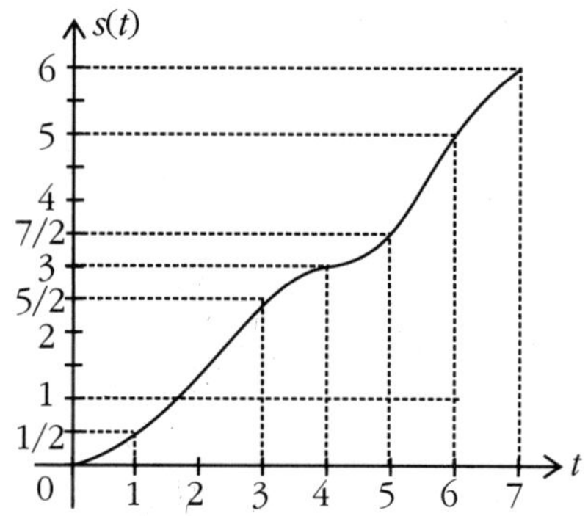

Now,
$$a_x=\frac{dv_x}{dt}=\begin{cases}1; & 0\le t<1\\ 0; & 1<t\le 3\\ -1; & 3<t\le 6\\ 2; & 6<t\le 7\end{cases}$$

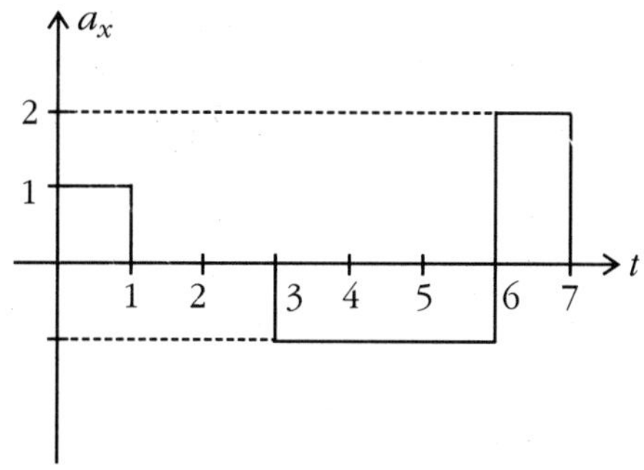

1.19 (a) Mean velocity

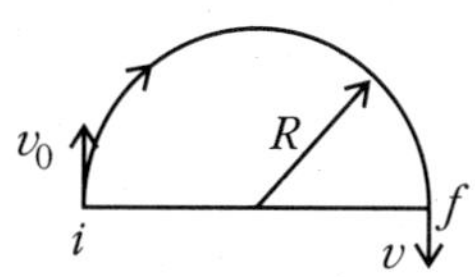

$$<v>=\frac{\text{Total distance covered}}{\text{Time elapsed}}$$
$$=\frac{s}{t}=\frac{\pi R}{\tau}=50\text{ cm/s} \tag{1}$$

(b) Modulus of mean velocity vector $|<\mathbf{v}>|=\dfrac{|\Delta\mathbf{r}|}{\Delta t}=\dfrac{2R}{\tau}=32\text{ cm/s}$ (2)

(c) Let the point moves from i to f along the half circle (see figure) and v_0 and v be the speed at the points respectively.

We have
$$\frac{dv}{dt}=w_t$$

or $v=v_0+w_t t$ (as w_t is constant, according to the problem)

So,
$$<v>=\frac{\int_0^t (v_0+w_t t)\,dt}{\int_0^t dt}=\frac{v_0+(v_0+w_t t)}{2}=\frac{v_0+v}{2} \tag{3}$$

So, from Eqs. (1) and (3)
$$\frac{v_0+v}{2}=\frac{\pi R}{\tau} \tag{4}$$

Now, the modulus of the mean vector of total acceleration

$$|\langle\mathbf{w}\rangle| = \frac{|\Delta\mathbf{v}|}{\Delta t} = \frac{|\mathbf{v}-\mathbf{v}_0|}{\tau} = \frac{v_0+v}{\tau} \quad \text{(see figure)} \tag{5}$$

Using Eq. (4) in Eq. (5), we get

$$|\langle\mathbf{w}\rangle| = \frac{2\pi R}{\tau^2} = 10 \text{ cm/s}^2 \quad \text{(on substituting values)}$$

1.20 (a) We have

$$\mathbf{r} = \mathbf{a}t(1 - 2\alpha t)$$

So,

$$\mathbf{v} = \frac{d\mathbf{r}}{dt} = \mathbf{a}(1-2\alpha t)$$

and

$$\mathbf{w} = \frac{d\mathbf{v}}{dt} = -2\alpha\,\mathbf{a}$$

(b) From the equation

$$\mathbf{r} = \mathbf{a}t(1 - \alpha t)$$

$$\mathbf{r} = 0, \text{ at } t = 0 \text{ and also at } t = \Delta t = \frac{1}{\alpha}.$$

So, the sought time $\Delta t = \dfrac{1}{\alpha}$

As

$$\mathbf{v} = \mathbf{a}(1 - 2\alpha t)$$

So,

$$v = |\mathbf{v}| = \begin{cases} a(1-2\alpha t) & \text{for } t \le \dfrac{1}{2\alpha} \\ a(2\alpha t - 1) & \text{for } t > \dfrac{1}{2\alpha} \end{cases}$$

Hence, the sought distance

$$s = \int v\,dt = \int_0^{1/2\alpha} a(1-2\alpha t)\,dt + \int_{1/2\alpha}^{1/\alpha} a(2\alpha t - 1)\,dt$$

Simplifying, we get, $s = \dfrac{a}{2\alpha}$

1.21 (a) As the particle leaves the origin at $t = 0$

So,

$$\Delta x = x = \int v_x\,dt \tag{1}$$

As

$$\mathbf{v} = \mathbf{v}_0\left(1 - \frac{t}{\tau}\right)$$

(where $\mathbf{v}_0$ is directed towards the +ve x-axis)

So,

$$v_x = v_0\left(1 - \frac{t}{\tau}\right) \tag{2}$$

From Eqs. (1) and (2),

$$x = \int_0^t v_0\left(1 - \frac{t}{\tau}\right)dt = v_0 t\left(1 - \frac{t}{2\tau}\right) \tag{3}$$

Hence x coordinate of the particle at $t = 6$ s

$$x = 10\times 6\left(1-\frac{6}{2\times 5}\right) = 24 \text{ cm} = 0.24 \text{ m}$$

Similarly at $t = 10$ s

$$x = 10\times 10\left(1-\frac{10}{2\times 5}\right) = 0$$

and at $t = 20$ s

$$x = 10\times 20\left(1-\frac{20}{2\times 5}\right) = -200 \text{ cm} = -2 \text{ m}$$

(b) At the moment the particle is at a distance of 10 cm from the origin, $x = \pm 10$ cm.

Putting $x = +10$ in Eq. (3),

$$10 = 10t\left(1-\frac{t}{10}\right) \quad \text{or} \quad t^2 - 10t + 10 = 0$$

So,
$$t = \frac{10 \pm \sqrt{100-40}}{2} = 5 \pm \sqrt{15} \text{ s}$$

Now, putting $x = -10$ in Eq. (3),

$$-10 = 10\left(1-\frac{t}{10}\right)$$

On solving, $t = 5 \pm \sqrt{35}$ s

As t cannot be negative, so,

$$t = (5+\sqrt{35}) \text{ s}$$

Hence the particle is at a distance of 10 cm from the origin at three moments is of time:

$$t = 5 \pm \sqrt{15} \text{ s}, 5+\sqrt{35} \text{ s}$$

(c) We have
$$\mathbf{v} = \mathbf{v}_0\left(1-\frac{t}{\tau}\right)$$

So,
$$v = |\mathbf{v}| = \begin{cases} v_0\left(1-\dfrac{t}{\tau}\right) & \text{for } t \le \tau \\ v_0\left(\dfrac{t}{\tau}-1\right) & \text{for } t > \tau \end{cases}$$

So,
$$s = \int_0^t v_0\left(1-\frac{t}{\tau}\right)dt \quad \text{for } t \le \tau = v_0 t(1 - t/2\tau) \tag{4}$$

and
$$s = \int_0^t v_0\left(1-\frac{t}{\tau}\right)dt + \int_\tau^t v_0\left(\frac{t}{\tau}-1\right)dt \quad \text{for } t > \tau$$
$$= v_0\tau[1+(1-t/\tau)^2]/2 \quad \text{for } t > \tau \tag{5}$$

$$s = \int_0^4 v_0\left(1-\frac{t}{\tau}\right)dt = \int_0^4 10\left(1-\frac{t}{5}\right)dt = 24 \text{ cm}$$

For $t = 8$ s

$$s = \int_0^5 10\left(1-\frac{t}{5}\right)dt + \int_5^8 10\left(\frac{t}{5}-1\right)dt$$

On integrating and simplifying, we get

$$s = 34 \text{ cm}$$

On the basis of Eqs. (3) and (4), $x(t)$ and $s(t)$ plots can be drawn as shown in the answer sheet.

1.22 (a) As particle is in unidirectional motion, it is directed along the x-axis all the time. At $t = 0$, $x = 0$.

So, $$\Delta x = x = s \quad \text{and} \quad w = \frac{dv}{dt} = \frac{vdv}{ds} = \frac{1}{2}\frac{d}{ds}(v^2)$$

According to the problem $v = \alpha\sqrt{x} = \alpha\sqrt{s}$

So, $$v^2 = \alpha^2 s$$

and $$w = \frac{1}{2}\frac{d}{ds}(v^2) = \frac{1}{2}\frac{d}{ds}(\alpha^2 s) = \frac{\alpha^2}{2} \tag{1}$$

As, $$w = \frac{dv}{dt} = \frac{\alpha^2}{2} \quad \text{so,} \quad dv = \frac{\alpha^2}{2}dt$$

On integrating, $$v = \frac{\alpha^2}{2}t = v(t) \tag{2}$$

(b) Let t be the time to cover first s m of the path. From Eq. (2)

$$s = \int v\,dt = \int_0^t \frac{\alpha^2}{2}t\,dt = \frac{\alpha^2}{2}\frac{t^2}{2}$$

Hence, $$t = \frac{2}{\alpha}\sqrt{s} \tag{3}$$

The mean velocity of particle

$$\langle v\rangle = \frac{\int v(t)\,dt}{\int dt} = \frac{\int_0^{2\sqrt{s}/\alpha}\frac{\alpha^2}{2}t\,dt}{2\sqrt{s}/\alpha} = \frac{\alpha\sqrt{s}}{2}$$

1.23 According to the problem

$$-\frac{v\,dv}{ds} = a\sqrt{v} \quad \text{(as } v \text{ decreases with time)}$$

or $$-\int_{v_0}^0 \sqrt{v}\,dv = a\int_0^s ds$$

On integrating we get, $$s = \frac{2}{3a}v_0^{3/2}$$

Again according to the problem

$$-\frac{dv}{dt} = a\sqrt{v} \quad \text{or} \quad -\frac{dv}{\sqrt{v}} = a\,dt$$

or
$$-\int_{v_0}^{0} \frac{dv}{\sqrt{v}} = a\int_0^t dt$$

Thus,
$$t = \frac{2\sqrt{v_0}}{a}$$

1.24 (a) As
$$\mathbf{r} = at\,\mathbf{i} - bt^2\,\mathbf{j}$$

So,
$$x = at,\ y = -bt^2$$

and therefore
$$y = \frac{-bx^2}{a^2}$$

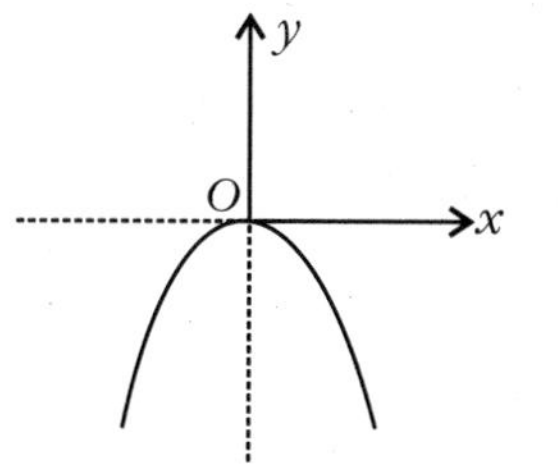

which is equation of a parabola, whose graph is shown in the figure.

(b) As
$$\mathbf{r} = at\,\mathbf{i} - bt^2\mathbf{j}$$

$$\mathbf{v} = \frac{d\mathbf{r}}{dt} = a\,\mathbf{i} - 2b\mathbf{j} \qquad (1)$$

So,
$$v = \sqrt{a^2 + (-2bt)^2} = \sqrt{a^2 + 4b^2t^2}$$

Differentiating Eq. (1) with respect to time, we get
$$\mathbf{w} = \frac{d\mathbf{v}}{dt} = -2b\mathbf{j}$$

So,
$$|\mathbf{w}| = w = 2b$$

(c)
$$\cos\alpha = \frac{\mathbf{v}\cdot\mathbf{w}}{vw} = \frac{(a\mathbf{i} - 2bt\mathbf{j})\cdot(-2b\mathbf{j})}{(\sqrt{a^2 + 4b^2t^2})\,2b}$$

or
$$\cos\alpha = \frac{2bt}{\sqrt{a^2 + 4b^2t^2}}$$

So,
$$\tan\alpha = \frac{a}{2bt} \quad \text{or} \quad \alpha = \tan^{-1}\left(\frac{a}{2bt}\right)$$

(d) The mean velocity vector
$$<\mathbf{v}> = \frac{\int \mathbf{v}\,dt}{\int dt} = \frac{\int_0^t (a\mathbf{i} - 2bt\mathbf{j})\,dt}{t} = a\mathbf{i} - bt\mathbf{j}$$

Hence,
$$|<\mathbf{v}>| = \sqrt{a^2 + (-bt)^2} = \sqrt{a^2 + b^2t^2}$$

1.25 (a) We have
$$x = at \quad \text{and} \quad y = at(1 - \alpha t) \qquad (1)$$

Hence, $y(x)$ becomes
$$y = \frac{ax}{a}\left(1 - \frac{ax}{a}\right) = x - \frac{a}{a}x^2 \quad \text{(parabola)}$$

(b) Differentiating Eq. (1) we get

$$v_x = a \quad \text{and} \quad v_y = a(1 - 2\alpha t) \tag{2}$$

So,
$$v = \sqrt{v_x^2 + v_y^2} = a\sqrt{1+(1-2\alpha t)^2}$$

Differentiating Eq. (2) with respect to time

$$w_x = 0 \quad \text{and} \quad w_y = -2a\alpha$$

So,
$$w = \sqrt{w_x^2 + w_y^2} = 2a\alpha \tag{3}$$

(c) From Eqs. (2) and (3) we have $\mathbf{v} = a\mathbf{i} + a(1 - 2\alpha t)\mathbf{j}$ and $\mathbf{w} = 2a\alpha\mathbf{j}$.

So at $t - t_0$, $\cos\frac{\pi}{4} = \frac{1}{\sqrt{2}} = \frac{\mathbf{v}\cdot\mathbf{w}}{vw} = \frac{-a(1-2\alpha t_0)2a\alpha}{a\sqrt{1+(1-2\alpha t_0)^2}\,2a\alpha}$

On simplifying,
$$1 - 2\alpha t_0 = \pm 1$$

As
$$t_0 \neq 0,\ t_0 = \frac{1}{\alpha}$$

1.26 Differentiating motion law: $x = a\sin\omega t$, $y = a(1 - \cos\omega t)$, with respect to time, $v_x = a\omega\cos\omega t$ and $v_y = a\omega\sin\omega t$.

So,
$$\mathbf{v} = a\omega\cos\omega t\mathbf{i} + a\omega\sin\omega t\mathbf{j} \tag{1}$$

and
$$v = a\omega = \text{constant} \tag{2}$$

Differentiating Eq. (1) with respect to time

$$\mathbf{w} = \frac{d\mathbf{v}}{dt} = -a\omega^2\sin\omega t\mathbf{i} + a\omega^2\cos\omega t\mathbf{j} \tag{3}$$

(a) The distance s traversed by the point during the time τ is given by

$$s = \int_0^\tau v\,dt = \int_0^\tau a\omega\,dt = a\omega\tau \quad \text{(using Eq. 2)}$$

(b) Taking scalar product of $\mathbf{v}$ and $\mathbf{w}$

We get,

$$\mathbf{v}\cdot\mathbf{w} = (a\omega\cos\omega t\mathbf{i} + a\omega\sin\omega t\mathbf{j})\cdot(a\omega^2\sin\omega t\,(-\mathbf{i}) + a\omega^2\cos\omega t\mathbf{j})$$

So,
$$\mathbf{v}\cdot\mathbf{w} = -a^2\omega^2\sin\omega t\cos\omega t + a^2\omega^3\sin\omega t\cos\omega t = 0$$

Thus, $\mathbf{v} \perp \mathbf{w}$, i.e., the angle between velocity vector and acceleration vector equals $\pi/2$.

1.27 According to the problem

$$\mathbf{w} = w(-\mathbf{j})$$

So,
$$w_x = \frac{dv_x}{dt} = 0 \quad \text{and} \quad w_y = \frac{dv_y}{dt} = -w \tag{1}$$

Differentiating equation of trajectory, $y = ax - bx^2$, with respect to time

$$\frac{dy}{dt} = \frac{adx}{dt} - 2bx\frac{dx}{dt} \tag{2}$$

So, $$\left.\frac{dy}{dt}\right|_{x=0} = \left.\frac{adx}{dt}\right|_{x=0}$$

Again differentiating with respect to time

$$\frac{d^2y}{dt^2} = \frac{ad^2x}{dt^2} - 2b\left(\frac{dx}{dt}\right)^2 - 2bx\frac{d^2x}{dt^2}$$

or $$-w = a(0) - 2b\left(\frac{dx}{dt}\right)^2 - 2bx(0) \quad \text{(using Eq. 1)}$$

or $$\frac{dx}{dt} = \sqrt{\frac{w}{2b}} \quad \text{(using Eq. 1)} \tag{3}$$

Using Eq. (3) in Eq. (2) $$\left.\frac{dy}{dt}\right|_{x=0} = a\sqrt{\frac{w}{2b}} \tag{4}$$

Hence, the velocity of the particle at the origin

$$v = \sqrt{\left(\frac{dx}{dt}\right)^2_{x=0} + \left(\frac{dy}{dt}\right)^2_{x=0}} = \sqrt{\frac{w}{2b} + a^2\frac{w}{2b}} \quad \text{(using Eqs. 3 and 4)}$$

Hence, $$v = \sqrt{\frac{w}{2b}(1+a^2)}$$

1.28 As the body is under gravity of constant acceleration **g**, its velocity vector and displacement vectors are:

$$\mathbf{v} = \mathbf{v}_0 + \mathbf{g}t \tag{1}$$

and $$\Delta\mathbf{r} = \mathbf{r} = \mathbf{v}_0 t + \frac{1}{2}\mathbf{g}t^2 \quad (\text{as } \mathbf{r} = 0 \text{ at } t = 0) \tag{2}$$

So, $\langle\mathbf{v}\rangle$ over the first t seconds

$$\langle\mathbf{v}\rangle = \frac{\Delta\mathbf{r}}{\Delta t} = \frac{\mathbf{r}}{t} = \mathbf{v}_0 + \frac{1}{2}\mathbf{g}t \tag{3}$$

Hence from Eq. (3), $\langle\mathbf{v}\rangle$ over the first τ seconds

$$\langle\mathbf{v}\rangle = \mathbf{v}_0 + \frac{\mathbf{g}}{2}\tau \tag{4}$$

For evaluating t, take

$$\mathbf{v}\cdot\mathbf{v} = (\mathbf{v}_0 + \mathbf{g}t)\cdot(\mathbf{v}_0 + \mathbf{g}t) = v_0^2 + 2(\mathbf{v}_0\cdot\mathbf{g})t + g^2t^2$$

or $$v^2 = v_0^2 + 2(\mathbf{v}_0\cdot\mathbf{g})t + g^2t^2$$

But we have $v = v_0$ at $t = 0$ and also at $t = \tau$ (also from energy conservation).

Hence using this property in Eq. (5),

$$v_0^2 = v_0^2 + 2(\mathbf{v}_0\cdot\mathbf{g})\tau + g^2\tau^2$$

As $$\tau \neq 0, \text{ so, } \tau = -\frac{2(\mathbf{v}_0\cdot\mathbf{g})}{g^2}$$

Putting this value of τ in Eq. (4), the average velocity over the time of flight

$$<\mathbf{v}> = \mathbf{v}_0 - \mathbf{g}\frac{(\mathbf{v}_0 \cdot \mathbf{g})}{g^2}$$

1.29 The body thrown in air with velocity v_0 at an angle α from the horizontal lands at point P on the Earth's surface at same horizontal level (see figure). The point of projection is taken as origin, so $\Delta x = x$ and $\Delta y = y$.

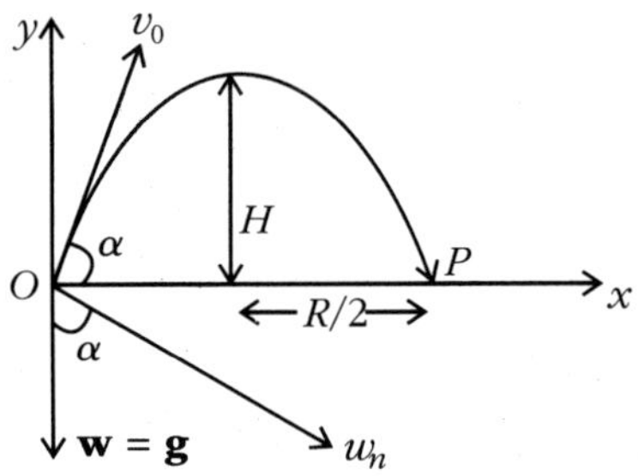

(a) From the equation $\quad \Delta y = v_{0y}t + \frac{1}{2}w_y t^2$

or $\quad 0 = v_0 \sin \alpha \tau - \frac{1}{2} g\tau^2$

As $\tau \neq 0$, so, time of motion $\tau = \frac{2v_0 \sin \alpha}{g}$

(b) At the maximum height of ascent, $v_y = 0$

so, from the equation

$$v_y^2 = v_{0y}^2 + 2w_y \Delta y$$

$$0 = (v_0 \sin \alpha)^2 - 2gH$$

Hence maximum height $\quad H = \frac{v_0^2 \sin^2 \alpha}{2g}$

During the time of motion the net horizontal displacement or horizontal range, will be obtained by the equation

$$\Delta x = v_{0x}t + \frac{1}{2}w_x \tau^2$$

or $\quad R = v_0 \cos \alpha\tau - \frac{1}{2}(0)\,\tau^2 = v_0 \cos \alpha\tau = \frac{v_0^2 \sin 2\alpha}{g}$

when $R = H$

$$\frac{v_0^2 \sin 2\alpha}{g} = \frac{v_0^2 \sin^2 \alpha}{2g} \quad \text{or} \quad \tan \alpha = 4, \quad \text{so,} \quad \alpha = \tan^{-1} 4$$

(c) For the body, $x(t)$ and $y(t)$ are

$$x = v_0 \cos \alpha t \qquad (1)$$

and $\quad y = v_0 \sin \alpha t - \frac{1}{2}gt^2 \qquad (2)$

Hence putting the value of t from Eq. (1) into Eq. (2) we get

$$y = v_0 \sin \alpha \left(\frac{x}{v_0 \cos \alpha}\right) - \frac{1}{2}g\left(\frac{x}{v_0 \cos \alpha}\right)^2 = x \tan \alpha - \frac{gx^2}{2v_0^2 \cos^2 \alpha}$$

Note: *One can use the formula of curvature radius of a trajectory $y(x)$, to solve part (d),*

$$R = \frac{[1 + (dy/dx)^2]^{3/2}}{|d^2y/dx^2|}$$

1.30 We have, $v_x = v_0 \cos\alpha,\ v_y = v_0 \sin\alpha - gt$

Thus, $v^2 = v_x^2 + v_y^2 = v_0^2 - 2gt\,v_0 \sin\alpha + g^2t^2$

Now, $$w_t = \frac{dv_t}{dt} = \frac{1}{2v_t}\frac{d}{dt}(v_t)^2 = \frac{1}{v}(g^2t - gv_0 \sin\alpha)$$

$$= -\frac{g}{v}(v_0 \sin\alpha - gt) = -g\frac{v_y}{v_t}$$

Hence, $$|w_t| = g\frac{|v_y|}{v}$$

Now, $$w_n = \sqrt{w^2 - w_t^2} = \sqrt{g^2 - g^2\frac{v_y^2}{v_t^2}}$$

or $$w_n = g\frac{v_x}{v_t} \quad \left[\text{where } v_x = \sqrt{v_t^2 - v_y^2} = \sqrt{v^2 - v_y^2}\right]$$

$$w_v = w_t = -g\frac{v_y}{v}$$

On the basis of obtained expressions or facts, the sought plots can be drawn as shown in the figure of answer sheet.

1.31 The ball strikes the inclined plane (Ox) at point O (origin) with velocity

$$v_0 = \sqrt{2gh} \tag{1}$$

As the ball classically rebounds, it recalls with same velocity v_0, at the same angle α from the normal on y axis (see figure). Let the ball strike the incline second time at P, which is at a distance l (say) from the point O, along the incline. From the equation

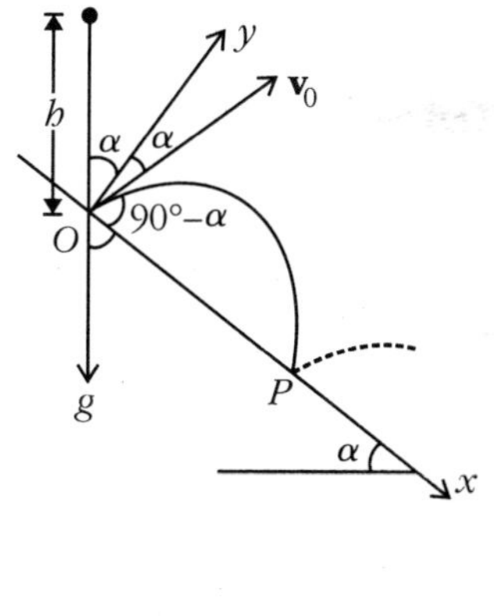

$$y = v_{0y}t + \frac{1}{2}w_yt^2$$

$$0 = v_0 \cos\alpha\tau - \frac{1}{2}g\cos\alpha\tau^2$$

where τ is the time of motion of ball in air while moving from O to P.

As $$\tau \neq 0, \quad \text{so,} \quad \tau = \frac{2v_0}{g} \tag{2}$$

Now from the equation.

$$x = v_{0x}t + \frac{1}{2}w_xt^2$$

$$l = v_0 \sin\alpha\tau + \frac{1}{2}g\sin\alpha\tau^2$$

So, $$l = v_0 \sin\alpha\left(\frac{2v_0}{g}\right) + \frac{1}{2}g\sin\alpha\left(\frac{2v_0}{g}\right) = \frac{4v_0^2 \sin\alpha}{g} \quad \text{(using Eq. 2)}$$

Hence the sought distance,

$$l = \frac{4(2gh)\sin\alpha}{g} = 8h\sin\alpha \quad \text{(using Eq. 1)}$$

1.32 Total time of motion

$$\tau = \frac{2v_0 \sin\alpha}{g} \quad \text{or} \quad \sin\alpha = \frac{\tau g}{2v_0} = \frac{9.8\tau}{2\times 240} \tag{1}$$

and horizontal range

$$R = v_0 \cos\alpha\tau \quad \text{or} \quad \cos\alpha = \frac{R}{v_0\tau} = \frac{5100}{240\tau} = \frac{85}{4\tau} \tag{2}$$

From Eqs. (1) and (2)

$$\frac{(9.8)^2\tau^2}{(480)^2} + \frac{(85)^2}{(4\tau^2)^2} = 1$$

On simplifying

$$\tau^4 - 2400\tau^2 + 1083750 = 1$$

Solving for τ^2, we get

$$\tau^2 = \frac{2400 \pm \sqrt{1425000}}{2} = \frac{2400 \pm 1194}{2}$$

Thus, $\tau = 42.39 \text{ s} = 0.71 \text{ min}$

and $\tau = 24.55 \text{ s} = 0.41 \text{ min}$ (depending on the angle α)

1.33 Let the shells collide at the point $P(x, y)$. If the first shell takes t s to collide with second and Δt be the time interval between the firings, then

$$x = v_0 \cos\theta_1 t = v_0 \cos\theta_2 (t - \Delta t) \tag{1}$$

and

$$y = v_0 \sin\theta_2 (t - \Delta t) - \frac{1}{2} g (t - \Delta t)^2 \tag{2}$$

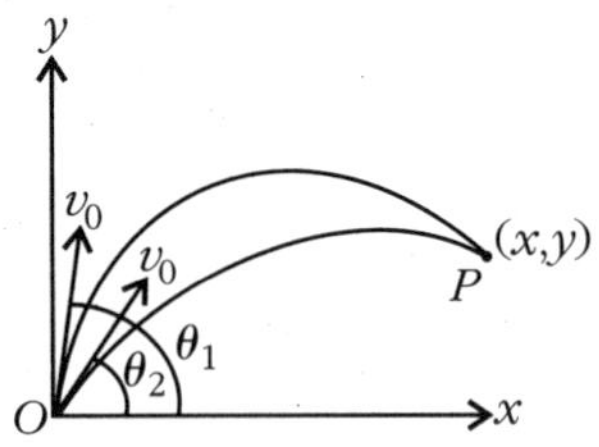

From Eq. (1)

$$t = \frac{\Delta t \cos\theta_2}{\cos\theta_2 - \cos\theta_1} \tag{3}$$

From Eqs. (2) and (3)

$$\Delta t = \frac{2v_0 \sin(\theta_1 - \theta_2)}{g(\cos\theta_2 + \cos\theta_1)} \text{ as } \Delta t \neq 0$$

$$= 11 \text{ s} \quad \text{(on substituting values)}$$

1.34 According to the problem

(a)

$$\frac{dy}{dt} = v_0 \quad \text{or} \quad dy = v_0\, dt$$

Integrating

$$\int_0^y dy = v_0 \int_0^t dt \quad \text{or} \quad y = v_0 t \tag{1}$$

Also, we have $\frac{dx}{dt} = ay$ or $dx = ay\,dt = av_0 t\,dt$ (using Eq. 1)

So, $$\int_0^x dx = av_0 \int_0^t t\,dt \quad \text{or} \quad x = \frac{1}{2}av_0 t^2 = \frac{1}{2}\frac{ay^2}{v_0} \quad \text{(using Eq. 1)}$$

(b) According to the problem

$$v_y = v_0 \quad \text{and} \quad v_x = ay \tag{2}$$

So, $$v = \sqrt{v_x^2 + v_y^2} = \sqrt{v_0^2 + a^2 y^2}$$

Therefore, $$w_t = \frac{dv}{dt} = \frac{a^2 y}{\sqrt{v_0^2 + ay^2}}\frac{dy}{dt} = \frac{a^2 y}{\sqrt{1+(ay/v_0)^2}}$$

Differentiating Eq. (2) with respect to time,

$$\frac{dv_y}{dt} = w_y = 0 \quad \text{and} \quad \frac{dv_x}{dt} = w_x = a\frac{dy}{dt} = av_0$$

So, $$w = |w_x| = av_0$$

Hence, $$w_n = \sqrt{w^2 - w_t^2} = \sqrt{a^2 v_0^2 - \frac{a^4 y^2}{1+(ay/v_0)^2}} = \frac{av_0}{\sqrt{1+(ay/v_0)^2}}$$

1.35 (a) The velocity vector of the particle

$$\mathbf{v} = a\mathbf{i} + bx\mathbf{j}$$

So, $$\frac{dx}{dt} = a \quad \text{and} \quad \frac{dy}{dt} = bx \tag{1}$$

From Eq. (1) $$\int_0^x dx = a\int_0^t dt \quad \text{or} \quad x = at \tag{2}$$

and $$dy = bx\,dt = bat\,dt$$

Integrating $$\int_0^y dy = ab\int_0^t t\,dt \quad \text{or} \quad y = \frac{1}{2}abt^2 \tag{3}$$

From Eqs. (2) and (3), we get,

$$y = \frac{b}{2a}(x^2) \tag{4}$$

(b) The curvature radius of trajectory $y(x)$ is

$$R = \frac{[1+(dy/dx)^2]^{3/2}}{|d^2y/dx^2|} \tag{5}$$

Let us differentiate the path equation $y = \frac{b}{2a}x^2$ with respect to x,

$$\frac{dy}{dx} = \frac{b}{a}x \quad \text{and} \quad \frac{d^2y}{dx^2} = \frac{b}{a} \tag{6}$$

From Eqs. (5) and (6), the sought curvature radius is

$$R = \frac{a}{b}\left[1+\left(\frac{b}{a}x\right)^2\right]^{3/2}$$

1.36 In accordance with the problem

$$w_t = \mathbf{a} \cdot \boldsymbol{\tau}$$

But, $$w_t = \frac{v\,dv}{ds} \quad \text{or} \quad v\,dv = w_t\,ds$$

So, $$v\,dv = (\mathbf{a} \cdot \boldsymbol{\tau})\,ds = \mathbf{a} \cdot ds\boldsymbol{\tau} = \mathbf{a} \cdot d\mathbf{r}$$

or $$v\,dv = a\mathbf{i} \cdot d\mathbf{r} = a\,dx \text{ (because } \mathbf{a} \text{ is directed towards the } x\text{-axis)}$$

So, $$\int_0^v v\,dv = a\int_0^x dx$$

Hence, $$v^2 = 2ax \quad \text{or} \quad v = \sqrt{2ax}$$

1.37 The velocity of the particle $v = at$.

So, $$\frac{dv}{dt} = w_t = a \tag{1}$$

and $$w_n = \frac{v^2}{R} = \frac{a^2t^2}{R} \quad \text{(using } v = at\text{)} \tag{2}$$

From $$s = \int v\,dt$$

$$2\pi R\eta = \int_0^t v\,dt = \int_0^t at\,dt$$

So, $$\frac{4\pi\eta}{a} = \frac{t^2}{R} \tag{3}$$

From Eqs. (2) and (3), $$w_n = 4\pi a\eta$$

Hence, $$w = \sqrt{w_t^2 + w_n^2}$$
$$= \sqrt{a^2 + (4\pi a\eta)^2} = a\sqrt{1+16\pi^2\eta^2} = 0.8 \text{ m/s}^2$$

1.38 (a) According to the problem

$$|w_t| = |w_n|$$

For $v(t)$, $$\frac{-dv}{dt} = \frac{v^2}{R}$$

Integrating this equation from $v_0 \le v \le v$ and $0 \le t \le t$, we get

$$-\int_{v_0}^{v} \frac{dv}{v^2} = \frac{1}{R}\int_0^t dt \quad \text{or} \quad v = \frac{v_0}{(1+v_0t/R)} \tag{1}$$

Now for $v(s)$, $$-\frac{v\,dv}{ds} = \frac{v^2}{R}$$

Integrating this equation from $v_0 \le v \le v$ and $0 \le s \le s$, we get

$$\int_{v_0}^{v} \frac{dv}{v} = -\frac{1}{R}\int_0^s ds \quad \text{or} \quad \ln\frac{v}{v_0} = -\frac{s}{R}$$

Hence,
$$v = v_0 e^{-s/R} \tag{2}$$

(b) The normal acceleration of the point

$$w_n = \frac{v_2}{R} = \frac{v_0^2 e^{-2s/R}}{R} \quad \text{(using Eq. 2)}$$

In accordance with the problem

$$|w_t| = |w_n| \quad \text{and} \quad w_t\mathbf{t} \perp w_n\mathbf{n}$$

So,
$$w = \sqrt{2}w_n = \sqrt{2}\frac{v_0^2}{R}e^{-2s/R} = \sqrt{2}\frac{v^2}{R}$$

1.39 From the equation
$$v = a\sqrt{s}$$

$$w_t = \frac{dv}{dt} = \frac{a}{2\sqrt{s}}\frac{ds}{dt} = \frac{a}{2\sqrt{s}}a\sqrt{s} = \frac{a^2}{2}$$

and
$$w_n = \frac{v^2}{R} = \frac{a^2 s}{R}$$

As w_t is a positive constant, the speed of the particle increases with time, and the tangential acceleration vector and velocity vector coincides in direction.

Hence the angle between **v** and **w** is equal to angle between $\mathbf{w}_t\mathbf{t}$ and **w** and α can be found by the formula

$$\tan\alpha = \frac{|w_n|}{|w_t|} = \frac{a^2 s/R}{a^2/2} = \frac{2s}{R}$$

1.40 Differentiating the equation
$$l = a\sin\omega t$$

$$\frac{dl}{dt} = v = a\omega\cos\omega t$$

So,
$$w_t = \frac{dv}{dt} = -a\omega^2\sin\omega t \tag{1}$$

and
$$w_n = \frac{v^2}{R} = \frac{a^2\omega^2\cos^2\omega t}{R} \tag{2}$$

(a) At the point $l = 0$, $\sin\omega t = 0$ and $\cos\omega t = \pm 1$ so, $\omega t = 0, \pi$, etc.

Hence, $w = w_n = \dfrac{a^2\omega^2}{R} = 2.6\ \text{m/s}^2$ (on substituting values)

Similarly at $l = \pm a$, $\sin\omega t = \pm 1$ and $\cos\omega t = 0$, so, $w_n = 0$.

Hence, $w = |w_t| = a\omega^2 = 3.2\ \text{m/s}^2$ (on substituting values)

(b) Using $\sin\omega t = l/a$ in Eqs. (1) and (2) we get

$$\omega_t = -\omega^2 l$$

and
$$w_n = \frac{\omega^2}{R}(a^2 - l^2)$$

Therefore, $$w_t = \sqrt{\omega^4 l^2 + \frac{\omega^4}{R^2}(a^2 - l^2)^2} = f(l)$$

On differentiating with respect to l and putting

$$\frac{dw_t}{dl} = 0$$

we get, $$l^2 = l_m^2 = a^2 - \frac{R^2}{2} = a^2\left(1 - \frac{R^2}{2a^2}\right)$$

So, $$l_m = \pm a\sqrt{1 - \frac{R^2}{2a^2}} = \pm 0.37 \text{ m} \quad \text{(on substituting values)}$$

The corresponding $w_{\min}$ at $l = l_m$ is

$$w_{\min} = \sqrt{\omega^4\left(a^2 - \frac{R^2}{2}\right) + \frac{\omega^4}{R^2}\left(\frac{R^2}{2}\right)^2}$$

$$= a\omega^2\sqrt{1 - \frac{R^2}{2a^2} + \frac{R^2}{4a^2}}$$

$$= a\omega^2\sqrt{1 - \left(\frac{R}{2a}\right)^2} = 2.5 \text{ m/s}^2 \quad \text{(on substituting values)}$$

1.41 As $w_t = a$ and at $t = 0$, the point is at rest.

So, $v(t)$ and $s(t)$ are

$$v = at \quad \text{and} \quad s = \frac{1}{2}at^2 \tag{1}$$

Let R be the curvature radius, then

$$w_n = \frac{v^2}{R} = \frac{a^2t^2}{R} = \frac{2as}{R} \quad \text{(using Eq. 1)}$$

But according to the problem

$$w_n = bt^4$$

So, $$bt^4 = \frac{a^2t^2}{R} \quad \text{or} \quad R = \frac{a^2}{bt^2} = \frac{a2}{2bs} \quad \text{(using Eq. 1)} \tag{2}$$

Therefore, $$w = \sqrt{w_t^2 + w_n^2} = \sqrt{a^2 + (2as/R)^2} = \sqrt{a^2 + (4bs^2/a^2)^2} \quad \text{(using Eq. 2)}$$

Hence, $$w = a\sqrt{1 + (4bs^2/a^3)^2}$$

1.42 (a) Let us differentiate twice the path equation $y(x)$ with respect to time,

$$\frac{dy}{dt} = 2ax\frac{dx}{dt}; \quad \frac{d^2y}{dt^2} = 2a\left[\left(\frac{dx}{dt}\right)^2 + x\frac{d^2x}{dt^2}\right]$$

Since the particle moves uniformly, its acceleration at all points of the path is normal and at the point $x = 0$ it coincides with the direction of derivative d^2y/dt^2.

Keeping in mind that at the point $x = 0$, $\left|\frac{dx}{dt}\right| = v$,

we get $$w = \left|\frac{d^2y}{dt^2}\right|_{x=0} = 2av^2 = w_n$$

So, $$w_n = 2av^2 = \frac{v^2}{R} \quad \text{or} \quad R = \frac{1}{2a}$$

Note: *We can also calculate it from the formula of problem 1.35(b).*

(b) Differentiating the equation of the trajectory with respect to time we see that

$$b^2x\frac{dx}{dt} + a^2y\frac{dy}{dt} = 0 \tag{1}$$

which implies that the vector $(b^2x\mathbf{i} + a^2y\mathbf{j})$ is normal to the velocity vector $\mathbf{v} = \frac{dx}{dt}\mathbf{i} + \frac{dy}{dt}\mathbf{j}$ which, of course, is along the tangent. Thus the former vector is along the normal and the normal component of acceleration is clearly

$$w_n = \frac{b^2x\frac{d^2x}{dt^2} + a^2y\frac{d^2y}{dt^2}}{(b^4x^2 + a^4y^2)^{1/2}} \quad \left(\text{on using } w_n = \frac{\mathbf{w}\cdot\mathbf{n}}{|\mathbf{n}|}\right)$$

At $x = 0$, $y = \pm b$, so at $x = 0$

$$w_n = \pm\frac{d^2y}{dt^2}\bigg|_{x=0}$$

Differentiating Eq. (1)

$$b^2\left(\frac{dx}{dt}\right)^2 + b^2x\left(\frac{d^2x}{dt^2}\right) + a^2\left(\frac{dy}{dt}\right)^2 + a^2y\left(\frac{d^2y}{dt}\right) = 0$$

Also from Eq. (1) $\quad \frac{dy}{dt} = 0$ at $x = 0$

So, $\quad \left(\frac{dx}{dt}\right) = \pm v \quad$ (since tangential velocity is constant $= v$)

Thus, $\quad \left(\frac{d^2y}{dt^2}\right) = \pm\frac{b}{a^2}v^2$

and $\quad |w_n| = \frac{bv^2}{a^2} = \frac{v^2}{R}$

This gives $\quad R = \frac{a^2}{b}$

1.43 Angular velocity of point A, with respect to centre C of the circle or turning rate of line CA taking the line OCX as reference line becomes

$$\omega_C = -\frac{d(2\theta)}{dt} = 2\left(\frac{-d\theta}{dt}\right) = 2\omega$$

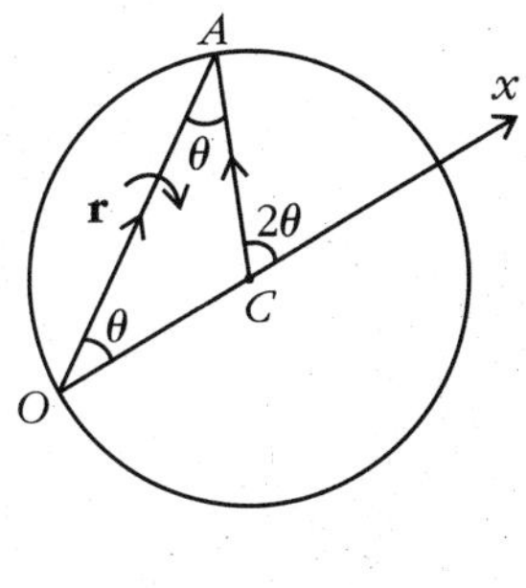

because angular speed of line OA is $\omega = -d\theta/dt$.

The turning rate of line CA is also the turning rate of velocity vector of point A, which is given by v_A/R.

So, $\quad v_A = \omega_C R = (2\omega)R = 0.4$ m/s $\quad$ (on substituting values)

Alternate:

Angular speed of a point relative to origin $= \dfrac{\text{Transverse velocity}}{\text{Magnitude of position vector}}$

So, angular speed of point A relative to origin O

$$= \frac{\text{Component velocity of point } A \text{ perpendicular to } OA}{\text{Length of line } OA}$$

$$\omega = \frac{v_A \cos\theta}{r}$$

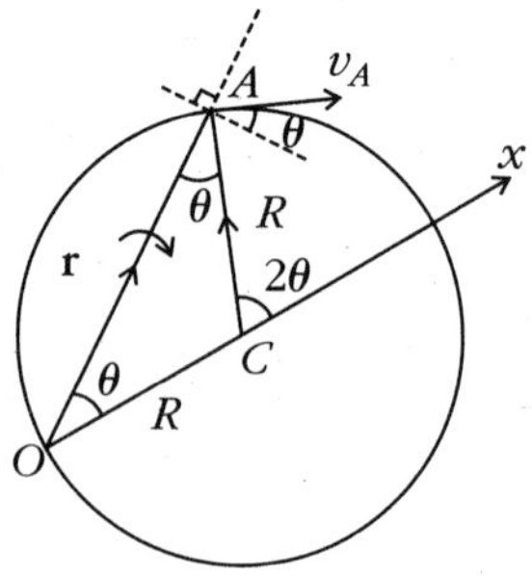

But from sine property of triangle OAC, $r = 2R\cos\theta$.

So, $$\omega = \frac{v_A}{2R} \quad \text{or} \quad v_A = 2\omega R$$

From both the methods the obtained speed v_A is constant, so the tangential acceleration of particle A is zero and

$$w = w_n = \frac{v_A^2}{R} = \frac{(2\omega R)^2}{R} = 4\omega^2 R = 0.32 \text{ m/s}^2 \quad \text{(on substituting values)}$$

1.44 Differentiating $\varphi(t)$ with respect to time

$$\frac{d\varphi}{dt} = \omega_z = 2at \qquad (1)$$

For fixed axis rotation, the speed of the point A

$$v = \omega R = 2atR \quad \text{or} \quad R = \frac{v}{2at} \qquad (2)$$

Differentiating with respect to time

$$w_t = \frac{dv}{dt} = 2aR = \frac{v}{t} \quad \text{(using Eq. 1)}$$

But, $$w_n = \frac{v^2}{R} = \frac{v^2}{v/2at} = 2atv \quad \text{(using Eq. 2)}$$

So, $$w = \sqrt{w_t^2 + w_n^2} = \sqrt{(v/t)^2 + (2atv)^2} = \frac{v}{t}\sqrt{1+4a^2t^4}$$

$$= 0.7 \text{ m/s}^2 \quad \text{(on substituting values)}$$

1.45 The shell acquires a constant angular acceleration at the same time as it accelerates linearly.

$$l = \frac{1}{2}wt^2 \text{ and } 2\pi n = \frac{1}{2}\beta t^2$$

So, $$\frac{w}{l} = \frac{\beta}{2\pi n}$$

(where w = linear acceleration and β = angular acceleration).

Then, $$\omega = \sqrt{2\beta\, 2\pi n} = \sqrt{2\frac{w}{l}(2\pi n)^2}$$

But, $$v^2 = 2wl$$

Hence, $$\omega = \frac{2\pi n v}{l} = 2.0 \times 10^3 \text{ rad/s} \quad \text{(on substituting values)}$$

1.46 Let us take the rotation axis as z-axis whose positive direction is associated with the positive direction of the co-ordinate φ, the rotation angle, in accordance with the right-hand screw rule (see figure).

(a) Differentiating $\varphi(t)$ with respect to time twice, we get

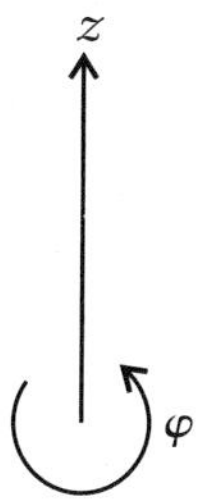

$$\frac{d\varphi}{dt} = a - 3bt^2 = \omega_z \tag{1}$$

$$\frac{d^2\varphi}{dt^2} = \frac{d\omega_z}{dt} = \beta_z = -6bt \tag{2}$$

From Eq. (1) the solid body comes to stop at

$$\Delta t = t = \sqrt{\frac{a}{3b}}$$

The angular velocity $\omega = a - 3bt^2$, for $0 \le t \le \sqrt{a/3b}$

So, $$\langle w \rangle = \frac{\int \omega\, dt}{\int dt} = \frac{\int_0^{\sqrt{a/3b}} (a - 3bt^2)\, dt}{\int_0^{\sqrt{a/3b}} dt}$$

$$= [at - bt^3]_0^{\sqrt{a/3b}} / \sqrt{a/3b} = 2a/3 = 4 \text{ rad/s}$$

Similarly, $\beta = |\beta_z| = 6bt$ for all values of t.

So, $$\langle \beta \rangle = \frac{\int \beta\, dt}{\int dt} = \frac{\int_0^{\sqrt{a/3b}} 6bt\, dt}{\int_0^{\sqrt{a/3b}} dt} = \sqrt{3ab} = 6 \text{ rad/s}^2$$

(b) From Eq. (2), $$\beta_z = -6bt$$

So, $$(\beta_z)_{t=\sqrt{a/3b}} = -6b\sqrt{\frac{a}{3b}} = -2\sqrt{3ab}$$

Hence, $$\beta = \left|(\beta_z)_{t=\sqrt{a/3b}}\right| = 2\sqrt{3ab} = 12 \text{ rad/s}^2$$

1.47 Angle α is related with $|w_t|$ and w_n by means of the formula

$$\tan\alpha = \frac{w_n}{|w_t|} \quad \text{where } w_n = \omega^2 R \quad \text{and} \quad |w_t| = \beta R \tag{1}$$

where R is the radius of the circle which an arbitrary point of the body circumscribes. From the given equation $\beta = d\omega/dt = at$ (here $\beta = d\omega/dt$ as β is positive for all values of t)

Integrating $$\int_0^{\omega} d\omega = a\int_0^{t} t\, dt \quad \text{or} \quad \omega = \frac{1}{2}at^2$$

So, $$w_n = \omega^2 R = \left(\frac{at^2}{2}\right)^2 R = \frac{a^2t^4}{4}R$$

and $$|w_t| = \beta R = atR$$

Putting the values of $|w_t|$ and w_n in Eq. (1), we get

$$\tan\alpha = \frac{a^2t^4R/4}{atR} = \frac{at^3}{4} \quad \text{or} \quad t = \left[\left(\frac{4}{a}\right)\tan\alpha\right]^{1/3}$$

$$t = 7\text{ s} \quad \text{(on substituting values)}$$

1.48 In accordance with the problem, $\beta_x < 0$.

Thus $$-\frac{d\omega}{dt} = k\sqrt{\omega} \quad \text{(where } k \text{ is proportionality constant)}$$

or $$-\int_{\omega_0}^{\omega}\frac{d\omega}{\sqrt{\omega}} = k\int_0^t dt \quad \text{or} \quad \sqrt{\omega} = \sqrt{\omega_0} - \frac{kt}{2} \tag{1}$$

When $\omega = 0$, total time of rotation $t = \tau = \dfrac{2\sqrt{\omega_0}}{k}$

Average angular velocity $$<\omega> = \frac{\int \omega\, dt}{\int dt} = \frac{\int_0^{2\sqrt{\omega_0}/k}\left(\omega_0 + \frac{k^2t^2}{4} - kt\sqrt{\omega_0}\right)dt}{2\sqrt{\omega_0}/k}$$

Hence, $$<\omega> = \frac{\left[\omega_0 t + \frac{k^2t^3}{12} - \frac{k}{2}\sqrt{\omega_0}\,t^2\right]_0^{2\sqrt{\omega_0}/k}}{2\sqrt{\omega_0}/k} = \frac{\omega_0}{3}$$

1.49 (a) We have $$\omega = \omega_0 - a\varphi = \frac{d\varphi}{dt}$$

Integrating this equation within its limit for $(\varphi)t$

$$\int_0^{\varphi}\frac{d\varphi}{\omega_0 - a\varphi} = \int_0^t dt \quad \text{or} \quad \ln\frac{\omega_0 - k\varphi}{\omega_0} = -kt$$

Hence, $$\varphi = \frac{\omega_0}{k}(1 - e^{-kt}) \tag{1}$$

(b) From the equation, $\omega = \omega_0 - k\varphi$ and Eq. (1) or by differentiating Eq. (1)

$$\omega = \omega_0 e^{-kt}$$

1.50 Let us choose the positive direction of z-axis (stationary rotation axis) along the vector β_0. In accordance with the equation

$$\frac{d\omega_z}{dt} = \beta_z \quad \text{or} \quad \omega_z\frac{d\omega_z}{d\varphi} = \beta_z$$

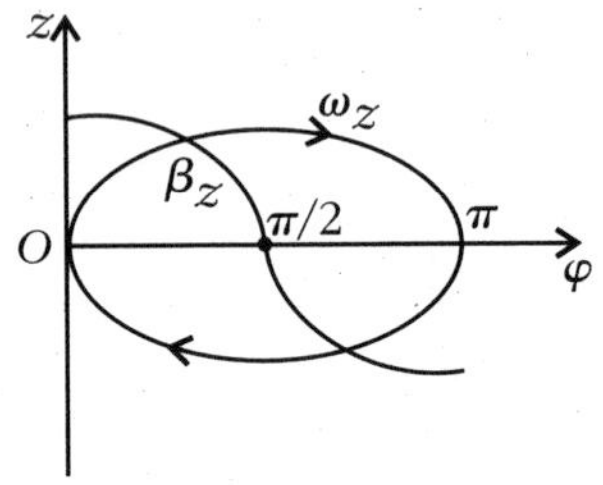

or $$\omega_z d\omega_z = \beta_z d\varphi = \beta\cos\varphi\, d\varphi$$

Integrating this equation within its limit for $\omega_z(\varphi)$

or $$\int_0^{\omega_z}\omega_z d\omega_z = \beta_0\sin\varphi$$

Hence, $$\omega_z = \pm\sqrt{2\beta_0\sin\varphi}$$

The plot $\omega_z(\varphi)$ is shown in the figure. It can be seen that as the angle φ grows, the vector $\boldsymbol{\omega}$ first increases, coinciding with the direction of the vector $\boldsymbol{\beta}_0(\omega_z > 0)$, reaches the maximum at $\varphi = \varphi/2$, then starts decreasing and finally turns into zero at $\varphi = \pi$. After that the body starts rotating in the opposite direction in a similar fashion ($\omega_z < 0$). As a result, the body will oscillate about the position $\varphi = \varphi/2$ with an amplitude equal to $\pi/2$.

1.51 A rotating disk moves along the x-axis, in plane motion in $x - y$ plane. For the calculation of velocity only, plane motion of a solid can be imagined to be in pure rotation about a point (say I) at a certain instant known as instantaneous centre of rotation. The axis is directed along $\boldsymbol{\omega}$ of the solid which passes through the point I at that instant and is known as instantaneous axis of rotation.

Therefore the velocity vector of an arbitrary point (P) of the solid can be represented as:

$$\mathbf{v}_P = \boldsymbol{\omega} \times \mathbf{r}_{PI} = \omega \times \boldsymbol{\rho}_{PI} \tag{1}$$

(where $\boldsymbol{\rho}_{PI}$ is normal location of point P relative to instantaneous rotation axis passing through point I).

So instantaneous rotation axis I is at the perpendicular distance $\rho_{PI} = v_P/\omega$ from point P. On the basis of Eq. (1) for the center of mass (C.M.) of the disk, velocity is

$$\mathbf{v}_C = \boldsymbol{\omega} \times \boldsymbol{\rho}_{CI} \tag{2}$$

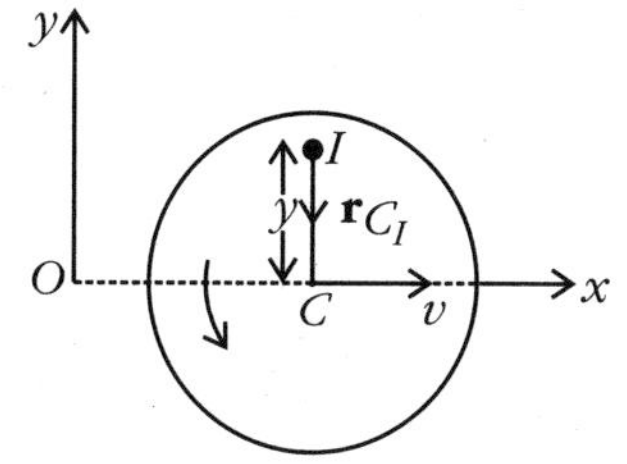

According to the problem $\mathbf{v}_C \uparrow\uparrow \mathbf{i}$ and $\boldsymbol{\omega} \uparrow\uparrow \mathbf{k}$ so to satisfy Eq. (2), $\boldsymbol{\rho}_{CI}$ is directed towards ($-\mathbf{j}$). Hence point I is at a distance $\boldsymbol{\rho}_{CI} = y$, above the centre of the disk along y-axis. Using all these facts in Eq. (2), we get

$$v_C = \omega y \quad \text{or} \quad y = \frac{v_C}{\omega} \tag{3}$$

(a) From the angular kinematical equation

$$\omega_z = \omega_{0z} + \beta_z t \tag{4}$$
$$\omega = \beta t$$

On the other hand

$$x = vt,\ t = x/v \quad \text{(where } x \text{ is the } x \text{ coordinate of the C.M.)} \tag{5}$$

From Eqs. (4) and (5),

$$\omega = \frac{\beta x}{v}$$

Using this value of ω in Eq. (3) we get

$$y = \frac{v_C}{\omega} = \frac{v}{\beta x/v} = \frac{v^2}{\beta x} \quad \text{(hyperbola)}$$

(b) As centre C moves with constant acceleration w, with zero initial velocity

So, $$x = \frac{1}{2}wt^2 \quad \text{and} \quad v_C = wt$$

Therefore, $$v_C = w\sqrt{\frac{2x}{w}} = \sqrt{2xw}$$

Hence, $$y = \frac{v_C}{\omega} = \frac{\sqrt{2wx}}{\omega} \quad \text{(parabola)}$$

1.52 (a) The general plane motion of a solid can be imagined as the combination of translation with C.M. and rotation about C.M.

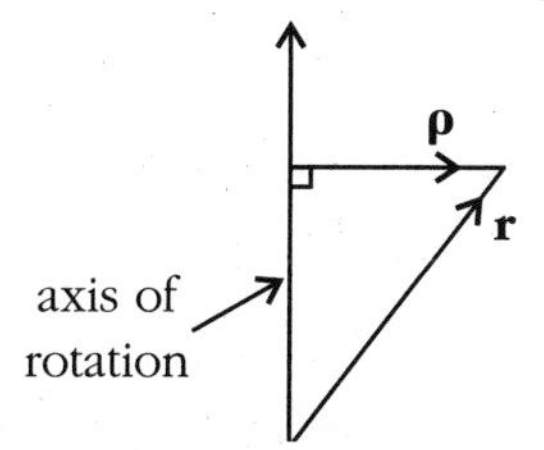

So, $$\mathbf{v}_A = \mathbf{v}_C + \mathbf{v}_{AC} = \mathbf{v}_C + \boldsymbol{\omega} \times \mathbf{r}_{AC} = \mathbf{v}_C + \boldsymbol{\omega} \times \boldsymbol{\rho}_{AC} \tag{1}$$

and $$\mathbf{w}_A = \mathbf{w}_C + \mathbf{w}_{AC} = \mathbf{w}_C + \boldsymbol{\omega} \times (\boldsymbol{\omega} \times \mathbf{r}_{AC}) + (\boldsymbol{\beta} \times \mathbf{r}_{AC})$$
$$= \mathbf{w}_C + \omega^2(-\boldsymbol{\rho}_{AC}) + (\boldsymbol{\beta} \times \boldsymbol{\rho}_{AC}) \tag{2}$$

where $\boldsymbol{\rho}$ is the component of $\mathbf{r}$ normal to the axis of rotation and directed away from it. In this problem $\rho_{AC} = r_{AC} = R$ and $v_C = v$.

Let the point A touch the horizontal surface at $t = 0$, further let us locate the point A at time t, when it makes an angle θ from vertical (see figure).

On the basis of Eqs. (1) and (2) the pictorial diagrams for velocity and acceleration are as follows:

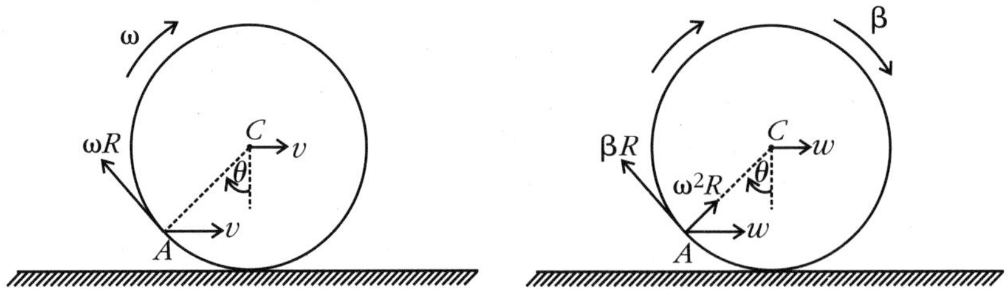

As the rolling is without slipping along a line, so, $v_C = \omega R$ and $w_C = \beta R$.

According to the problem $v_C = v$ (constant), so $\omega = v/R$, $w_C = 0$ and $\beta = 0$. Using these facts, $\mathbf{w}_A = v^2/R = 2.0\ \text{m/s}^2$ and the vector $\mathbf{w}$, is directed toward centre C of the wheel:

$$\begin{aligned} v_A &= \sqrt{v^2 + (\omega R)^2 + 2v(\omega R)\cos(\pi - \theta)} \\ &= \sqrt{v^2 + v^2 + 2v^2\cos(\pi - \theta)} = v\sqrt{2(1 - \cos\theta)} \\ &= 2v\sin(\theta/2) = 2v\sin(\omega t/2) \end{aligned}$$

Hence, distance covered by the point A during time interval $2\pi/\omega$

$$\begin{aligned} s &= \int v_A\, dt = \int_0^{2\pi/\omega} 2v\sin(\omega t/2)\, dt \\ &= \frac{8v}{\omega} = 8R = 4.0\ \text{m} \quad \text{(on substituting values)} \end{aligned}$$

Note: *One can easily find v_A, assuming the body to rotate about the instantaneous centre of rotation of zero velocity (not of zero acceleration), which is the contact point of the rolling body in this case.*

1.53 As the ball rolls without slipping on the rigid surface along a line so, contact point O has zero velocity, which is possible when the ball rotates in clockwise sense if it's centre moves towards right and such that $v_C = \omega R$, and $w_C = w = \beta R$. As motion starts at $t = 0$ with constant acceleration w so at time t, the velocity of mass centre C becomes $v_C = wt$ and $\omega = v_C/R = wt/R$.

(a) The contact point O becomes instantaneous centre of rotation, thus, the velocity of any arbitrary point P (say) of the ball can be obtained by the relation

$$\mathbf{v}_P = \boldsymbol{\omega} \times \boldsymbol{\rho}_{PO} \tag{1}$$

Using Eq. (1), the pictorial diagram to find the velocities of the points A and B of the ball is shown below

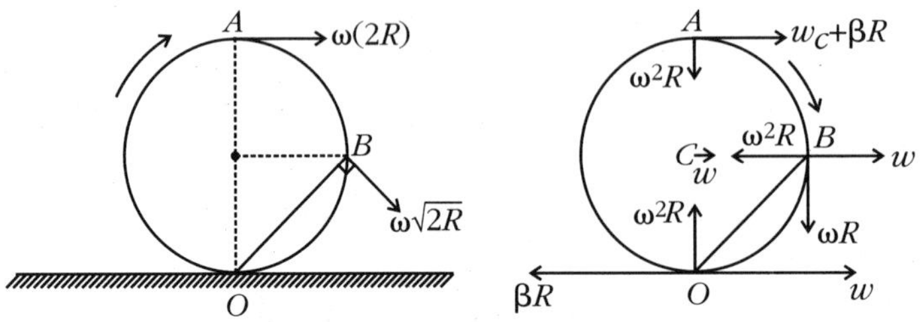

Hence, $v_A = 2v_C = 2wt = 10\ \text{cm/s}$

and $$v_B = \sqrt{2}\, v_C = \sqrt{2}\, wt = 7.1 \text{ cm/s}$$

(b) One can write for the any arbitrary point P

$$\mathbf{w}_P = \mathbf{w}_C + \mathbf{w}_{PC} = \mathbf{w}_C + \boldsymbol{\omega}^2(-\boldsymbol{\rho}_{PC}) + (\boldsymbol{\beta} \times \boldsymbol{\rho}_{PC}) \qquad (2)$$

Taking into account Eq. (2), one can easily get the acceleration of the points O, A and B on using the pictorial diagram.

So, $$w_0 = w^2t^2/R = 2.5 \text{ cm/s}^2$$

$$w_A = \sqrt{4w^2 + \frac{w^4t^4}{R^2}} = 2w\sqrt{1 + \left(\frac{wt^2}{2R}\right)^2} = 5.6 \text{ cm/s}^2$$

and $$w_B = \sqrt{\left(w - \frac{w^2t^2}{R}\right)^2 + w^2} = 2.5 \text{ m/s}^2$$

1.54 In the frame of horizontal plane the direction of velocity at a point is along the tangent of the curve at that point, so normal acceleration at that point is perpendicular to the direction of velocity vector and directed inwards. We make the pictorial diagram as in solution of problem 1.53.

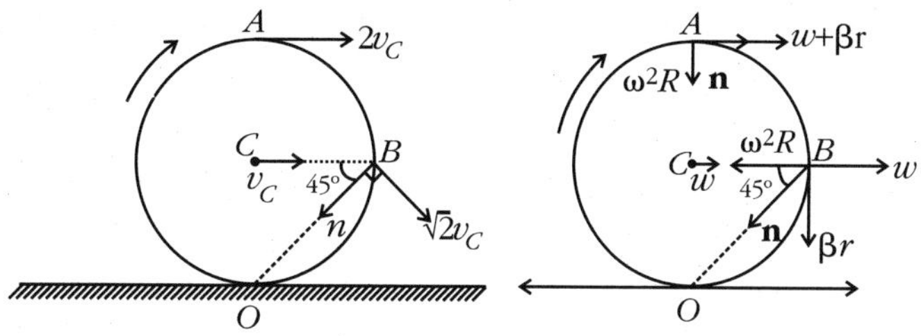

As an arbitrary point of the cylinder follows a curve, its normal acceleration and radius of curvature are related by the well-known equation

$$w_n = \frac{v^2}{R}$$

So, for point A, $$w_{A(n)} = \frac{v_A^2}{R_A}$$

or $$\omega^2 r = \frac{(2v_C)^2}{R_A}$$

which gives $$R_A = 4r \quad (\text{on using } v_C = \omega r)$$

Similarly for point B, $$w_{B(n)} = \frac{v_B^2}{R_B}$$

or $$\omega^2 r \cos 45° + \beta r \cos 45° - w \cos 45° = \frac{\left(\sqrt{2}v_C\right)^2}{R_B}$$

or $$\omega^2 r \cos 45° = \frac{\left(\sqrt{2}v_C\right)^2}{R_B}, \quad \text{as } w = \beta r \text{ due to pure rolling}$$

$$R_B = 2\sqrt{2}\frac{v_C^2}{\omega^2 r} = 2\sqrt{2}r \quad (\text{again on using } v_C = \omega r)$$

1.55 In rotating frame attached with solid 2, from the relation

$$\mathbf{v}' = \mathbf{v} - (\boldsymbol{\omega} \times \mathbf{r})$$
$$= (\boldsymbol{\omega}_1 \times \mathbf{r}) - (\boldsymbol{\omega}_2 \times \mathbf{r}) = (\boldsymbol{\omega}_1 - \boldsymbol{\omega}_2) \times \mathbf{r}$$

Hence sought relative angular velocity

$$\boldsymbol{\omega}_{12} = \boldsymbol{\omega}_1 - \boldsymbol{\omega}_2$$

As $\boldsymbol{\omega}_1 \perp \boldsymbol{\omega}_2$ So, $|\boldsymbol{\omega}_{12}| = \sqrt{\omega_1^2 + \omega_2^2} = 5$ rad/s (On substituting values)

If frame K' rotates with angular velocity $\boldsymbol{\omega}$ with respect to frame K, the relation between the time derivatives of any vector $\mathbf{a}$, seen from different frames is:

$$\left.\frac{d\mathbf{a}}{dt}\right|_K = \left.\frac{d\mathbf{a}}{dt}\right|_{K'} + \boldsymbol{\omega} \times \mathbf{a}$$

If the frame attached with the intersection point of two axes (about which solids are rotating) is K and the frame attached with rotating solid with angular velocity $\boldsymbol{\omega}_2$ is K'

Then, $$\left(\frac{d\boldsymbol{\omega}_1}{dt}\right)_K = \left(\frac{d\boldsymbol{\omega}_1}{dt}\right)_{K'} + \boldsymbol{\omega}_2 \times \boldsymbol{\omega}_1$$

However $\left(\frac{d\boldsymbol{\omega}}{dt}\right)_K = 0$, as the first body rotates with constant angular velocity in space,

and $$\left(\frac{d\boldsymbol{\omega}_1}{dt}\right)_{K'} = \boldsymbol{\beta}_{12} \quad \text{(the sought angular velocity)}$$

Hence, $$\boldsymbol{\beta}_{12} = \boldsymbol{\omega}_1 \times \boldsymbol{\omega}_2$$

So, $$|\boldsymbol{\beta}_{12}| = \omega_1\omega_2 = \text{rad/s}^2 \quad \text{(On substituting values)}$$

1.56 (a) We have $$\boldsymbol{\omega} = at\mathbf{i} + bt^2\mathbf{j} \tag{1}$$

So, $$\omega = \sqrt{(at)^2 + (bt^2)^2}, \text{ thus, } \omega|_{t=10\,s} = 7.81 \text{ rad/s}$$

Differentiating Eq. (1) with respect to time

$$\boldsymbol{\beta} = \frac{d\boldsymbol{\omega}}{dt} = a\mathbf{i} + 2bt\mathbf{j} \tag{2}$$

So, $$\beta = \sqrt{a^2 + (2bt)^2}$$

and $$\beta|_{t=10s} = 1.3 \text{ rad/s}^2$$

(b) $$\cos\alpha = \frac{\boldsymbol{\omega}\cdot\boldsymbol{\beta}}{\omega\beta} = \frac{(at\mathbf{i} + bt^2\mathbf{j})\cdot(a\mathbf{i} + 2bt\mathbf{j})}{\sqrt{(at)^2 + (bt^2)^2}\,\sqrt{a^2 + (2bt)^2}}$$

Putting the values of (*a*) and (*b*) and taking $t = 10$ s, we get

$$\alpha = 17°$$

1.57 (a) Let the axis of the cone (OC) rotate in anticlockwise sense with constant angular velocity $\boldsymbol{\omega}'$ and the cone itself about it's own axis (OC) in clockwise sense with angular velocity $\boldsymbol{\omega}_0$. Then the resultant angular velocity of the cone

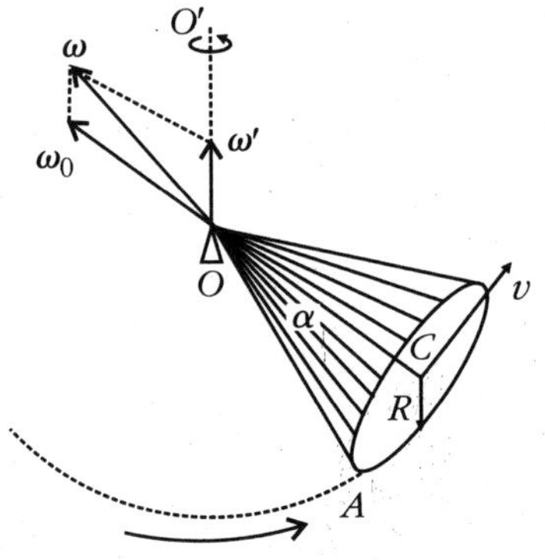

$$\boldsymbol{\omega} = \boldsymbol{\omega} + \boldsymbol{\omega}_0$$

As the rolling is pure the magnitudes of the vectors $\boldsymbol{\omega}'$ and $\boldsymbol{\omega}_0$ can be easily found from the figure as

$$\omega' = \frac{v}{R\cot\alpha} \quad \text{and} \quad \omega_0 = \frac{v}{R}$$

As $\boldsymbol{\omega} \perp \boldsymbol{\omega}_0$, hence

$$\omega = \sqrt{\omega'^2 + \omega_0^2} \sqrt{\left(\frac{v}{R\cot\alpha}\right)^2 + \left(\frac{v}{R}\right)^2}$$

$$= \frac{v}{R\cos\alpha} = 2.32 \text{ rad/s}$$

(b) Vector of angular acceleration

$$\boldsymbol{\beta} = \frac{d\boldsymbol{\omega}}{dt} = \frac{d(\boldsymbol{\omega}' + \boldsymbol{\omega}_0)}{dt} = \frac{d\boldsymbol{\omega}_0}{dt} \quad (\text{as } \boldsymbol{\omega}' = \text{constant})$$

If any vector **a** (say) rotates with angular velocity $\boldsymbol{\omega}$ keeping its value constant then $d\mathbf{a}/dt = \boldsymbol{\omega} \times \mathbf{a}$. The vector $\boldsymbol{\omega}_0$ keeping its magnitude constant rotates about the OO' axis with the angular velocity $\boldsymbol{\omega}'$. So, $d\boldsymbol{\omega}_0/dt = (\boldsymbol{\omega}' \times \boldsymbol{\omega}_0)$. Hence, $\boldsymbol{\beta} = \boldsymbol{\omega}' \times \boldsymbol{\omega}_0$. The magnitude of the vector $\boldsymbol{\beta}$ is equal to $\beta = \omega'\omega_0$ (as $\boldsymbol{\omega}' \perp \boldsymbol{\omega}_0$).

So, $$\beta = \frac{v}{R\cot\alpha}\left(\frac{v}{R}\right)$$

$$= \frac{v^2}{R^2}\tan\alpha = 2.3 \text{ rad/s}^2, \text{ on putting the values of } v, R \text{ and } \alpha.$$

Alternate:

Vector $\boldsymbol{\omega}_0$ is turning keeping the magnitude constant. To find $d\boldsymbol{\omega}_0$, let us make a vector diagram as shown in the figure. The circle shown in the figure has the radius ω_0 and the tail of $\boldsymbol{\omega}_0$ and of $\boldsymbol{\omega}_0 + d\boldsymbol{\omega}_0$ coincides at the centre of the circle.

Let $\boldsymbol{\omega}_0$ turn by the small angle $d\varphi$ in the differential time interval dt, so, $d\varphi = \omega' dt$. Thus, $|d\boldsymbol{\omega}_0| \approx$ arc length $= \omega_0 d\varphi = \omega_0\omega' dt$. As $\boldsymbol{\omega}_0$ is along radial line and $d\boldsymbol{\omega}_0$ is along the tangent in the turning sense of $\boldsymbol{\omega}_0$ so $d\boldsymbol{\omega}_0 \perp \boldsymbol{\omega}_0$. In vector form

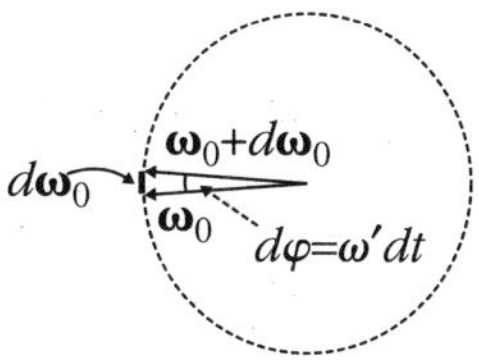

$$d\boldsymbol{\omega}_0 = (\boldsymbol{\omega}' \times \boldsymbol{\omega}_0)\, dt$$

Hence, $$\boldsymbol{\beta} = \frac{d\boldsymbol{\omega}_0}{dt} = (\boldsymbol{\omega}' \times \boldsymbol{\omega}_0)$$

1.58 The axis AB acquired the angular velocity

$$\boldsymbol{\omega}' = \boldsymbol{\beta}_0 t$$

Using the facts of the solution of problem 1.57, the angular velocity of the body is

$$\omega = \sqrt{\omega_0^2 + \omega'^2}$$

$$= \sqrt{\omega_0^2 + \beta_0^2\, t^2} = 0.6 \text{ rad/s}$$

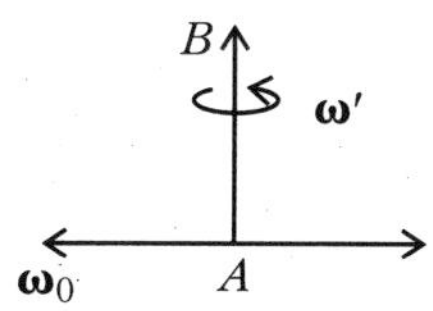

The angular acceleration,

$$\boldsymbol{\beta} = \frac{d\boldsymbol{\omega}'}{dt} = \frac{d(\boldsymbol{\omega}' + \boldsymbol{\omega}_0)}{dt} = \frac{d\boldsymbol{\omega}'}{dt} + \frac{d\boldsymbol{\omega}_0}{dt}$$

But, $$\frac{d\boldsymbol{\omega}_0}{dt} = \boldsymbol{\omega}' \times \boldsymbol{\omega}_0 \quad \text{and} \quad \frac{d\boldsymbol{\omega}'}{dt} = \boldsymbol{\beta}_0$$

So, $$\boldsymbol{\beta} = (\boldsymbol{\beta}_0 + \boldsymbol{\omega}' \times \boldsymbol{\omega}_0) = \boldsymbol{\beta}_0 + (\boldsymbol{\beta}_0 t \times \boldsymbol{\omega}_0) \quad (\text{because } \boldsymbol{\omega}' = \boldsymbol{\beta}_0 t)$$

As, $$\boldsymbol{\beta}_0 \perp \boldsymbol{\omega}_0 \text{ so, } \boldsymbol{\beta} = \sqrt{(\omega_0\beta_0 t)^2 + \beta_0^2} = \beta_0\sqrt{1 + (\omega_0 t)^2} = 0.2 \text{ rad/s}^2$$

1.2 The Fundamental Equation of Dynamics

1.59 Let R be the constant upward thrust on the aerostat of mass m, coming down with a constant acceleration w. Applying Newton's second law of motion for the aerostat in projection form

$$F_y = mw_y \tag{1}$$

$$mg - R = mw$$

Now, if Δm be the mass to be dumped, then using the equation $F_y = mw_y$

$$R - (m - \Delta m)g = (m - \Delta m)w \tag{2}$$

From Eqs. (1) and (2), we get

$$\Delta m = \frac{2mw}{g + w}$$

1.60 Let us write the fundamental equation of dynamics for all the three blocks in terms of projections, having taken the positive direction of x and y axes as shown in the figure and using the fact that kinematical relation between the accelerations is such that the blocks move with same value of acceleration (say w)

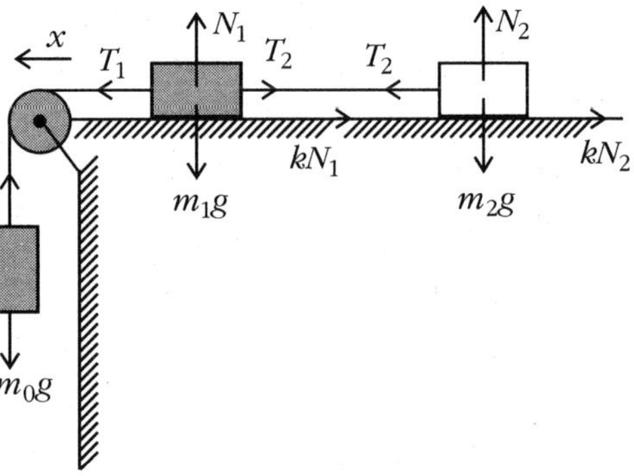

$$m_0g - T_1 = m_0w \tag{1}$$

$$T_1 - T_2 - km_1g = m_1w \tag{2}$$

and

$$T_2 - km_2g = m_2w \tag{3}$$

The simultaneous solution of Eqs. (1), (2) and (3) yields,

$$w = g\frac{[m_0 - k(m_1 + m_2)]}{m_0 + m_1 + m_2}$$

$$T_2 = \frac{(1+k)m_0}{m_0 + m_1 + m_2}m_2g$$

As the block m_0 moves down with acceleration w, so in vector form

$$\mathbf{w} = \frac{[m_0 - k(m_1 + m_2)]\mathbf{g}}{m_0 + m_1 + m_2}$$

1.61 (a) Let us indicate the positive the direction of x-axis along the incline (see figure). The figures show the force diagram for the blocks.

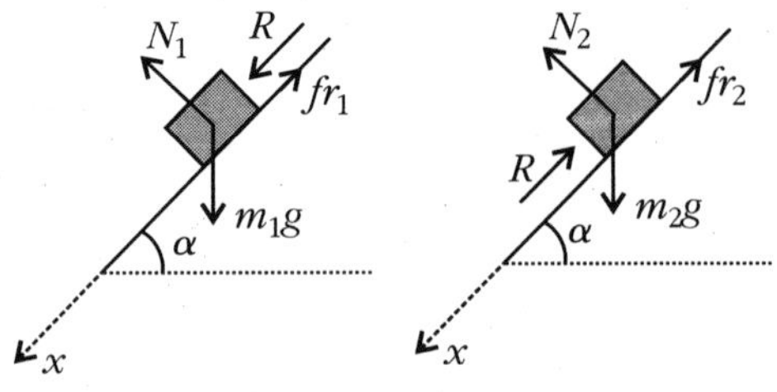

Let R be the force of interaction between the bars and they are obviously sliding down with the same constant acceleration w.

Newton's second law of motion in projection form along x-axis for the blocks gives

$$m_1g\sin\alpha - k_1m_1g\cos\alpha + R = m_1w \tag{1}$$

$$m_2g\sin\alpha - R - k_2m_2g\cos\alpha = m_2w \tag{2}$$

Solving Eqs. (1) and (2) simultaneously, we get

$$w = g\sin\alpha - g\cos\alpha\frac{k_1m_1 + k_2m_2}{m_1 + m_2} \quad (3)$$

and

$$R = \frac{m_1m_2(k_1 - k_2)g\cos\alpha}{m_1 + m_2}$$

(b) When the blocks just slide down the plane, $w = 0$, then from Eq. (3)

$$g\sin\alpha - g\cos\alpha\frac{k_1m_1 + k_2m_2}{m_1 + m_2} = 0$$

or

$$(m_1 + m_2)\sin\alpha = (k_1m_1 + k_2m_2)\cos\alpha$$

Hence,

$$\tan\alpha = \frac{(k_1m_1 + k_2m_2)}{m_1 + m_2}$$

1.62 Case 1: When the body is launched up.

Let k be the coefficient of friction, u the velocity of projection and l the distance traversed along the incline. Retarding force on the block $= mg\sin\alpha + kmg\cos\alpha$ and hence the retardation $= g\sin\alpha + kg\cos\alpha$.

Using the equation of particle kinematics along the incline,

$$0 = u^2 - 2(g\sin\alpha + kg\cos\alpha)l$$

or

$$l = \frac{u^2}{2(g\sin\alpha + kg\cos\alpha)} \quad (1)$$

and

$$0 = u - (g\sin\alpha + kg\cos\alpha)t$$

or

$$u = (g\sin\alpha + kg\cos\alpha)t \quad (2)$$

Using Eq. (2) in (1) we get

$$l = \frac{1}{2}(g\sin\alpha + kg\cos\alpha)t^2 \quad (3)$$

Case 2: When the block comes downward.

The net force on the body $= mg\sin\alpha - kmg\cos\alpha$ and hence its acceleration $= g\sin\alpha - kg\cos\alpha$.

Let, t' be the time required, then

$$l = \frac{1}{2}(g\sin\alpha - kg\cos\alpha)t'^2 \quad (4)$$

From Eqs. (3) and (4)

$$\frac{t^2}{t'^2} = \frac{\sin\alpha - k\cos\alpha}{\sin\alpha + k\cos\alpha}$$

But,

$$\frac{t}{t'} = \frac{1}{\eta} \quad \text{(according to the question)}$$

Hence, on solving we get

$$k = \frac{(\eta^2 - 1)}{(\eta^2 + 1)}\tan\alpha = 0.16$$

1.63 At the initial moment, obviously the tension in the thread connecting m_1 and m_2 equals the weight of m_2.

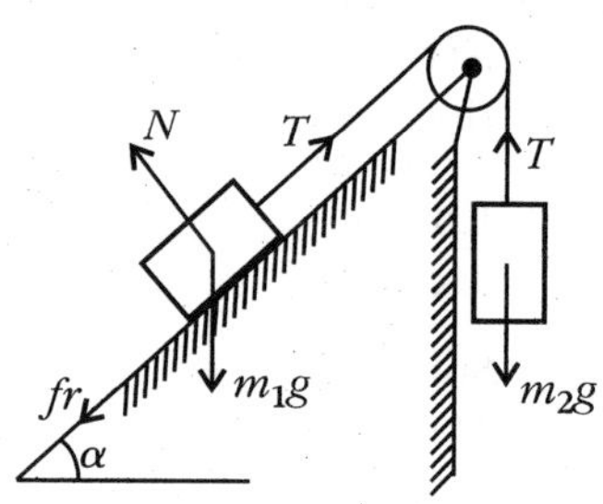

(a) For the block m_2 to come down or the block m_1 to go up, the conditions

$$m_2g - T \geq 0 \quad \text{and} \quad T - m_1g\sin\alpha - fr \geq 0$$

where T is tension and fr is friction which in the limiting case equals $km_1g\cos\alpha$.

Then
$$m_2g - m_1g\sin\alpha > km_1\,g\cos\alpha$$

or
$$\frac{m_2}{m_1} > (k\cos\alpha + \sin\alpha)$$

(b) Similarly in the case

$$m_1g\sin\alpha - m_2g > fr_{\text{lim}}$$

or
$$m_1g\sin\alpha - m_2g > km_1g\cos\alpha$$

or
$$\frac{m_2}{m_1} < (\sin\alpha - k\cos\alpha)$$

(c) For this case; neither kind of motion is possible, and fr need not be limiting.

Hence,
$$(k\cos\alpha + \sin\alpha) > \frac{m_2}{m_1} > (\sin\alpha - k\cos\alpha)$$

1.64 From the conditions obtained in the previous problem, first we will check whether the mass m_2 goes up or down.

Here, $m_2/m_1 = \eta > \sin\alpha + k\cos\alpha$, (substituting the values). Hence the mass m_2 will come down with an acceleration (say w). From the free-body diagram of previous problem,

$$m_2g - T = m_2w \tag{1}$$

and
$$T - m_1g\sin\alpha - km_1g\cos\alpha = m_1w \tag{2}$$

Adding Eqs. (1) and (2), we get,

$$m_2g - m_1g\sin\alpha - km_1g\cos\alpha = (m_1 + m_2)w$$

$$w = \frac{(m_2/m_1 - \sin\alpha - k\cos\alpha)g}{(1 + m_2/m_1)} = \frac{(\eta - \sin\alpha - k\cos\alpha)g}{1+\eta}$$

Substituting all the values, $w = 0.048\,g \approx 0.05\,g$

As m_2 moves down with acceleration of magnitude $w = 0.05\,g > 0$, thus in vector form, acceleration of m_2 is given by

$$\mathbf{w}_2 = \frac{(\eta - \sin\alpha - k\cos\alpha)\mathbf{g}}{1+\eta} = 0.05\,\mathbf{g}$$

1.65 Let us write Newton's second law in projection form along positive x-axis for the plank and the bar

$$fr = m_1w_1, \quad F - f_r = m_2w_2 \tag{1}$$

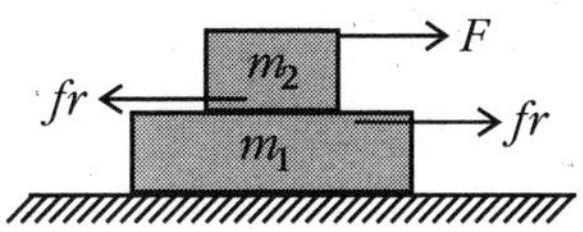

At the initial moment, *fr* represents the static friction, and as the force *F* grows so does the friction force *fr*, but up to its limiting value, i.e., $fr = fr_{s(\max)} = kN = km_2g$.

Unless this value is reached, both bodies move as a single body with equal acceleration. But as soon as the force *fr* reaches the limit, the bar starts sliding over the plank, i.e., $w_2 \geq w_1$.

Substituting here the values of w_1 and w_2 taken from Eq. (1) and taking into account that $fr = km_2g$, we obtain,

$$(at - m_2g)/m_2 \geq \frac{km_2}{m_1}g$$

(where the sign "=" corresponds to the moment $t = t_0$).

Hence, $$t_0 = \frac{kgm_2(m_1 + m_2)}{am_1}$$

If $t \leq t_0$, then $$w_1 = w_2 = \frac{at}{m_1 + m_2}$$

and if $t < t_0$, then $$w_1 = \frac{km_2g}{m_1} = \text{constant},\ w_2 = \frac{(at - km_2g)}{m_2}$$

On this basis $w_1(t)$ and $w_2(t)$ plots are as shown in the figure of answer sheet.

1.66 Let us designate the *x*-axis (see figure) and apply $F_x = mw_x$ for body *A*

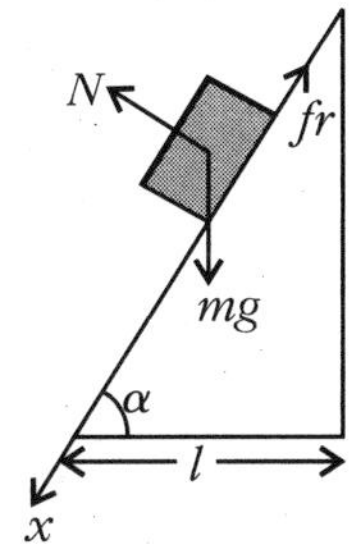

$$mg \sin \alpha - kmg \cos \alpha = mw$$

or $$w = g \sin \alpha - kg \cos \alpha$$

Now, from kinematical equation:

$$l \sec \alpha = 0 + (1/2)wt^2$$

or $$t = \sqrt{2l \sec \alpha/(\sin \alpha - k \cos \alpha)g}$$

$$= \sqrt{2l/(\sin 2\alpha/2 - k \cos^2 \alpha)g} \quad \text{(using Eq. 1)}$$

For $t_{\min}$, $$\frac{d\left(\frac{\sin 2\alpha}{2} - k \cos^2\alpha\right)}{d\alpha} = 0$$

i.e., $$\frac{2 \cos 2\alpha}{2} + 2k \cos \alpha \sin \alpha = 0$$

or $$\tan 2\alpha = -\frac{1}{k} \Rightarrow \alpha = 49°$$

and putting the values of α, k and l in Eq. (2) we get $t_{\min} = 1$ s.

1.67 Let us fix the $x - y$ co-ordinate system to the wedge, taking the *x*-axis up along the incline and the *y*-axis perpendicular to it (see figure).

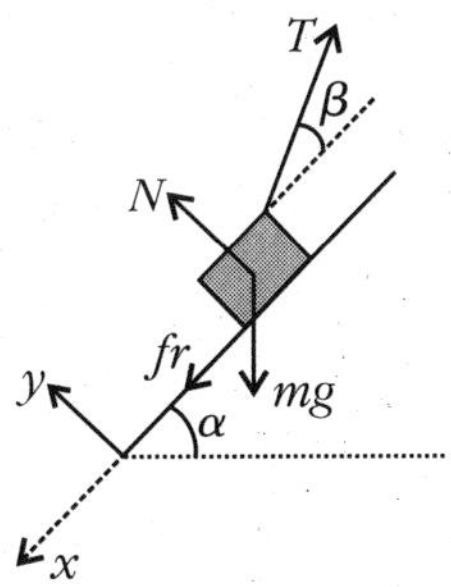

Now, we draw the free body diagram for the bar.

Let us apply Newton's second law in projection form along *x*- and *y*-axes for the bar

$$T\cos\beta - mg\sin\alpha - fr = 0 \quad (1)$$

$$T\sin\beta + N - mg\cos\alpha = 0$$

or $$N = mg\cos\alpha - T\sin\beta \tag{2}$$

But, $fr = kN$ and using Eq. (2) in Eq. (1), we get

$$T = mg\sin\alpha + kmg\cos\alpha/(\cos\beta + k\sin\beta) \tag{3}$$

For T_{min} the value of $(\cos\beta + k\sin\beta)$ should be maximum

So, $$\frac{d(\cos\beta + k\sin\beta)}{d\beta} = 0 \quad \text{or} \quad \tan\beta = k$$

Putting this value of β in Eq. (3), we get

$$T_{min} = \frac{mg(\sin\alpha + k\cos\alpha)}{\sqrt{1+k^2}}$$

1.68 First of all let us draw the free body diagram for the small body of mass m and indicate x-axis along the horizontal plane and y-axis, perpendicular to it, as shown in the figure. Let the block break off the plane at $t = t_0$, i.e., $N = 0$.

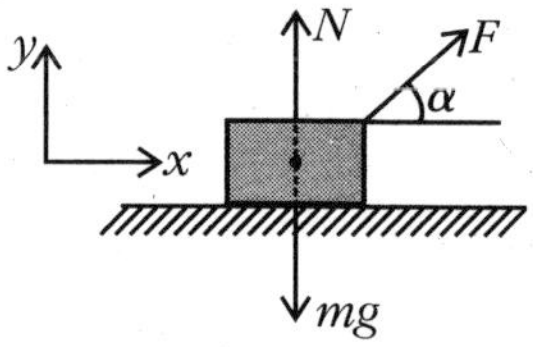

So, $$N = mg - at_0\sin\alpha = 0$$

or $$t_0 = \frac{mg}{a\sin\alpha} \tag{1}$$

From $F_x = mw_x$, for the body under investigation

$$\frac{mdv_x}{dt} = at\cos\alpha$$

Integrating within the limits for $v(t)$

$$m\int_0^v dv_x = a\cos\alpha\int_0^t tdt \quad \text{(using Eq. 1)}$$

So, $$v = \frac{ds}{dt} = \frac{a\cos\alpha}{2m}t^2 \tag{2}$$

Integrating, Eq. (2) for $s(t)$ we get

$$s = \frac{a\cos\alpha}{2m}\frac{t^3}{3} \tag{3}$$

Using the value of $t = t_0$ from Eq. (1), into Eqs. (2) and (3)

$$v = \frac{mg^2\cos\alpha}{2a\sin^2\alpha} \quad \text{and} \quad s = \frac{m^2g^3\cos\alpha}{6a^2\sin^3\alpha}$$

1.69 Newton's second law of motion in projection form, along horizontal or x-axis, i.e., $F_x = mw_x$ gives

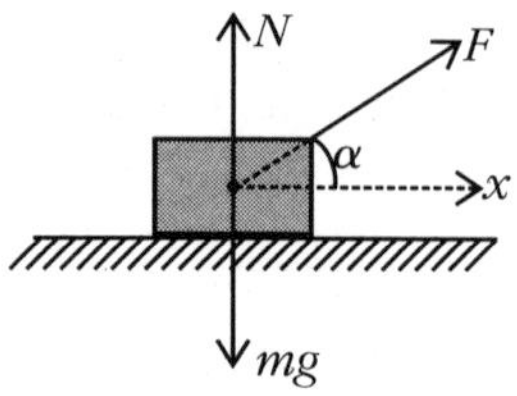

$$F\cos(as) = mv\frac{dv}{ds} \quad (\text{as } \alpha = as)$$

or $$F\cos(as)\,ds = mvdv$$

Integrating, over the limits for $v(s)$

$$\frac{F}{m}\int_0^x \cos(ax)\,dx = \frac{v^2}{2}$$

or $$v = \sqrt{\frac{2F \sin \alpha}{ma}} \quad (\text{as } \alpha = as)$$

$$= \sqrt{2g \sin \alpha / 3a} \quad (\text{using } F = mg/3)$$

This is the sought relationship.

1.70 From the Newton's second law in projection form

For the bar,

$$T - 2kmg = (2m)w \tag{1}$$

For the motor,

$$T - kmg = mw' \tag{2}$$

Now, from the equation of kinematics in the frame of bar or motor

$$l = \frac{1}{2}(w + w')t^2 \tag{3}$$

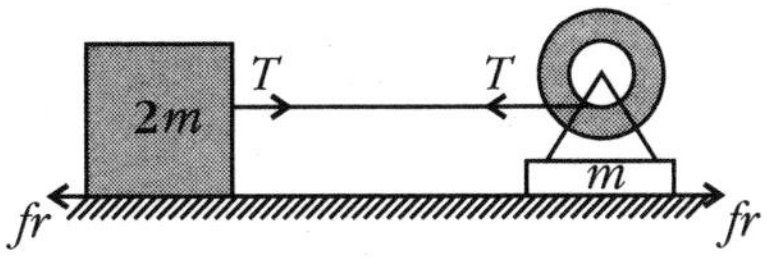

From Eqs. (1), (2) and (3) we get on eliminating T and w'

$$t = \sqrt{2l/(kg + 3w)}$$

1.71 (a) Let us write Newton's second law in vector form $\mathbf{F} = m\mathbf{w}$, for both the blocks (in the frame of ground)

$$\mathbf{T} + m_1\mathbf{g} = m_1\mathbf{w}_1 \tag{1}$$

$$\mathbf{T} + m_2\mathbf{g} = m_2\mathbf{w}_2 \tag{2}$$

These two equations contain three unknown quantities $\mathbf{w}_1$, $\mathbf{w}_2$ and $\mathbf{T}$. The third equation is provided by the kinematic relationship between the accelerations

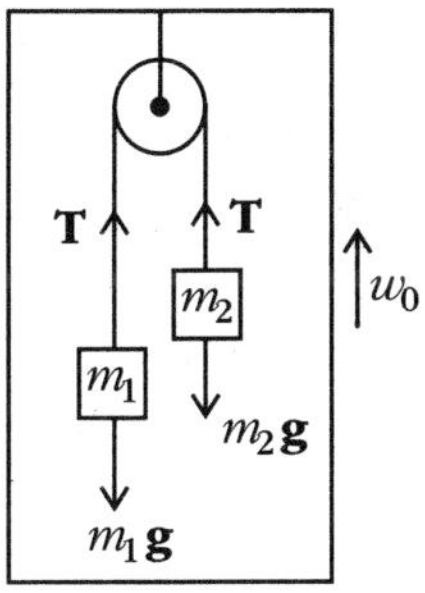

$$\mathbf{w}_1 = \mathbf{w}_0 + \mathbf{w}', \quad \mathbf{w}_2 = \mathbf{w}_0 - \mathbf{w}' \tag{3}$$

where $\mathbf{w}$ is the acceleration of the mass m_1 with respect to the pulley or elevator car.

Summing up term wise the left-hand and the right-hand sides of these kinematical equations, we get

$$\mathbf{w}_1 + \mathbf{w}_2 = 2\mathbf{w}_0 \tag{4}$$

The simultaneous solution of Eqs. (1), (2) and (4) yields

$$\mathbf{w}_1 = \frac{(m_1 - m_2)\mathbf{g} + 2m_2\mathbf{w}_0}{m_1 + m_2}$$

Using this result in Eq. (3), we get,

$$\mathbf{w}_1' = \frac{m_1 - m_2}{m_1 + m_2}(\mathbf{g} - \mathbf{w}_0) \quad \text{and} \quad \mathbf{T} = \frac{2m_1m_2}{m_1 + m_2}(\mathbf{w}_0 - \mathbf{g})$$

Using the results in Eq. (3) we get, $\mathbf{w} = \dfrac{m_1 - m_2}{m_1 + m_2}(\mathbf{g} - \mathbf{w}_0)$

(b) Obviously the force exerted by the pulley on the ceiling of the car

$$\mathbf{F} = -2\mathbf{T} = \frac{4m_1m_2}{m_1 + m_2}(\mathbf{g} - \mathbf{w}_0)$$

Note: *One could also solve this problem in the frame of elevator car.*

1.72 Let us write Newton's second law for both, bar 1 and body 2 in terms of projection having taken the positive direction of x_1 and x_2 as shown in the figure and assuming that body 2 starts sliding, say, upward along the incline.

$$T_1 - m_1 g \sin\alpha = m_1 w_1 \quad (1)$$

$$m_2 g - T_2 = m_2 w \quad (2)$$

For the pulley, moving in vertical direction from the equation $F_x = mw_x$

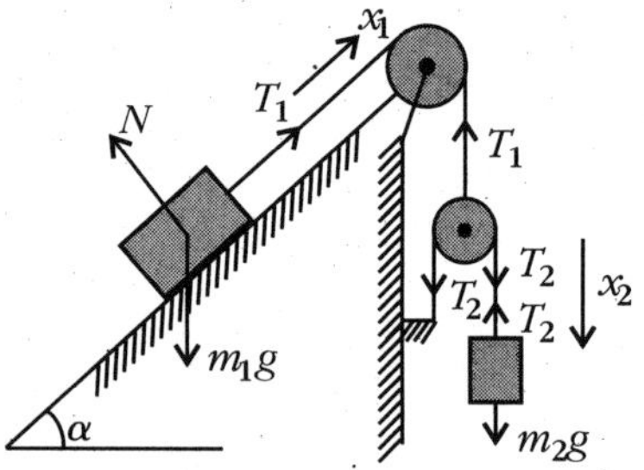

$$2T_2 - T_1 = (m_p)w_1 = 0$$

(as mass of the pulley $m_p = 0$)

or $$T_1 = 2T_2 \quad (3)$$

As the length of the threads are constant, the kinematic relationship of accelerations becomes

$$w = 2w_1 \quad (4)$$

Simultaneous solutions of all these equations yields

$$w = \frac{2g(2m_2/m_1 - \sin\alpha)}{(4m_2/m_1 + 1)} = \frac{2g(2\eta - \sin\alpha)}{4\eta + 1}$$

As $\eta > 1$, w is directed vertically downward, and hence in vector form

$$\mathbf{w} = \frac{2(2\eta - \sin\alpha)}{4\eta + 1}\mathbf{g}$$

1.73 Let us write Newton's second law for masses m_1 and m_2 and moving pulley in vertical direction along positive x-axis (see figure):

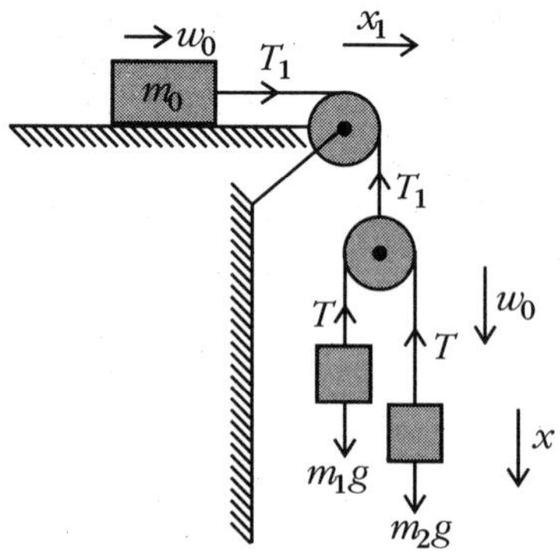

$$m_1 g - T = m_1 w_{1x}$$

$$m_2 g - T = m_2 w_{2x}$$

$$T_1 - 2T = 0 \quad (\text{as } m = 0)$$

or $$T_1 = 2T$$

Again using Newton's second law in projection form for mass m_0 along positive x_1 direction (see figure), we get

$$T_1 = m_0 w_0$$

The kinematic relationship between the accelerations of masses given in terms of projection on the x-axis

$$w_{1x} + w_{2x} = 2w_0$$

Simultaneous solution of the obtained five equations yields, in vector form

$$\mathbf{w}_1 = \frac{[4m_1m_2 + m_0(m_1 - m_2)]\mathbf{g}}{4m_1m_2 + m_0(m_1 + m_2)}$$

1.74 As the thread is not tied with m, so if there were no friction between the thread and the ball m, the tension in the thread would be zero and as a result both bodies will have free fall motion. Obviously in the given problem it is the friction force exerted by the ball on the thread, which becomes the tension in the thread. From the condition or language of the problem $w_M > w_m$ and as both are directed downward so, relative acceleration of $M = w_M - w_m$ and is directed downward. Kinematical equation for the ball in the frame of rod in projection form along upward direction gives

$$l = \frac{1}{2}(w_M - w_m)t^2 \quad (1)$$

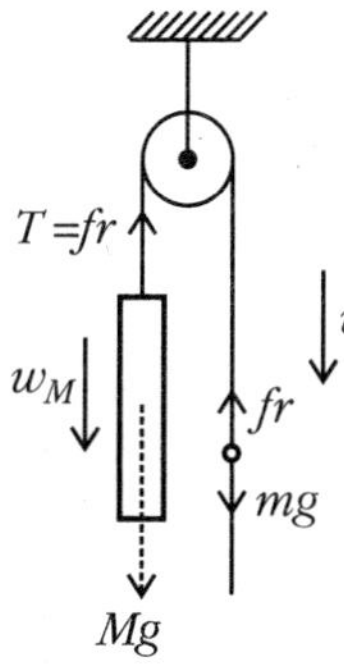

Newton's second law in projection form along vertically down direction for both, rod and ball gives,

$$Mg - fr = Mw_M \quad (2)$$

$$mg - fr = mw_m \quad (3)$$

Multiplying Eq. (2) by m and Eq. (3) by M and then subtracting Eq. (3) from Eq. (2) and after using Eq. (1), we get

$$fr = \frac{2lMm}{(M-m)t^2}$$

1.75 Suppose, the ball goes up with acceleration w_1 and the rod comes down with the acceleration w_2.

As the length of the thread is constant, $2w_1 = w_2$ (1)

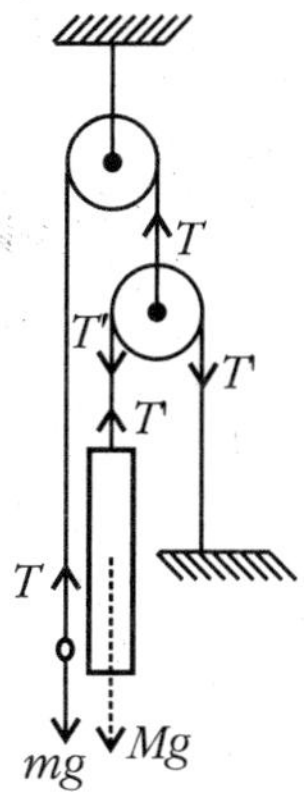

Newton's second law in projection form, vertically upwards for the ball and vertically downwards for the rod respectively gives,

$$T - mg = mw_1 \quad (2)$$

and $$Mg - T' = Mw_2 \quad (3)$$

but, $$T = 2T' \text{ (because pulley is massless)} \quad (4)$$

From Eqs. (1), (2), (3) and (4)

$$w_1 = \frac{(2M-m)g}{m+4M} = \frac{(2-\eta)g}{\eta+4} \text{ (upwards)}$$

and $$w_2 = \frac{2(2-\eta)g}{(\eta+4)} \text{ (downwards)}$$

From kinematical equation in projection form, we get

$$l = \frac{1}{2}(w_1 + w_2)t^2$$

as w_1 and w_2 are in the opposite direction.

Putting the values of w_1 and w_2, the sought time becomes

$$t = \sqrt{2l(\eta+4)/3(2-\eta)g} = 1.4 \text{ s}$$

1.76 Using Newton's second law in projection form along x-axis for the body 1 and along negative x-axis for the body 2, respectively, we get

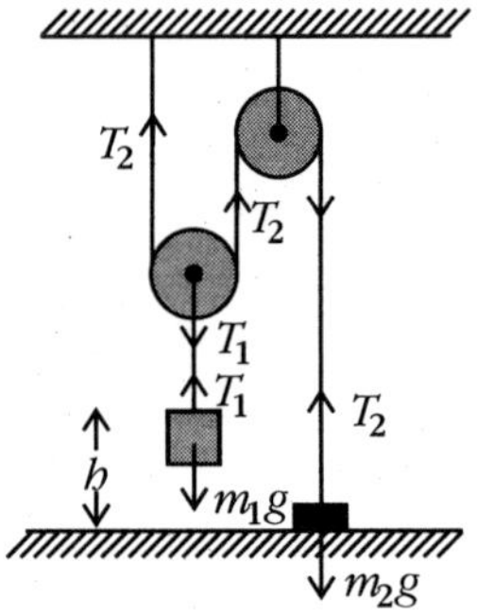

$$m_1g - T_1 = m_1w_1 \quad (1)$$

$$T_2 - m_2g = m_2w_2 \quad (2)$$

For the pulley lowering in downward direction from along x-axis, Newton's law gives

$$T_1 - 2T_2 = 0 \quad \text{(as pulley is massless)}$$

or

$$T_1 = 2T_2 \quad (3)$$

As the length of the thread is constant, so

$$w_2 = 2w_1 \quad (4)$$

The simultaneous solution of above equations yields

$$w_2 = \frac{2(m_1 - 2m_2)g}{4m_2 + m_1} = \frac{2(\eta - 2)}{\eta + 4}g \quad \left(\text{as} \frac{m_1}{m_2} = \eta\right) \quad (5)$$

Obviously during the time interval in which the body 1 comes to the horizontal floor covering the distance h, the body 2 moves upward the distance $2h$. At the moment when the body 2 is at the height $2h$ from the floor its velocity is given by the expression

$$v_2^2 = 2w_2(2h) = 2\left[\frac{2(\eta - 2)g}{\eta + 4}\right]2h = \frac{8h(\eta - 2)g}{\eta + 4}$$

After the body m_1 touches the floor, the thread becomes slack or the tension in the thread zero, thus as a result body 2 is only under gravity for it's subsequent motion.

Owing to the velocity v_2 at that moment or at the height $2h$ from the floor, the body 2 further goes up under gravity by the distance,

$$h' = \frac{v_2^2}{2g} = \frac{4h(\eta - 2)}{\eta + 4}$$

Thus the sought maximum height attained by body 2 is

$$H = 2h + h' = 2h + \frac{4h(\eta - 2)}{(\eta + 4)}$$

$$= \frac{6\eta h}{\eta + 4} = 0.6 \text{ m} \quad \text{(on substituting values)}$$

1.77 Let us draw free body diagram of each body, i.e., of rod A and of wedge B and also draw the kinematical diagram for accelerations, after analysing the directions of motion of A and B. The kinematic relationship of accelerations is

$$\tan\alpha = \frac{w_A}{w_B} \quad (1)$$

Let us write Newton's second law for both bodies in terms of projections having taken positive directions of y- and x-axes as shown in the figure:

$$m_Ag - N\cos\alpha = m_AW_A \quad (2)$$

and

$$N\sin\alpha = m_Bw_B \quad (3)$$

Simultaneous solution of Eqs. (1), (2) and (3) yields

$$w_A = \frac{m_A g \sin\alpha}{m_A \sin\alpha + m_B \cot\alpha \cos\alpha} = \frac{g}{(1+\eta \cot^2\alpha)}$$

and
$$w_B = \frac{w_A}{\tan\alpha} = \frac{g}{(\tan\alpha + \eta\cot\alpha)}$$

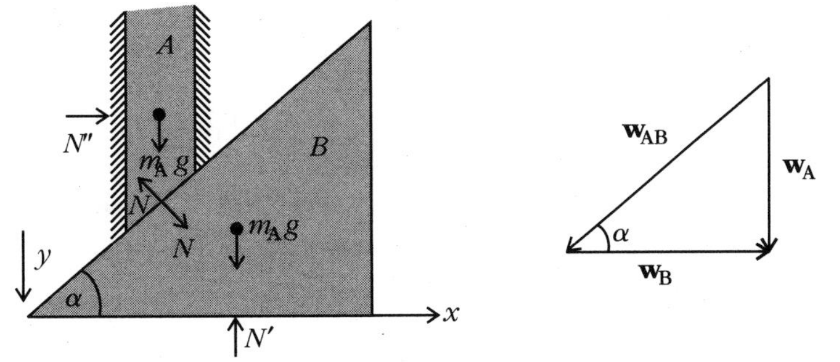

Note: *We may also solve this problem using conservation of mechanical energy instead of Newton's second law.*

1.78 Let us draw free body diagram of each body and fix the coordinate system, as shown in the figure. After analysing the motion of M and m on the basis of force diagrams, let us draw the kinematic diagram for accelerations (see figure).

As the length of thread is constant so,

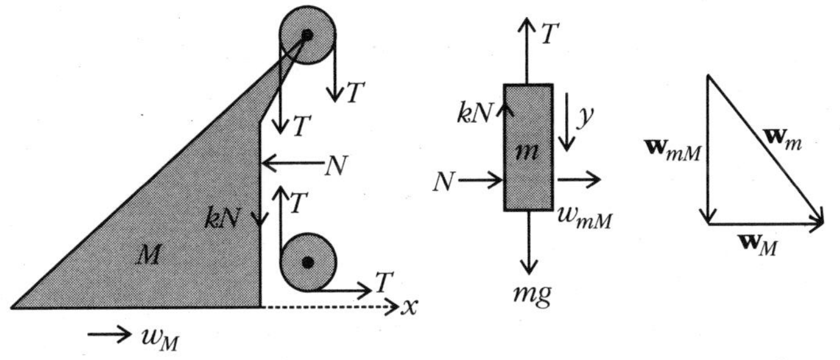

$ds_{mM} = ds_M$ and as $\mathbf{v}_{mM}$ and $\mathbf{v}_M$ do not change their directions that's why

$$|\mathbf{w}_{mM}| = |\mathbf{w}_M| = w \quad \text{(say)}$$

As
$$\mathbf{w}_m = \mathbf{w}_{mM} + \mathbf{w}_M \tag{1}$$

So, from the triangle law of vector addition

$$w_M = \sqrt{2}w$$

From equation $F_x = mw_x$, for the wedge and the block,

$$T - N = Mw \tag{2}$$

and
$$N = mw \tag{3}$$

Now, from equation $F_y = mw_y$, for the block,

$$mg - T - kN = mw \tag{4}$$

Simultaneous solution of Eqs. (2), (3) and (4) yields

$$w = \frac{mg}{(km + 2m + M)} = \frac{g}{(k + 2 + M/m)}$$

Hence, using Eq. (1)

$$w_m = \frac{g\sqrt{2}}{(2 + k + M/m)}$$

1.79 Bodies 1 and 2 will remain at rest with respect to bar A for $w_{min} \le w \le w_{max}$ where w_{min} is the sought minimum acceleration of the bar. Beyond these limits there will be a relative motion between bar and the bodies. For $0 \le w \le w_{min}$, the tendency of body 1 in relation to the bar A is to move towards right and is in the opposite sense for $w \ge w_{max}$. On the basis of above argument the static friction on 2 by A is directed upward and on 1 by A is directed towards left for the purpose of calculating w_{min}.

Let us write Newton's second law for bodies 1 and 2 in terms of projection along positive x-axis (see figure).

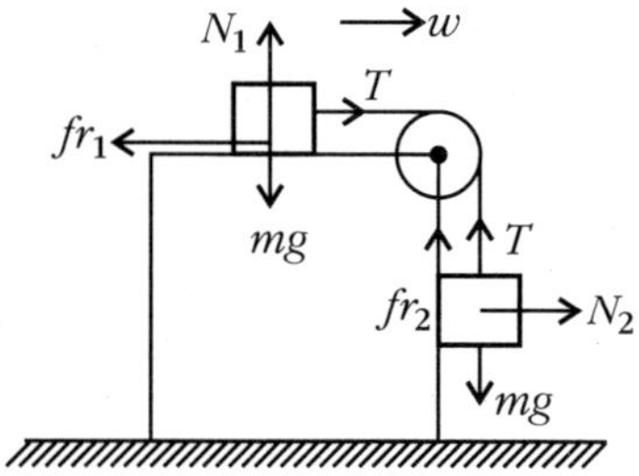

$$T - fr_1 = mw \quad \text{or} \quad fr_1 = T - mw \qquad (1)$$

$$N_2 = mw \qquad (2)$$

As body 2 has no acceleration in vertical direction, so

$$fr_2 = mg - T \qquad (3)$$

From Eqs. (1) and (3)

$$(fr_1 + fr_2) = m(g - w) \qquad (4)$$

But, $\quad fr_1 + fr_2 \le k(N_1 + N_2)$

or $\quad fr_1 + fr_2 \le k(mg + mw) \qquad (5)$

From Eqs. (4) and (5)

$$m(g - w) \le mk(g + w), \quad \text{or} \quad w \ge w \ge \frac{g(1-k)}{(1+k)}$$

Hence, $\quad w_{min} = \dfrac{g(1-k)}{(1+k)}$

1.80 On the basis of the initial argument of the solution of problem 1.79, the tendency of bar 2 with respect to 1 will be to move up along the plane.

Let us fix x-y coordinate system in the frame of ground as shown in the figure.

From second law of motion in projection form along y- and x-axes

$$mg \cos\alpha - N = mw \sin\alpha$$

or $\quad N = m(g \cos\alpha - w \sin\alpha) \qquad (1)$

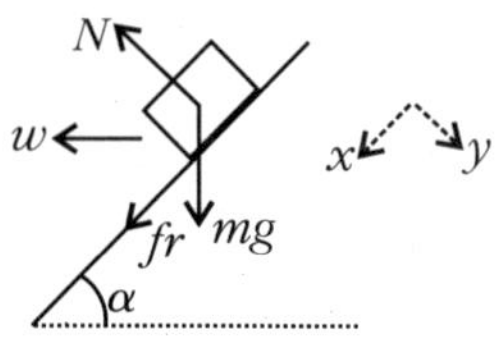

$$mg \sin\alpha + fr = mw \cos\alpha$$

or $\quad fr = m(w \cos\alpha - g \sin\alpha) \qquad (2)$

but $\quad fr \le kN$

So from Eqs. (1) and (2), $(w\cos\alpha - g\sin\alpha) \le k(g\cos\alpha + w\sin\alpha)$

or $\quad w(\cos\alpha - k\sin\alpha) \le g(k\cos\alpha + \sin\alpha)$

or $$w \le g\frac{(k\cos\alpha + \sin\alpha)}{\cos\alpha - k\sin\alpha}$$

So, the sought maximum acceleration of the wedge is

$$w_{max} = \frac{(k\cos\alpha + \sin\alpha)g}{\cos\alpha - k\sin\alpha} = \frac{(k\cot\alpha + 1)g}{\cot\alpha - k} \quad \text{where } \cot\alpha > k$$

1.81 Let us draw the force diagram of each body, and on this basis we observe that the prism moves towards right (say) with an acceleration w_1 and the bar 2 of mass m_2 moves down the plane with respect to 1, say with acceleration $\mathbf{w}_{21}$, then $\mathbf{w}_2 = \mathbf{w}_{21} + \mathbf{w}_1$ (see figure).

Let us write Newton's second law of both bodies in projection form along positive y_2 and x_1 axes as shown in the figure.

$$m_2 g\cos\alpha - N = m_2 w_{2(y_2)} = m_2[w_{21(y_2)} + w_{1(y_2)}] = m_2[0 + w_1 \sin\alpha]$$

or $$m_2 g\cos\alpha - N = m_2 w_1 \sin\alpha \quad (1)$$

and $$N\sin\alpha = m_1 w_1 \quad (2)$$

Solving Eqs. (1) and (2), we get

$$w_1 = \frac{m_2 g\sin\alpha\cos\alpha}{m_1 + m_2\sin^2\alpha} = \frac{g\sin\alpha\cos\alpha}{(m_1/m_2) + \sin^2\alpha}$$

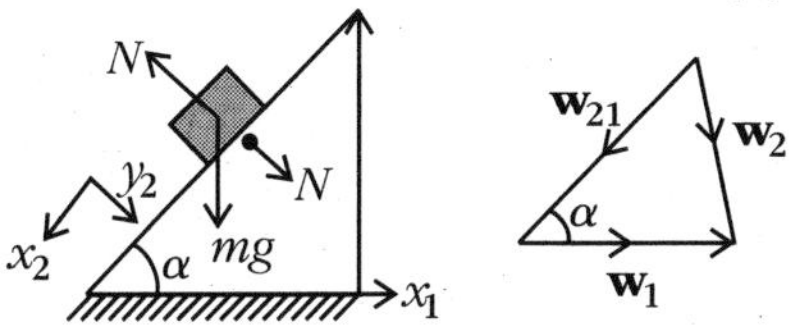

1.82 To analyse the kinematical relations between the bodies, sketch the force diagram of each body as shown in the figure.

On the basis of force diagram, it is obvious that the wedge M will move towards right and the block will move down along the wedge. As the length of the thread is constant, the distance traveled by the block on the wedge must be equal to the distance traveled by the wedge on the floor. Hence, $ds_{mM} = ds_M$. As $\mathbf{v}_{mM}$ and $\mathbf{v}_M$ do not change their directions and acceleration that's why $\mathbf{w}_{mM} \uparrow\uparrow \mathbf{v}_{mM}$ and $\mathbf{w}_M \uparrow\uparrow \mathbf{v}_M$ and $w_{mM} = w$ (say) and accordingly the diagram of kinematical dependence is shown in the figure.

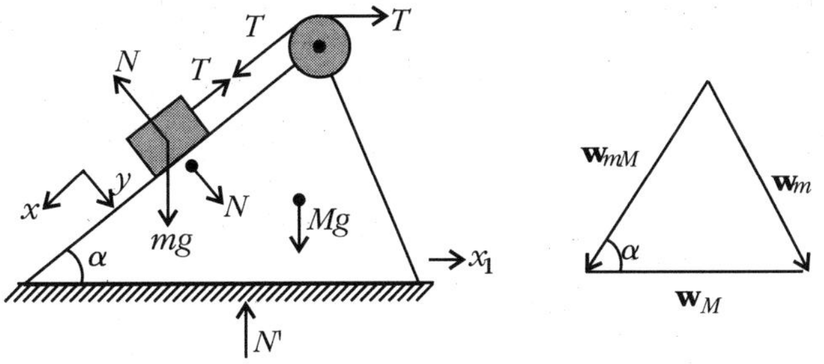

As $\mathbf{w}_m = \mathbf{w}_{mM} + \mathbf{w}_M$, so from triangle law of vector addition

$$w_M = \sqrt{w_M^2 + w_M^2 - 2\,w_{mM}w_M\cos\alpha} = w\sqrt{2(1-\cos\alpha)} \quad (1)$$

From $F_x = mw_x$, (for the wedge),

$$T - T\cos\alpha + N\sin\alpha = Mu \quad (2)$$

For the bar m let us fix x-y coordinate system in the frame of ground. Newton's law in projection form along x- and y-axes (see figure) gives

$$mg\sin\alpha - T = mw_{m(x)} = m[w_{mM(x)} + w_{M(x)}]$$

$$= m[w_{mM} + w_M\cos(\pi - \alpha)] = mw(1 - \cos\alpha) \quad (3)$$

$$mg\cos\alpha - N = mw_{m(y)} = m[w_{mM(y)} + w_{m(y)}] = m[0 + w\sin\alpha] \quad (4)$$

Solving the above equations simultaneously, we get

$$w = \frac{mg \sin \alpha}{M + 2m(1 - \cos\alpha)}$$

Note: *We can study the motion of the block m in the frame of wedge also, alternately we may solve this problem using conservation of mechanical energy.*

1.83 Let us sketch the diagram for the motion of the particle of mass m along the circle of radius R as shown in the figure.

For a particle, modulus of change in linear momentum $|\Delta \mathbf{p}| = m|\Delta \mathbf{v}|$.

(a) In this case $|\Delta \mathbf{v}| = \sqrt{2}v$ (see figure), and the time taken in describing quarter of the circle,

$$\Delta t = \frac{\pi R}{2v}$$

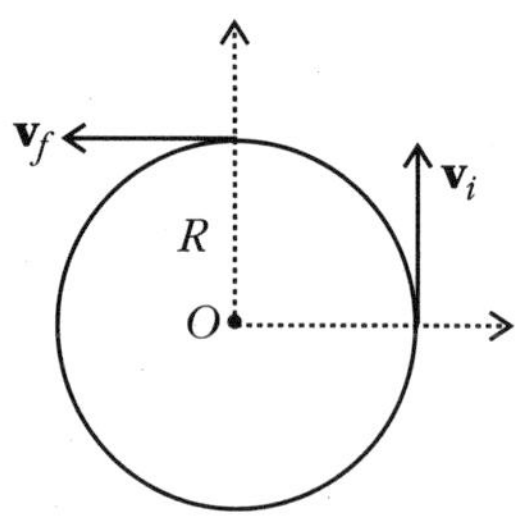

Hence, $$|\langle \mathbf{F} \rangle| = \frac{|\Delta \mathbf{p}|}{\Delta t} = \frac{m|\Delta \mathbf{v}|}{\Delta t} = \frac{\sqrt{2}\, mv}{\pi R/2v} = \frac{2\sqrt{2}\, mv^2}{\pi R}$$

(b) In this case $\mathbf{v}_i = 0$ so, $\mathbf{v}_f = \mathbf{v}(t)$. Thus $|\Delta \mathbf{v}| = |\mathbf{v}(t)| = v(t) = w_t t$.

Hence, $$|\langle \mathbf{F} \rangle| = \frac{|\Delta \mathbf{p}|}{\Delta t} = \frac{m|\Delta \mathbf{v}|}{\Delta t} = \frac{m|\mathbf{v}(t)|}{t} = mw_t$$

1.84 While moving in a loop, normal reaction exerted by the flyer on the loop at different points and uncompensated weight if any contribute to the weight of flyer at those points.

(a) When the aircraft is at the lowermost point, Newton's second law of motion in projection form, $F_n = mw_n$ gives

$$N - mg = \frac{mv^2}{R}$$

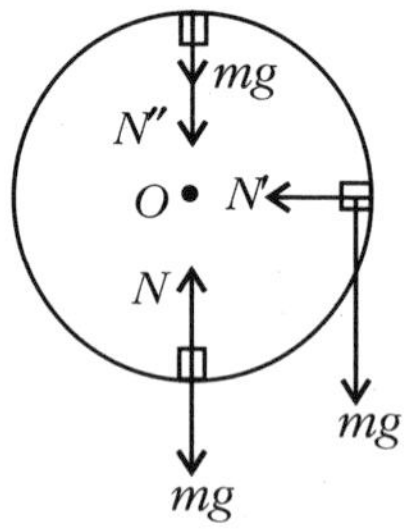

or $$N = mg + \frac{mv^2}{R} = 2.09 \text{ kN}$$

(b) When it is at the upper most point, again from $F_n = mw_n$, we get

$$N'' + mg = \frac{mv^2}{R}$$

$$N'' = \frac{mv^2}{R} - mg = 0.7 \text{ kN}$$

(c) When the aircraft is at the middle point of the loop, again from $F_n = mw_n$

$$N' = \frac{mv^2}{R} = 1.4 \text{ kN}$$

The uncompensated weight is mg. Thus, effective weight $= \sqrt{N'^2 + m^2g^2} = 1.56$ kN acts obliquely.

1.85 (a) Let us depict the forces acting on the small sphere m, (at an arbitrary position when the thread makes an angle θ from the vertical) and write equation $\mathbf{F} = m\mathbf{w}$ via projection on the unit vectors $\mathbf{t}$ and $\mathbf{n}$. From $F_t = mw_t$, we have

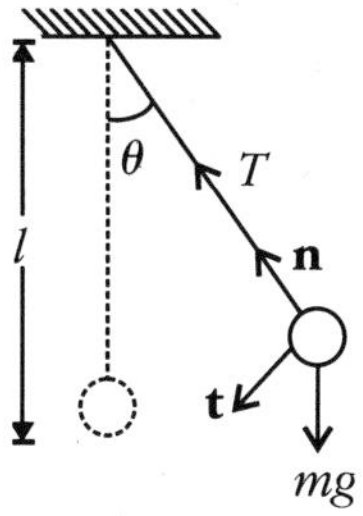

$$mg\sin\theta = m\frac{dv}{dt}$$

$$= m\frac{vdv}{dt} = m\frac{vdv}{l(-d\theta)}$$

(as vertical is reference line of angular position)

or $$vdv = -gl\sin\theta\, d\theta$$

Integrating both the sides we get

$$\int_0^v vdv = -gl\int_{\pi/2}^{\theta}\sin\theta\, d\theta$$

or $$\frac{v^2}{2} = gl\cos\theta$$

Hence, $$\frac{v^2}{l} = 2g\cos\theta = w_n \tag{1}$$

Note: *Eq. (1) can be easily obtained by the conservation of mechanical energy.*

From $$F_n = mw_n$$

$$T - mg\cos\theta = \frac{mv^2}{l}$$

Using Eq. (1) we have

$$T = 3mg\cos\theta \tag{2}$$

Again from equation $F_t = mw_t$,

$$mg\sin\theta = mw_t \quad \text{or} \quad w_t = g\sin\theta \tag{3}$$

Hence, $$w = \sqrt{w_t^2 + w_n^2} = \sqrt{(g\sin\theta)^2 + (2g\cos\theta)^2} \quad \text{(using Eqs. 1 and 3)}$$

$$= g\sqrt{1 + 3\cos^2\theta}.$$

(b) Vertical component of velocity, $v_y = v\sin\theta$

So, $$v_y^2 = v^2\sin^2\theta = 2gl\cos\theta\sin^2\theta \quad \text{(using Eq. 1)}$$

For maximum v_y or v_y^2, $$\frac{d(\cos\theta\sin^2\theta)}{d\theta} = 0$$

which yields $$\cos\theta = \frac{1}{\sqrt{3}}$$

Therefore from Eq. (2), $$T = 3mg\frac{1}{\sqrt{3}} = \sqrt{3}\, mg$$

(c) We have $\mathbf{w} = w_t\mathbf{t} + w_n\mathbf{n}$ thus $w_y = w_{t(y)} + w_{n(y)}$

But in accordance with the problem $w_y = 0$

So, $$w_{t(y)} + w_{n(y)} = 0$$

or $$-g\sin\theta\sin\theta + 2g\cos^2\theta = 0$$

or $$\cos\theta = \frac{1}{\sqrt{3}} \quad \text{or} \quad \theta = 54.7°$$

1.86 The ball has only normal acceleration at the lowest position and only tangential acceleration at either of the extreme positions. Let v be the speed of the ball at its lowest position and l be length of the thread, then according to the problem

$$\frac{v^2}{l} = g\sin\alpha \tag{1}$$

where α is the maximum deflection angle.

From Newton's law in projection form:

$$F_t = mw_t$$

$$-mg\sin\theta = mv\frac{dv}{ld\theta}$$

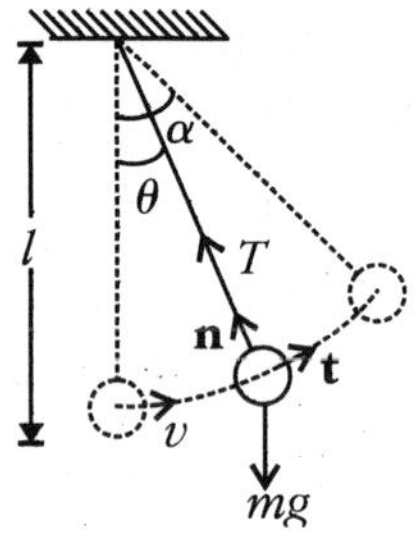

or $$-gl\sin\theta\, d\theta = vdv$$

On integrating both the sides within their limits,

$$-gl\int_0^{\alpha}\sin\theta d\theta = \int_v^0 vdv$$

or $$v^2 = 2gl(1 - \cos\alpha) \tag{2}$$

Note: *Eq. (2) can easily be obtained by the conservation of mechanical energy of the ball in the uniform field of gravity.*

From Eqs. (1) and (2) with $\theta = \alpha$

$$\sin\alpha = 2(1 - \cos\alpha)$$

or $$\sin\alpha + 2\cos\alpha = 2$$

On solving we get,

$$\alpha \cong 53°$$

1.87 Let us depict the forces acting on the body A (which are the force of gravity mg and the normal reaction N) and write equation $\mathbf{F} = m\mathbf{w}$ via projection on the unit vectors $\mathbf{t}$ and $\mathbf{n}$ (see figure).

From $F_t = mw_t$

$$mg\sin\theta = m\frac{dv}{dt}$$

$$= m\frac{vdv}{ds} = m\frac{vdv}{Rd\theta}$$

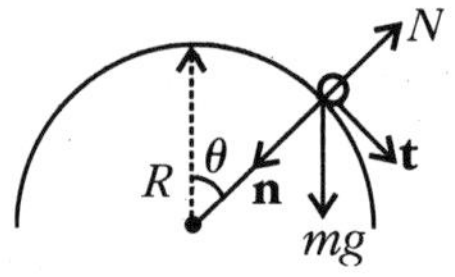

or $$gR\sin\theta\, d\theta = vdv$$

Integrating both sides for obtaining $v(\theta)$, we get

$$\int_0^{\theta} gR\sin\theta\, d\theta' = \int_0^{v} vdv$$

or $$v^2 = 2gR(1 - \cos\theta) \tag{1}$$

From $F_n = mw_n$

$$mg\cos\theta - N = m\frac{v^2}{R} \tag{2}$$

At the moment the body loses contact with the surface, $N = 0$, and therefore Eq. (2) becomes

$$v^2 = gR\cos\theta \tag{3}$$

where v and θ correspond to the moment when the body loses contact with the surface.

Solving Eqs. (1) and (3), we obtain

$$\cos\theta = \frac{2}{3} \quad \text{or} \quad \theta = \cos^{-1}(2/3) \cong 48° \quad \text{and} \quad v = \sqrt{2gR/3}$$

1.88 At first draw the free body diagram of the device as, shown. The forces, acting on the sleeve are its weight, acting vertically downward, spring force, along the length of the spring and normal reaction by the rod, perpendicular to its length.

Let F be the spring force, and Δl be the elongation.

From $F_n = mw_n$

$$N\sin\theta + F\cos\theta = m\omega^2 r \tag{1}$$

where $r\cos\theta = (l_0 + \Delta l)$

Similarly, from $F_l = mw_l$

$$N\cos\theta - F\sin\theta = 0 \quad \text{or} \quad N = F\sin\theta/\cos\theta \tag{2}$$

From Eqs. (1) and (2)

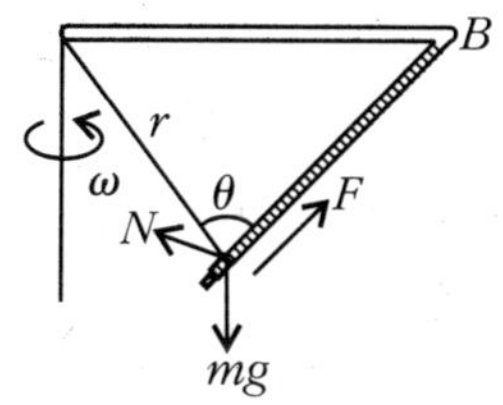

$$F(\sin\theta/\cos\theta)\cdot\sin\theta + F\cos\theta = m\omega^2 r$$

$$= m\omega^2(l_0 + \Delta l)/\cos\theta$$

On putting $F = \kappa\Delta l$,

$$\kappa\Delta l\sin^2\theta + \kappa\Delta l\cos^2\theta = m\omega^2(l_0 + \Delta l)$$

On solving, we get

$$\Delta l = m\omega^2\frac{l_0}{\kappa - m\omega^2} = \frac{l_0}{(\kappa/m\,\omega^2 - 1)}$$

It is independent of the direction of rotation.

1.89 According to the question, the cyclist moves along the circular path and the centripetal force is provided by the frictional force. Thus from the equation

$$F_n = mw_n$$

$$fr = \frac{mv^2}{r} \quad \text{or} \quad kmg = \frac{mv^2}{r}$$

or

$$k_0\left(1 - \frac{r}{R}\right)g = \frac{v^2}{r} \quad \text{or} \quad v^2 = k_0(r - r^2/R)g \tag{1}$$

For v_{max}, we should have $\dfrac{d\left(r - \dfrac{r^2}{R}\right)}{dr} = 0$

or $$1-\frac{2r}{R}=0, \quad \text{so} \quad r=\frac{R}{2}$$

Hence, $$v_{\max}=\frac{1}{2}\sqrt{k_0 gR}$$

1.90 As initial velocity is zero thus,

$$v^2 = 2w_t s = f(s) \tag{1}$$

As $w_t > 0$ the speed of the car increases with time or distance. Till the moment, sliding starts, the static friction provides the required centripetal accelaration the car.

Thus, $$fr = mw, \quad \text{but} \quad fr \leq kmg$$

So, $$w^2 \leq k^2 g^2 \quad \text{or} \quad w_t^2 + \left(\frac{v^2}{R}\right)^2 \leq k^2 g^2$$

or $$v^2 \leq R(k^2 g^2 - w_t^2)$$

Hence, $$v_{\max}^2 = R\sqrt{(k^2 g^2 - w_t^2)}$$

So, from Eq. (1), the sought distance $s = \dfrac{v_{\max}^2}{2w_t} = \dfrac{R}{2}\sqrt{\left(\dfrac{kg}{w_t}\right) - 1} = 60 \text{ m}$

1.91 Since the car follows a sinusoidal curve, so the maximum velocity at which it can ride without sliding is the point of minimum radius of curvature. Obviously in this case the static friction between the car and the road is limiting. Hence, from the equation $F_n = mw_n$

$$\frac{mv^2}{R} \leq kmg$$

or $$v \leq \sqrt{kRg}$$

So, $$v_{\max} = \sqrt{kR_{\min} g} \tag{1}$$

We know that the radius of curvature for a curve at any point (x, y) is given as

$$R = \left|\frac{[1+(dy/dx)^2]^{3/2}}{(d^2y)/dx^2}\right| \tag{2}$$

For the given curve,

$$\frac{dy}{dx} = \frac{a}{\alpha}\cos\left(\frac{x}{\alpha}\right) \quad \text{and} \quad \frac{d^2y}{dx^2} = \frac{-a}{\alpha^2}\sin\frac{x}{\alpha}$$

Substituting this value in Eq. (2), we get

$$R = \frac{[1+(a^2/\alpha^2)\cos^2(x/\alpha)]^{3/2}}{(a/\alpha^2)\sin(x/\alpha)}$$

For the minimum R, $$\frac{x}{\alpha} = \frac{\pi}{2}$$

and therefore, corresponding radius of curvature

$$R_{\min} = \frac{\alpha^2}{a} \tag{3}$$

Hence, from Eqs. (1) and (3)

$$v_{\max} = \alpha\sqrt{kg/a}$$

1.92 The sought tensile stress acts on each element of the chain. Hence divide the chain into small, similar elements so that each element may be assumed as a particle. We consider one such element of mass dm, which subtends angle $d\alpha$ at the centre. The chain moves along a circle of known radius R with known angular speed ω and certain forces act on it. We have to find one of these forces.

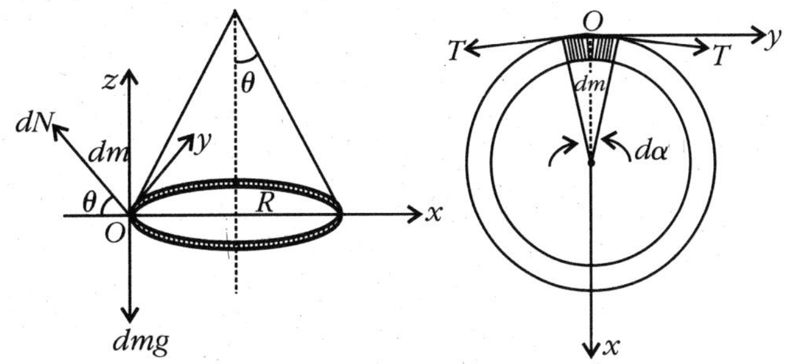

From Newton's second law in projection form, $F_x = mw_x$, we get

$$2T\sin(d\alpha/2) - dN\cos\theta = dm\,\omega^2 R$$

and from $F_z = mw_z$, we get

$$dN\sin\theta = (dm)g$$

Then putting $dm = m d\alpha/2\pi$ and $\sin(d\alpha/2) = d\alpha/2$ and solving, we get

$$T = \frac{m(\omega^2 R + g\cot\theta)}{2\pi}$$

1.93 Let us consider a small element of the thread and draw free body diagram for this element.

(a) Applying Newton's second law of motion in projection form, $F_n = mw_n$ for this element,

$$(T + dT)\sin(d\theta/2) + T\sin(d\theta/2) - dN = dm\omega^2 R = 0$$

or $$2T\sin(d\theta/2) = dN \quad [\text{neglecting the term } (dT\sin d\theta/2)]$$

or $$T\,d\theta = dN\left(\text{as } \sin\frac{d\theta}{2} \approx \frac{d\theta}{2}\right) \quad (1)$$

Also, $$dfr = kdN = (T + dT) - T = dT \quad (2)$$

From Eqs. (1) and (2)

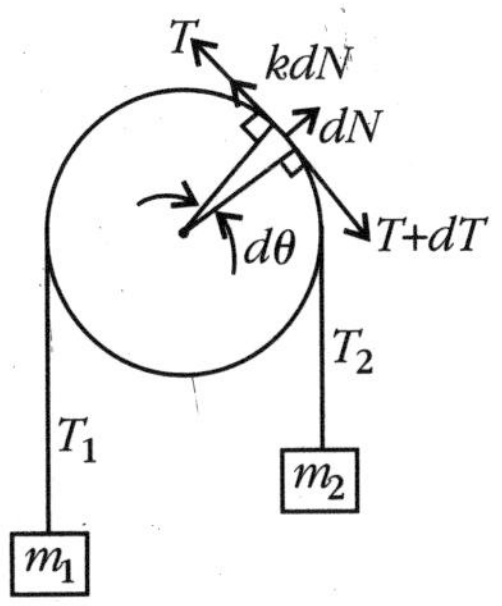

$$kT\,d\theta = dT \quad \text{or} \quad \frac{dT}{T} = k\,d\theta$$

Integrating θ from $\theta = 0$ to $\theta = \pi$, we get

$$\ln\frac{T_2}{T_1} = k\pi \quad (3)$$

So, $$k = \frac{1}{\pi}\ln\frac{T_2}{T_1} = \frac{1}{\pi}\ln\eta_0 \quad \left(\text{as } \frac{T_2}{T_1} = \frac{m_g g}{m_1 g} = \frac{m_2}{m_1} = \eta_0\right)$$

(b) When $m_2/m_1 = \eta > \eta_0$, the blocks will move with same value of acceleration (say w) and clearly m_2 moves downward. From Newton's second law in projection form (downward for m_2 and upward for m_1), we get

$$m_2 g - T_2 = m_2 w \tag{4}$$

and
$$T_2 - m_1 g = m_1 w \tag{5}$$

Also
$$\frac{T_2}{T_1} = \eta_0 \tag{6}$$

Simultaneous solution of Eqs. (4), (5) and (6) yields

$$w = \frac{(m_2 - \eta_0 m_1)g}{(m_2 + \eta_0 m_1)} = \frac{(\eta - \eta_0)}{(\eta + \eta_0)} g \quad \left(\text{as} \frac{m_2}{m_1} = \eta\right)$$

1.94 The force with which the cylinder wall acts on the particle will provide centripetal force necessary for the motion of the particle, and since there is no acceleration acting in the horizontal direction, horizontal component of the velocity will remain constant throughout the motion.

So,
$$v_x = v_0 \cos\alpha$$

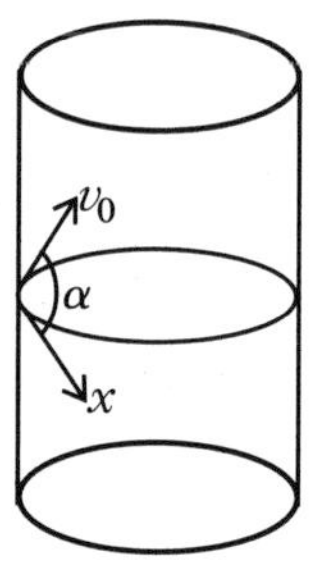

Using, $F_n = mw_n$, for the particle of mass m

$$N = \frac{mv_x^2}{R} = \frac{mv_0^2 \cos^2\alpha}{R}$$

which is the required normal force.

1.95 Obviously, the radius vector describing the position of the particle relative to the origin of the coordinate is,

$$\mathbf{r} = x\mathbf{i} + y\mathbf{j} = a\sin\omega t\mathbf{i} + b\cos\omega t\mathbf{j}$$

Differentiating twice with respect to time

$$\mathbf{w} = \frac{d^2\mathbf{r}}{dt^2} = -\boldsymbol{\omega}^2 (a\sin\omega t\mathbf{i} + b\cos\omega t\mathbf{j}) = -\omega^2\mathbf{r} \tag{1}$$

Thus,
$$\mathbf{F} = m\mathbf{w} = -m\omega^2\mathbf{r} \quad \text{(using Eq. 1)}$$

1.96 (a) We have
$$\Delta\mathbf{p} = \int \mathbf{F}dt = \int_0^t m\mathbf{g}\,dt = m\mathbf{g}t \tag{1}$$

(b) Using the solution of problem 1.28 (b), the total time of motion,

$$\tau = -\frac{2(\mathbf{v}_0 \cdot \mathbf{g})}{\mathbf{g}^2}$$

Hence, using $t = \tau$ in Eq. (1)

$$|\Delta\mathbf{p}| = mg\tau$$
$$= \frac{-2m(\mathbf{v}_0 \cdot \mathbf{g})}{g}$$

1.97 From the equation of the given time dependence force $\mathbf{F} = \mathbf{a}t(\tau - t)$ at $\mathbf{t} = \tau$, the force vanishes.

(a) Thus,

$$\Delta\mathbf{p} = \mathbf{p} = \int_0^\tau \mathbf{F}dt$$

or

$$\mathbf{p} = \int_0^\tau \mathbf{a}t(\tau - t)\,dt = \frac{\mathbf{a}\tau^3}{6}$$

But,

$$\mathbf{p} = m\mathbf{v} \quad \text{so,} \quad \mathbf{v} = \frac{\mathbf{a}\tau^3}{6m}$$

(b) Again from the equation $\mathbf{F} = m\mathbf{w}$

$$\mathbf{a}t(\tau - t) = m\frac{d\mathbf{v}}{dt}$$

or

$$\mathbf{a}(t\tau - t^2)\,dt = md\mathbf{v}$$

Integrating within the limits for $\mathbf{v}(t)$

$$\int_0^t \mathbf{a}(t\tau - t^2)dt = m\int_0^{\mathbf{v}} d\mathbf{v}$$

or

$$\mathbf{v} = \frac{\mathbf{a}}{m}\left(\frac{\tau t^2}{2} - \frac{t^3}{3}\right) = \frac{\mathbf{a}t^2}{m}\left(\frac{\tau}{2} - \frac{t}{3}\right)$$

Thus,

$$v = \frac{at^2}{m}\left(\frac{\tau}{2} - \frac{t}{3}\right) \quad \text{for } t \le \tau$$

Hence, distance covered during the time interval $t = \tau$ is given by

$$s = \int_0^\tau v dt = \int_0^\tau \frac{at^2}{m}\left(\frac{\tau}{2} - \frac{t}{3}\right)dt = \frac{a}{m}\frac{\tau^4}{12}$$

1.98 We have

$$F = F_0 \sin\omega t$$

or

$$m\frac{d\mathbf{v}}{dt} = \mathbf{F}_0 \sin\omega t \quad \text{or} \quad md\mathbf{v} = \mathbf{F}_0 \sin\omega t\, dt$$

On integrating,

$$m\mathbf{v} = \frac{-\mathbf{F}_0}{\omega}\cos\omega t + C \quad \text{(where } C \text{ is integration constant)}$$

When $t = 0$, $v = 0$, so $C = \dfrac{\mathbf{F}_0}{m\omega}$

Hence,

$$\mathbf{v} = \frac{-\mathbf{F}_0}{m\omega}\cos\omega t + \frac{\mathbf{F}_0}{m\omega}$$

As $\cos\omega t \le 1$ so,

$$\mathbf{v} = \frac{\mathbf{F}_0}{m\omega}(1 - \cos\omega t)$$

Thus,

$$s = \int_0^t v dt = \frac{F_0 t}{m\omega} - \frac{F_0 \sin\omega t}{m\omega^2} = \frac{F_0}{m\omega^2}(\omega t - \sin\omega t).$$ (see figure in the answer sheet)

1.99 According to the problem, the force acting on the particle of mass m is, $\mathbf{F} = \mathbf{F}_0 \cos \omega t$.

So, $$m\frac{d\mathbf{v}}{dt} = \mathbf{F}_0 \cos \omega t \quad \text{or} \quad d\mathbf{v} = \frac{\mathbf{F}_0}{m} \cos \omega t \, dt$$

Integrating, within the limits

$$\int_0^{\mathbf{v}} d\mathbf{v} = \frac{\mathbf{F}_0}{m} \int_0^t \cos \omega t \, dt \quad \text{or} \quad \mathbf{v} = \frac{\mathbf{F}_0}{m\omega} \sin \omega t \tag{1}$$

It is clear from Eq. (1), that after starting at $t = 0$, the particle comes to rest for the first time at $t = \pi/\omega$.

From Eq. (1)

$$v = |\mathbf{v}| = \frac{\mathbf{F}_0}{m\omega} \sin \omega t \quad \text{for} \quad t \le \frac{\pi}{\omega} \tag{2}$$

Thus during the time interval $t = \pi/\omega$, the sought distance

$$s = \frac{F_0}{m\omega} \int_0^{\pi/\omega} \sin \omega t \, dt = \frac{2F}{m\omega^2}$$

From Eq. (1)

$$v_{\text{max}} = \frac{F_0}{m\omega} \quad \text{as} \, |\sin \omega t| \le 1$$

1.100 (a) From the problem $\mathbf{F} = -r\mathbf{v}$ so, $m\frac{d\mathbf{v}}{dt} = -r\mathbf{v}$

Thus, $$m\frac{dv}{dt} = -rv \quad [\text{as } d\mathbf{v} \uparrow\downarrow \mathbf{v}]$$

or $$\frac{dv}{v} = -\frac{r}{m} dt$$

On integrating $$\ln v = -\frac{r}{m} t + C$$

But at $t = 0$, $v = v_0$ so, $C = \ln v_0$.

So, $$\ln \frac{v}{v_0} = -\frac{r}{m} t \quad \text{or} \quad v = v_0 e^{-(r/m)t}$$

Thus, for $t \to \infty$, $v = 0$.

(b) We have $m\frac{dv}{dt} = -rv$ so, $dv = \frac{-r}{m} ds$

Integrating within the given limits to obtain $v(s)$, we get

$$\int_{v_0}^{v} dv = -\frac{r}{m} \int_0^s ds \quad \text{or} \quad v = v_0 - \frac{rs}{m} \tag{1}$$

Thus for $v = 0$, $s = s_{\text{total}} = \frac{mv_0}{r}$

(c) We have $$\frac{mdv}{dt} = -rv \quad \text{or} \quad \frac{dv}{v} = \frac{-r}{m} dt$$

or $$\int_0^{v_0/\eta} \frac{dv}{v} = \frac{-r}{m} \int_0^t dt \quad \text{or} \quad \ln \frac{v_0}{\eta v_0} = -\frac{r}{m} t$$

So, $$t = \frac{-m\ln(1/\eta)}{r} = \frac{m\ln\eta}{r}$$

Now, average velocity over this time interval

$$<v> = \frac{\int v dt}{\int dt} = \frac{\int_0^{\frac{m}{r}\ln\eta} v_0 e^{-\frac{rt}{m}} dt}{m/r\ln\eta} = \frac{v_0(\eta-1)}{\eta\ln\eta}$$

1.101 According to the problem

$$m\frac{dv}{dt} = -kv^2 \quad \text{or} \quad m\frac{dv}{v^2} = -kdt$$

Integrating, within the limits

$$\int_{v_0}^{v} \frac{dv}{v^2} = -\frac{k}{m}\int_0^t dt \quad \text{or} \quad t = \frac{m}{k}\frac{(v_0 - v)}{v_0 v} \tag{1}$$

To find the value of k, rewrite

$$mv\frac{dv}{ds} = -kv^2 \quad \text{or} \quad \frac{dv}{v} = -\frac{k}{m}ds$$

On integrating

$$\int_{v_0}^{v} \frac{dv}{v} = -\frac{k}{m}\int_0^b ds$$

So, $$k = \frac{m}{b}\ln\frac{v_0}{v} \tag{2}$$

Putting the value of k from Eq. (2) in Eq. (1), we get

$$t = \frac{b(v_0 - v)}{v_0 v \ln\frac{v_0}{v}}$$

1.102 From Newton's second law for the bar in projection form, $F_x = mw_x$, along x direction, we get

$$mg\sin\alpha - kmg\cos\alpha = mw$$

or $$v\frac{dv}{dx} = g\sin\alpha - ax\, g\cos\alpha \quad (\text{as } k = ax)$$

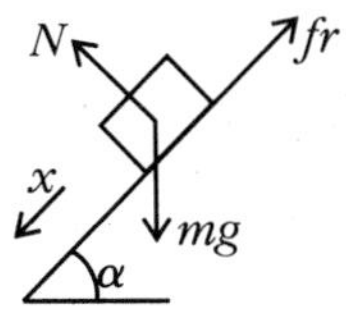

or $$vdv = (g\sin\alpha - ax\, g\cos\alpha)\, dx$$

or $$\int_0^v vdv = g\int_0^x (\sin\alpha - x\cos\alpha)\, dx$$

So, $$\frac{v^2}{2} = g(x\sin\alpha - \frac{x^2}{2}a\cos\alpha) \tag{1}$$

From Eq. (1),

$$v = 0 \text{ at either } x = 0 \quad \text{or} \quad x = \frac{2}{a}\tan\alpha$$

As the motion of the bar is unidirectional it stops after going through a distance of $2/a \tan\alpha$.

From Eq. (1), for v_{max},

$$\frac{d}{dx}(x \sin\alpha - \frac{x^2}{2} a\cos\alpha) = 0 \quad \text{or} \quad x = \frac{1}{a}\tan\alpha$$

Hence, the maximum velocity will be at the distance, $x = \tan\alpha/a$.

Putting this value of x in Eq. (1), the maximum velocity,

$$v_{max} = \sqrt{\frac{g\sin\alpha\tan\alpha}{a}}$$

1.103 Since the applied force is proportional to the time and the frictional force also exists, the motion does not start just after applying the force. The body starts its motion when F equals the limiting friction. Let the motion start after time t_0, then

$$F = at_0 = kmg \quad \text{or} \quad t_0 = \frac{kmg}{a}$$

So, for $t \le t_0$, the body remains at rest and for $t > t_0$ obviously

$$\frac{mdv}{dt} = a(t - t_0) \quad \text{or} \quad mdv = a(t - t_0)dt$$

Integrating, and noting $v = 0$ at $t = t_0$, we have for $t > t_0$

$$\int_0^v mdv = a\int_0^t (t - t_0)\, dt \quad \text{or} \quad v = \frac{a}{2m}(t - t_0)^2$$

Thus,

$$s = \int vdt = \frac{a}{2m}\int_{t_0}^t (t - t_0)^2\, dt = \frac{a}{6m}(t - t_0)^3$$

1.104 While going upward, from Newton's second law, in vertical direction

$$m\frac{vdv}{ds} = -(mg + kv^2) \quad \text{or} \quad \frac{vdv}{(g + kv^2/m)} = -ds$$

At the maximum height h, the speed $v = 0$, so

$$\int_{v_0}^0 \frac{vdv}{g + (kv^2/m)} = -\int_0^h ds$$

Integrating and solving, we get

$$h = \frac{m}{2k}\ln\left(1 + +\frac{kv_0^2}{mg}\right) \tag{1}$$

When the body falls downward, the net force acting on the body in downward direction equals $(mg - kv^2)$.

Hence net acceleration, in downward direction, according to second law of motion

$$\frac{vdv}{ds} = g - \frac{kv^2}{m} \quad \text{or} \quad \frac{vdv}{g - (kv^2/m)} = ds$$

Thus,

$$\int_0^{v'} \frac{vdv}{g - (kv^2/m)} = \int_0^h ds$$

Integrating and putting the value of h from Eq. (1), we get

$$v' = \frac{v_0}{\sqrt{1 + kv_0^2/mg}}$$

1.105 Let us fix $x - y$ coordinate system to the given plane, taking x-axis in the direction along which the force vector was oriented at the moment $t = 0$, then the fundamental equation of dynamics expressed via the projection on x and y-axes gives,

$$F\cos\omega t = m\frac{dv_x}{dt}$$

and

$$F\sin\omega t = m\frac{dv_y}{dt}$$

(a) Using the condition $v(0) = 0$, we obtain $v_x = \frac{F}{m\omega}\sin\omega t$

and

$$v_y = \frac{F}{m\omega}(1-\cos\omega t)$$

Hence,

$$v = \sqrt{v_x^2 + v_y^2} = \left(\frac{2F}{m\omega}\right)\left|\sin\left(\frac{\omega t}{2}\right)\right|$$

(b) It is seen from this that the velocity v turns into zero after the time interval Δt, which can be found from the relation, $\omega\frac{\Delta t}{2} = \pi$. Consequently, the sought distance is

$$s = \int_0^{\Delta t} v\,dt = \frac{8F}{m\omega^2}$$

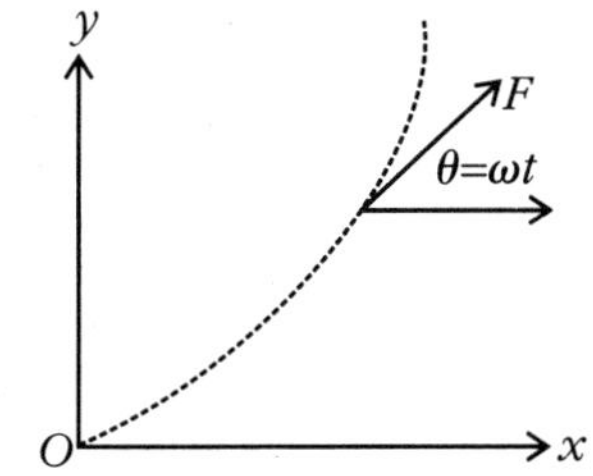

Average velocity,

$$<v> = \frac{\int v\,dt}{\int dt}$$

So,

$$<v> = \int_0^{2\pi/\omega} \frac{2F}{m\omega}\sin\left(\frac{\omega t}{2}\right)dt/(2\pi\omega) = \frac{4F}{\pi m\omega}$$

1.106 The acceleration of the disk along the plane is determined by the projection of the force of gravity on this plane $F_x = mg\sin\alpha$ and the friction force $fr = kmg\cos\alpha$. In our case, $k = \tan\alpha$ and therefore

$$fr = F_x = mg\sin\alpha$$

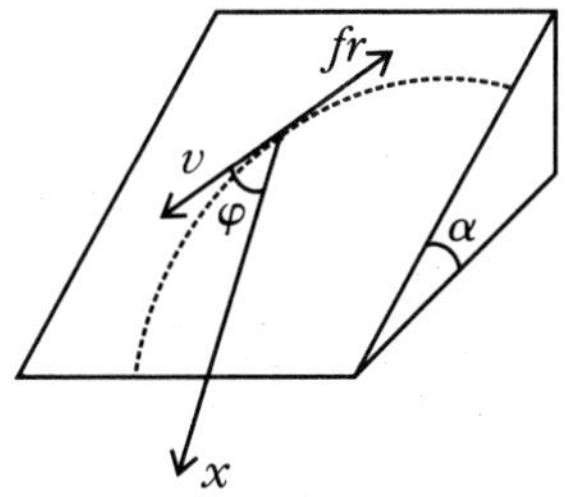

Let us find the projection of acceleration on the direction of the tangent to the trajectory and on the x-axis

$$mw_t = F_x\cos\varphi - fr = mg\sin\alpha\,(\cos\varphi - 1)$$

$$mw_x = F_x - fr\cos\varphi = mg\sin\alpha\,(1 - \cos\varphi)$$

It is seen from this that $w_t = -w_x$, which means that the velocity v and its projection v_x differ only by a constant value C which does not change with time, i.e., $v = -v_x + C$, where $v_x = v\cos\varphi$. The constant C is found from the initial condition $v = v_0$, $U_x = 0$ since $\varphi = \pi/2$ initially. Finally, we obtain

$$C = v_0$$

$$v = \frac{v_0}{(1+\cos\varphi)}$$

In the course of time $\varphi \to 0$ and $v \to v_0/2$. (Motion then is unaccelerated.)

1.107 Let us consider an element of length ds at an angle φ from the vertical diameter. As the speed of this element is zero at initial instant of time, its centripetal acceleration is zero, and hence, $dN - \lambda ds\, g\cos\varphi = 0$,

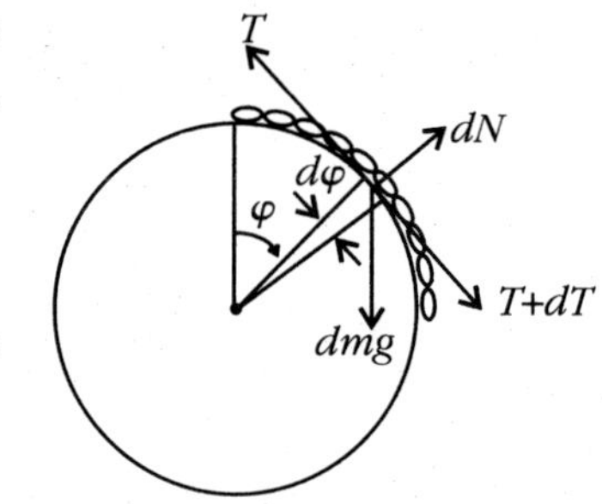

where λ is the linear mass density of the chain. Let T and $T + dT$ be the tension at the upper and the lower ends of ds, then we have from $F_t = mw_t$,

$$(T + dT) + \lambda\, ds\, g \sin\varphi - T = \lambda\, ds\, w_t$$

or
$$dT + \lambda\, R\, d\varphi\, g \sin\varphi = \lambda\, ds\, w_t$$

If we sum the above equation for all elements, the term $\int dT = 0$, because there is no tension at the free ends, so

$$\lambda g R \int_0^{l/R} \sin\varphi\, d\varphi = \lambda \omega_t \int ds = \lambda l w_t$$

Hence,
$$\omega_t = \frac{gR}{l}\left(1 - \cos\frac{l}{R}\right)$$

As $w_n = a$ at initial moment, so

$$w = |w_t| = \frac{gR}{l}\left(1 - \cos\frac{l}{R}\right)$$

1.108 In the problem, we require the velocity of the body, relative to the sphere, which itself moves with an acceleration w_0 in horizontal direction (say towards left). Hence it is advisable to solve the problem in the frame of sphere (*a translating frame here*).

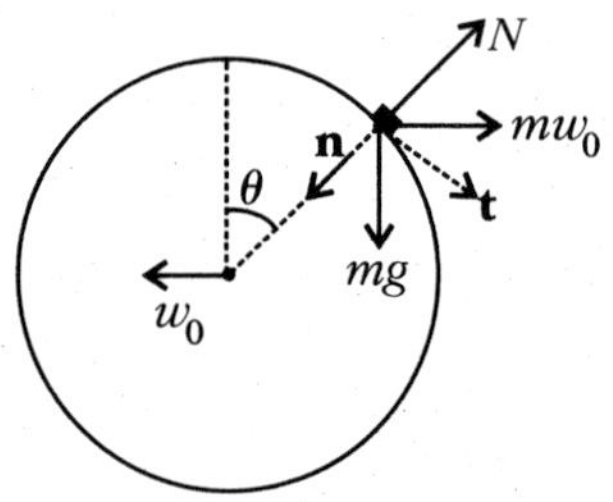

At an arbitrary moment, when the body is at an angle θ with the vertical, we sketch the force diagram for the body and write the second law of motion in projection form $F_n = mw_n$ as,

$$mg\cos\theta - N - mw_0 \sin\theta = \frac{mv^2}{R} \tag{1}$$

At the break off point, $N = 0$, $\theta = \theta_0$ and let $v = v_0$, so Eq. (1) becomes,

$$\frac{v_0^2}{R} = g\cos\theta_0 - w_0 \sin\theta_0 \tag{2}$$

From $F_t = mw_t$

$$mg\sin\theta - mw_0\cos\theta = m\frac{vdv}{ds} = m\frac{v\,dv}{Rd\theta}$$

or
$$vdv = R(g\sin\theta + w_0\cos\theta)\,d\theta$$

Integrating
$$\int_0^{v_0} v\,dv = \int_0^{\theta_0} R(g\sin\theta + w_0\cos\theta)\,d\theta$$

$$\frac{v_0^2}{2R} = g(1 - \cos\theta_0) + w_0 \sin\theta_0 \tag{3}$$

Note that Eq. (3) can also be obtained by the work-energy theorem $A = \Delta T$ (in the frame of sphere). Therefore,

$$mgR(1 - \cos\theta_0) + mw_0 R\sin\theta_0 = \frac{1}{2}mv_0^2$$

[here $mw_0 R\sin\theta_0$ is the work done by the pseudoforce $(-m\mathbf{w}_0)$]

or
$$\frac{v_0^2}{2R} = g(1 - \cos\theta_0) + w_0 \sin\theta_0$$

Solving Eqs. (2) and (3), we get

$$v_0 = \sqrt{\frac{2gR}{3}} \quad \text{and} \quad \theta_0 = \cos^{-1}\left[\frac{2+k\sqrt{5+9k^2}}{3(1+k^3)}\right] \quad \left(\text{where } k = \frac{w_0}{g}\right)$$

Hence,

$$\theta_0\Big|_{w_0=g} = 17°$$

1.109 This is not central force problem unless the path is a circle about the said point. Rather here F_t (tangential force) vanishes. Thus the equation of motion becomes

$$v_t = v_0 = \text{constant}$$

and

$$\frac{mv_0^2}{r} = \frac{A}{r^n} \quad (\text{for } r = r_0)$$

We can consider the latter equation as the equilibrium under two forces. When the motion is perturbed, we write $r = r_0 + x$ and the net force acting on the particle is

$$-\frac{A}{(r_0+x)^n} + \frac{mv_0^2}{r_0+x} = \frac{-A}{r_0^n}\left(1-\frac{nx}{r_0}\right) + \frac{mv_0^2}{r_0}\left(1-\frac{x}{r_0}\right) = -\frac{mv_0^2}{r_0^2}(1-n)x$$

This opposes the displacement x, if $n < 1$. (mv_0^2/r is an outward directed centrifugal force while $-A/r^n$ is the inward directed external force.)

1.110 Let us observe the behaviour of the sleeve in the frame fixed to the rotating rod bent into the shape of half circle. We resolve all the forces into tangential and normal components, then the net downward tangential force on the sleeve is

$$mg\sin\theta\left(1-\frac{\omega^2 R}{g}\cos\theta\right)$$

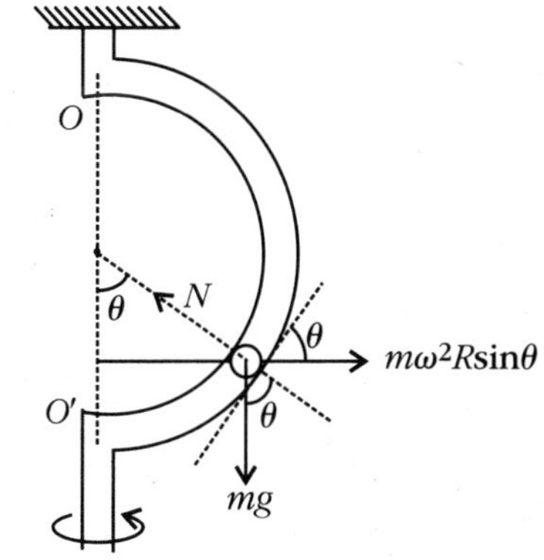

This vanishes for $\theta = 0$ and for $\theta = \theta_0 = \cos^{-1}(g/\omega^2 R)$, which is real if $\omega^2 R > g$.

If $\omega^2 R < g$, then $[1-(\omega^2 R/g)\cos\theta]$ is always positive for small values of θ and hence the net tangential force near $\theta = 0$ opposes any displacement away from it, and $\theta = 0$ is then stable.

If $\omega^2 R > g$, then $[1-(\omega^2 R/g)\cos\theta]$ is negative for small θ near $\theta = 0$, and $\theta = 0$ is then unstable.

However $\theta = \theta_0$ is stable because the force tends to bring the sleeve near the equilibrium position $\theta = \theta_0$.

If $\omega^2 R = g$, the two positions coincide and becomes a stable equilibrium point.

1.111 Define the axes as shown with z along the local vertical, x due east and y due north. (We assume we are in the northern hemisphere). Then the Coriolis force has the components.

$$\begin{aligned}\mathbf{F}_{\text{cor}} &= -2m(\boldsymbol{\omega} \times \mathbf{v}) \\ &= 2m\omega\,[v_y\cos\theta - v_z\sin\theta)\mathbf{i} - v_x\cos\theta\mathbf{j} + v_x\cos\theta\mathbf{k}] \\ &= 2m\omega\,(v_y\cos\theta - v_z\sin\theta)\mathbf{i}\end{aligned}$$

Since v_x is small when the direction in which the gun is fired is due north. Thus the equations of motion (neglecting centrifugal forces) are

$$\ddot{x} = 2\omega(v_y \sin\varphi - v_z \cos\varphi), \quad \ddot{y} = 0 \quad \text{and} \quad \ddot{z} = -g$$

Integrating, we get $v_y = v$ (constant),

$$\dot{z} = -gt \quad \text{and} \quad \dot{x} = 2\omega v \sin\varphi t + \omega g t^2 \cos\varphi$$

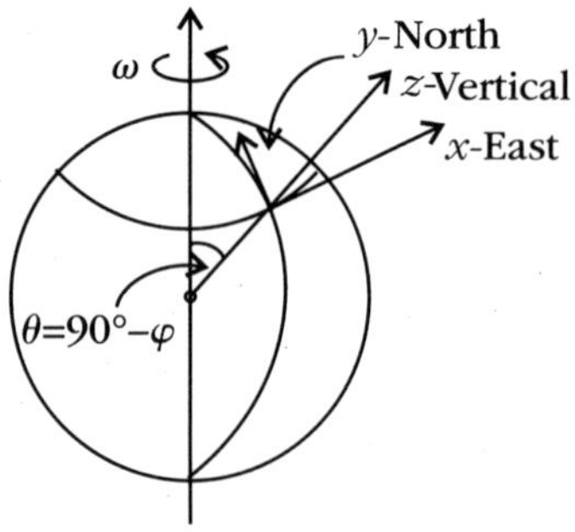

Finally,

$$x = \omega v t^2 \sin\varphi + \frac{1}{3}\omega g t^3 \cos\varphi$$

Now, $v \gg gt$ in the present case. So,

$$x = \omega v \sin\varphi\left(\frac{s}{v}\right)^2 \approx \omega \sin\varphi \frac{s^2}{v}$$

$$= 7 \text{ cm (to the east)}$$

1.112 For the observer attached in the frame of discs

$$\mathbf{R} + \mathbf{F}_{cor} + m\mathbf{g} + \mathbf{F}_{cf} = 0$$

So, $$\mathbf{R} = -(\mathbf{F}_{cor} + m\mathbf{g} + \mathbf{F}_{cf})$$

As $$\mathbf{F}_{cor} \perp m\mathbf{g} \perp \mathbf{E}_{cf}$$

So, $$|\mathrm{R}| = \sqrt{(2mv'\omega)^2 + (mg)^2 + (m\omega^2 r)^2} = m\sqrt{g^2 + \omega^4 r^2 + (2v'\omega)} = 8 \text{ N.}$$

1.113 The sleeve is free to slide along the rod *AB*. Thus in the frame of rod, the centrifugal force acts on it in the outward direction along the rod.

So, we have $$m\dot{v} = m\omega^2 r$$

Where $$v = \frac{dr}{dt}$$

But $$\dot{v} = \frac{vdv}{dr} = \frac{d}{dr}\left(\frac{1}{2}v^2\right)$$

so, $$\frac{1}{2}v^2 = \frac{1}{2}\omega^2 r^2 + \text{constant}$$

or, $$v^2 = v_0^2 + \omega^2 r^2$$

v_0 being the initial velocity when $r = 0$. The Coriolis force is then

$$F_{cor} = 2m\omega\sqrt{v_0^2 + \omega^2 r^2} = 2m\omega^2 r\sqrt{1 + (v_0^2/\omega^2 r^2)} = 2.83 \text{ N} \quad \text{(on putting the values)}$$

1.114 The disc *OBAC* is rotating with angular velocity ω about the axis OO' passing through the edge point *O*. The equation of motion in rotating frame is,

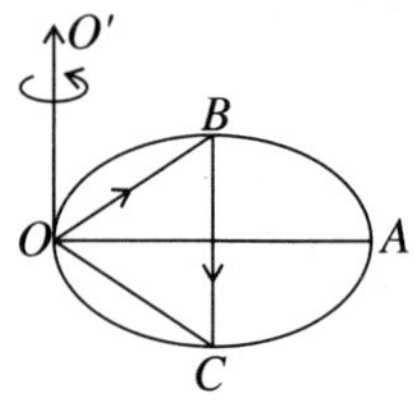

$$m\mathbf{w}' = \mathbf{F} + m\omega^2\mathbf{R} + 2m(\mathbf{v}' \times \omega) = \mathbf{F} + \mathbf{F}_{in}$$

where $\mathbf{F}_{in}$ is the resultant inertial force (pseudo-force) which is the vector sum of centrifugal and Coriolis forces.

(a) At A, F_{in} vanishes.

Thus $$0 = -2m\omega^2 R\mathbf{n} + 2mv'\omega\mathbf{n}$$

where **n** is the inward drawn unit vector to the centre, from the point in question, here A.

Thus, $$v' = \omega R$$

So, $$w = \frac{v'^2}{\rho} = \frac{v'^2}{R} = \omega^2 R$$

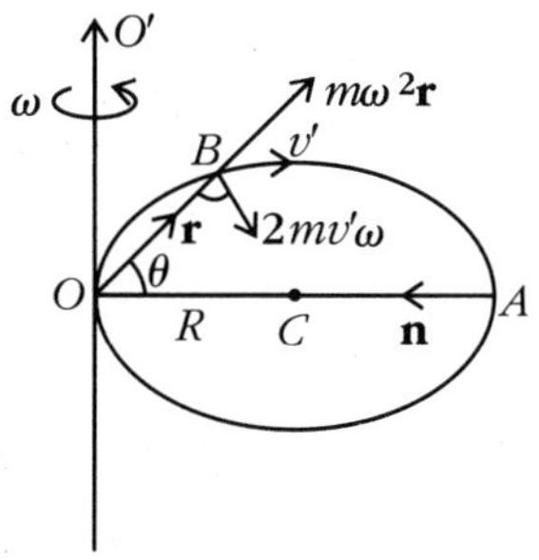

(b) At B; $\mathbf{F}_{in} = m\omega^2\mathbf{r} + 2mv'\omega\mathbf{u_{BC}}$

(where $\mathbf{u_{BC}}$ is unit vector, directed from B to C).

$$v' = \omega r = \text{constant} \quad \text{(From question)}$$

$$\mathbf{F}_{in} = m\omega^2\mathbf{r} + 2m\omega^2\,\mathbf{u_{BC}}$$

From geometry,

$$r = 2R\cos\theta \quad \text{(see fig.)}$$

so, $$|\mathbf{F}_{in}| = m\omega^2\sqrt{4R^2 - r^2}$$

1.115 The equation of motion in the rotating coordinate system is,

$$m\mathbf{w} = \mathbf{F} + m\omega^2\boldsymbol{\rho} + 2m\,(\mathbf{v}' \times \omega)$$

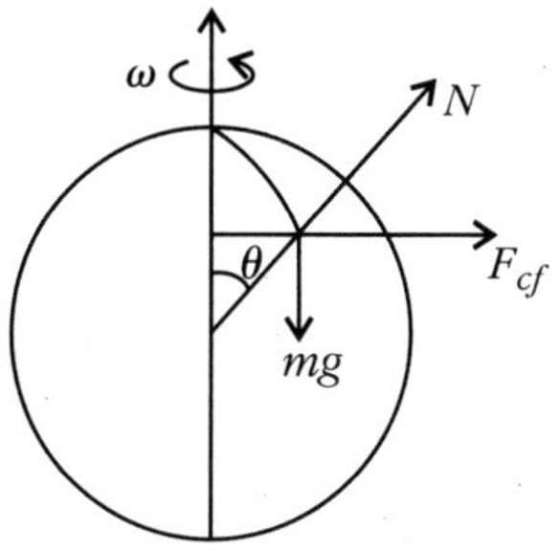

Now, $$\mathbf{v}' = R\dot{\theta}\mathbf{e}_\theta + R\sin\theta\,\dot{\varphi}\,\mathbf{e}_\varphi$$

and $$\boldsymbol{\omega} = \omega\cos\theta\,\mathbf{e}_r - \omega\sin\theta\,\mathbf{e}_\theta$$

$$\frac{1}{2m}\mathbf{F}_{cor} = \begin{vmatrix} \mathbf{e}_r & \mathbf{e}_\theta & \mathbf{e}_\varphi \\ 0 & R\dot{\theta} & (R\sin\theta)\dot{\varphi} \\ \omega\cos\theta & -\omega\sin\theta & 0 \end{vmatrix}$$

$$= \mathbf{e}_r(\omega R\sin^2\theta\dot{\varphi}) + [\omega R\sin\theta\cos\theta\dot{\varphi}]\mathbf{e}_\theta - (\omega R\dot{\theta}\cos\theta)\mathbf{e}_\varphi$$

Now on the sphere

$$\mathbf{w}' = (-R\dot{\theta}^2 - R\sin^2\theta\,\dot{\varphi}^2)\mathbf{e}_r + (R\ddot{\theta} - R\sin\theta\cos\theta\dot{\varphi}^2)\mathbf{e}_\theta$$
$$+ (R\sin\theta\ddot{\varphi} + 2R\cos\theta\dot{\theta}\dot{\varphi})\mathbf{e}_\varphi$$

Thus the equations of motion are:

$$m(-R\dot{\theta}^2 - R\sin^2\theta\,\dot{\varphi}^2) = N - mg\cos\theta + m\omega^2 R\sin^2\theta + 2m\omega R\sin^2\theta\dot{\varphi}$$
$$m(R\ddot{\theta} - R\sin\theta\cos\theta\,\ddot{\varphi}) = mg\sin\theta + m\omega^2 R\sin\theta\cos\theta + 2m\omega R\sin\theta\cos\theta\,\dot{\varphi}$$
$$m(R\sin\theta\ddot{\varphi} + 2R\cos\theta\dot{\theta}\,\dot{\varphi}) = -2m\omega R\dot{\theta}\cos\theta$$

From the third equation, we get, $\dot{\varphi} = -\omega$

A result that is easy to understand by considering the motion in non-rotating frame.

Eliminating $\dot{\varphi}$ we get, $$mR\dot{\theta}^2 = mg\cos\theta - N$$
$$mR\ddot{\theta} = mg\sin\theta$$

Integrating the last equation

$$\frac{1}{2}mR\dot{\theta}^2 = mg\,(1 - \cos\theta) \qquad (1)$$

Hence $$N = (3 - 2\cos\theta)mg$$

So the body must fly off for $\theta = \theta_0 = \cos^{-1}\frac{2}{3}$, exactly as if the sphere were non-rotating.

Now, at this point F_{cf} = centrifugal force $= m\omega^2 R \sin\theta_0 = \sqrt{\frac{5}{9}}\, m\omega^2 R$

$$F_{cor} = 2m\sqrt{\omega^2 R^2 \dot{\theta}^2 \cos^2\theta + (\omega^2 R)^2 \sin^2\theta}$$

Putting the value of θ from Eq. (1)

$$\theta = \theta_0$$

We get, $$F_{cor} = \frac{2m\omega^2 R}{3}\sqrt{5 + \frac{8g}{3\omega^2 R}} = 17 \text{ N} \quad \text{(on substituting values)}$$

1.116 (a) When the train is moving along a meridian only the Coriolis force has a lateral component and its magnitude (see the previous problem) is,

$$2m\omega v \cos\theta = 2m\omega v \sin\varphi \quad \text{(here we have put } R\dot{\theta} \to v\text{)}$$

So, $$F_{lateral} = 2 \times 2000 \times 10^3 \times \frac{2\pi}{86400} \times \frac{54000}{3600} \times \frac{\sqrt{3}}{2}$$

$$= 3.77 \text{ kN} \quad \text{(we write } \varphi \text{ for the latitude)}$$

(b) The resultant of the inertial forces acting on the train is,

$$\mathbf{F}_{in} = -2m\omega R\dot{\theta}\cos\theta\,\mathbf{e}_\varphi + (m\omega^2 R\sin\theta\cos\theta + 2m\,\omega R\sin\theta\cos\theta\,\dot{\varphi})\mathbf{e}_\theta + (m\omega^2 R\sin^2\theta + 2m\,\omega R\sin^2\theta\,\dot{\varphi})\mathbf{e}_r$$

This vanishes if $\dot{\theta} = 0,\ \dot{\varphi} = -\frac{1}{2}\omega$

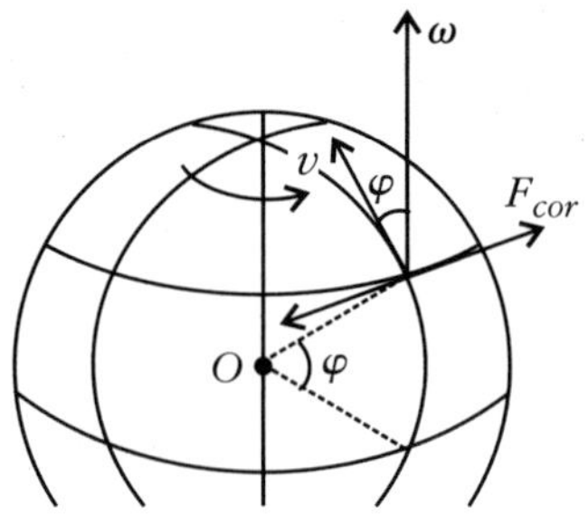

Thus, $\mathbf{v} = v_\varphi \mathbf{e}_\varphi, v_\varphi = -\frac{1}{2}\omega R\sin\theta = -\frac{1}{2}\omega R\cos\varphi$

where (we write φ for the latitude here)

Thus the train must move from the east to west along the 60th parallel with a speed,

$$\frac{1}{2}\omega R\cos\varphi = \frac{1}{4} \times \frac{2\pi}{8.64} \times 10^{-4} \times 6.37 \times 10^6 = 115.8 \text{ m/s} \approx 420 \text{ km/h}$$

1.117 We go to the equation given in problem 1.111. Here $v_y = 0$ so we can take $y = 0$; thus, we get for the motion in the xz plane.

$$\ddot{x} = -2\omega v_z \cos\varphi$$

and $$\ddot{z} = -g$$

Integrating, $$z = -\frac{1}{2}gt^2$$

$$\dot{x} = \omega g \cos\varphi t^2$$

So $$x = \frac{1}{3}\omega g\cos\varphi t^3 = \frac{1}{3}\omega g\cos\varphi\left(\frac{2h}{g}\right)^{3/2} = \frac{2\omega h}{3}\cos\varphi\sqrt{\frac{2h}{g}}$$

There is thus a displacement to the east of magnitude

$$\frac{2}{3} \times 500 \times \frac{2\pi}{8.64} \times 10^{-4} \times 1 \times \sqrt{\frac{2 \times 500}{9.8}} \approx 26 \text{ cm}$$

1.3 Law of Conservation of Energy, Momentum and Angular Momentum

1.118 As **F** is constant so the sought work done

$$A = \mathbf{F}\cdot\Delta\mathbf{r} = \mathbf{F}\cdot(\mathbf{r}_2 - \mathbf{r}_1)$$

or $$A = (3\mathbf{i}+4\mathbf{j})\cdot[(2\mathbf{i}-3\mathbf{j})-(\mathbf{i}+2\mathbf{j})] = (3\mathbf{i}+4\mathbf{j})\cdot(\mathbf{i}-5\mathbf{j}) = -17\,\text{J}$$

1.119 As locomotive is in unidirectional motion, its acceleration

$$w = \frac{dv}{dt} = \frac{1}{2}\frac{dv^2}{ds} = \frac{a^2}{2}$$

Hence, force acting on the locomotive $F = mw = \dfrac{ma^2}{2}$

Let, $v = 0$ at $t = 0$, then the distance covered during the first t seconds

$$s = \frac{1}{2}wt^2 = \frac{1}{2}\frac{a^2}{2}t^2 = \frac{a^2}{4}t^2$$

Hence the sought work,

$$A = Fs = \frac{ma^2}{2}\frac{(a^2t^2)}{4} = \frac{ma^4t^2}{8}$$

1.120 We have $$T = \frac{1}{2}mv^2 = as^2 \quad \text{or} \quad v^2 = \frac{2as^2}{m} \tag{1}$$

Differentiating Eq. (1) with respect to time

$$2vw_t = \frac{4as}{m}v \quad \text{or} \quad w_t = \frac{2as}{m} \tag{2}$$

Hence net acceleration of the particle

$$w = \sqrt{w_t^2 + w_n^2} = \sqrt{\left(\frac{2as}{m}\right)^2 + \left(\frac{2as^2}{mR}\right)^2} = \frac{2as}{m}\sqrt{1+(s/R)^2}$$

Hence the sought, force $F = mw = 2as\sqrt{1+(s/R)^2}$

1.121 Let **F**, make an angle θ with the horizontal at an arbitrary instant of time (see figure). Newton's second law in projection form along the direction of the force gives

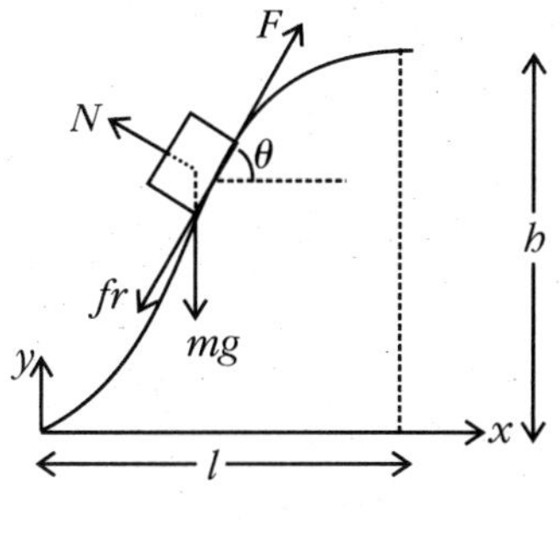

$F = kmg\cos\theta + mg\sin\theta$ (because there is no acceleration of the body)

As $\mathbf{F}\uparrow\uparrow d\mathbf{r}$, the differential work done by the force **F**

$$dA = \mathbf{F}\cdot d\mathbf{r} = F\,ds \quad (\text{where } ds = |d\mathbf{r}|)$$
$$= kmg\,ds(\cos\theta) + mg\,ds\sin\theta$$
$$= kmg\,dx + mg\,dy$$

Hence, $$A = kmg\int_0^l dx + mg\int_0^h dy$$
$$= kmgl + mgh = mg(kl + h)$$

1.122 Let s be the distance covered by the disk along the incline, from the equation of increment of mechanical energy of the disk in the field of gravity: $\Delta T + \Delta U = A_{fr}$

$$0 + (-mgs \sin\alpha) = -kmg \cos\alpha s - kmgl$$

or
$$s = \frac{kl}{\sin\alpha - k\cos\alpha} \tag{1}$$

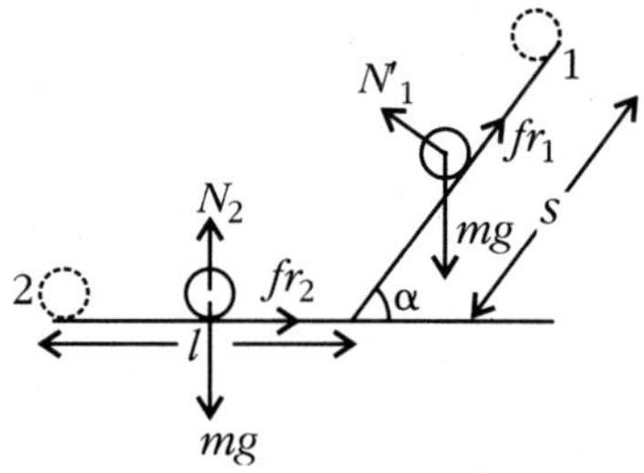

Hence the sought work

$$A_{fr} = -kmg[s\cos\alpha + l]$$

$$A_{fr} = -\frac{klmg}{1 - k\cot\alpha} \quad \text{(using Eq. 1)}$$

On putting the values

$$A_{fr} = -0.05 \text{ J}$$

1.123 Let x be the compression in the spring when the bar m_2 is about to shift. Therefore at this moment spring force on m_2 is equal to the limiting friction between the bar m_2 and horizontal floor. Hence

$$\kappa x = km_2 g \quad \text{[where } \kappa \text{ is the spring constant (say)]} \tag{1}$$

For the block m_1 from work-energy theorem:

$A = \Delta T = 0$ for minimum force. (A here includes the work done in stretching the spring.)

So,
$$Fx - \frac{1}{2}\kappa x^2 - km_1 gx = 0 \quad \text{or} \quad \kappa\frac{x}{2} = F - km_1 g \tag{2}$$

From Eqs. (1) and (2),
$$F = kg\left(m_1 + \frac{m_2}{2}\right)$$

1.124 From the initial condition of the problem, the limiting friction between the chain lying on the horizontal table equals the weight of the overhanging part of chain, i.e., $\lambda\eta lg = k\lambda(1 - \eta)lg$ (where λ is the linear mass density of the chain).

So,
$$k = \frac{\eta}{1 - \eta} \tag{1}$$

Let (at an arbitrary moment of time) length of the chain on the table be x. So the net friction force between the chain and the table, at this moment

$$fr = kN = k\lambda xg \tag{2}$$

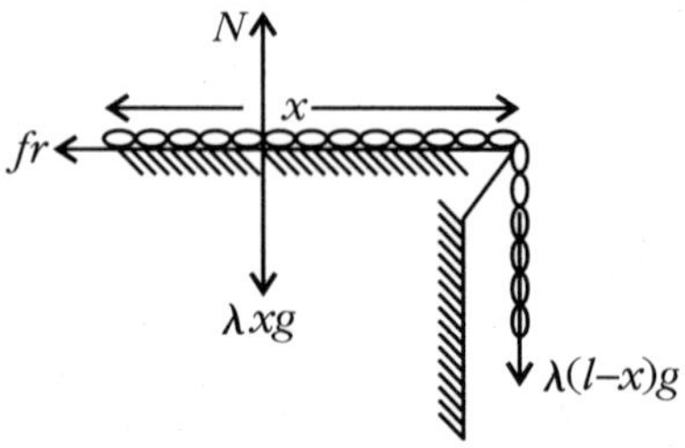

The differential work done by the friction forces

$$dA = \mathbf{fr} \cdot d\mathbf{r} = -fr\, ds$$

$$= -k\lambda xg(-dx) = \lambda g\left(\frac{\eta}{1 - \eta}\right) xdx \tag{3}$$

(Note that here we have written $ds = -dx$, because ds is essentially a positive term and as the length of the chain decreases with time, dx is negative.)

Hence, the sought work done

$$A = \int_{(1-\eta)l}^{0} \lambda g \frac{\eta}{1 - \eta} xdx = -(1 - \eta)\eta \frac{mgl}{2} = -1.3 \text{ J}$$

1.125 The velocity of the body, t seconds after the beginning of the motion becomes $\mathbf{v} = \mathbf{v}_0 + \mathbf{g}t$. The power developed by the gravity ($m\mathbf{g}$) at that moment, is

$$P = m\mathbf{g} \cdot \mathbf{v} = m(\mathbf{g} \cdot \mathbf{v}_0 + g^2 t) = mg(gt - v_0 \sin \alpha)$$

As $m\mathbf{g}$ is a constant force, so the average power

$$<P> = \frac{A}{\tau} = \frac{m\mathbf{g} \cdot \Delta\mathbf{r}}{\tau}$$

where $\Delta\mathbf{r}$ is the net displacement of the body during time of flight.

As, $m\mathbf{g} \perp \Delta\mathbf{r}$, so $<P> = 0$

1.126 We have $w_n = v^2/R = at^2$ or $v = \sqrt{aR}\, t$

where t is defined to start from the beginning of motion from rest.

So, $$w_t = \frac{dv}{dt} = \sqrt{aR}$$

Instantaneous power, $P = \mathbf{F} \cdot \mathbf{v} = m(w_t\mathbf{t} + w_n\mathbf{n}) \cdot (\sqrt{aR}\, t\mathbf{t})$, where $\mathbf{t}$ and $\mathbf{n}$ are unit vectors along the direction of tangent (velocity) and normal, respectively.

So, $$P = mw_t\sqrt{aR}\, t = maRt$$

Hence the sought average power

$$<P> = \frac{\int_0^t P\, dt}{\int_0^t dt} = \frac{\int_0^t ma\, Rt\, dt}{t}$$

Hence, $$<P> = \frac{maRt^2}{2t} = \frac{maRt}{2}$$

1.127 Let the body m acquire the horizontal velocity v_0 along positive x-axis at the point O.

(a) Velocity of the body t seconds after the beginning of the motion,

$$\mathbf{v} = \mathbf{v}_0 + \mathbf{w}t = (v_0 - kg\, t)\,\mathbf{i} \tag{1}$$

Instantaneous power $P = \mathbf{F} \cdot \mathbf{v} = (-kmg\,\mathbf{i}) \cdot (v_0 - kgt)\,\mathbf{i} = -kmg\,(v_0 - kgt)$

From Eq. (1), the time of motion $\tau = v_0/kg$.

Hence sought average power during the time of motion

$$<P> = \frac{\int_0^\tau -kmg(v_0 - kgt)\, dt}{\tau} = -\frac{kmg\, v_0}{2} = -2\text{ W} \quad \text{(on substituting values)}$$

(b) From $$F_x = mw_x$$

$$-kmg = mw_x = mv_x\frac{dv_x}{dx}$$

or $$v_x dv_x = -kg\, dx = -\alpha gx\, dx$$

To find $v(x)$, let us integrate the above equation

$$\int_{v_0}^{v} v_x dv_x = -\alpha g \int_0^x x dx \quad \text{or} \quad v^2 = v_0^2 - \alpha g x^2$$

Now, $$\mathbf{P} = \mathbf{F} \cdot \mathbf{v} = -m\alpha x g\sqrt{v_0^2 - \alpha g x^2} \tag{2}$$

For maximum power,

$$\frac{d}{dt}\left(\sqrt{v_0^2 x^2 - \lambda g x^4}\right) = 0 \text{ which yields } x = \frac{v_0}{\sqrt{2\alpha g}}$$

Putting this value of x in Eq. (2), we get

$$P_{\max} = -\frac{1}{2} m v_0^2 \sqrt{\alpha g}$$

1.128 Centrifugal force of inertia is directed outward along radial line, thus the sought work

$$A = \int_{r_1}^{r_2} m\omega^2 r dr = \frac{1}{2} m\omega^2 (r_2^2 - r_1^2) = 0.20 \text{ J} \quad \text{(on substituting values)}$$

1.129 Since the springs are connected in series, the combination may be treated as a single spring of spring constant.

So, $$\kappa = \frac{\kappa_1 \kappa_2}{\kappa_1 + \kappa_2}$$

From the equation of increment of mechanical energy,

$$\Delta T + \Delta U_{\text{spring}} = A_{\text{ext}}$$

$$0 + \frac{1}{2}\kappa \Delta l^2 = A_{\text{ext}} \quad \text{or} \quad A_{\text{ext}} = \frac{1}{2}\left(\frac{\kappa_1 \kappa_2}{\kappa_1 + \kappa_2}\right)\Delta l^2$$

1.130 First, let us find the total height of ascent. At the beginning and at the end of the path, velocity of the body is equal to zero, and therefore the increment of the kinetic energy of the body is also equal to zero. On the other hand, according to work-energy theorem, ΔT is equal to the algebraic sum of the work A performed by all the forces, i.e. by the force F and gravity, over this path. However, since $\Delta T = 0$ then $A = 0$. Taking into account that the upward direction is assumed to coincide with the positive direction of the y-axis, we can write

$$A = \int_0^h (\mathbf{F} + m\mathbf{g}) \cdot d\mathbf{r} = \int_0^h (F_y - mg)\, dy$$

$$= mg \int_0^h (1 - 2ay)\, dy = mgh(1 - ah) = 0 \quad (\text{when } h = 1/a)$$

The work performed by the force F over the first half of the ascent is

$$A_F = \int_0^{h/2} F_y dy = 2mg \int_0^{h/2} (1 - ay)\, dy = \frac{3mg}{4a}, \quad \text{where } h = 1/a$$

The corresponding increment of the potential energy is

$$\Delta U = \frac{mgh}{2} = \frac{mg}{2a}$$

1.131 From the equation $F_r = -\frac{dU}{dr}$, we get $F_r = \left[-\frac{2a}{r^3} + \frac{b}{r^2}\right]$

(a) We have at $r = r_0$, the particle is in equilibrium position, i.e, $F_r = 0$.

So, $$r_0 = \frac{2a}{b}$$

To check whether the position is steady (the position of stable equilibrium), we have to satisfy

$$\frac{d^2U}{dr^2} > 0$$

We have $$\frac{d^2U}{dr^2} = \left[\frac{6a}{r^4} - \frac{2b}{r^3}\right]$$

Putting the value of $r = r_0 = 2a/b$, we get

$$\frac{d^2U}{dr^2} = \frac{b^4}{8a^3} \quad \text{(as } a \text{ and } b \text{ are positive constants)}$$

So, $$\frac{d^2U}{dr^2} = \frac{b^2}{8a^3} > 0$$

which indicates that the potential energy of the system is minimum, hence this position is steady.

(b) We have $$F_r = -\frac{dU}{dr} = \left[-\frac{2a}{r^3} + \frac{b}{r^2}\right]$$

For F to be maximum, $$\frac{dF_r}{dr} = 0$$

So, $$r = \frac{3a}{b} \quad \text{and} \quad F_{r(\max)} = \frac{-b^3}{27a^2}$$

As F_r is negative, the force is attractive.

1.132 (a) We have

$$F_x = -\frac{\partial U}{\partial x} = -2\alpha x \quad \text{and} \quad F_y = \frac{-\partial U}{\partial y} = -2\beta y$$

So, $$\mathbf{F} = -2\alpha x\mathbf{i} - 2\beta y\mathbf{i} \quad \text{and} \quad F = 2\sqrt{\alpha^2x^2 + \beta^2y^2}$$

For a central force, $\mathbf{r} \times \mathbf{F} = 0$.

Here, $$\mathbf{r} \times \mathbf{F} = (x\mathbf{i} + y\mathbf{j}) \times (-2\alpha x\mathbf{i} - 2\beta y\mathbf{j})$$
$$= -2\beta xy\mathbf{k} + 2\alpha xy(\mathbf{k}) \neq 0$$

Hence the force is not a central force.

(b) As $U = \alpha x^2 + \beta y^2$

So, $$F_x = -\frac{\partial U}{\partial x} = -2\alpha x \quad \text{and} \quad F_y = \frac{-\partial U}{\partial y} = -2\beta y$$

So, $$F = \sqrt{F_x^2 + F_y^2} = \sqrt{4\alpha^2x^2 + 4\beta^2y^2}$$

According to the problem

$$F = 2\sqrt{\alpha^2 x^2 + \beta^2 y^2} = C \text{ (constant)}$$

or
$$\alpha^2 x^2 + \beta^2 y^2 = \frac{C^2}{2}$$

or
$$\frac{x^2}{\beta^2} + \frac{y^2}{\alpha^2} = \frac{C^2}{2\alpha^2\beta^2} = k \text{ (say)}$$

Therefore, the surface for which F is constant is an ellipse.

For an equipotential surface U is constant.

So,
$$\alpha x^2 + \beta y^2 = C_0 \quad \text{(constant)}$$

or
$$\frac{x^2}{\sqrt{\beta^2}} + \frac{y^2}{\sqrt{\alpha^2}} = \frac{C_0}{\alpha\beta} = K_0 \quad \text{(constant)}$$

Hence, the equipotential surface is also an ellipse.

1.133 Let us calculate the work performed by the forces of each field over the path from a certain point 1 (x_1, y_1) to another point 2 (x_2, y_2)

(i)
$$dA = \mathbf{F}\cdot d\mathbf{r} = ay\mathbf{i}\cdot d\mathbf{r} = aydx \quad \text{or} \quad A = a\int_{x_1}^{x_2} ydx$$

(ii)
$$dA = \mathbf{F}\cdot d\mathbf{r} = (ax\mathbf{i} + by\mathbf{j})\cdot d\mathbf{r} = axdx + bydy$$

Hence,
$$A = \int_{x_1}^{x_2} axdx + \int_{y_1}^{y_2} bydy$$

In the first case, the integral depends on the function of type $y(x)$, i.e., on the shape of the path. Consequently, the first field of force is not potential. In the second case, both the integrals do not depend on the shape of the path. They are defined only by the coordinate, of the initial and final points of the path, therefore the second field of force is potential.

1.134 Let s be the sought distance, then from the equation of increment of mechanical energy,

$$\Delta T + \Delta U_{gr} = A_{fr}$$

$$\left(0 - \frac{1}{2}mv_0^2\right) + (mg\, s\sin\alpha) = -kmg\cos\alpha s$$

or
$$s = \frac{v_0^2/2g}{(\sin\alpha + k\cos\alpha)}$$

Hence,
$$A_{fr} = -kmg\cos\alpha s = \frac{-km\, v_0^2}{2(k + \tan\alpha)}$$

1.135 Velocity of the body at height h, $v_h = \sqrt{2g(H-h)}$, horizontally (from the figure given in the problem book). Time taken in falling through the distance h is given by

$$t = \sqrt{\frac{2h}{g}} \quad \text{(as initial vertical component of the velocity is zero)}$$

Now,
$$s = v_h t = \sqrt{2g(H-h)} \times \sqrt{\frac{2h}{g}} = \sqrt{4(Hh - h^2)}$$

For s_{max},
$$\frac{d}{ds}(Hh - h^2) = 0, \text{ which yields } h = \frac{H}{2}$$

Putting this value of h in the expression obtained for s, we get

$$S_{max} = H$$

1.136 To complete a smooth vertical track of radius R, the minimum height at which a particle starts, must be equal to $\frac{5}{2}R$. Let us first prove this.

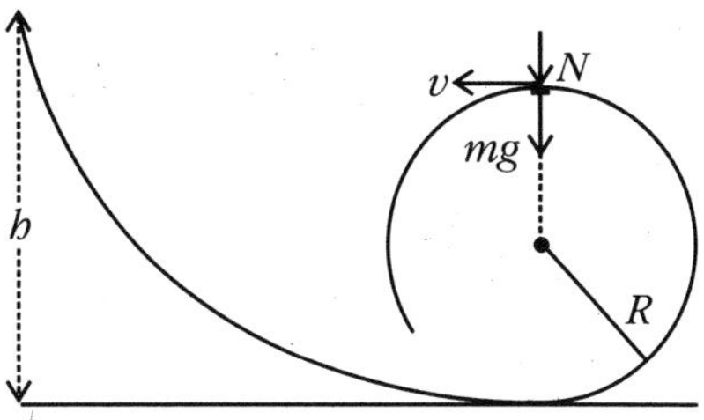

In the figure there is a smooth track and a small body is released at height h. We have to find minimum value for h so that the body can complete smooth vertical track of radius R. As the track is smooth and the body is under the action of only two forces, normal reaction and gravitational pull. The mechanical energy of the block in the gravitational field is conversed, because normal reaction does zero work whereas the gravitational force is conservative. From the conservation of mechanical energy between the point of release and the upper most point of the circular track

$$\Delta T + \Delta U_{gr} = 0$$

$$\left(\frac{1}{2}mv^2 - 0\right) - mg(h - 2R) = 0$$

which yields $$v^2 = 2g\,(h - 2R) \qquad (1)$$

To complete the vertical circular track the condition is that the normal reaction at the upper most point of the circular track $N \geq 0$.

From Newton's second law in projection form towards the centre of the track

$$F_n = mw_n \quad \text{(where } w_n \text{ is the normal acceleration)}$$

$$N + mg = m\frac{v^2}{R}$$

$$N = m\left(\frac{v^2}{R} - g\right)$$

As $N \geq 0$ at upper most point of the circular track

So, $$m\left(\frac{v^2}{R} - g\right) \geq 0$$

or $$v^2 \geq gR \qquad (2)$$

From Eqs. (1) and (2)

$$2g(h - 2R) \geq gR$$

Hence, $$h \geq \frac{5}{2}R$$

Thus in our problem, the body could not reach the upper most point of the vertical track of radius $h/2$. Let the particle A leave the track at some point with speed v (see figure). Now from energy conservation for the body A in the field of gravity

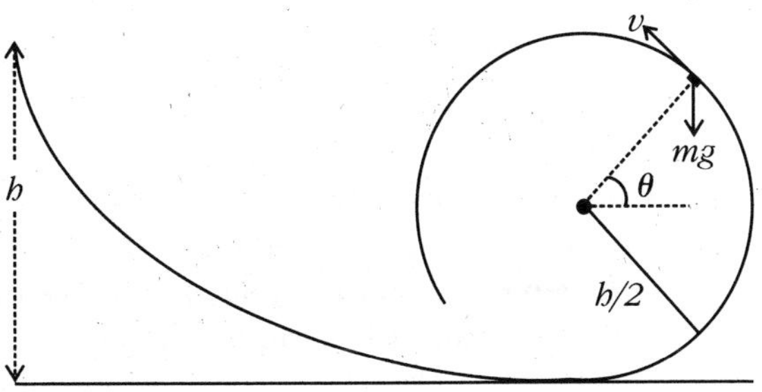

$$mg\left[h - \frac{h}{2}(1 + \sin\theta)\right] = \frac{1}{2}mv^2$$

or $$v^2 = gh(1 - \sin\theta) \qquad (3)$$

At the break off point the normal reaction $N = 0$, so from Newton's second law at this point

$$F_n = mw_n$$

or
$$mg\sin\theta = \frac{mv^2}{(h/2)}$$

So,
$$v^2 = \frac{gh}{2}\sin\theta \tag{4}$$

From Eqs. (3) and (4),
$$\sin\theta = \frac{2}{3} \quad \text{and} \quad v = \sqrt{\frac{gh}{3}}$$

After leaving the track, the body A comes in air and further goes up and at maximum height of its trajectory in air and its velocity (say v') becomes horizontal (Fig.). Hence, the sought velocity of A at this point.

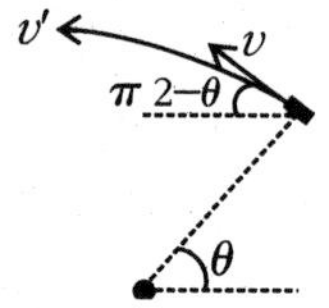

$$v' = v\cos\left(\frac{\pi}{2} - \theta\right) = v\sin\theta = \frac{2}{3}\sqrt{\frac{gh}{3}}$$

1.137 Let the point of suspension be shifted with velocity v_A in the horizontal direction towards left, then in the rest frame of point of suspension the ball starts with same velocity horizontally towards right. Let us work in this frame. From Newton's second law in projection form towards the point of suspension at the upper most point (say B):

$$mg + T = \frac{mv_B^2}{l} \quad \text{or} \quad T = \frac{mv_B^2}{l} - mg \tag{1}$$

Condition required to complete the vertical circle is that $T \geq 0$. (2)

But,
$$\frac{1}{2}mv_A^2 = mg(2l) + \frac{1}{2}mv_B^2$$

So,
$$v_B^2 = v_A^2 - 4gl \tag{3}$$

From Eqs. (1), (2) and (3)

$$T = \frac{m(v_A^2 - 4gl)}{l} - mg \geq 0 \quad \text{or} \quad v_A \geq \sqrt{5gl}$$

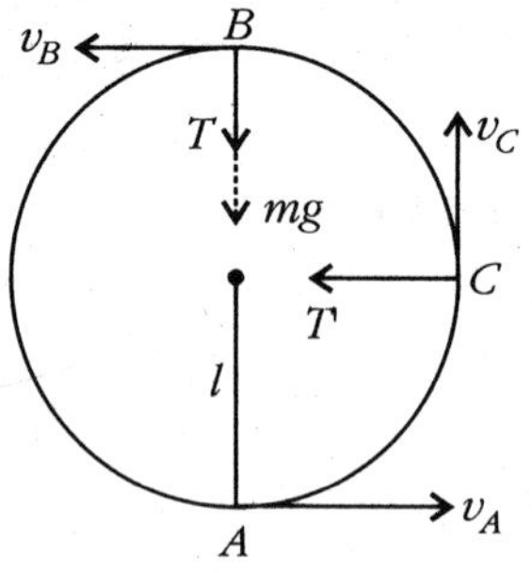

Thus
$$v_{A(\min)} = \sqrt{5gl}$$

From the equation $F_n = mw_n$ at point C

$$T' = \frac{mv_C^2}{l} \tag{4}$$

Again from energy conservation

$$\frac{1}{2}mv_A^2 = \frac{1}{2}mv_C^2 + mgl \tag{5}$$

From Eqs. (4) and (5)

$$T = 3mg$$

1.138 Since the tension is always perpendicular to the velocity vector, the work done by the tension force will be zero. Hence, according to the work energy theorem, the kinetic energy or velocity of the disk will remain constant during its motion. Hence, the sought time $t = s/v_0$, where s is the total distance traversed by the small disk during its motion.

Now, at an arbitrary position (see figure)

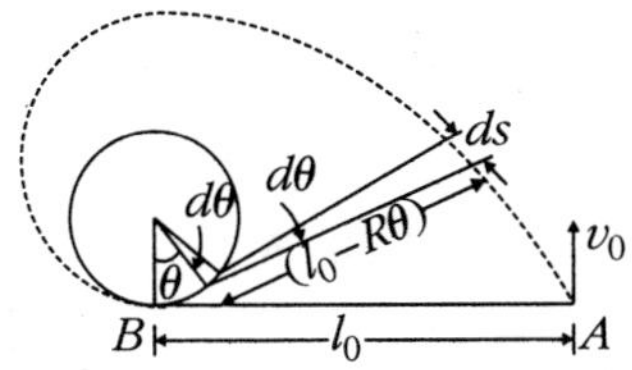

$$ds = (l_0 - R\theta)\, d\theta$$

So,
$$s = \int_0^{l_0/R} (l_0 - R\theta)\, d\theta$$

or
$$s = \frac{l_0^2}{R} - \frac{Rl_0^2}{2R^2} = \frac{l_0^2}{2R}$$

Hence, the required time
$$t = \frac{l_0^2}{2Rv_0}$$

Note: *It should be clearly understood that the only uncompensated force acting on the disk A in this case is the tension T, of the thread. It is easy to see that there is no point here, relative to which the moment of force T is invariable in the process of motion. Hence conservation of angular momentum is not applicable here.*

1.139 Suppose that Δl is the elongation of the rubber cord. Then, from energy conservation

$$\Delta U_{gr} + \Delta U_{el} = 0 \text{ (as } \Delta T = 0)$$

or
$$-mg(l + \Delta l) + \frac{1}{2}\kappa \Delta l^2 = 0$$

or
$$\frac{1}{2}\kappa \Delta l^2 - mg\,\Delta l - mgl = 0$$

or
$$\Delta l = \frac{mg \pm \sqrt{(mg)^2 + 4 \times \frac{\kappa}{2} mgl}}{2 \times \frac{\kappa}{2}} = \frac{mg}{\kappa}\left[1 \pm \sqrt{1 + \frac{2\kappa l}{mg}}\right]$$

Since the value of $\sqrt{1 + \frac{2\kappa l}{mg}}$ is certainly greater than 1, hence negative sign is avoided.

So,
$$\Delta l = \frac{mg}{\kappa}\left(1 + \sqrt{1 + \frac{2\kappa l}{mg}}\right)$$

1.140 When the thread *PA* is burnt, obviously the speed of the bars will be equal at any instant of time until it breaks off. Let v be the speed of each block and θ be the angle, which the elongated spring makes with the vertical at the moment when the bar *A* breaks off the plane. At this stage the elongation in the spring is

$$\Delta l = l_0 \sec\theta - l_0 = l_0 (\sec\theta - 1) \tag{1}$$

Since the problem is concerned with position and there are no forces other than conservative forces, the mechanical energy of the system (both bars + spring) in the field of gravity is conserved, i.e., $\Delta T + \Delta U = 0$.

So,
$$2\left(\frac{1}{2}mv^2\right) + \frac{1}{2}kl_0^2(\sec\theta - 1)^2 - mgl_0 \tan\theta = 1 \tag{2}$$

From Newton's second law in projection form along vertical direction:

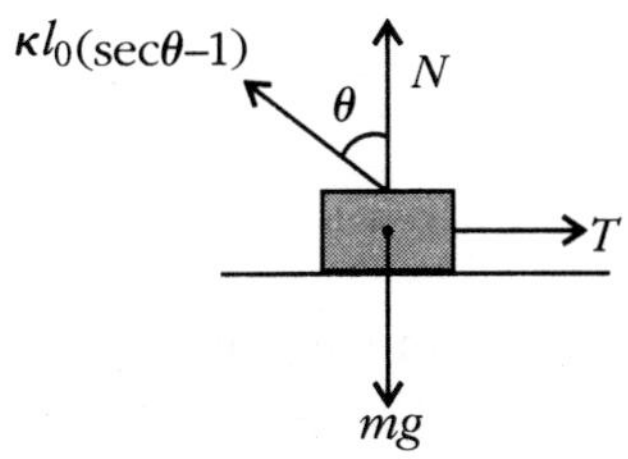

$$mg = N + \kappa l_0 (\sec\theta - 1)\cos\theta$$

But, at the moment of break off, $N = 0$.

Hence,
$$\kappa l_0 (\sec\theta - 1)\cos\theta = mg$$

or
$$\cos\theta = \frac{\kappa l_0 - mg}{k l_0} \quad (3)$$

Taking $\kappa = \frac{5mg}{l_0}$, simultaneous solution of Eqs. (2) and (3) yields

$$v = \sqrt{\frac{19 g l_0}{32}} = 1.7 \text{ m/s}$$

1.141 Obviously the elongation in the cord, $\Delta l = l_0 (\sec\theta - 1)$, at the moment the sliding first starts and at the moment horizontal projection of spring force equals the limiting friction.

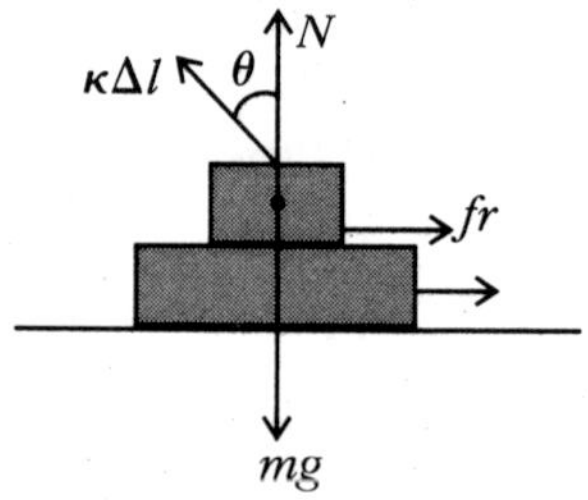

So,
$$\kappa \Delta l \sin\theta = kN \quad (1)$$

(where κ is the elastic constant)

From Newton's law in projection form along vertical direction:

$$\kappa\Delta l \cos\theta + N = mg$$

or
$$N = mg - \kappa\Delta l \cos\theta \quad (2)$$

From Eqs. (1) and (2),

$$\kappa\Delta l \sin\theta = k(mg - \kappa\Delta l \cos\theta)$$

or
$$\kappa = \frac{kmg}{\Delta l \sin\theta + k\Delta l \cos\theta}$$

From the equation of the increment of mechanical energy: $\Delta U + \Delta T = A_{fr}$

or
$$\left(\frac{1}{2}\kappa\Delta l^2\right) = A_{fr}$$

or
$$\frac{kmg\,\Delta l^2}{2\Delta l(\sin\theta + k\cos\theta)} = A_{fr}$$

Thus,
$$A_{fr} = \frac{kmgl_0(\sec\theta - 1)}{2(\sin\theta - k\cos\theta)} = 0.09 \text{ J} \quad \text{(on substituting values)}$$

1.142 Let the deformation in the spring be Δl, when the rod AB has attained the angular velocity ω. From the second law of motion in projection form $F_n = mw_n$.

$$\kappa\Delta l = m\omega^2 (l_0 + \Delta l) \text{ or } \Delta l = \frac{m\omega^2 l_0}{\kappa - m\omega^2}$$

From the energy equation,
$$A_{ext} = \frac{1}{2}mv^2 + \frac{1}{2}\kappa\Delta l^2$$
$$= \frac{1}{2}m\omega^2(l_0 + \Delta l)^2 + \frac{1}{2}\kappa\Delta l^2$$
$$= \frac{1}{2}m\omega^2\left(l_0 + \frac{m\omega^2 l_0}{\kappa - m\omega^2}\right)^2 + \frac{1}{2}\kappa\left(\frac{m\omega^2 l_0^2}{\kappa - m\omega^2}\right)^2$$

On solving
$$A_{ext} = \frac{\kappa}{2}\frac{l_0^2\eta(1+\eta)}{(1-\eta)^2}, \quad \text{where } \eta = \frac{m\omega^2}{\kappa}$$

1.143 We know that acceleration of centre of mass of the system is given by the expression.

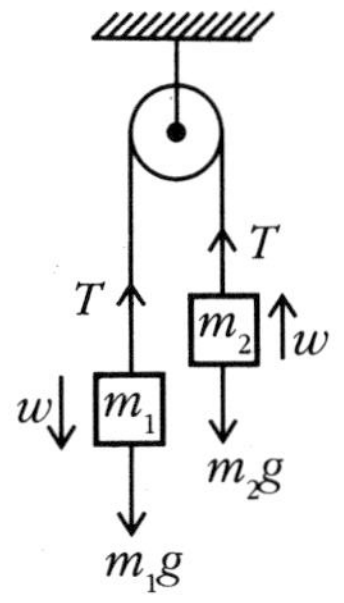

$$\mathbf{w}_C = \frac{m_1\mathbf{w}_1 + m_2\mathbf{w}_2}{m_1 + m_2}$$

Since $$\mathbf{w}_1 = -\mathbf{w}_2$$

$$\mathbf{w}_C = \frac{(m_1 - m_2)\mathbf{w}_1}{m_1 + m_2} \quad (1)$$

Now from Newton's second law $\mathbf{F} = m\mathbf{w}$, for bodies m_1 and m_2, respectively,

$$\mathbf{T} + m_1\mathbf{g} = m_1\mathbf{w}_1 \quad (2)$$

and $$\mathbf{T} + m_2\mathbf{g} = m_2\mathbf{w}_2 = -m_2\mathbf{w}_1 \quad (3)$$

Solving Eqs. (2) and (3) $$\mathbf{w}_1 = \frac{(m_1 - m_2)\mathbf{g}}{m_1 + m_2} \quad (4)$$

Thus, from Eqs. (1) and (4), $$\mathbf{w}_C = \frac{(m_1 - m_2)^2\,\mathbf{g}}{(m_1 + m_2)^2}$$

1.144 As the closed system consisting of two particles m_1 and m_2 is initially at rest, the centre of mass of the system will remain at rest. Further as $m_2 = m_1/2$, the centre of mass of the system divides the line joining m_1 and m_2 at all the moments of time in the ratio 1:2. In addition to it, the total linear momentum of the system at all the times is zero. So, $\mathbf{p}_1 = -\mathbf{p}_2$ and therefore the velocities of m_1 and m_2 are also directed in opposite sense. Bearing in mind all these things, the sought trajectory is as shown in the figure.

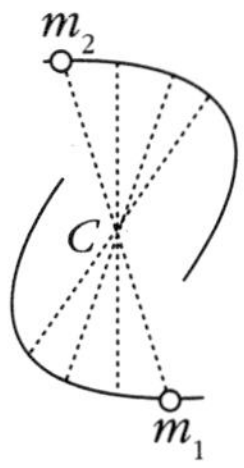

1.145 First of all, it is clear that the chain does not move in the vertical direction during the uniform rotation. This means that the vertical component of the tension T balances gravity. As for the horizontal component of the tension T, it is constant in magnitude and permanently directed toward the rotation axis. It follows from this that the centre of mass of the chain, the point C, travels along horizontal circle of radius ρ (say). Therefore we have

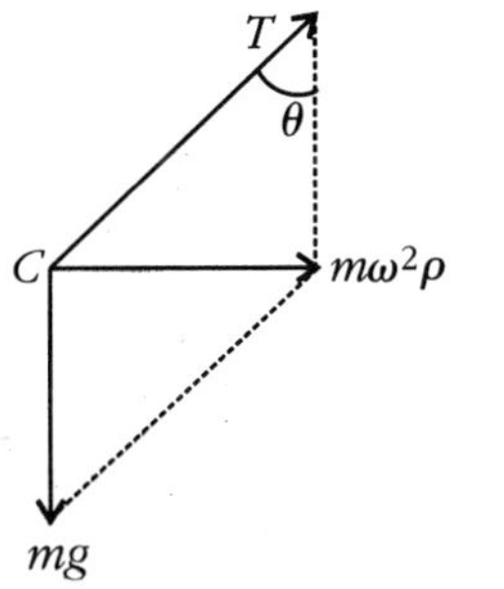

$$T\cos\theta = mg \quad \text{and} \quad T\sin\theta = m\omega^2\rho$$

Thus $$\rho = \frac{g\tan\theta}{\omega^2} = 0.8 \text{ cm}$$

and $$T = \frac{mg}{\cos\theta} = 5 \text{ N} \quad \text{(on substituting values)}$$

1.146 (a) Let us draw free body diagram and write Newton's second law in terms of projection along vertical and horizontal directions, respectively,

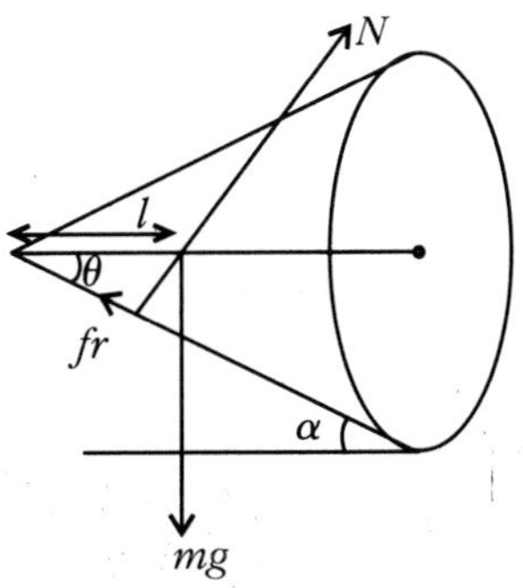

$$N\cos\alpha - mg + fr\sin\alpha = 0 \quad (1)$$

$$fr\cos\alpha - N\sin\alpha = m\omega^2 l \quad (2)$$

From Eqs. (1) and (2)

$$fr\cos\alpha - \frac{\sin\alpha}{\cos\alpha}(-fr\sin\alpha + mg) = m\omega^2 l$$

So, $$fr = mg\left(\sin\alpha + \frac{\omega^2 l}{g}\cos\alpha\right) \quad (3)$$

$$= 6 \text{ N} \quad \text{(on substituting values)}$$

(b) For rolling, without sliding,

$$fr \leq kN$$

but, $N = mg\cos\alpha - m\omega^2 l\sin\alpha$

$$mg\left(\sin\alpha + \frac{\omega^2 l}{g}\cos\alpha\right) \leq k(mg\cos\alpha - m\,\omega^2 l\sin\alpha) \quad \text{(using Eq. 3)}$$

Rearranging, we get

$$m\omega^2 l(\cos\alpha + k\sin\alpha) \leq (k\,mg\cos\alpha - mg\sin\alpha)$$

Thus, $$\omega \leq \sqrt{g(k - \tan\alpha)/(1 + k\tan\alpha)l} = 2 \text{ rad/s}$$

1.147 (a) Total kinetic energy in frame K' is

$$T = \frac{1}{2}m_1(\mathbf{v}_1 - \mathbf{V})^2 + \frac{1}{2}m_2(\mathbf{v}_2 - \mathbf{V})^2$$

This is minimum with respect to variation in $\mathbf{V}$, when

$$\frac{\delta T}{\delta \mathbf{V}} = 0, \quad \text{i. e.,} \quad m_1(\mathbf{v}_1 - \mathbf{V}) + m_2(\mathbf{v}_2 - \mathbf{V}) = 0$$

or $$\mathbf{V} = \frac{m_1\mathbf{v}_1 + m_2\mathbf{v}_2}{m_1 + m_2} = \mathbf{v}_C$$

Hence, it is the frame of center of mass in which kinetic energy of a system is minimum.

(b) Linear momentum of the particle 1 in the K' or C frame

$$\tilde{\mathbf{p}}_1 = m_1(\mathbf{v}_1 - \mathbf{v}_C) = \frac{m_1 m_2}{m_1 + m_2}(\mathbf{v}_1 - \mathbf{v}_2)$$

or $$\tilde{\mathbf{p}}_1 = \mu(\mathbf{v}_1 - \mathbf{v}_2), \quad \text{where} \quad \mu = \frac{m_1 m_2}{m_1 + m_2} = \text{reduced mass}$$

Similarly, $$\tilde{\mathbf{p}}_2 = \mu(\mathbf{v}_2 - \mathbf{v}_1)$$

So, $$|\tilde{\mathbf{p}}_1| = |\tilde{\mathbf{p}}_2| = \tilde{p} = \mu v_{rel}, \quad \text{where} \quad v_{rel} = |\mathbf{v}_1 - \mathbf{v}_2|$$

Now the total kinetic energy of the system in the C frame is

$$\tilde{T} = \tilde{T}_1 + \tilde{T}_2 = \frac{\tilde{p}^2}{2m_1} + \frac{\tilde{p}^2}{2m_2} = \frac{\tilde{p}^2}{2\mu}$$

Hence, $$\tilde{T} = \frac{1}{2}\mu v_{rel}^2 = \frac{1}{2}\mu|\mathbf{v}_1 - \mathbf{v}_2|^2$$

1.148 To find the relationship between the values of the mechanical energy of a system in the K and C reference frames, let us begin with the kinetic energy T of the system. The velocity of the i-th particle in the K frame may be represented as $\mathbf{v}_i = \tilde{\mathbf{v}}_i + \mathbf{v}_C$. Now we can write

$$T = \sum \frac{1}{2} m_i v_i^2 = \sum \frac{1}{2} m_i (\tilde{\mathbf{v}}_i + \mathbf{v}_C)\cdot(\tilde{\mathbf{v}}_i + \mathbf{v}_C)$$
$$= \sum \frac{1}{2} m_i \tilde{v}_i^2 = \mathbf{v}_C \sum m_i \tilde{\mathbf{v}}_i + \sum \frac{1}{2} m_i v_C^2$$

Since in the C frame $\sum m_i \tilde{\mathbf{v}}_i = 0$, the previous expression takes the form

$$T = \tilde{T} + \frac{1}{2} m \tilde{v}_C^2 = \tilde{T} + \frac{1}{2} m V^2 \quad \text{(since according to the problem } v_C = V) \tag{1}$$

Since the internal potential energy U of a system depends only on its configuration, the magnitude U is the same in all reference frames. Adding U to the left and right hand sides of Eq. (1), we obtain the sought relationship,

$$E = \tilde{E} + \frac{1}{2} m V^2$$

1.149 As initially $U = \tilde{U} = 0$, so, $\tilde{E} = \tilde{T}$. From the solution of problem 1.147(b)

$$\tilde{T} = \frac{1}{2} \mu |\mathbf{v}_1 - \mathbf{v}_2|^2$$

As $\qquad \mathbf{v}_1 \perp \mathbf{v}_2$

Thus, $\qquad \tilde{T} = \frac{1}{2} \frac{m_1 m_2}{m_1 + m_2} (v_1^2 + v_2^2)$

1.150 Velocities of masses m_1 and m_2 after t seconds are, respectively,

$$\mathbf{v}_1' = \mathbf{v}_1 + \mathbf{g}t \quad \text{and} \quad \mathbf{v}_2' = \mathbf{v}_2 + \mathbf{g}t$$

Hence the final momentum of the system

$$\mathbf{p} = m_1 \mathbf{v}_1 + m_2 \mathbf{v}_2 = m_1 \mathbf{v}_1 + m_1 \mathbf{v}_2 + (m_1 + m_2)\mathbf{g}t$$
$$= \mathbf{p}_0 + m\mathbf{g}t \quad \text{(where } \mathbf{p}_0 = m_1 \mathbf{v}_1 + m_2 \mathbf{v}_2 \text{ and } m = m_1 + m_2)$$

Radius vector, is given by

$$\mathbf{r}_C = \mathbf{v}_C t + \frac{1}{2} \mathbf{w}_C t^2$$
$$= \frac{(m_1 \mathbf{v}_1 + m_2 \mathbf{v}_2)t}{(m_1 + m_2)} + \frac{1}{2} \mathbf{g}t^2 = \mathbf{v}_0 t + \frac{1}{2} \mathbf{g}t^2, \quad \text{where } \mathbf{v}_0 = \frac{m_1 \mathbf{v}_1 + m_2 \mathbf{v}_2}{m_1 + m_2}$$

1.151 After releasing, bar 2 acquires velocity v_2, obtained by the energy conservation:

$$\frac{1}{2} m_2 v_2^2 = \frac{1}{2} \kappa x^2 \quad \text{or} \quad v_2 = x\sqrt{\kappa / m_2}$$

Thus, the sought velocity of C.M.

$$v_C = \frac{0 + m_2 x \sqrt{\kappa / m_2}}{m_1 + m_2} = \frac{x\sqrt{m_2 \kappa}}{m_1 + m_2}$$

1.152 Let us consider both blocks and spring as the physical system. The centre of mass of the system moves with acceleration $a = F/(m_1 + m_2)$ towards right. Let us work in the frame of centre of mass. As this frame is a non-inertial frame (accelerated with respect to the ground) we have to apply a pseudo force $m_1 a$ towards left on the block m_1 and $m_2 a$ towards left on the block m_2.

As the center of mass is at rest in this frame, the blocks move in opposite directions and come to instantaneous rest at some instant. The elongation of the spring will be maximum or minimum at this instant. Assume that the block m_1 is displaced by the distance x_1 and the block m_2 through a distance x_2 from the initial positions.

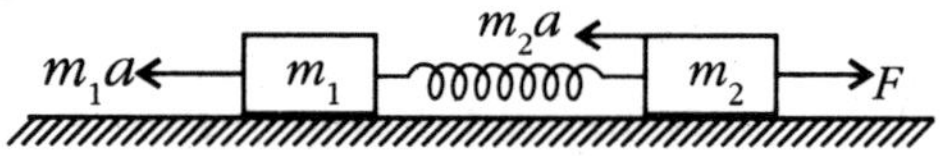

From the equation of increment of mechanical energy in C.M. frame

$$\Delta\tilde{T} + \Delta U = A_{ext}$$

where A_{ext} also includes the work done by the pseudo forces.

Here, $$\Delta\tilde{T} = 0,\ \Delta U = \frac{1}{2}k(x_1 + x_2)^2 \text{ and}$$

$$W_{ext} = \left(F - \frac{m_2 F}{m_1 + m_2}\right)x_2 + \frac{m_1 F}{m_1 + m_2}x_1 = \frac{m_1 F(x_1 + x_2)}{m_1 + m_2}$$

or $$\frac{1}{2}k(x_1 + x_2)^2 = \frac{m_1(x_1 + x_2)F}{m_1 + m_2}$$

So, $$x_1 + x_2 = 0 \quad \text{or} \quad x_1 + x_2 = \frac{2m_1 F}{k(m_1 + m_2)}$$

Hence, the maximum and minimum separations between the blocks are $l_0 + \dfrac{2m_1 F}{k(m_1 + m_2)}$ and l_0, respectively.

Alternate:

Let us link the inertial frame with horizontal floor and point the coordinate axis as shown in figure.

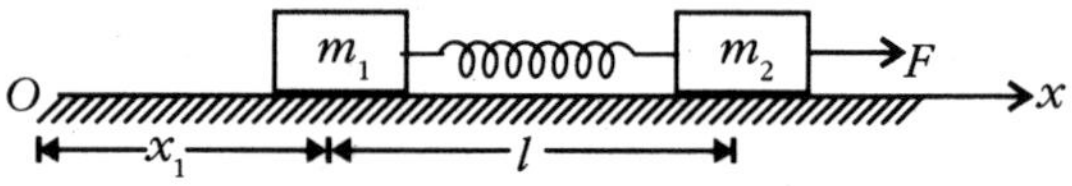

At an arbitrary instant of time the separation between the blocks is l and the x coordinate of the block m_1 is x_1. From Newton's second law for the block m_1,

$$m_1 \frac{d^2 x_1}{dt^2} = k(l - l_0) \tag{1}$$

Similarly for block m_2,

$$m_2 \frac{d^2(x_1 + l)}{dt^2} = F - k(l - l_0) \tag{2}$$

Multiplying Eq. (2) by m_1 and Eq. (1) by m_2 and further subtracting Eq. (1) from Eq. (2), we get

$$m_1 m_2 \frac{d^2(x_1 + l)}{dt^2} - m_1 m_2 \frac{d^2 x_1}{dt^2} = m_1 F - (m_1 + m_2)k(l - l_0)$$

or $$m_1 m_2 \frac{d^2 l}{dt^2} = m_1 F - (m_1 + m_2)k(l - l_0) \tag{3}$$

or
$$m_1 m_2 \frac{d^2(l-l_0)}{dt^2} = m_1 F - (m_1 + m_2)k(l - l_0)$$

or
$$\frac{d^2(l-l_0)}{dt^2} + \frac{k(m_1+m_2)(l-l_0)}{m_1 m_2} = \frac{F}{m_2} \tag{4}$$

Putting
$$\frac{k(m_1+m_2)}{m_1 m_2} = \omega^2 \quad \text{and} \quad \frac{F}{m_2} = A$$

and comparing with differential equation $\frac{d^2x}{dt^2} + \omega^2 x = A$, we get

$$l - l_0 = \frac{A}{\omega^2} + B\sin(\omega t + \delta) \tag{5}$$

But at time $t = 0$,
$$l - l_0 = 0 \quad \text{and} \quad \frac{d(l-l_0)}{dt} = 0$$

So, $B\omega\cos(\omega t + \delta)|_{t=0} = 0$ and hence $\delta = \pi/2$.

Therefore, Eq. (5) becomes

$$l - l_0 = \frac{A}{\omega^2} + B\cos\omega t \tag{6}$$

But at $t = 0$,
$$l - l_0 = 0, \text{ so, } B = \frac{-A}{\omega^2}$$

Hence, Eq. (6) becomes

$$l - l_0 = \frac{A}{\omega^2}(1 - \cos\omega t) \tag{7}$$

But $(l - l_0)$ will be maximum, when $\cos\omega t = -1$

Therefore, from Eq. (7)

$$l - l_0 = \frac{2A}{\omega^2} = \frac{2Fm_1 m_2}{km_2(m_1+m_2)}$$

or
$$l = l_0 + \frac{2Fm_1}{k(m_1+m_2)}$$

Similarly $(l - l_0)$ will be minimum when $\cos\omega t = 1$

Therefore, from Eq. (7)

$$l - l_0 = 0 \quad \text{or} \quad l = l_0$$

1.153 (a) The initial compression in the spring Δl must be such that after burning of the thread, the upper cube rises to a height that produces a tension in the spring that is atleast equal to the weight of the lower cube. Actually, the spring will first go from its compressed state to its natural length and then get elongated beyond this natural length. Let l be the maximum elongation produced under these circumstances.

Then
$$\kappa l = mg \tag{1}$$

Now, from energy conservation,

$$\frac{1}{2}\kappa\Delta l^2 = mg(\Delta l + l) + \frac{1}{2}\kappa l^2 \tag{2}$$

(because at maximum elongation of the spring, the speed of upper cube becomes zero).

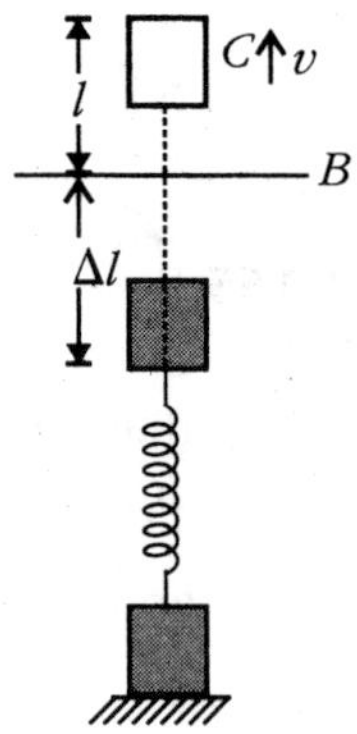

From Eqs. (1) and (2),

$$\Delta l^2 - \frac{2mg\Delta l}{\kappa} - \frac{3m^2g^2}{\kappa^2} = 0 \quad \text{or} \quad \Delta l = \frac{3mg}{\kappa}, \frac{-mg}{\kappa}$$

Therefore, acceptable solution of Δl equals $3mg/\kappa$.

(b) Let v be the velocity of upper cube at the position (say, at 2) when the lower block breaks off the floor, then from energy conservation,

$$\frac{1}{2}mv^2 = \frac{1}{2}\kappa(\Delta l^2 - l^2) - mg(l + \Delta l) \quad \left(\text{where } l = \frac{mg}{\kappa} \text{ and } \Delta l = 7\frac{mg}{\kappa}\right)$$

or
$$v^2 = 32\frac{mg^2}{\kappa} \tag{3}$$

The displacement of C.M. (of two blocks systems) in the interval in which upper block reaches position 2 is

$$\Delta y_{C_1} = \frac{m \cdot 0 + m(7mg/\kappa + mg/\kappa)}{2m} = \frac{4mg}{\kappa}$$

When the upper block reaches position 2, the lower block just leaves the floor with zero initial velocity. Now, treat two blocks system like a single particle of mass $2m$ projected vertically upward in a uniform gravitational field with the velocity of C.M.

$$v_C = \frac{mv + 0}{2m} = \frac{v}{2}$$

So, the further vertical displacement of C.M. of two block system

$$\Delta y_{C_2} = \frac{v_C^2}{2g} = \frac{(v/2)^2}{2g} = \frac{4mg}{\kappa} \quad \text{(using Eq. (3))}$$

Hence, the net displacement of the C.M. of the system, in upward direction

$$\Delta y_C = \Delta y_{C_1} + \Delta y_{C_2} = \frac{8mg}{\kappa}$$

1.154 Due to ejection of mass from a moving system (which moves due to inertia) in a direction perpendicular to it, the velocity of moving system does not change. The momentum change being adjusted by the forces on the rails. Hence in our problem velocities of buggies change only due to the entrance of the man coming from the other buggy. From conservation of linear momentum for the system (buggy 1 + man coming from buggy 2) in the direction of motion of buggy 1,

$$Mv_1 - mv_2 = 0 \tag{1}$$

Similarly from conservation of linear momentum for the system (buggy 2 + man coming from buggy 1) in the direction of motion of buggy 2,

$$Mv_2 - mv_1 = (M + m)v \tag{2}$$

Solving Eqs. (1) and (2), we get

$$v_1 = \frac{mv}{M - m} \quad \text{and} \quad v_2 = \frac{Mv}{M - m}$$

As
$$\mathbf{v}_1 \uparrow\downarrow \mathbf{v} \quad \text{and} \quad \mathbf{v}_1 \uparrow\uparrow \mathbf{v}$$

So, $$\mathbf{v}_1 = \frac{-m\mathbf{v}}{(M-m)} \quad \text{and} \quad \mathbf{v}_2 = \frac{M\mathbf{v}}{(M-m)}$$

1.155 From momentum conservation, for the system (rear buggy + man)

$$(M+m)\mathbf{v}_0 = m(\mathbf{u}+\mathbf{v}_R) + M\mathbf{v}_R \tag{1}$$

From momentum conservation, for the system (front buggy + man coming from rear buggy)

$$M\mathbf{v}_0 + m(\mathbf{u}+\mathbf{v}_R) = (M+m)\mathbf{v}_F$$

So, $$\mathbf{v}_F = \frac{M\mathbf{v}_0}{M+m} + \frac{m}{M+m}(\mathbf{u}+\mathbf{v}_R)$$

Putting the value of $\mathbf{v}_R$ from Eq. (1), we get

$$\mathbf{v}_F = \mathbf{v}_0 + \frac{mM}{(M+m)^2}\mathbf{u}$$

1.156 (i) Let $\mathbf{v}_1$ be the velocity of the buggy after both men jump off simultaneously. For the closed system (two men + buggy), from the conservation of linear momentum,

$$M\mathbf{v}_1 + 2m(\mathbf{u}+\mathbf{v}_1) = 0$$

or $$\mathbf{v}_1 = \frac{-2m\mathbf{u}}{M+2m} \tag{1}$$

(ii) Let $\mathbf{v}$ be the velocity of buggy with man, when one man jumps off the buggy. For the closed system (buggy with one man + other man) from the conservation of linear momentum:

$$0 = (M+m)\mathbf{v} + m\,(\mathbf{u}+\mathbf{v}) \tag{2}$$

Let $\mathbf{v}_2$ be the sought velocity of the buggy when the second man jumps off the buggy; then from conservation of linear momentum of the system (buggy + one man):

$$(M+m)\mathbf{v} = Mv_2 + m(\mathbf{u}+\mathbf{v}_2) \tag{3}$$

Solving Eqs. (2) and (3) we get

$$\mathbf{v}_2 = \frac{m(2M+3m)\mathbf{u}}{(M+m)(M+2m)} \tag{4}$$

From Eqs. (1) and (4)

$$\frac{v_2}{v_1} = 1 + \frac{m}{2(M+m)} > 1$$

Hence, $$v_2 > v_1$$

1.157 The descending part of the chain is in free fall, it has speed $v = \sqrt{2gy}$ at the instant when the chain descended a distance y. The length of the chain which lands on the floor, during the differential time interval dt following this instant is vdt.

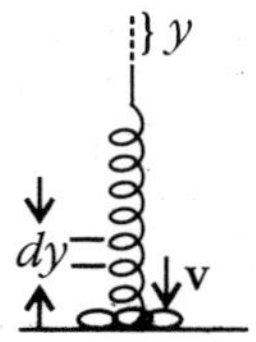

For the incoming chain element on the floor, from $dp_y = F_y dt$ (where y-axis is directed down)

$$0 - (\lambda\, vdt)v = F_y dt$$

or $$F_y = -\lambda v^2 = -2\lambda g y$$

Hence, the force exerted on the falling chain equals λv^2 and is directed upward. Therefore from third law the force exerted by the falling chain on the table at the same instant of time becomes λv^2 and is directed downward.

Since a length of chain of weight (λyg) already lies on the table and the table is at rest, the total force on the floor is $(2\lambda yg) + (\lambda yg) = (3\lambda yg)$ or the weight of a length $3y$ of chain.

1.158 Velocity of the ball, with which it hits the slab, $v = \sqrt{2gh}$.

After first impact $v' = ev$ (upward) but according to the problem

$$v' = \frac{v}{\eta'} \quad \text{so} \quad e = \frac{1}{\eta} \tag{1}$$

and momentum imparted to this slab equals

$$mv - (-mv') = mv(1 + e)$$

Similarly, velocity of the ball after second impact is

$$v'' = ev' = e^2 v$$

and momentum imparted is

$$m(v' + v'') = m(1 + e)ev$$

Again, momentum imparted during third impact is

$$m(1 + e)e^2 v, \text{ and so on}$$

Hence, net momentum imparted $= mv(1+e) + mve(1+e) + \cdots$

$$= mv(1+e)(1+e+e^2+\cdots)$$

$$= mv\frac{(1+e)}{1-e} \quad \text{(from summation of G.P.)}$$

$$= m\sqrt{2gh}\frac{(1+1/\eta)}{(1-1/\eta)} = m\sqrt{2gh}(\eta+1)/(\eta-1) \quad \text{(using Eq.(1))}$$

$$= 0.2 \text{ kg m/s (on substituting values)}$$

1.159 (a) Since the resistance of water is negligibly small the resultant of all external forces acting on the system "a man and a raft" is equal to zero. This means that the position of the C.M. of the given system does not change in the process of motion.

i.e., $$\mathbf{r}_C = \text{constant} \quad \text{or} \quad \Delta\mathbf{r}_C = 0, \sum m_i \Delta\mathbf{r}_i = 0$$

or $$m(\Delta\mathbf{r}_{mM} + \Delta\mathbf{r}_M) + M\Delta\mathbf{r}_M = 0$$

Thus, $$m(\mathbf{l}' + \mathbf{l}) + M\mathbf{l} = 0 \quad \text{or} \quad \mathbf{l} = -\frac{m\mathbf{l}'}{m+M}$$

(b) As net external force on "man-raft" system is equal to zero, therefore the momentum of this system does not change.

So, $$0 = \mathbf{m}[\mathbf{v}'(t) + \mathbf{v}_2(t)] + M\mathbf{v}_2(t)$$

or $$\mathbf{v}_2(t) = -\frac{m\mathbf{v}'(t)}{m+M} \tag{1}$$

Thus, the sought force on the raft

$$M\frac{d\mathbf{v}_2(t)}{dt} = -\frac{Mm}{m+M}\frac{d\mathbf{v}'(t)}{dt}$$

Note: *We may get the result of part (a), if we integrate Eq. (1) over the time of motion of man or raft*

1.160 The displacement of the C.M. of the system, man (m), ladder $(M - m)$ and the counterweight (M), is described by radius vector

$$\Delta \mathbf{r}_C = \frac{\Sigma m_i \Delta \mathbf{r}_i}{\Sigma m_i} = \frac{M\Delta \mathbf{r}_M + (M-m)\Delta \mathbf{r}_{(M-m)} + m\Delta \mathbf{r}_m}{2M} \qquad (1)$$

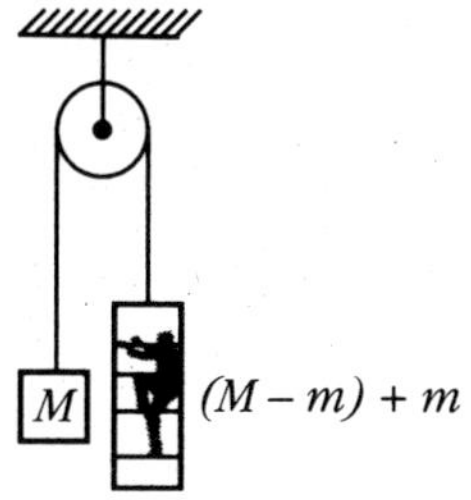

But, $$\Delta \mathbf{r}_M = -\Delta \mathbf{r}_{(M-m)}$$

and $$\Delta \mathbf{r}_m = -\Delta \mathbf{r}_{m(M-m)} + \Delta \mathbf{r}_{(M-m)} \qquad (2)$$

Using Eq. (2) in Eq. (1) we get

$$\Delta \mathbf{r}_C = \frac{ml}{2M}$$

Alternate:

Initially all the bodies of the system are at rest, and therefore the increments of linear momentum of the bodies in their motion are equal to the momentum themselves. The rope tension is the same both on the left and on the right-hand side, and consequently the momentum of the counter-balancing mass ($\mathbf{p}_1$) and the ladder with the man ($\mathbf{p}_2$) are equal at any instant of time, i.e., $\mathbf{p}_1 = \mathbf{p}_2$,

or $$M\mathbf{v}_1 = m\mathbf{v} + (M - m)\mathbf{v}_2$$

where $\mathbf{v}_1$, $\mathbf{v}$, and $\mathbf{v}_2$ are the velocities of the mass, the man, and the ladder, respectively. Taking into account that $\mathbf{v}_2 = -\mathbf{v}_1$ and $\mathbf{v} = \mathbf{v}_2 + \mathbf{v}'$, where $\mathbf{v}'$ is the man's velocity relative to the ladder, we obtain

$$\mathbf{v}_1 = (m/2M)\mathbf{v}'$$

On the other hand, the momentum of the whole system

$$\mathbf{p} = \mathbf{p}_1 + \mathbf{p}_2 = 2\mathbf{p}_1 \quad \text{or} \quad 2M\mathbf{V}_C = 2M\mathbf{v}_1 \qquad (1)$$

where $\mathbf{V}_C$ is the velocity of the centre of inertia of the system. With allowance made for Eq. (1) we get

$$\mathbf{V}_C = \mathbf{v}_1 = (m/2M)\mathbf{v}'$$

Finally, the sought displacement is

$$\Delta \mathbf{r}_C = \int \mathbf{v}_C dt = (m/2M)\int \mathbf{v}' dt = (m/2M)\Delta \mathbf{r}'$$

1.161 Velocity of cannon as well as that of shell equals $\sqrt{2gl \sin\alpha}$ down the inclined plane taken as the positive x-axis. From the linear impulse momentum theorem in projection form along x-axis for the system (cannon + shell),

$$\Delta p_x = F_x \Delta t$$

i.e., $$p\cos\alpha - M\sqrt{2gl\sin\alpha} = Mg\sin\alpha\,\Delta t \text{ (as mass of the shell is negligible)}$$

or $$\Delta t = \frac{p\cos\alpha - M\sqrt{2gl\sin\alpha}}{Mg\sin\alpha}$$

1.162 From conservation of momentum, for the system "bullet + body" along the initial direction of bullet

$$mv_0 = (m+M)v \quad \text{or} \quad v = \frac{mv_0}{m+M} \qquad (1)$$

where v_0 is the initial velocity of the bullet and v is combined velocity, after the collision.

From conservation of mechanical energy of the system (bullet + body) in the field of gravity:

$$\frac{1}{2}(m+M)v^2 = (M+m)gl(1-\cos\theta) \quad \text{or} \quad v^2 = 2gl(1-\cos\theta) \tag{2}$$

(a) From Eqs. (1) and (2),

$$v_0 = \frac{(m+M)}{m}\sqrt{2gl(1-\cos\theta)} = \frac{2(M+m)}{m}\sqrt{gl}\,\sin(\theta/2)$$

As $m \ll M$,

$$v_0 \cong \frac{2M}{m}\sqrt{gl}\,\sin(\theta/2)$$

(b) Fraction of initial kinetic energy turned into heat

$$= \frac{T_i - T_f}{T_i} = 1 - \frac{T_f}{T_i} = 1 - \frac{\frac{1}{2}(m+M)v^2}{\frac{1}{2}mv_0^2}$$

$$= 1 - \frac{m}{(m+M)} \cong \left(1 - \frac{m}{M}\right) \quad \text{(using Eq. (1))}$$

1.163 When the disk breaks off the body M, its velocity towards right (along x-axis) equals the velocity of the body M, and let the disk's velocity in upward direction (along y-axis) at that moment be v'_y.

From conservation of momentum, along x-axis for the system (disk + body)

$$mv = (m+M)v'_x \quad \text{or} \quad v'_x = \frac{mv}{m+M} \tag{1}$$

And from energy conservation, for the same system in the field of gravity:

$$\frac{1}{2}mv^2 = \frac{1}{2}(m+M)v_x'^2 + \frac{1}{2}mv_y'^2 + mgh'$$

where h' is the height of break-off point from initial level.

So,

$$\frac{1}{2}mv^2 = \frac{1}{2}(m+M)\frac{m^2v^2}{(m+M)^2} + \frac{1}{2}mv_y'^2 + mgh' \quad \text{(using Eq. (1))}$$

or

$$v_y'^2 = v^2 - \frac{mv^2}{(m+m)} - 2gh'$$

Also, if h'' is the height of the disk from the break-off point, then,

$$v_y'^2 = 2gh''$$

So,

$$2g(h''+h') = v^2 - \frac{mv^2}{(M+m)}$$

Hence, the total height, raised from the initial level is

$$h' + h'' = \frac{Mv^2}{2g(M+m)}$$

1.164 (a) When the disk slides and comes to the plank, it has a velocity equal to $v = \sqrt{2gh}$. Due to friction between the disk and the plank the disk slows down and after some time the disk moves with the plank with velocity v' (say).

From the momentum conservation for the system (disk + plank) along horizontal towards right:

$$mv = (m+M)v' \quad \text{or} \quad v' = \frac{mv}{m+M}$$

Now from the equation of the increment of total mechanical energy of a system:

$$\frac{1}{2}(M+m)v'^2 - \frac{1}{2}mv^2 = A_{fr}$$

or

$$\frac{1}{2}(M+m)\frac{m^2v^2}{(m+M)^2} - \frac{1}{2}mv^2 = A_{fr}$$

so,

$$\frac{1}{2}v^2\left[\frac{m^2}{(M+m)} - m\right] = A_{fr}$$

Hence, $A_{fr} = -\left(\frac{mM}{m+M}\right)gh = -\mu gh \quad \left(\text{where } \mu = \frac{mM}{m+M} = \text{reduce mass}\right).$

(b) We look at the problem from a frame in which the hill is moving (together with the disk on it) to the right with speed u. Then in this frame the speed of the disk when it just gets onto the plank is, by the law of addition of velocities, $v = u + \sqrt{2gh}$. Similarly the common speed of the plank and the disk when they move together is

$$v = u + \frac{m}{m+M}\sqrt{2gh}$$

Then as above

$$A_{fr} = \frac{1}{2}(m+M)v^2 - \frac{1}{2}mv^2 - \frac{1}{2}Mu^2$$

$$= \frac{1}{2}(m+M)\left\{u^2 + \frac{2m}{m+M}u\sqrt{2gh} + \frac{m^2}{(m+M)^2}2gh\right\}$$

$$-\frac{1}{2}(m+M)u^2 - \frac{1}{2}m2u\sqrt{2gh} - mgh$$

We see that A_{fr} is independent of u and is in fact just $-\mu gh$ as in (a). Thus the result obtained does not depend on the choice of reference frame.

Do note however that it will be incorrect to apply "conservation of energy" formula in the frame in which the hill is moving. The energy carried by the hill is not negligible in this frame. See also the next problem.

1.165 In a frame moving relative to the Earth, one has to include the kinetic energy of the Earth as well as Earth's acceleration to be able to apply conservation of energy to the problem. In a reference frame falling to the Earth with velocity v_0, the stone is initially going up with velocity v_0 and so is the Earth. The final velocity of the stone is $0 = v_0 - gt$ and that of the Earth is $v_0 + (m/M)gt$ (M is the mass of the Earth), from Newton's third law, where t = time of fall. From conservation of energy

$$\frac{1}{2}mv_0^2 + \frac{1}{2}Mv_0^2 + mgh = \frac{1}{2}M\left(v_0 + \frac{m}{M}v_0\right)^2$$

Hence,

$$\frac{1}{2}v_0^2\left(m + \frac{m^2}{M}\right) = mgh$$

Neglecting m/M in comparison with 1, we get

$$v_0^2 = 2gh \quad \text{or} \quad v_0 = \sqrt{2gh}$$

The point in this is that in the Earth's rest frame the effect of Earth's acceleration is of order m/M and can be neglected but in a frame moving with respect to the Earth, the effect of Earth's acceleration must be kept because it is of order one (i.e., large).

1.166 From conservation of momentum for the closed system "both colliding particles"

$$m_1\mathbf{v}_1 + m_2\mathbf{v}_2 = (m_1 + m_2)\mathbf{v}$$

or
$$\mathbf{v} = \frac{m_1\mathbf{v}_1 + m_2\mathbf{v}_2}{m_1 + m_2} = \frac{1(3\mathbf{i} - 2\mathbf{j}) + 2(4\mathbf{j} - 6\mathbf{k})}{3} = \mathbf{i} + 2\mathbf{j} - 4\mathbf{k}$$

Hence,
$$|\mathbf{v}| = \sqrt{1 + 4 + 16}\ \text{m/s} = 4.6\ \text{m/s}$$

1.167 For perfectly inelastic collision, in the C.M. frame, final kinetic energy of the colliding system (both spheres) becomes zero. Hence initial kinetic energy of the system in C.M. frame completely turns into the internal energy (Q) of the formed body.

Hence,
$$Q = T_i = \frac{1}{2}\mu|\mathbf{v}_1 - \mathbf{v}_2|^2$$

Now from energy conservation
$$\Delta T = -Q = -\frac{1}{2}\mu|\mathbf{v}_1 - \mathbf{v}_2|^2$$

In laboratory frame the same result is obtained as

$$\Delta\tilde{T} = \frac{1}{2}\frac{(m_1\mathbf{v}_1 + m_2\mathbf{v}_2)^2}{m_1 + m_2} - \frac{1}{2}m_1|\mathbf{v}_1|^2 + \frac{1}{2}m_2|\mathbf{v}_2|^2$$

$$= -\frac{1}{2}\mu|\mathbf{v}_1 - \mathbf{v}_2|^2$$

1.168 (a) From conservation of linear momentum

$$\mathbf{p}_1 = \mathbf{p}_1' + \mathbf{p}_2' \quad \text{or} \quad \mathbf{p}_1 - \mathbf{p}_1' = \mathbf{p}_2'$$

As
$$\mathbf{p}_1' \perp \mathbf{p}_1 \quad \text{so,} \quad p_1^2 + p_1'^2 = p_2'^2 \tag{1}$$

From conservation of kinetic energy
$$\frac{p_1^2}{2m_1} = \frac{p_1'^2}{2m_1} + \frac{p_2'^2}{2m_2} \tag{2}$$

Using Eq. (1) in Eq. (2) we get
$$\frac{p_1^2}{2m_1} = \frac{p_1'^2}{2m_1} + \frac{(p_1^2 + p_1'^2)}{2m_2}$$

which yields
$$\frac{p_1'^2}{p_1^2} = \left(\frac{m_2 - m_1}{m_2 + m_1}\right) \tag{3}$$

So, sought fraction of kinetic energy
$$\eta = 1 - \frac{T_1'}{T_1} = 1 - \frac{p_1'^2/2m_1}{p_1^2/2m_1} = 1 - \frac{p_1'^2}{p_1^2}$$

Hence,
$$\eta = 1 - \left(\frac{m_2 - m_1}{m_2 + m_1}\right) = \frac{2m_1}{m_1 + m_2}$$

(b) For two particle closed system momentum of each particle in their C.M. frame are always equal and opposite. In C.M. frame the kinetic energy of the two particle system $\tilde{T} = \tilde{p}^2/2\mu$ where μ is the reduced mass. In perfectly elastic collision, the kinetic energy of the system is conserved.

So,
$$\frac{\tilde{p}'^2}{2\mu} = \frac{\tilde{p}^2}{2\mu'} \quad \text{which gives} \quad \tilde{p}' = \tilde{p}$$

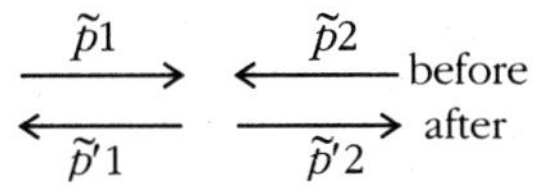

Being head-on collision, both the particles have to keep their motion along the same straight line before and after the collision. On collision, momentum vector of each particle has to change due to the reaction force by the other particle. So only choice left to each of the particle is to reverse the direction of its momentum after the collision, keeping the magnitude constant, i.e., $\tilde{\mathbf{p}}'_i = -\tilde{\mathbf{p}}_i$, where $i = 1, 2$.

The same can be said about the velocity of each particle in the C.M. frame

$$\tilde{\mathbf{v}}'_i = -\tilde{\mathbf{v}}_i$$

but $$\mathbf{v}'_i = \mathbf{V}_C + \tilde{\mathbf{v}}'_i = \mathbf{V}_C - \tilde{\mathbf{v}}_i = \mathbf{V}_C - (\mathbf{v}_i - \mathbf{V}_C) = 2\mathbf{V}_C - \mathbf{v}_i$$

Hence the velocity of i-th particle after collision:

$$\tilde{\mathbf{v}}_i = 2\mathbf{V}_C - \mathbf{v}_i \quad (\text{where } i = 1, 2)$$

So velocity of particle 1 just after the collision from above relation is

$$\mathbf{v}'_1 = 2\mathbf{V}_C - \mathbf{v}_1 = 2\left(\frac{m_1\mathbf{v}_1}{m_1 + m_2}\right) - \mathbf{v}_1 = \frac{(m_1 - m_2)\mathbf{v}_1}{m_1 + m_2}$$

Therefore, $$\frac{v_1'^2}{v_1^2} = \frac{(m_1 - m_2)^2}{(m_1 + m_2)^2}$$

Thus sought fraction $$= 1 - \frac{T'_1}{T_1} = 1 - \frac{v_1'^2}{v_1^2} = 1 - \frac{(m_1 - m_2)^2}{(m_1 + m_2)^2} = \frac{4m_1m_2}{(m_1 + m_2)^2}$$

1.169 (a) Being perfectly elastic head on collision, the velocity of i-th particle after collision

$$\mathbf{v}'_i = 2\mathbf{V}_C - \mathbf{v}_i \quad (\text{where } i = 1, 2) \quad (\text{see previous problem})$$

So, $$\mathbf{v}'_i = 2\left(\frac{m_1\mathbf{v}_1}{m_1 + m_2}\right) - \mathbf{v}_1 = \frac{(m_1 - m_2)\mathbf{v}_1}{m_1 + m_2} \quad \text{and} \tag{1}$$

$$\mathbf{v}'_i = 2\left(\frac{m_1\mathbf{v}_1}{m_1 + m_2}\right) \tag{2}$$

According to the problem

$$\mathbf{v}'_2 = -\mathbf{v}'_1$$

so, $$2\left(\frac{m_1\mathbf{v}_1}{m_1 + m_2}\right) = \frac{(m_2 - m_1)\mathbf{v}_1}{m_1 + m_2} \quad \text{which gives } 2m_1 = (m_2 - m_1)$$

Hence, $$\frac{m_1}{m_2} = \frac{1}{3}$$

(b) From conservation of linear momentum in the direction perpendicular to initial motion, direction of striking particle 1 gives

$$p'_1 \sin 30° = p'_2 \sin 30°$$

So, $$p'_1 = p'_2 \tag{1}$$

From conservation of linear momentum

$$\mathbf{p}_1 = \mathbf{p}'_1 + \mathbf{p}'_2 \quad \text{or,} \quad \mathbf{p}_1 - \mathbf{p}'_1 + \mathbf{p}'_2$$

so, $$p_1^2 + p_1'^2 - 2p_1 p_1' \cos 30° = p_2'^2 \qquad (2)$$

On using Eq. (1) in Eq. (2) we get on manipulation

$$p_1' = \frac{p_1}{2\cos 30°} = \frac{p_1}{\sqrt{3}} \qquad (3)$$

From conservation of kinetic energy

$$\frac{p_1^2}{2m_1} = \frac{p_1'^2}{2m_1} + \frac{p_2'^2}{2m_2}$$

On using Eq. (1) and Eq. (3), we get on solving

$$\frac{m_1}{m_2} = 2$$

1.170 At the moment of maximum deformation the velocity of colliding bodies along their common normal **n** (here the line joining the mass centres of the balls) must be equal so

$$v_{1n}' = v_{2n}' = v_n' \text{ (say)} \qquad (1)$$

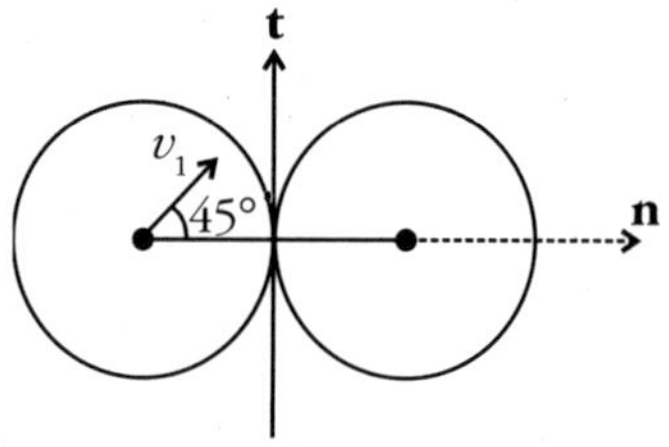

As the balls are smooth so, velocity of each ball along their common tangent **t** remains constant.

So, $$v_{1t}' = v_{1t} = v_1 \sin 45° \quad \text{and} \quad v_{2t}' = v_{2t} = 0 \qquad (2)$$

Now, from conservation of linear momentum

$$mv_1 \cos 45° = 2m\, v_n' \quad \text{so,} \quad v_n' = \frac{v_1}{2\sqrt{2}} \qquad (3)$$

From energy conservation, gain in internal potential energy is due to loss of kinetic energy of the system. Initial kinetic energy of the system is the kinetic energy T_1 of striking ball only so, the sought fraction $\eta = 1 - T'_{\text{system}}/T_1$.

But, $$T'_{\text{system}} = \frac{1}{2} m [v_{1n}'^2 + v_{1t}'^2] + \frac{1}{2} m [v_{2n}'^2 + v_{2t}'^2]$$

On using Eq. (1), (2) and (3), we get

$$T'_{\text{system}} = \frac{1}{2} m \frac{3v_1^2}{4}$$

So, $$\frac{T'_{\text{system}}}{T_1} = \frac{1}{2} m \frac{3v_1^2}{4} \Big/ \frac{1}{2} m v_1^2 = \frac{3}{4}$$

Hence, sought fraction $\eta = 1 - \frac{3}{4} = \frac{1}{4}$.

1.171 From the conservation of linear momentum of the shell just before and after its fragmentation

$$3\mathbf{v} = \mathbf{v}_1 + \mathbf{v}_2 + \mathbf{v}_3 \qquad (1)$$

where $\mathbf{v}_1$, $\mathbf{v}_2$ and $\mathbf{v}_3$ are the velocities of its fragments.

From the energy conservation

$$3\eta v^2 = v_1^2 + v_2^2 + v_3^2 \qquad (2)$$

Now, $$\tilde{\mathbf{v}} \text{ or } \mathbf{v}_{iC} = \mathbf{v}_i - \mathbf{v}_C = \mathbf{v}_i - \mathbf{v} \quad (3)$$

where $\mathbf{v}_C = \mathbf{v}$ = velocity of the C.M. of the fragments of the shell. Obviously in the C.M. frame the linear momentum of a system is equal to zero so

$$\tilde{\mathbf{v}}_1 + \tilde{\mathbf{v}}_2 + \tilde{\mathbf{v}}_3 = 0 \quad (4)$$

Using Eqs. (3) and (4) in Eq. (2), we get

$$3\eta v^2 = (\mathbf{v} + \tilde{\mathbf{v}}_1)^2 + (\mathbf{v} + \tilde{\mathbf{v}}_2)^2 + (\mathbf{v} - \tilde{\mathbf{v}}_1 - \tilde{\mathbf{v}}_2)^2 = 3v^2 + 2\tilde{\mathbf{v}}_1^2 + 2\tilde{\mathbf{v}}_2^2 + 2\tilde{\mathbf{v}}_1 \cdot \tilde{\mathbf{v}}_2$$

or $$2\tilde{v}_1^2 + 2\tilde{v}_1\tilde{v}_2 \cos\theta + 2\tilde{v}_2^2 + 3(1-\eta)v^2 = 0 \quad (5)$$

If we have had used $\tilde{\mathbf{v}}_2 = -\tilde{\mathbf{v}}_1 - \tilde{\mathbf{v}}_3$, then Eq. (5) would contain $\tilde{v}_3$ instead of $\tilde{v}_2$ and so on. The problem being symmetrical we can look for the maximum of any one. Obviously it will be the same for each.

For $$\tilde{v}_2^2 \cos^2\theta \geq 8(2\tilde{v}_2^2 + 3(1-\eta)v^2) \text{ or } 6(\eta - 1)v^2 \geq (4 - \cos^2\theta)\tilde{v}_2^2$$

So, $$\tilde{v}_2 \leq v\sqrt{\frac{6(\eta-1)}{4-\cos^2\theta}} \text{ or } \tilde{v}_{2(\max)} = \sqrt{2(\eta-1)}v$$

Hence, $v_{2(\max)} = |\mathbf{v} + \tilde{\mathbf{v}}_2|_{\max} = v + \sqrt{2(\eta-1)}v = v(1+\sqrt{2(\eta-1)}) = 1$ km/s.

Thus, owing to the symmetry

$$v_{1(\max)} = v_{2(\max)} = v_{3(\max)} = v(1+\sqrt{2(\eta-1)}) = 1 \text{ km/s}$$

1.172 From the conservation of momentum

$$mv_1 = mv_1' + mv_2' \quad \text{or} \quad v_1 = v_1' + v_2' \quad (1)$$

From the given condition,

$$\eta = \frac{T-T'}{T} = 1 - \frac{T'}{T} = 1 - \frac{v_1'^2 + v_2'^2}{v_1^2}$$

or $$v_1'^2 + v_2'^2 = (1-\eta)v_1^2 \quad (2)$$

From Eqs. (1) and (2)

$$v_1'^2 + (v_1 - v_1')^2 = (1-\eta)v_1^2$$

or $$2v_1'^2 - 2v_1'v_1 + \eta v_1^2 = 0$$

So, $$v_1' = \frac{2v_1 \pm \sqrt{4v_1^2 - 8\eta\, v_1^2}}{4}$$

$$= \frac{1}{2}\left[v_1 \pm \sqrt{v_1^2 - 2\eta\, v_1^2}\right] = \frac{1}{2}v_1(1 \pm \sqrt{1-2\eta})$$

Positive sign gives the velocity of the second particle which lies ahead. The negative sign is correct for v_1. So, $v_1' = 1/2v_1(1-\sqrt{1-2\eta}) = 5$ m/s will continue moving in the same direction.

Note that $v_1' = 0$ if $\eta = 0$.

Alternate:

As internal energy is independent of reference frame, so loss in kinetic energy of the system in the C.M. frame is the same.

So, $$\tilde{T} - \tilde{T}' = T - T'$$

or $$\left(1 - \frac{\tilde{T}'}{\tilde{T}}\right) = \frac{T}{\tilde{T}}\left(1 - \frac{T'}{T}\right)$$

Using $\left(1 - \frac{T'}{\tilde{T}}\right) = \eta,\ T = \frac{1}{2}mv^2, \text{and}\ \tilde{T} = \frac{1}{2}\mu v_{rel}^2 = \frac{1}{2}\left(\frac{m}{2}\right)v^2$, where m is the mass of each particle, we get

$$\left(1 - \frac{T'}{\tilde{T}}\right) = 2\eta$$

Now, according to the problem

$$\tilde{T} - \tilde{T}' = 2\eta\tilde{T}$$

$$\frac{\tilde{p}^2}{2\mu} - \frac{\tilde{p}'^2}{2\mu} = 2\eta\frac{\tilde{p}^2}{2\mu}$$

which yields $$\tilde{p}' = \tilde{p}\sqrt{(1 - 2\eta)}$$

Being head-on collision, final momentum in C.M. frame of any particle should be reversed, so

$$\tilde{\mathbf{p}}_1' = -\tilde{\mathbf{p}}_1\sqrt{(1 - 2\eta)}$$

or $$\tilde{\mathbf{v}}_1' = -\tilde{\mathbf{v}}_1\sqrt{(1 - 2\eta)},$$

But, $$\mathbf{v}_1' = \tilde{\mathbf{v}}_1' + \mathbf{v}_C = -\tilde{\mathbf{v}}_1\sqrt{(1 - 2\eta)} + \mathbf{v}_C = -(\mathbf{v}_1 - \mathbf{v}_C)\sqrt{(1 - 2\eta)} + \mathbf{v}_C$$

Thus, $$\mathbf{v}_1' = \mathbf{v}_C[1 + \sqrt{1 - 2\eta}] - \mathbf{v}_1\sqrt{1 - 2\eta}$$

Using $$\mathbf{v}_C = \frac{m\mathbf{v}}{2m} = \frac{\mathbf{v}}{2}$$ (where $\mathbf{v}$ is the velocity of striking particle)

$$\mathbf{v}_1' = \frac{\mathbf{v}}{2}(1 - \sqrt{1 - 2\eta}),$$ putting $\mathbf{v}_1 = \mathbf{v}$

1.173 From conservation of linear momentum in the initial motion direction of striking particle 1 and perpendicular to it gives

$$mv_1 = Mv_2' \cos\theta \quad \text{and} \quad mv_1' = Mv_1' = Mv_2' \sin\theta, \text{ respectively}$$

Thus, $$v_1' = v_1 \tan\theta$$

Percentage change in the kinetic energy of the system $= \dfrac{T'_{\text{system}} - T_{\text{system}}}{T_{\text{system}}} \times 100$

$$= \left(\frac{T'_{\text{system}}}{T_{\text{system}}} - 1\right) \times 100$$

$$= \left[\frac{\frac{1}{2}mv_1^2 \tan^2\theta + \frac{1}{2}M\left(\frac{mv_1}{M\cos\theta}\right)^2}{\frac{1}{2}mv_1^2} - 1\right] \times 100 = \left(\tan^2\theta + \frac{m}{M}\sec^2\theta - 1\right) \times 100$$

Putting the values of θ and m/M, we get % of change in kinetic energy = –40%.

1.174 (a) Let the particles m_1 and m_2 move with velocities $\mathbf{v}_1$ and $\mathbf{v}_2$, respectively. On the basis of solution of problem 1.147(b):

$$\tilde{p} = \mu v_{rel} = \mu|\mathbf{v}_1 - \mathbf{v}_2|$$

As $$\mathbf{v}_1 \perp \mathbf{v}_2$$

so, $$\tilde{p} = \mu\sqrt{v_1^2 + v_2^2} \text{ where } \mu = \frac{m_1 m_2}{m_1 + m_2}$$

(b) Again from solution of problem 1.147(b):

$$\tilde{T} = \frac{1}{2}\mu v_{rel}^2 = \frac{1}{2}\mu|\mathbf{v}_1 - \mathbf{v}_2|^2$$

So, $$\tilde{T} = \frac{1}{2}\mu(v_1^2 + v_2^2)$$

1.175 From conservation of momentum

$$\mathbf{p}_1 = \mathbf{p}_1' + \mathbf{p}_2 \text{ or } \mathbf{p}_1 - \mathbf{p}_1' = \mathbf{p}_2$$

So, $p_1^2 - 2p_1p_1'\cos\theta_1 + p_1'^2 = p_2'^2$, where θ is the angle between $\mathbf{p}_1$ and $\mathbf{p}_1'$.

From conservation of energy

$$\frac{p_1^2}{2m_1} = \frac{p_1'^2}{2m_1} + \frac{p_2'^2}{2m_2}$$

Eliminating p_2' we get

$$0 = p_1'^2\left(1 + \frac{m_2}{m_1}\right) - 2p_1'p_1\cos\theta_1 + p_1^2\left(1 - \frac{m_2}{m_1}\right)$$

This quadratic equation for p_1' has a real solution in terms of p_1 and θ_1 only if

$$4\cos^2\theta_1 \geq 4\left(1 - \frac{m_2^2}{m_1^2}\right)$$

or $$\sin^2\theta_1 \leq \frac{m_2^2}{m_1^2} \text{ or } \sin\theta_1 \geq -\frac{m_2}{m_1}$$

This clearly implies (since only + sign makes sense) that

$$\sin\theta_{1\max} = \frac{m_2}{m_1}$$

Alternate:

The solution of the problem becomes standard in the frame of C.M., which is moving with velocity $\mathbf{v}_C = m_1\mathbf{v}_1/(m_1 + m_2)$ in the frame of laboratory. In the frame of C.M., the momenta of the particles must always be equal and opposite.

So, $$\tilde{\mathbf{p}}_1 = -\tilde{\mathbf{p}}_2,$$

but $|\tilde{\mathbf{p}}_1| = |\tilde{\mathbf{p}}_2| = \tilde{p} = \mu v_{rel} = \mu v_1$ (where, $\mu = m_1m_2/(m_1 + m_2)$ is the reduced mass of the system).

The initial kinetic energy of the system in the frame of C.M. $\tilde{T}_i = \tilde{p}^2/2\mu$.

As the collision is perfectly elastic and the reduced mass of the system is constant.

$$\tilde{T}_i = \tilde{T}_f \quad \text{or} \quad \frac{\tilde{p}^2}{2\mu} = \frac{\tilde{p}'^2}{2\mu}$$

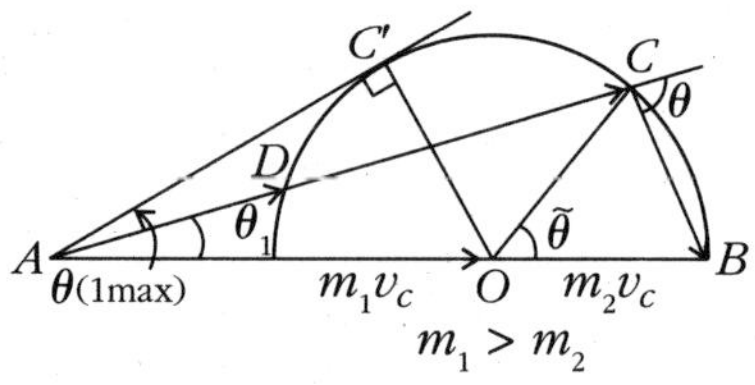

Hence,
$$\tilde{p} = \tilde{p}'$$

In the frame of laboratory from the conservation of linear momentum

$$\mathbf{p}_1' + \tilde{\mathbf{p}}_2' = \mathbf{p}_1 = (m_1 + m_2)\mathbf{v}_C$$

Now let us draw the so-called vector diagram (see figure) of momenta. First we depict the vector $\mathbf{p}_1$ as the section AB and the $\vec{\mathbf{p}}_1'$ and $\vec{\mathbf{p}}_2'$ as sections AC and CB each of which represents according to

$$\mathbf{p}_1' = \tilde{\mathbf{p}}_1' + m_1\mathbf{v}_C$$

and
$$\mathbf{p}_2' = \tilde{\mathbf{p}}_2' + m_2\mathbf{v}_C$$

where
$$\vec{\tilde{\mathbf{p}}}_1 = \overrightarrow{OC} \quad \text{and} \quad \vec{\tilde{\mathbf{p}}}_2 = \overrightarrow{CO}$$

Note that this diagram is valid regardless of the angle $\tilde{\theta}$. The point C (centre of mass) therefore can be located only on the circle of radius $\tilde{p}$ having its centre at the point O, which divides the section AB into two parts in the ratio AO: $OB = m_1$: m_2. In our case, this circle passes through the point B, the end point of vector $\tilde{\mathbf{p}}_1$ since the section $OB = m_2 v_C = \mu v_1 = \tilde{p}$. This circle is the locus of all possible locations of the apex C of the momenta triangle ABC, whose side AC and CB represent the possible momentum of particles $\mathbf{p}_1'$ and $\mathbf{p}_2'$ after the collision in the frame of laboratory.

Depending on the particle mass ratio ($m_1 = m_2$) the point A, the beginning of the vector $\mathbf{p}_1$ can be located inside the given circle, on it, or outside it. In our case point A will lie outside the circle and it is also another interesting fact that the particle m_1 can be scattered by the same angle θ_1, where it possesses the momentum AC or AD (figure). The same is true for particle m_2. The maximum scattering in the laboratory $\theta_{1\,max}$ corresponds to the case when the velocity vector in the frame of laboratory become a tangent (AC') to the circle (see figure).

It then follows that

$$\sin\theta_{1\,max} = \frac{OC'}{AO} = \frac{\tilde{p}}{m_1 v_C} = \frac{(m_1 m_2/m_1 + m_2)v_1}{(m_1^2/m_1 + m_2)v_1} = \frac{m_2}{m_1}$$

Note: *It is clear that* θ_m *is limited to* $\pi/2$ *for* $m_1 = m_2$, *for the case* $m_2 > m_1$, *point A will be inside the sphere, and all angles of scattering from* 0 *to* $\pi/2$ *are permitted.*

1.176 From the symmetry of the problem, the velocity of the disk A will be directed either in the initial direction or opposite to it just after the impact. Let the velocity of the disk A after the collision be v' and be directed towards right after the collision. It is also clear from the symmetry of the problem that the disks B and C have equal speed (say v''), in the directions shown. From the condition of the problem,

$$\cos\theta = \frac{\eta\dfrac{d}{2}}{d} = \frac{\eta}{2} \quad \text{so,} \quad \sin\theta = \sqrt{4 - \eta^2/2} \tag{1}$$

For the three disk system, from the conservation of linear momentum in the symmetry direction (towards right)

$$mv = 2mv''\sin\theta + mv' \quad \text{or} \quad v = 2v''\sin\theta + v' \tag{2}$$

From the definition of the coefficient of restitution, we have for disks A, B and C

$$e = \frac{v'' - v'\sin\theta}{v\sin\theta - 0}$$

But $e = 1$, for perfectly elastic collision.

So,
$$v\sin\theta = v'' - v'\sin\theta \tag{3}$$

From Eqs. (2) and (3)

$$v' = \frac{v(1-2\sin^2\theta)}{(1+2\sin^2\theta)}$$
$$= \frac{v(\eta^2-2)}{6-\eta^2} \quad \text{(using Eq. (1))}$$

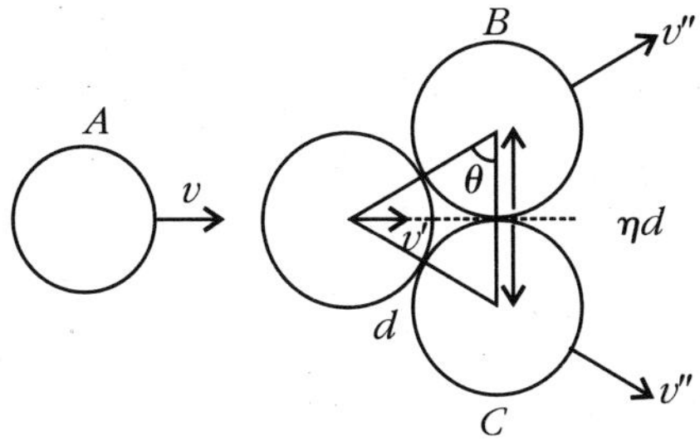

Hence, we have,

$$v' = \frac{v(\eta^2-2)}{6+\eta^2}$$

Therefore, the disk A will recoil if $\eta < \sqrt{2}$ and stop if $\eta = \sqrt{2}$.

Note: *One can write the equations of momentum conservation along the direction perpendicular to the initial direction of disk A and the conservation of kinetic energy instead of the equation of restitution.*

1.177 (a) Let a molecule come with velocity $\mathbf{v}_1$ to strike another stationary molecule and just after collision their velocities become $\mathbf{v}_1'$ and $\mathbf{v}_2'$, respectively. As the mass of each molecule is same, conservation of linear momentum and conservation of kinetic energy for the system (both molecules), respectively, gives

$$\mathbf{v}_1 = \mathbf{v}_1' + \mathbf{v}_2'$$

and
$$v_1^2 = v_1'^2 + v_2'^2$$

From the property of vector addition it is obvious from the obtained equations that

$$\mathbf{v}_1 \perp \mathbf{v}_2 \quad \text{or} \quad \mathbf{v}_1 \cdot \mathbf{v}_2 = 0$$

(b) Due to the loss of kinetic energy in inelastic collision $v_1^2 > v_1'^2 + v_2'^2$ so, $\mathbf{v}_1 \cdot \mathbf{v}_2 > 0$ and therefore angle of divergence < 90°.

1.178 Suppose that at time t, the rocket has the mass m and the velocity $\mathbf{v}$, relative to the reference frame, employed. Now consider the inertial frame moving with the velocity that the rocket has at the given moment. In this reference frame, the momentum increment that the rocket and ejected gas system acquires during time dt is,

$$d\mathbf{p} = m d\mathbf{v} + \mu dt\, \mathbf{u} = \mathbf{F}\, dt$$

or
$$m\frac{d\mathbf{v}}{dt} = \mathbf{F} - \mu\mathbf{u}$$

or
$$m\mathbf{w} = \mathbf{F} - \mu\mathbf{u}$$

1.179 According to the question, $\mathbf{F} = 0$ and $\mu = -dm/dt$ so the equation for this system becomes,

$$m\frac{d\mathbf{v}}{dt} = \frac{dm}{dt}\mathbf{u}$$

As $d\mathbf{v} \uparrow\downarrow \mathbf{u}$, so, $m\, dv = -u dm$.

Integrating within the limits

$$\frac{1}{u}\int_0^v dv = -\int_{m_0}^0 \frac{dm}{m} \quad \text{or} \quad \frac{v}{u} = \ln\frac{m_0}{m}$$

Thus,
$$v = v \ln\frac{m_0}{m}$$

As $d\mathbf{v} \uparrow\downarrow \mathbf{u}$, so in vector from $\mathbf{v} = -\mathbf{u} \ln m_0/m$.

1.180 According to the question, **F** (external force) = 0

So, $$m\frac{d\mathbf{v}}{dt} = \frac{dm}{dt}\mathbf{u}$$

As, $$d\mathbf{v} \uparrow\downarrow \mathbf{u}$$

So, in scalar form $$mdv = -udm$$

or $$\frac{wdt}{u} = -\frac{dm}{m}$$

Integrating within the limit for $m(t)$

$$\frac{wt}{u} = -\int_{m_0}^{m}\frac{dm}{m} \quad \text{or} \quad \frac{v}{u} = -\ln\frac{m}{m_0}$$

Hence, $$m = m_0 e^{-(wt/u)}$$

1.181 As **F** = 0, from the equation of dynamics of a body with variable mass,

$$m\frac{d\mathbf{v}}{dt} = \mathbf{u}\frac{dm}{dt} \quad \text{or} \quad d\mathbf{v} = \mathbf{u}\frac{dm}{m} \tag{1}$$

Now $dv \uparrow\downarrow \mathbf{u}$ and since $\mathbf{u} \perp \mathbf{v}$, we must have $|d\mathbf{v}| = v_0 d\alpha$ (because v_0 is constant), where $d\alpha$ is the angle by which the spaceship turns in time dt.

So, $$-u\frac{dm}{m} = v_0 d\alpha \quad \text{or} \quad d\alpha = -\frac{u}{v_0}\frac{dm}{m}$$

or $$\alpha = -\frac{u}{v_0}\int_{m_0}^{m}\frac{dm}{m} = \frac{u}{v_0}\ln\left(\frac{m_0}{m}\right)$$

1.182 We have $$\frac{dm}{dt} = -\mu \quad \text{or} \quad dm = -\mu dt$$

Integrating $$\int_{m_0}^{m} dm = -\mu\int_0^t dt \quad \text{or} \quad m = m_0 - \mu t$$

As **u** = 0 so, from the equation of variable mass system

$$(m_0 - \mu t)\frac{d\mathbf{v}}{dt} = \mathbf{F} \quad \text{or} \quad \frac{d\mathbf{v}}{dt} = \mathbf{w} = \frac{\mathbf{F}}{(m_0 - \mu t)}$$

or $$\int_0^{\mathbf{v}} d\mathbf{v} = \mathbf{F}\int_0^t \frac{dt}{(m_0 - \mu t)}$$

Hence, $$\mathbf{v} = \frac{\mathbf{F}}{\mu}\ln\left(\frac{m_0}{m_0 - \mu t}\right)$$

1.183 Let the car be moving in a reference frame to which the hopper is fixed and at any instant of time, let its mass be m and velocity **v**.

Then from the general equation, for variable mass system

$$m\frac{d\mathbf{v}}{dt} = \mathbf{F} + \mathbf{u}\frac{dm}{dt}$$

We write the equation, for our system as,

$$m\frac{d\mathbf{v}}{dt} = \mathbf{F} - \mathbf{v}\frac{dm}{dt} \quad \text{as} \quad \mathbf{u} = -\mathbf{v}$$

So, $$\frac{d}{dt}(m\mathbf{v}) = \mathbf{F}$$

and $$\mathbf{v} = \frac{\mathbf{F}t}{m} \quad \text{on integration}$$

But $$m = m_0 + \mu t$$

so, $$\mathbf{v} = \frac{\mathbf{F}t}{m_0\left(1+\dfrac{\mu t}{m_0}\right)}$$

Thus, the sought acceleration is

$$\mathbf{w} = \frac{d\mathbf{v}}{dt} = \frac{\mathbf{F}}{m_0\left(1+\dfrac{\mu t}{m_0}\right)^2}$$

1.184 Let the length of the chain inside the smooth horizontal tube at an arbitrary instant be x. From the equation,

$$m\mathbf{w} = \mathbf{F} + \mathbf{u}\frac{dm}{dt}$$

as $\mathbf{u} = 0$, $\mathbf{F} \uparrow\uparrow \mathbf{w}$, for the chain inside the tube

$$\lambda x w = T, \quad \text{where } \lambda = \frac{m}{l} \tag{1}$$

Similarly for the overhanging part,

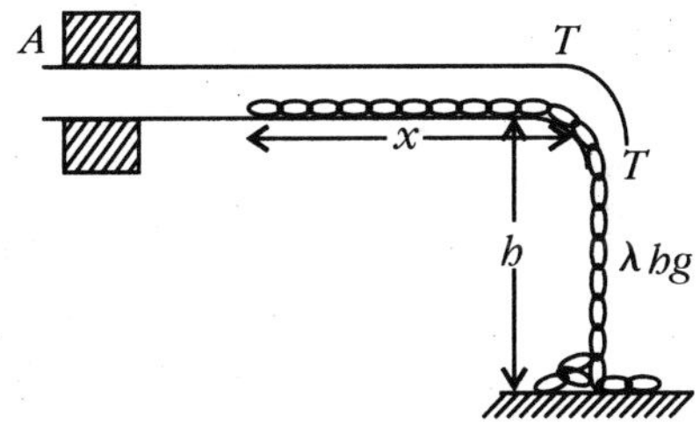

$$\mathbf{u} = 0$$

Thus, $$mw = F$$

or $$\lambda h w = \lambda h g - T \tag{2}$$

From Eqs. (1) and (2)

$$\lambda(x+h)w = \lambda hg \quad \text{or} \quad (x+h)v\frac{dv}{ds} = hg$$

or $$(x+h)v\frac{dv}{(-dx)} = gh$$

(As the length of the chain inside the tube decreases with time, $ds = -dx$.)

or $$vdv = -gh\frac{dx}{x+h}$$

Integrating, $$\int_0^v vdv = -gh\int_{(l-h)}^{0}\frac{dx}{x+h}$$

or $$\frac{v^2}{2} = gh\ln\left(\frac{l}{h}\right) \quad \text{or} \quad v = \sqrt{2gh\ln\left(\frac{l}{h}\right)}$$

1.185 The angular momentum of the particle relative to point O is given as function of a time as $\mathbf{M} = \mathbf{a} + \mathbf{b}t^2$. So, force moment relative to point O is given by

$$\mathbf{N} = \frac{d\mathbf{M}}{dt} = 2\mathbf{b}t$$

As force moment $\mathbf{N} = 2\mathbf{b}t$, so vector $\mathbf{N}$ coincides with vector $\mathbf{b}$.

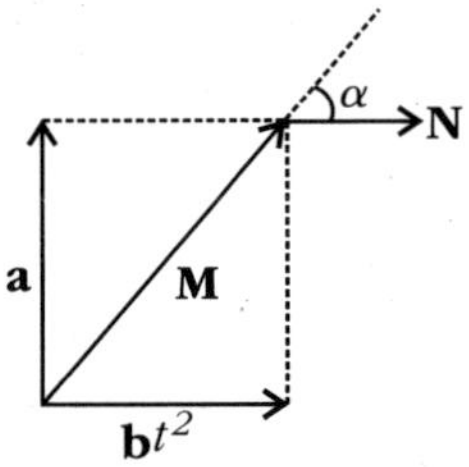

Let us depict the vectors $\mathbf{N}$ and $\mathbf{M}$ at an arbitrary instant of time t, when angle between $\mathbf{M}$ and $\mathbf{N}$ is α (see figure). It is obvious from the figure that $\tan\alpha = a/bt^2$, therefore $\alpha = 45°$, the time $t = t_0 = \sqrt{a/b}$.

Thus, $\mathbf{N} = 2\mathbf{b}\sqrt{a/b}$, when vector $\mathbf{N}$ forms angle $\alpha = 45°$ with vector $\mathbf{M}$.

Alternate:

Let the angle between $\mathbf{M}$ and $\mathbf{N}$, $\alpha = 45°$ at $t = t_0$.

Then
$$\frac{1}{\sqrt{2}} = \frac{\mathbf{M}\cdot\mathbf{N}}{|\mathbf{M}|\,|\mathbf{N}|} = \frac{(\mathbf{a}+\mathbf{b}t_0^2)\cdot(2\mathbf{b}t_0)}{\sqrt{a^2+b^2t_0^4}\,2bt_0}$$
$$= \frac{2b^2t_0^3}{\sqrt{a^2+b^2t_0^4}\,2bt_0} = \frac{bt_0^2}{\sqrt{a^2+b^2t_0^4}}$$

So,
$$2b^2t_0^4 = a^2 + b^2t_0^4 \quad \text{or} \quad t_0 = \sqrt{\frac{a}{b}} \quad \text{(as } t_0 \text{ cannot be negative)}$$

Hence,
$$\mathbf{N} = 2\mathbf{b}\sqrt{\frac{a}{b}}$$

1.186
$$\mathbf{M}(t) = \mathbf{r}\times\mathbf{p} = \left(\mathbf{v}_0 t + \frac{1}{2}\mathbf{g}t^2\right)\times m(\mathbf{v}_0 + \mathbf{g}t)$$
$$= mv_0\,gt^2\sin\left(\frac{\pi}{2}+\alpha\right)(-\mathbf{k}) + \frac{1}{2}mv_0gt^2\sin\left(\frac{\pi}{2}+\alpha\right)(\mathbf{k})$$
$$= \frac{1}{2}mv_0gt^2\cos\alpha(-k)$$

Thus
$$M(t) = \frac{mv_0gt^2\cos\alpha}{2}$$

Thus, angular momentum at maximum height, i.e., at t

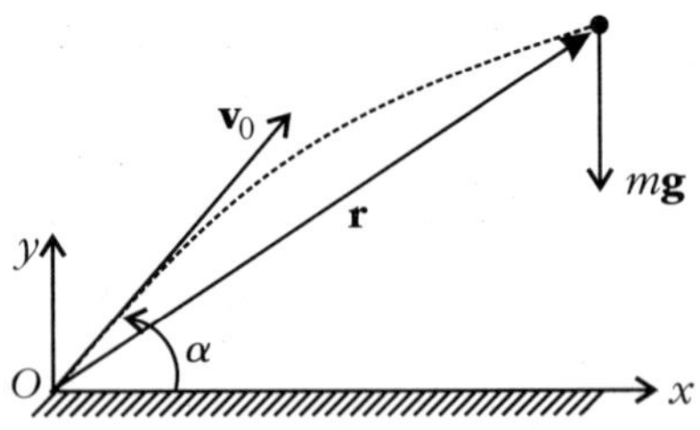

$$\frac{\tau}{2} = \frac{v_0\sin\alpha}{g}$$
$$M\left(\frac{\tau}{2}\right) = \left(\frac{mv_0^3}{2g}\right)\sin^2\alpha\cos\alpha$$
$$= 37\ \text{kg m}^2/\text{s}$$

Alternate:

$$\mathbf{M}(0) = 0 \quad \text{so} \quad \mathbf{M}(t) = \int_0^t \mathbf{N}\,dt = \int_0^t (\mathbf{r}\times m\mathbf{g})dt$$
$$= \int_0^t \left[\left(\mathbf{v}_0 t + \frac{1}{2}\mathbf{g}t^2\right)\times m\mathbf{g}\right]dt = (\mathbf{v}_0\times m\mathbf{g})\frac{t^2}{2}$$

1.187 (a) The disk experiences gravity, the force of reaction of the horizontal surface, and the force R of reaction of the wall at the moment of the impact against it. The first two forces counter-balance each other, leaving only the force **R**. It's moment relative to any point of the line along which the vector **R** acts or along normal to the wall is equal to zero and therefore the angular momentum of the disk relative to any of these points does not change in the given process.

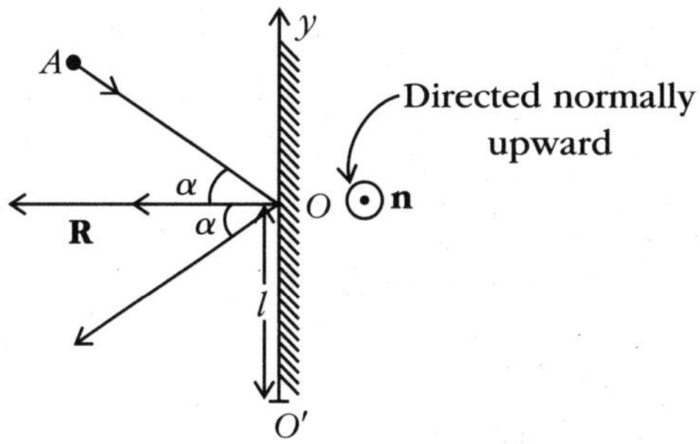

(b) From
$$\vec{\mathbf{M}} = \vec{\mathbf{r}} \times \vec{\mathbf{p}}$$
$$\Delta\vec{\mathbf{M}} = \Delta\vec{\mathbf{r}} \times \vec{\mathbf{p}} + \vec{\mathbf{r}} \times \Delta\vec{\mathbf{p}}$$

During the course of collision with wall the position of disk is same and is equal to $\mathbf{r}_{00'}$. Obviously the increment in momentum

$$\Delta\mathbf{M} = \mathbf{r}_{00'} \times \Delta\mathbf{p} = 2mvl\cos\alpha\,\mathbf{n} \quad (\text{since } \Delta\vec{\mathbf{r}} = 0)$$

Here, $|\Delta\mathbf{M}| = 2mvl\cos\alpha$

1.188 (a) The ball is under the influence of forces **T** and ***mg*** at all the moments of time, while moving along a horizontal circle. Obviously, the vertical component of **T** balances ***mg*** and so the net moment of these two forces about any point becomes zero. The horizontal component of **T**, which provides the centripetal acceleration to ball is already directed toward the centre (*C*) of the horizontal circle, thus its moment about the point *C* equals zero at all the moments of time. Hence the net moment of the force acting on the ball about point *C* equals zero and that's why the angular momentum of the ball is conserved about the horizontal circle.

(b) Let α be the angle which the thread forms with the vertical.

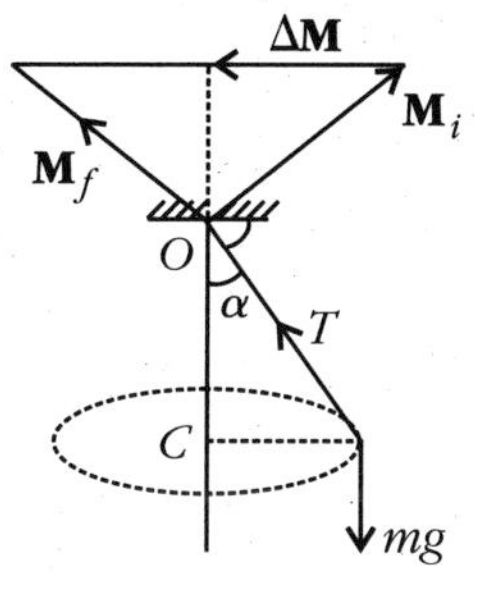

Now from equation of particle dynamics

$$T\cos\alpha = mg \quad \text{and} \quad T\sin\alpha = m\omega^2 l\sin\alpha$$

Hence on solving,

$$\cos\alpha = \frac{g}{\omega^2 l} \qquad (1)$$

As $|\mathbf{M}|$ is constant in magnitude, so from the figure

$$|\Delta\mathbf{M}| = 2M\cos\alpha$$

where
$$M = |\mathbf{M}_i| = |\mathbf{M}_f|$$
$$= |\mathbf{r}_{bo} \times m\mathbf{v}| = mvl \ (\text{as } \mathbf{r}_{bo} \perp \mathbf{v})$$

Thus,
$$|\Delta M| = 2mvl\cos\alpha = 2\,m\omega l^2 \sin\cos\alpha$$
$$= \frac{2mgl}{\omega}\sqrt{1-\left(\frac{g}{\omega^2 t}\right)^2} \quad (\text{using Eq. (1)})$$

1.189 During the free fall, time $\tau = \sqrt{2h/g}$, the reference point *O* moves in horizontal direction by the distance $V\tau$. In the translating frame as $\mathbf{M}_i(O) = 0$, so

$$\Delta\mathbf{M} = \mathbf{M}_f = \mathbf{r}_f \times m\mathbf{v}_f$$
$$= \left(\frac{1}{2}\mathbf{g}\tau^2 - \mathbf{V}\tau\right) \times m(\mathbf{g}\tau - \mathbf{V})$$
$$= \frac{1}{2}m(\mathbf{g} \times \mathbf{V})\tau^2$$

So, $|\mathbf{M}_f| = \frac{1}{2} mgV\left(\frac{2h}{g}\right)$ (using the value of τ and taking into account $\mathbf{V} \perp \mathbf{g}$)

Hence, $|\Delta \mathbf{M}| = mVh$

1.190 The Coriolis force is $2m(\mathbf{v}' \times \omega)$. Here ω is along the z-axis (vertical). The moving disk is moving with velocity v_0 which is constant, the motion is along the x-axis, then the Coriolis force is along y-axis and has the magnitude $2mv_0\omega$. At time t, the distance of the centre of moving disk from O is $v_0 t$ (along x-axis). Thus the torque N due to the Coriolis force is $N = 2mv_0\omega \cdot v_0 t$ along the z-axis. Hence, equating this to dM/dt, we get

$$\frac{dm}{dt} = 2mv_0^2\,\omega t$$

or

$$M = mv_0^2\,\omega t^2 + \text{constant}$$

The constant is irrelevant and may be put equal to zero if the disk is originally set in motion from the point O.

This discussion is approximate. The Coriolis force which causes the disk to swerve from straight line motion and thus causes deviation from the above formula will be substantial for large t.

1.191 If $\dot{r}$ = radial velocity of the particle then the total energy of the particle at any instant is

$$\frac{1}{2}m\dot{r}^2 + \frac{M^2}{2mr^2} + kr^2 = E \qquad (1)$$

where the second term is the kinetic energy of angular motion about the centre O.

Then the extreme values of r are determined by $\dot{r} = 0$ and solving the resulting quadratic equation

$$k(r^2)^2 - Er^2 + \frac{M^2}{2m} = 0$$

we get

$$r^2 = \frac{E \pm \sqrt{E^2 - 2M^2k/m}}{2k}$$

From this we see that

$$E = k(r_1^2 + r_2^2)$$

So,

$$\frac{1}{2}mv_2^2 + kr_2^2 = k(r_1^2 + r_1^2)$$

Hence,

$$m = \frac{2kr_1^2}{v_2^2}$$

Note: *Eq. (1) can be derived from the standard expression for kinetic energy and angular momentum in plane polar coordinates:*

$$T = \frac{1}{2}m\dot{r}^2 + \frac{1}{2}mr^2\dot{\theta}^2 \quad \text{and} \quad M = mr^2\dot{\theta}, \text{ respectively}$$

1.192 The swinging sphere experiences two forces: the gravitational force and the tension of the thread. Now, it is clear from the condition, given in the problem, that the moment of these forces about the vertical axis, passing through the point of suspension $N_Z = 0$. Consequently, the angular momentum M_Z of the sphere relative to the given axis (z) is constant.

Thus,

$$mv_0(l\sin\theta) = mvl \qquad (1)$$

where m is the mass of sphere and v is its velocity in the position, when the thread forms an angle $\pi/2$ with the vertical. Mechanical energy is also conserved, as the sphere is under the influence of only one other force, i.e., tension, which does not perform any work, as it is always perpendicular to the velocity.

So, $$\frac{1}{2}mv_0^2 + mgl\cos\theta = \frac{1}{2}mv^2 \quad (2)$$

From Eqs. (1) and (2), we get

$$v_0 = \sqrt{2gl/\cos\theta}$$

1.193 Forces, acting on the mass m are shown in the figure. As $\mathbf{N} = m\mathbf{g}$, the net torque of these two forces about any fixed point must be equal to zero. Tension T, acting on the mass m is a central force, which is always directed towards the centre O. Hence the moment of force T is also zero about the point O and therefore the angular momentum of the particle m is conserved about O.

So, $$mr^2\dot{\theta} = mr_0^2\omega_0$$

or $$\omega = \dot{\theta} = \frac{\omega_0 r_0^2}{r^2}$$

From $$F_r = ma_r$$

$$-T = m(\ddot{r} - r\dot{\theta}^2)$$

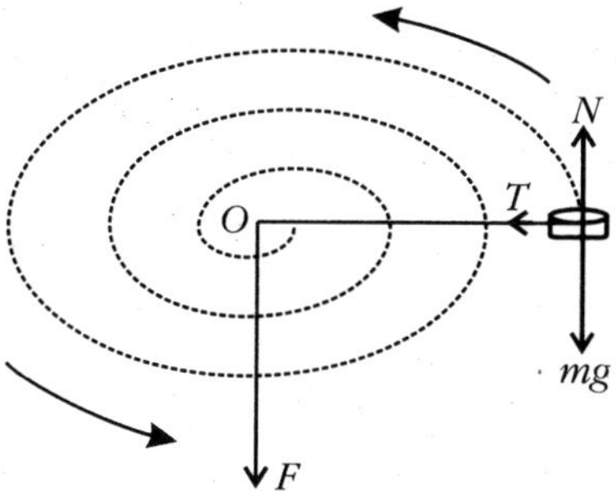

As the thread is pulled with constant speed so, $\dot{r}$ = constant and $\ddot{r} = 0$ (where dot stands for time derivative).

Thus $$T = F = mr\dot{\theta}^2$$

Hence, the sought tension is

$$F = mr\dot{\theta}^2 = mr\left(\frac{\omega_0 r_0^2}{r^2}\right)^2 = \frac{m\omega_0^2 r_0^4}{r^3}$$

1.194 On the given system the weight of the body m is the only force whose moment is effective about the axis of pulley. Let us take the sense of ω of the pulley at an arbitrary instant as the positive sense of axis of rotation (z-axis).

As $$M_z(0) = 0 \quad \text{so} \quad \Delta M_z = M_z(t) = \int N_z\,dt$$

So, $$M_z(t) = \int_0^t mgR\,dt = mgRt$$

1.195 Let the point of contact of sphere at initial moment ($t = 0$) be at O. At an arbitrary moment, the forces acting on the sphere are shown in the figure.

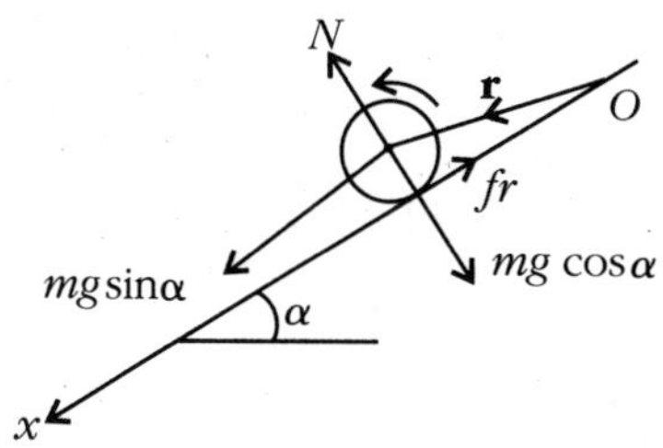

As $N = mg\cos\alpha$ and both (N and $mg\cos\alpha$) pass through the same line, their net torque about any point is zero. The force of static friction passes through the point O, thus its moment about point O becomes zero. Hence, $mg\sin\alpha$ is the only force which has effective torque about point O, and is given by $|\mathbf{N}| = mgR\sin\alpha$ normally emerging from the plane of figure.

As $\quad \mathbf{M}(t=0)=0, \text{ so } \Delta\mathbf{M}=\mathbf{M}(t)=\int \mathbf{N}\,dt$

Hence, $\quad \mathbf{M}(t)=Nt=mg\,R\sin\alpha\,t$

1.196 Let position vectors of the particles of the system be $\mathbf{r}_i$ and $\mathbf{r}_i'$ with respect to the points O and O' respectively, then we have

$$\mathbf{r}_i=\mathbf{r}_i'+\mathbf{r}_0 \tag{1}$$

(where $\mathbf{r}_0$ is the radius vector of O' with respect to O).

Now, the angular momentum of the system relative to the point O can be written as follows:

$$\mathbf{M}=\Sigma\,\mathbf{r}_i\times\mathbf{p}_i=\Sigma(\mathbf{r}_i'\times\mathbf{p}_i)+\Sigma\int(\mathbf{r}_0\times\mathbf{p}_i) \quad \text{(using Eq. (1))}$$

or

$$\mathbf{M}=\mathbf{M}'+(\mathbf{r}_0\times\mathbf{p}) \quad (\text{where } \mathbf{p}=\Sigma\,\mathbf{p}_i) \tag{2}$$

From Eq. (2), if the total linear momentum of the system, $\mathbf{p}=0$, then its angular momentum does not depend on the choice of the point O.

Note: *In the C.M. frame, the system of particles, as whole is at rest.*

1.197 On the basis of solution of problem 1.196, we have concluded that; "in the C.M. frame, the angular momentum of system of particles is independent of the choice of the point, relative to which it is determined" and in accordance with the problem this is denoted by $\mathbf{M}$.

We denote the angular momentum of the system of particles, relative to the point O, by $\mathbf{M}$. Since the internal and proper angular momentum $\tilde{\mathbf{M}}$, in the C.M. frame, does not depend on the choice of the point O', this point may be taken coincident with the point O of the K-frame, at a given moment of time. Then at that moment, the radius vectors of all the particles, in both reference frames, are equal ($\mathbf{r}_i'=\mathbf{r}_i$) and the velocities are related by the equation

$$\mathbf{v}_i=\tilde{\mathbf{v}}_i+\mathbf{v}_C$$

where $\mathbf{v}_C$ is the velocity of C.M. frame, relative to the K-frame. Consequently, we may write

$$\mathbf{M}=\Sigma\,m_i(\mathbf{r}_i\times\mathbf{v}_i)=\Sigma\,m_i(\mathbf{r}_i\times\tilde{\mathbf{v}}_i)+\Sigma\,m_i(\mathbf{r}_i\times\mathbf{v}_C)$$

or $\quad \mathbf{M}=\tilde{\mathbf{M}}+m(\mathbf{r}_C\times\mathbf{v}_C), \quad \text{as} \quad \Sigma\,m_i\mathbf{r}_i=m\mathbf{r}_C \quad (\text{where } m=\Sigma\,m_i)$

or $\quad \mathbf{M}=\tilde{\mathbf{M}}+(\mathbf{r}_C\times m\mathbf{v}_C)=\tilde{\mathbf{M}}+(\mathbf{r}_C\times\mathbf{p})$

1.198 Taking striking ball as particle 1 and upper ball of the dumbbell as particle 2 the velocity acquired by particle 2 on collsion, from the formula

$$\mathbf{v}_i'=2\mathbf{v}_C-\mathbf{v}_i \quad \text{(see solution of problem 1.168)}$$

$$\mathbf{v}_i'=2\left(\frac{m\mathbf{v}_0}{m+m/2}\right)-0=\frac{4}{3}\mathbf{v}_0$$

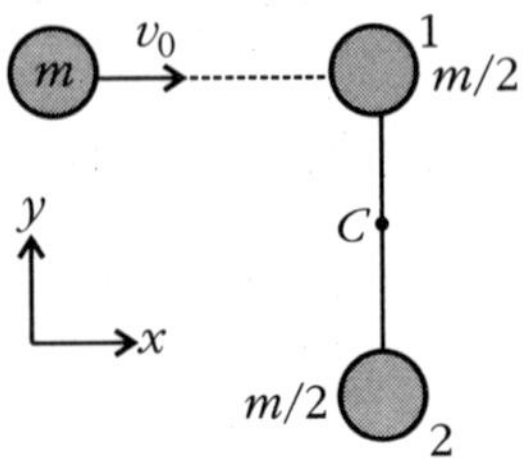

To find the proper angular momentum of the dumbbell, let us denote upper ball of the dumbbell as particle 1 and lower ball as particle 2.

Then proper angular momentum vector is

$$\tilde{\mathbf{M}}=\mathbf{r}_{1C}\times m_1\mathbf{v}_{1C}+\mathbf{r}_{2C}\times m_2\mathbf{v}_{2C}=\mathbf{r}_{1C}\times\tilde{\mathbf{p}}_1+\mathbf{r}_{2C}\times\tilde{\mathbf{p}}_2$$

In C.M. frame net linear momentum of any system is zero, so, $\tilde{\mathbf{p}}_1 = \tilde{\mathbf{p}}_2$.

So, $$\tilde{\mathbf{M}} = (\mathbf{r}_{1C} - \mathbf{r}_{2C}) \times \tilde{\mathbf{p}}_1 = \mathbf{r}_{12} \times \tilde{\mathbf{p}}_1$$

We have $$\tilde{\mathbf{p}}_1 = \mu(\mathbf{v}_1 - \mathbf{v}_2) \quad \text{(see solution of problem 1.147)}$$

$$= \frac{m}{4}\left(\frac{4v_0}{3}\mathbf{i} - 0\right) = \frac{mv_0}{3}\mathbf{i}$$

Hence, $$\tilde{\mathbf{M}} = [\mathbf{r}_{12} \times \tilde{\mathbf{p}}_1] = \left[l\mathbf{j} \times \frac{mv_0}{3}\mathbf{i}\right] = \frac{mv_0 l}{3}(-\mathbf{k})$$

Hence, $$\tilde{M} = \frac{mv_0 l}{3}$$

1.199 In the C.M. frame of the system (both the disks + spring), the linear momentum of the disks are related by the relation, $\tilde{\mathbf{p}}_1 = -\tilde{\mathbf{p}}_2$, at all the moments of time.

Where, $$\tilde{p}_1 = \tilde{p}_2 = \tilde{p} = \mu v_{rel}$$

And the total kinetic energy of the system,

$$T = \frac{1}{2}\mu v_{rel}^2 \quad \text{[See solution of problem 1.147(b)]}$$

Bearing in mind that at the moment of maximum deformation of the spring, the projection of $\mathbf{v}_{rel}$ along the length of the spring becomes zero, i.e., $v_{rel(x)} = 0$.

The conservation of mechanical energy of the considered system in the C.M. frame gives

$$\frac{1}{2}\left(\frac{m}{2}\right)v_0^2 = \frac{1}{2}\kappa x^2 + \frac{1}{2}\left(\frac{m}{2}\right)v_{rel(y)}^2 \tag{1}$$

Now from the conservation of angular momentum of the system about the C.M.,

$$\frac{1}{2}\left(\frac{l_0}{2}\right)\left(\frac{m}{2}v_0\right) = 2\left(\frac{l_0 + x}{2}\right)\frac{m}{2}v_{rel(y)}$$

or $$v_{rel(y)} = \frac{v_0 l_0}{(l_0 + x)} = v_0\left(1 + \frac{x}{l_0}\right)^{-1} = v_0\left(1 - \frac{x}{l_0}\right) \quad (\text{as } x \ll l_0) \tag{2}$$

Using Eq. (2) in Eq.(1), $$\frac{1}{2}mv_0^2\left[1 - \left(1 - \frac{x}{l_0}\right)^2\right] = \kappa x^2$$

or $$\frac{1}{2}mv_0^2\left[1 - \left(1 - \frac{2x}{l_0} + \frac{x^2}{l_0^2}\right)\right] = \kappa x^2$$

or $$\frac{mv_0^2 x}{l_0} = \kappa x^2 \quad [\text{neglecting } x^2/l_0^2]$$

As $$x \neq 0, \text{ thus } x = \frac{mv_0^2}{\kappa l_0}$$

1.4 Universal Gravitation

1.200 We have $\frac{Mv^2}{r} = \frac{\gamma M M_S}{r^2}$ or $r = \frac{\gamma M_S}{v^2}$ (here M_S is the mass of the Sun)

Thus $$\omega = \frac{v}{r} = \frac{v}{\gamma M_S / v^2} = \frac{v^3}{\gamma M_S}$$

So, $$T = \frac{2\pi\gamma M_S}{v^3} = \frac{2\pi \times 6.67 \times 10^{-11} \times 1.97 \times 10^{30}}{(34.9 \times 10^3)^3} = 1.94 \times 10^7 \text{ s} = 225 \text{ days}$$

(The answer is incorrectly written in terms of the planetary mass M in answer sheet.)

1.201 For any planet

$$MR\omega^2 = \frac{\gamma M M_S}{R^2} \quad \text{or} \quad \omega = \sqrt{\frac{\gamma M_S}{R^3}}$$

So, $$T = \frac{2\pi}{\omega} = 2\pi R^{3/2} / \sqrt{\gamma M_S}$$

(a) Thus, $$\frac{T_J}{T_E} = \left(\frac{R_J}{R_E}\right)^{3/2}$$

So, $$\frac{R_J}{R_E} = \left(\frac{T_J}{T_E}\right)^{2/3} = (12)^{2/3} = 5.24$$

(b) $$V_J^2 = \frac{\gamma M_S}{R_J} \quad \text{and} \quad R_J = \left(T \frac{\sqrt{\gamma M_S}}{2\pi}\right)^{2/3}$$

So, $$V_J^2 = \frac{(\gamma M_S)^{2/3} (2\pi)^{2/3}}{T^{2/3}} \quad \text{or} \quad V_J = \left(\frac{2\pi\gamma M_S}{T}\right)^{2/3}$$

where $T = 12$ years, M_S = mass of the Sun.

Putting the values, we get $V_J = 12.97$ km/s

$$\text{Acceleration} = \frac{V_J^2}{R_j} = \left(\frac{2\pi\gamma M_S}{T}\right)^{2/3} \times \left(\frac{2\pi}{T\sqrt{\gamma M_S}}\right)^{2/3}$$
$$= \left(\frac{2\pi}{T}\right)^{4/3} (\gamma M_S)^{1/3}$$
$$= 2.15 \times 10^{-4} \text{ km/s}^2$$

1.202 Semi-major axis = $(r + R)/2$. It is sufficient to consider the motion be along a circle of semi-major axis $(r + R)/2$ for T does not depend on eccentricity.

Hence, $$T = \frac{[2\pi(r+R)/2]^{3/2}}{\sqrt{\gamma M_S}}$$

$$= \pi\sqrt{(r+R)^3/2\gamma M_S} \quad \text{(again } M_S \text{ is the mass of the Sun)}$$

1.203 We can think of the body as moving in a very elongated orbit of maximum distance R and minimum distance 0 so semi-major axis $= R/2$. Hence if τ is the time of fall then

$$\left(\frac{2\tau}{T}\right)^2 = \left(\frac{R/2}{R}\right)^3 \quad \text{or} \quad \tau^2 = \frac{T^2}{32}$$

or $$\tau = T/4\sqrt{2} = 365/4\sqrt{2} = 64.5 \text{ days}$$

1.204 $T = 2\pi R^{3/2}/\sqrt{\gamma M_S}$. If the distances are scaled down, $R^{3/2}$ decreases by a factor $\eta^{3/2}$ and so does M_S. Hence, T does not change.

1.205 The double star can be replaced by a single star of mass $m_1 m_2/(m_1 + m_2)$ moving about the centre of mass subjected to the force $\gamma m_1 m_2/r^2$. Then

$$T = \frac{2\pi r^{3/2}}{\sqrt{\gamma m_1 m_2 \Big/ \dfrac{m_1 m_2}{m_1 + m_2}}}$$

$$= \frac{2\pi r^{3/2}}{\sqrt{\gamma M}}$$

So, $$r^{3/2} = \frac{T}{2\pi}\sqrt{\gamma M}$$

or $$r = \left(\frac{T}{2\pi}\right)^{2/3} (\gamma M)^{1/3} = \sqrt[3]{\gamma M (T/2\pi)^2}$$

1.206 (a) The gravitational potential due to m_1 at the point of location of m_2

$$V_2 = \int_r^\infty \mathbf{G}\cdot d\mathbf{r} = \int_r^\infty -\frac{\gamma m_1}{x^2}\,dx = -\frac{\gamma m_1}{r}$$

So, $$U_{21} = m_2 V_2 = -\frac{\gamma m_1 m_2}{r}$$

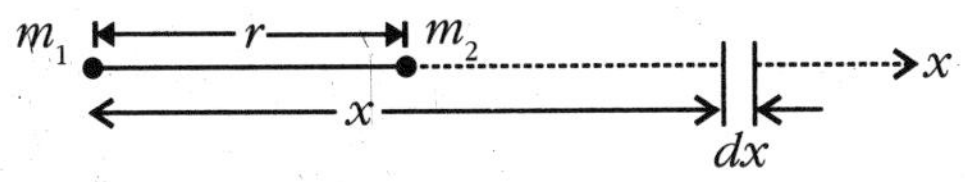

Similarly, $$U_{12} = -\frac{\gamma m_1 m_2}{r}$$

Hence, $$U_{12} = U_{21} = U = -\frac{\gamma m_1 m_2}{r}$$

(b) Choose the location of the point mass as the origin. Then the potential energy dU of an element of mass $dM = M/l\ dx$ of the rod in the field of the point mass is

$$dU = -\gamma m \frac{M}{l}\,dx\,\frac{1}{x}$$

where x is the distance between the element and the point. (Note that the rod and the point mass are on a straight line.) Then if a is the distance of the nearer end

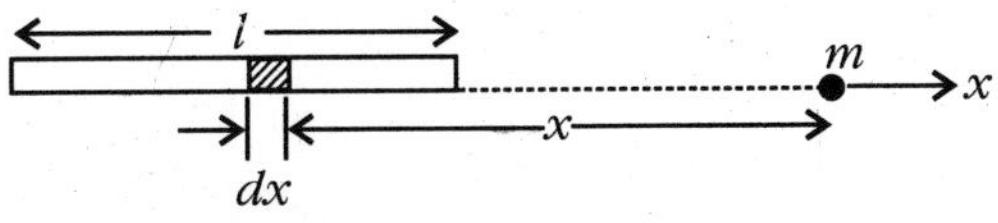

of the rod from the point mass,

$$U = -\gamma \frac{mM}{l} \int_a^{a+l} \frac{dx}{x} = -\gamma m \frac{M}{l} \ln\left(1 + \frac{l}{a}\right)$$

The force of interaction is

$$F = -\frac{\partial U}{\partial a} = \gamma \frac{mM}{l} \times \frac{1}{1 + \frac{l}{a}} \left(-\frac{l}{a^2}\right) = -\frac{\gamma mM}{a(a+l)}$$

Minus sign means attraction.

1.207 As the planet is under central force (gravitational interaction), its angular momentum is conserved about the Sun (which is situated at one of the focii of the ellipse).

So, $$mv_1 r_1 = mv_2 r_2 \quad \text{or} \quad v_1^2 = \frac{v_2^2 r_2^2}{r_1^2} \tag{1}$$

From the conservation of mechanical energy of the system (Sun + planet),

$$-\frac{\gamma M_S m}{r_1} + \frac{1}{2} m v_1^2 = \frac{\gamma M_S m}{r_2} + \frac{1}{2} m v_2^2$$

or $$-\frac{\gamma M_S}{r_1} + \frac{1}{2} v_2^2 \frac{r_2^2}{r_1^2} = -\left(\frac{\gamma M_S}{r_2}\right) + \frac{1}{2} v_2^2 \quad \text{(using Eq. (1))}$$

Thus, $$v_2 = \sqrt{2\gamma M_S r_1 / r_2 (r_1 + r_2)} \tag{2}$$

Hence, $$M = m v_2 r_2 = m\sqrt{2\gamma M_S r_1 r_2 / (r_1 + r_2)}$$

1.208 From the previous problem, if r_1, r_2 are the maximum and minimum distances from the Sun to the planet and v_1, v_2 are the corresponding velocities, then, say,

$$E = \frac{1}{2} m v_2^2 - \frac{\gamma m M_S}{r_2}$$

$$= \frac{\gamma m M_S}{r_1 + r_2} \cdot \frac{r_1}{r_2} - \frac{\gamma m M_S}{r_2} = \frac{-\gamma m M_S}{r_1 + r_2} = \frac{\gamma m M_S}{2a} \quad \text{(using Eq. (2) of problem 1.207)}$$

where $2a$ = major axis = $r_1 + r_2$.

The same result can also be obtained directly by writing an equation analogous to Eq. (1) of problem 1.191

$$E = \frac{1}{2} m \dot{r}^2 - \frac{M^2}{2mr^2} - \frac{\gamma m M_S}{r}$$

(Here M is angular momentum of the planet and m is its mass.) For extreme position $\dot{r} = 0$ and we get the quadratic equation

$$Er^2 + \gamma m M_S r - \frac{M^2}{2m} = 0$$

The sum of the two roots of this equation is

$$r_1 + r_2 = -\frac{\gamma m M_S}{E}$$

Thus, $$E = -\frac{\gamma m M_S}{2a} = \text{constant}$$

1.209 From the conservation of angular momentum about the Sun.

$$m v_0 r_0 \sin\alpha = m v_1 r_1 = m v_2 r_2 \quad \text{or} \quad v_1 r_1 = v_2 r_2 = v_0 r_0 \sin\alpha \tag{1}$$

From conservation of mechanical energy,

$$\frac{1}{2} m v_0^2 - \frac{\gamma M_S m}{r_0} = \frac{1}{2} m v_1^2 - \frac{\gamma M_S m}{r_1}$$

or $$\frac{v_0^2}{2} - \frac{\gamma M_S}{r_0} = \frac{v_0^2 r_0^2 \sin^2\alpha}{2 r_1^2} - \frac{\gamma M_S}{r_1} \quad \text{(using Eq. 1)}$$

or $$\left(v_0^2 - \frac{2\gamma M_S}{r_0}\right) r_1^2 + 2\gamma M_S r_1 - v_0^2\, r_0^2 \sin\alpha = 0$$

So, $$r_1 = \frac{-2\gamma M_S \pm \sqrt{4\gamma^2 M_S^2 + 4(v_0^2 r_0^2 \sin^2\alpha)\left(v_0^2 - \frac{2\gamma M_S}{r_0}\right)}}{2\left(v_0^2 - \frac{2\gamma M_S}{r_0}\right)}$$

$$= \frac{1 \pm \sqrt{1 - \frac{v_0^2 r_0^2 \sin^2\alpha}{\gamma M_S}\left(\frac{2}{r_0} - \frac{v_0^2}{r_0 M_S}\right)}}{\left(\frac{2}{r_0} - \frac{v_0^2}{\gamma M_S}\right)} = \frac{r_0[1 \pm \sqrt{1-(2-\eta)\eta \sin^2\alpha}]}{(2-\eta)}$$

where $\eta = v_0^2\, r_0 / \gamma M_S$.

1.210 At the minimum separation with the Sun, the cosmic body's velocity is perpendicular to its position vector relative to the Sun. If $r_{\min}$ be the sought minimum distance, from conservation of angular momentum about the Sun (C),

$$m v_0 l = m v r_{\min} \quad \text{or} \quad v = \frac{v_0 l}{r_{\min}} \tag{1}$$

From conservation of mechanical energy of the system (Sun 1 cosmic body),

$$\frac{1}{2} m v_0^2 = -\frac{\gamma M_S m}{r_{\min}} + \frac{1}{2} m v^2$$

So, $$\frac{v_0^2}{2} = -\frac{\gamma M_S}{r_{\min}} + \frac{v_0^2 l^2}{2 r_{\min}^2} \quad \text{(using Eq. 1)}$$

or $$v_0^2 r_{\min}^2 + 2\gamma M_S r_{\min} - v_0^2\, l^2 = 0$$

So, $$r_{\min} = \frac{-2\gamma M_S \pm \sqrt{4\gamma^2 M_S^2 + 4 v_0^2 v_0^2 l^2}}{2 v_0^2} = \frac{-\gamma M_S \pm \sqrt{\gamma^2 M_S^2 + v_0^4 l^2}}{v_0^2}$$

Hence, taking positive root

$$r_{\min} = (\gamma M_S / v_0^2)\left[\sqrt{1 + \left(\frac{l v_0^2}{\gamma M_S}\right)^2} - 1\right]$$

1.211 (a) Suppose that the sphere has a radius equal to a. We may imagine that the sphere is made up of concentric thin spherical shells (layers) with radii ranging from 0 to a, and each spherical layer is made up of elementary bands (rings). Let us first calculate potential due to an elementary band of a spherical layer at the point of location of the point mass m (say point P) (see figure). As all the points of the band are located at the distance l from the point P, so,

$$\partial\varphi = -\frac{\gamma\partial M}{l} \quad \text{(where } \partial M \text{ is mass of the band)} \tag{1}$$

$$\partial M = \left(\frac{dM}{4\pi a^2}\right)(2\pi a \sin\theta)(a d\theta)$$

$$= \left(\frac{dM}{2}\right)\sin\theta d\theta \tag{2}$$

and $$l^2 = a^2 + r^2 - 2\,ar\cos\theta \tag{3}$$

Differentiating Eq. (3), we get

$$l dl = ar\sin\theta d\theta \tag{4}$$

Hence using above equations

$$\partial\varphi = -\left(\frac{\gamma dM}{2ar}\right)dl \tag{5}$$

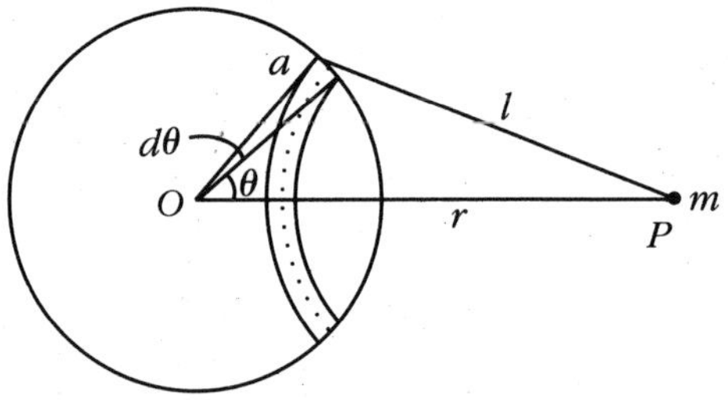

Now integrating Eq. (5) over the whole spherical layer

$$d\varphi = \int\partial\varphi = -\frac{\gamma dM}{2ar}\int_{r-a}^{r+a} dl$$

So, $$d\varphi = -\frac{\gamma dM}{r} \tag{6}$$

Equation (6) demonstrates that the potential produced by a thin uniform spherical layer outside the layer is such as if the whole mass of the layer were concentrated at its centre. Hence the potential due to the sphere at point P is given by

$$\varphi = \int d\varphi = -\frac{\gamma}{r}\int dM = -\frac{\gamma M}{r} \tag{7}$$

This expression is similar to that of Eq. (6).

Hence the sought potential energy of gravitational interaction of the particle m and the sphere,

$$U = m\varphi = -\frac{\gamma Mm}{r}$$

(b) Using the equation $$G_r = -\frac{\partial\varphi}{\partial r}$$

$$G_r = -\frac{\gamma M}{r^2} \quad \text{(using Eq. 7)}$$

So, $$\mathbf{G} = -\frac{\gamma M}{r^3}\mathbf{r} \quad \text{and} \quad \mathbf{F} = m\mathbf{G} = -\frac{\gamma m M}{r^3}\mathbf{r} \tag{8}$$

1.212 (The problem has already a clear hint in the answer sheet of the problem book. Here we adopt a different method.)

Let m be the mass of the spherical layer, which is imagined to be made up of rings. At a point inside the spherical layer at distance r from the centre, the gravitational potential due to a ring element of radius a equals

$$d\varphi = -\frac{\gamma m}{2ar}dl \text{ (see Eq. (5) of solution of problem 1.211)}$$

So,
$$\varphi = \int d\varphi = -\frac{\gamma m}{2ar}\int_{a-r}^{a+r} dl = -\frac{\gamma m}{a}$$

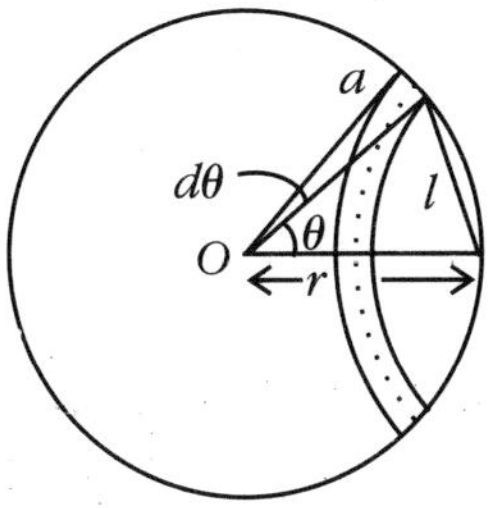

Hence,
$$G_r = -\frac{\partial \varphi}{\partial r} = 0$$

Hence, gravitational field strength as well as field force becomes zero inside a thin spherical layer.

1.213 One can imagine that the uniform hemisphere is made up of thin hemispherical layers of radii ranging from 0 to R. Let us consider such a layer (see figure). Potential at point O due to this layer is

$$d\varphi = -\frac{\gamma dm}{r} = -\frac{3\gamma m}{r^3}rdr, \quad \text{where } dm = \frac{M}{(2/3)\pi R^3}\left(\frac{4\pi r^2}{2}\right)dr$$

(This is because all points of each hemispherical shell are equidistant from O).

Hence,
$$\varphi = \int d\varphi = -\frac{3\gamma M}{R^3}\int_0^R rdr = -\frac{3\gamma M}{2R}$$

Hence, the work done by the gravitational field force on the particle of mass m, to move it to infinity is given by the formula

$$A = m\varphi, \quad \text{since } \varphi = 0 \text{ at infinity}$$

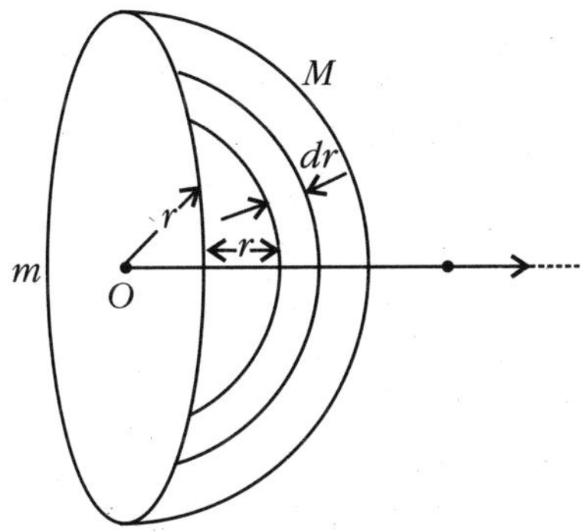

Hence the sought work,

$$A_{0\to\infty} = -\frac{3\gamma m M}{2R}$$

(The work done by the external agent is $-A$.)

1.214 In the solution of problem 1.211, we have obtained φ and **G** due to a uniform sphere, at a distance r from it's centre outside it. We have from Eqs. (7) and (8) of problem 1.211,

$$\varphi = -\frac{\gamma M}{r} \quad \text{and} \quad \mathbf{G} = -\frac{\gamma M}{r^3}\mathbf{r} \qquad \text{(A)}$$

In accordance with Eq. (1) of the solution of 1.212, potential due to a spherical shell of radius a, at any point, inside it becomes

$$\varphi = \frac{-\gamma M}{a} = \text{constant} \quad \text{and} \quad G_r = -\frac{\partial \varphi}{\partial r} = 0 \qquad \text{(B)}$$

For a point (say P) which lies inside the uniform solid sphere, the potential φ at that point may be represented as a sum

$$\varphi_{\text{inside}} = \varphi_1 + \varphi_2$$

where φ_1 is the potential of a solid sphere having radius r and φ_2 is the potential of the layer of radii r and R. In accordance with Eq. (A)

$$\varphi_1 = -\frac{\gamma}{r}\left(\frac{M}{(4/3)\pi R^3}\frac{4}{3}\pi r^3\right) = -\frac{\gamma M}{R^3}r^2$$

The potential φ_2 produced by the layer (thick shell) is the same at all points inside it. The potential φ_2 is easiest to calculate, for the point positioned at the layer's centre. Using Eq. (B)

$$\varphi_2 = -\gamma\int_r^R \frac{dM}{r} = -\frac{3}{2}\frac{\gamma M}{R^3}(R^2 - r^2)$$

where $$dM = \frac{M}{(4/3)\pi R^3}4\pi r^2 dr = \left(\frac{3M}{R^3}\right)r^2 dr$$

is the mass of a thin layer between the radii r and $r + dr$.

Thus, $$\varphi_{\text{inside}} = \varphi_1 + \varphi_2 = \left(\frac{\gamma M}{2R}\right)\left(3 - \frac{r^2}{R^2}\right) \qquad \text{(C)}$$

From Eq. (B) $$G_r = \frac{-\partial\varphi}{\partial r}$$

$$G_r = \frac{\gamma M r}{R^3}$$

or $$\mathbf{G} = -\frac{\gamma M}{R^3}\mathbf{r} = -\gamma\frac{4}{3}\pi\rho\mathbf{r}$$

where $\rho = \dfrac{M}{4/3\,\pi R^3}$ is the density of the sphere.

The plots $\varphi(r)$ and $G(r)$ for a uniform sphere of radius R are shown in the figure of answer sheet.

Alternate:

Like Gauss's theorem of electrostatics, one can derive Gauss's theorem for gravitation in the form $\mathbf{G} \cdot d\mathbf{S} = -4\pi\gamma m_{\text{enclosed}}$. For calculation of $\mathbf{G}$ at a point inside the sphere at a distance r from its centre, let us consider a Gaussian surface of radius r. Then

$$G_r 4\pi r^2 = -4\pi\gamma\left(\frac{M}{r^3}\right)r^3 \quad \text{or} \quad G_r = -\frac{\gamma M}{R^3}r$$

Hence, $$\mathbf{G} = \frac{\gamma M}{R^3}\mathbf{r} = -\gamma\frac{4}{3}\pi\rho\mathbf{r} \quad \left(\text{as } \rho = \frac{M}{(4/3)\pi R^3}\right)$$

So, $$\varphi = \int_r^\infty G_r dr = \int_r^R -\frac{\gamma M}{R^3}r\,dr + \int_R^\infty -\frac{\gamma M}{r^2}dr$$

Integrating and summing up, we get

$$\varphi = -\frac{\gamma M}{2R}\left(3 - \frac{r^2}{R^2}\right)$$

From Gauss' theorem for a point outside it,

$$G_r 4\pi r^2 = -4\pi\gamma M \quad \text{or} \quad G_r = -\frac{\gamma M}{r^2}$$

Thus,
$$\varphi(r) = \int_r^\infty G_r dr = -\frac{\gamma M}{r}$$

1.215 Treating the cavity as negative mass of density $-\rho$ in a uniform sphere density $+\rho$ and using the superposition principle, the sought field strength is

$$\mathbf{G} = \mathbf{G}_1 + \mathbf{G}_2$$

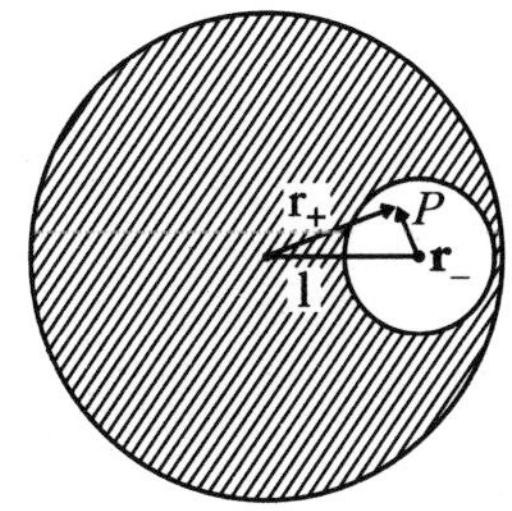

or
$$\mathbf{G} = -\frac{4}{3}\pi\gamma\rho\mathbf{r}_+ + -\frac{4}{3}\gamma\pi(-\rho)\mathbf{r}_-$$

where $\mathbf{r}_+$ and $\mathbf{r}_-$ are the position vectors of an arbitrary point P inside the cavity with respect to centre of sphere and cavity, respectively.

Thus,
$$\mathbf{G} = -\frac{4}{3}\pi\gamma\rho(\mathbf{r}_+ - \mathbf{r}_-) = -\frac{4}{3}\pi\gamma\rho\mathbf{l}$$

1.216 We partition the solid sphere into thin spherical layers and consider a layer of thickness dr lying at a distance r from the centre of the ball. Each spherical layer presses on the layers within it. The considered layer is attracted to the part of the sphere lying within it (the outer part does not act on the layer). Hence for the considered layer

$$dp\, 4\pi r^2 = dF$$

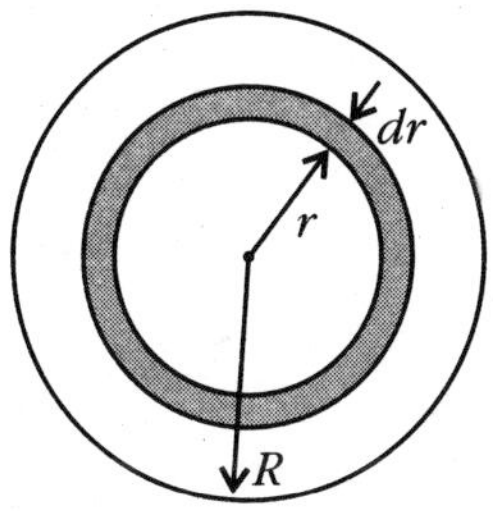

or
$$dp 4\pi r^2 = \frac{\gamma(4/3\pi r^3\rho)(4\pi r^2 dr\rho)}{r^2}$$

(where ρ is the mean density of sphere)

or
$$dp = \frac{4}{3}\pi\gamma\rho^2 r dr$$

Thus,
$$p = \int_r^R dp = \frac{2\pi}{3}\gamma\rho^2(R^2 - r^2)$$

(the pressure must vanish at $r = R$)

or
$$p = \frac{3}{8}(1-(r^2/R^2))\gamma M^2/\pi R^4, \quad \text{putting } \rho = M/(4/3)\pi R^3$$

Putting $r = 0$, we have the pressure at sphere's centre, and treating it as the Earth where mean density is equal to $\rho = 5.5 \times 10^3$ kg/m^3 and $R = 64 \times 10^2$ km we have, $p = 1.73 \times 10^{11}$ Pa or 1.72×10^6 atm.

1.217 (a) Since the potential at each point of a spherical surface (shell) is constant and is equal to $\varphi = -\gamma m/R$ [as we have in Eq. (1) of solution of problem 1.212].

We obtain in accordance with the equation

$$U = \frac{1}{2}\int dm\varphi = \frac{1}{2}\varphi\int dm$$
$$= \frac{1}{2}\left(-\frac{\gamma m}{R}\right)m = -\frac{\gamma m^2}{2R}$$

(The factor 1/2 is needed otherwise contribution of different mass elements is counted twice.)

(b) In this case the potential inside the sphere depends only on r (see Eq. (C) of the solution of problem 1.214)

$$\varphi = -\frac{3\gamma M}{2R}\left(1 - \frac{r^2}{3R^2}\right)$$

Here dm is the mass of an elementary spherical layer confined between the radii r and $r + dr$ given by

$$dm = (4\pi r^2\, dr\rho) = \left(\frac{3m}{R^3}\right) r^2 dr$$

$$U = \frac{1}{2}\int dm\,\varphi$$

$$= \frac{1}{2}\int_0^R \left(\frac{3m}{R^3}\right) r^2\, dr\left(-\frac{3\gamma m}{2R}\left(1 - \frac{r^2}{3R^2}\right)\right)$$

After integrating, we get

$$U = -\frac{3}{5}\frac{\gamma m^2}{R}$$

1.218 Let $\omega = \sqrt{\frac{\gamma M_E}{r^3}}$ = circular frequency of the satellite in the outer orbit,

$\omega_0 = \sqrt{\frac{\gamma M_E}{(r - \Delta r)^3}}$ = circular frequency of the satellite in the inner orbit.

So, relative angular velocity $= \omega_0 \pm \omega$, where $-ve$ sign is to be taken when the satellites are moving in the same sense and $+ve$ sign if they are moving in opposite sense.

Hence, time between closest approaches

$$\frac{2\pi}{\omega_0 \pm \omega} = \frac{2\pi}{\sqrt{\gamma M_E}}\frac{r^{3/2}}{3\Delta r/2r + \delta} = \begin{cases} 4.5 \text{ days } (\delta = 0) \\ 0.84 \text{ hour } (\delta = 2) \end{cases}$$

where δ is 0 in the first case and 2 in the second case.

1.219 $\omega_1 = \frac{\gamma M}{R^2} = \frac{6.67 \times 10^{-11} \times 5.96 \times 10^{24}}{(6.37 \times 10^6)^2} = 9.8 \text{ m/s}^2$

$$\omega_2 = \omega^2 R = \left(\frac{2\pi}{T}\right)^2 R = \left(\frac{2 \times 22}{24 \times 3600 \times 7}\right)^2 6.37 \times 10^6 = 0.034 \text{ m/s}^2$$

and $\omega_3 = \frac{\gamma M_S}{R^2_{\text{mean}}} = \frac{6.67 \times 10^{-11} \times 1.97 \times 10^{30}}{(149.50 \times 10^6 \times 10^3)^2} = 5.9 \times 10^{-3} \text{ m/s}^2$

Then, $\omega_1 : \omega_2 : \omega_3 = 1 : 0.0034 : 0.0006$

1.220 Let h be the sought height in the first case, so

$$\frac{99}{100}g = \frac{\gamma M}{(R + h)^2}$$

$$= \frac{\gamma M}{R^2(1 + h/R)^2} = \frac{g}{(1 + h/R)^2}$$

or

$$\frac{99}{100} = \left(1 + \frac{h}{R}\right)^{-2}$$

From the statement of the problem, it is obvious that in this case $h << R$

Thus, $$\frac{99}{100} = \left(1 - \frac{2h}{R}\right) \quad \text{or} \quad h = \frac{R}{200} = \left(\frac{6400}{200}\right)\text{km} = 32\text{ km}$$

In the other case, if h' be the sought height, then

$$\frac{g}{2} = g\left(1 + \frac{h'}{R}\right)^{-2} \quad \text{or} \quad \frac{1}{2} = \left(1 + \frac{h'}{R}\right)^{-2}$$

From the language of the problem, in this case h' is not very small in comparison with R. Therefore in this case we cannot use the approximation adopted in the previous case.

Here, $$\left(1 + \frac{h'}{R}\right)^2 = 2$$

So, $$\frac{h'}{R} = \pm\sqrt{2} - 1$$

As $-ve$ sign is not acceptable

$$h' = (\sqrt{2} - 1)R = (\sqrt{2} - 1)6400\text{ km} = 2650\text{ km}$$

1.221 Let the mass of the body be m and let it go upto a height h.

From conservation of mechanical energy of the system

$$-\frac{\gamma Mm}{R} + \frac{1}{2}mv_0^2 = \frac{-\gamma Mm}{(R+h)} + 0$$

Using $\gamma M/R^2 = g$, in above equation and on solving, we get

$$h = \frac{Rv_0^2}{2gR - v_0^2}$$

1.222 Gravitational pull provides the required centripetal acceleration to the satellite. Thus if h be the sought distance, we have

$$\frac{mv^2}{(R+h)} = \frac{\gamma mM}{(R+h)^2} \quad \text{or} \quad (R+h)v^2 = \gamma M$$

or $$Rv^2 + hv^2 = gR^2 \quad \text{as} \quad g = \frac{\gamma M}{R^2}$$

Hence, $$h = \frac{gR^2 - Rv^2}{v^2} = R\left[\frac{gR}{v^2} - 1\right]$$

1.223 A satellite that hovers above the Earth's equator and co-rotates with it moving from west to east with the diurnal angular velocity of the Earth appears stationary to an observer on the Earth. It is called geostationary. For this calculation we may neglect the annual motion of the Earth as well as all other influences. Then, by Newton's law,

$$\frac{\gamma Mm}{r^2} = m\left(\frac{2\pi}{T}\right)^2 r$$

where M = mass of the Earth, T = 86400 = period of daily rotation of the Earth and r = distance of the satellite from the centre of the Earth.

Then,
$$r = \sqrt[3]{\gamma M \left(\frac{T}{2\pi}\right)^2}$$

Substitution of $M = 5.96 \times 10^{24}$ kg, gives $r = 4.220 \times 10^4$ km.

Thus instantaneous velocity with respect to an inertial frame fixed to the centre of the Earth at that moment will be

$$\left(\frac{2\pi}{T}\right)r = 3.07 \text{ km/s}$$

and the acceleration will be the centripetal acceleration

$$\left(\frac{2\pi}{T}\right)^2 r = 0.223 \text{ m/s}^2$$

1.224 We know from the previous problem that a satellite moving west to east at distance $R = 2.00 \times 10^4$ km from the centre of the Earth will be revolving round the Earth with an angular velocity faster than the Earth's diurnal angular velocity. Let

$$\omega = \text{angular velocity of the satellite,}$$

$$\omega_0 = 2\pi/T = \text{angular velocity of the Earth.}$$

Then
$$\omega - \omega_0 = 2\pi/\tau$$

is the relative angular velocity with respect to Earth. Now by Newton's law

$$\frac{\gamma M}{R^2} = \omega^2 R$$

So,
$$M = \frac{R^3}{\gamma}\left(\frac{2\pi}{\tau} + \frac{2\pi}{T}\right)^2$$
$$= \frac{4\pi^2 R^3}{\gamma T^2}\left(1 + \frac{T}{\tau}\right)^2$$

Substitution gives

$$m = 6.27 \times 10^{24} \text{ kg}$$

1.225 The velocity of the satellite in the inertial space fixed frame is $\sqrt{\gamma M/R}$ east to west. With respect to the Earth fixed frame, from the $\mathbf{v}' = \mathbf{v} - (\boldsymbol{\omega} \times \mathbf{r})$, the velocity is

$$v' = \frac{2\pi R}{T} + \sqrt{\frac{\gamma M}{R}} = 7.03 \text{ km/s}$$

Here M is the mass of the Earth and T is its period of rotation about its own axis. If the satellite were moving from west to east, velocity would be

$$-\frac{2\pi R}{T} + \sqrt{\frac{\gamma m}{R}}$$

To find the acceleration we note the formula

$$m\mathbf{w}' = \mathbf{F} + 2m(\mathbf{v}' \times \boldsymbol{\omega}) + m\omega^2 \mathbf{R}$$

Here $\mathbf{F} = (\gamma M m/R^3)\mathbf{R}$ and $\mathbf{v}' \times \boldsymbol{\omega}$ is directed towards the centre of the Earth.

Thus, $$\mathbf{w}' = \frac{\gamma M}{R^2} + 2\left(\frac{2\pi R}{T} + \sqrt{\frac{\gamma M}{R}}\right)\frac{2\pi}{T} - \left(\frac{2\pi}{T}\right)^2 R \text{ (toward the Earth's rotation axis)}$$

$$= \frac{\gamma M}{R^2} + \frac{2\pi}{T}\left[\frac{2\pi R}{T} + 2\sqrt{\frac{\gamma M}{R}}\right] = 4.94 \text{ m/s}^2 \text{ on substitution}$$

1.226 From the well known relationship between the velocities of a particle with respect to a space fixed frame (K) and rotating frame (K'),

$$\mathbf{v} = \mathbf{v}' + (\boldsymbol{\omega} \times \mathbf{r})$$

we get, $$v_1' = v - \left(\frac{2\pi}{T}\right)R$$

Thus kinetic energy of the satellite in the Earth's frame

$$T_1' = \frac{1}{2}mv_1'^2 = \frac{1}{2}m\left(v - \frac{2\pi R}{T}\right)^2 \tag{1}$$

Obviously when the satellite moves in opposite sense compared to the rotation of the Earth, its velocity relative to the same frame would be

$$v_2' = v + \left(\frac{2\pi}{T}\right)R$$

and kinetic energy

$$T_2' = \frac{1}{2}mv_2'^2 = \frac{1}{2}m\left(v + \frac{2\pi R}{T}\right)^2 \tag{2}$$

From Eqs. (1) and (2)

$$\frac{T_2'}{T_1'} = \frac{(v + 2\pi R/T)^2}{(v - 2\pi R/T)^2} \tag{3}$$

Now from Newton's second law

$$\frac{\gamma Mm}{R^2} = \frac{mv^2}{R} \quad \text{or} \quad v = \sqrt{\frac{\gamma M}{R}} = \sqrt{gR} \tag{4}$$

Using Eqs. (3) and (4)

$$\frac{T_2'}{T_1'} = \frac{(\sqrt{gR} + 2\pi R/T)^2}{(\sqrt{gR} - 2\pi R/T)^2} = 1.27 \text{ nearly (using Appendices)}$$

1.227 For a satellite in a circular orbit about any massive body, the following relation holds between kinetic, potential and total energy

$$T = -E, \; U = 2E \tag{1}$$

Thus, since total mechanical energy must decrease due to resistance of the cosmic dust, the kinetic energy will increase and the satellite will "fall". We see then, by work energy theorem

$$dT = -dE = -dA_{fr}$$

So, $$mv\,dv = \alpha v^2 v\,dt \quad \text{or} \quad \frac{\alpha\,dt}{m} = \frac{dv}{v^2}$$

Now from Newton's law at an arbitrary radius r from the Moon's centre,

$$\frac{v^2}{r} = \frac{\gamma M}{r^2} \quad \text{or} \quad v = \sqrt{\frac{\gamma M}{r}} \text{ (where } M \text{ is the mass of the moon)}$$

Then, $$v_i = \sqrt{\frac{\gamma M}{\eta R}} \quad \text{and} \quad v_f = \sqrt{\frac{\gamma M}{R}} \quad \text{(where } R = \text{Moon's radius)}$$

So, $$\int_{v_i}^{v_f} \frac{dv}{v^2} = \frac{\alpha}{m}\int_0^{\tau} dt = \frac{\alpha\tau}{m}$$

or $$\tau = \frac{m}{\alpha}\left(\frac{1}{v_i} - \frac{1}{v_f}\right) = \frac{m}{\alpha\sqrt{\gamma M/R}}(\sqrt{\eta} - 1)$$

$$= \frac{m}{\alpha\sqrt{gR}}(\sqrt{\eta} - 1) \quad \text{(where } g \text{ is Moon's gravity)}$$

The averaging implied by Eq. (1) (for non-circular orbits) makes the result approximate.

1.228 From Newton's second law

$$\frac{\gamma Mm}{R^2} = \frac{mv_0^2}{R} \quad \text{or} \quad v_0 = \sqrt{\frac{\gamma M}{R}} = 1.67\,\text{km/s} \tag{1}$$

From conservation of mechanical energy

$$\frac{1}{2}mv_e^2 - \frac{\gamma Mm}{R} = 0 \quad \text{or} \quad v_e = \sqrt{\frac{2\gamma M}{R}} = 2.37\,\text{km/s} \tag{2}$$

In Eqs. (1) and (2), M and R are the mass of the Moon and its radius. In Eq. (1) if M and R represent the mass of the Earth and its radius, then, using appendices, we can easily get orbital and escape velocities for Earth as

$$v_0 = 7.9 \text{ km/s} \quad \text{and} \quad v_e = 11.2 \text{ km/s}$$

1.229 In a parabolic orbit, $E = 0$

So, $$\frac{1}{2}mv_i^2 - \frac{\gamma Mm}{R} = 0 \quad \text{or} \quad v_i = \sqrt{2}\sqrt{\frac{\gamma M}{R}}$$

where M = mass of the Moon, R = its radius and this is just the escape velocity. On the other hand in orbit,

$$mv_f^2/R = \frac{\gamma Mm}{R^2} \quad \text{or} \quad v_f = \sqrt{\frac{\gamma M}{R}}$$

Thus, $$\Delta v = (1 - \sqrt{2})\sqrt{\frac{\gamma M}{R}} = -0.70\,\text{km/s}$$

1.230 From solution of problem 1.228 for the Earth surface

$$v_0 = \sqrt{\frac{\gamma M}{R}} \quad \text{and} \quad v_e = \sqrt{\frac{2\gamma M}{R}}$$

Thus the sought additional velocity

$$\Delta v = v_e - v_0 = \sqrt{\frac{\gamma M}{R}}(\sqrt{2} - 1) = \sqrt{gR}(\sqrt{2} - 1)$$

This "kick" in velocity must be given along the direction of motion of the satellite in its orbit.

1.231 Let r be the sought distance then

$$\frac{\gamma\eta M}{(nR-r)^2}=\frac{\gamma M}{r^2} \quad \text{or} \quad \eta r^2=(nR-r)^2$$

or
$$\sqrt{\eta}r=(nR-r) \quad \text{or} \quad r=\frac{nR}{\sqrt{\eta}+1}=3.8\times 10^4 \text{ km}$$

1.232 Between the Earth and the Moon, the potential energy (P.E.) of the spaceship will have a maximum at the point where the attractions of the Earth and the moon balance each other. This maximum P.E. is approximately zero. We can also neglect the contribution of either body to the P.E. of the spaceship sufficiently near the other body. Then the minimum energy that must be imparted to the spaceship to cross the maximum of the P.E. is clearly (using E to denote the Earth)

$$\frac{\gamma M_E m}{R_E} \tag{1}$$

With this energy the spaceship will cross over the hump in the P.E. and coast down the hill of P.E. towards the moon and crash land on it. What the problem seeks is the minimum energy required for soft-landing. That requires the use of rockets to lowering and braking of the spaceship, and since the kinetic energy of the gases ejected from the rocket will always be positive, the total energy required for soft-landing is greater than that required for crash-landing. To calculate this energy we assume the rockets are used fairly close to the moon when the spaceship has nearly attained its terminal velocity on the moon $\sqrt{2\gamma M_0/R_0}$ where M_0 is the mass of the moon and R_0 is its radius. In generai $dE=vdp$ and since the speed of the ejected gases is not less than the speed of the rocket, and momentum transferred to the gases must equal the momentum of the spaceship the energy E of the gas ejected is not less than the kinetic energy of spaceship

$$\frac{\gamma M_0 m}{R_0} \tag{2}$$

Adding Eqs. (1) and (2) we get the minimum work done on the ejected gases to bring about the soft landing.

$$A_{\min}=\gamma m\left(\frac{M_E}{R_E}+\frac{M_0}{R_0}\right)$$

On substitution, we get 1.3 × 10⁸ kJ.

1.233 Assume first that the attraction of the Earth can be neglected. Then the minimum velocity, that must be imparted to the body to escape from the Sun's pull, is, as in solution of problem 1.230, equal to

$$(\sqrt{2}-1)v_1$$

where $v_1^2=\gamma M_S/r$, $r=$ radius of the Earth's orbit, $M_S=$ mass of the Sun.

In the actual case near the Earth, the pull of the Sun is small and does not change much over distances, which are several times the radius of the Earth. The velocity v_3 in question is that which overcomes the Earth's pull with sufficient velocity to escape the Sun's pull.

Thus,
$$\frac{1}{2}mv_3^2-\frac{\gamma M_E m}{R}=\frac{1}{2}m\left(\sqrt{2}-1\right)^2 v_1^2$$

where $R=$ radius of the Earth and $M_E=$ mass of the Earth.

Writing $v_1^2=\gamma M_S/R$, we get

$$v_3=\sqrt{2v_2^2+(\sqrt{2}-1)^2v_1^2}=16.6\,\text{km/s}$$

1.5 Dynamics of a Solid Body

1.234 Since, motion of the rod is purely translational, net torque about the C.M. of the rod should be equal to zero.

Thus, $$F_1\frac{l}{2} = F_2\left(\frac{l}{2} - a\right) \quad \text{or} \quad \frac{F_1}{F_2} = 1 - \frac{a}{l/2} \tag{1}$$

For the translational motion of rod,

$$F_1 - F_2 = mw_C \quad \text{or} \quad 1 - \frac{F_1}{F_2} = \frac{mw_C}{F_2} \tag{2}$$

From Eqs. (1) and (2)

$$\frac{a}{l/2} = \frac{mw_C}{F_2} \quad \text{or} \quad l = \frac{2aF_2}{mw_C} = 1\text{ m}$$

1.235 Sought moment $\mathbf{N} = \mathbf{r} \times \mathbf{F} = (a\mathbf{i} + b\mathbf{j}) \times (A\mathbf{i} + B\mathbf{j})$

$$= aB\mathbf{k} + Ab(-\mathbf{k}) = (aB - Ab)\mathbf{k}$$

and arm of the force $$l = \frac{N}{F} = \frac{aB - Ab}{\sqrt{A^2 + B^2}}$$

1.236 Relative to point O, the net moment of force is given by

$$N = \mathbf{r}_1 \times \mathbf{F}_1 + \mathbf{r}_2 \times \mathbf{F}_2 = (a\mathbf{i} \times A\mathbf{j}) + (b\mathbf{j} \times B\mathbf{i})$$

$$= aA\vec{\mathbf{k}} + bB(-\vec{\mathbf{k}}) \tag{1}$$

Resultant of the external force

$$\mathbf{F} = \mathbf{F}_1 + \mathbf{F}_2 = A\mathbf{j} + B\mathbf{i} \tag{2}$$

$\mathbf{N} \cdot \mathbf{F} = 0$ (as $\mathbf{N} \perp \mathbf{F}$), so the sought arm l of the force $\mathbf{F}$ is given by

$$l = \frac{N}{F} = \frac{aA - bB}{\sqrt{A^2 + B^2}}$$

1.237 For coplanar forces, about any point in the same plane,

$$\sum \mathbf{r}_i \times \mathbf{F}_i = \mathbf{r} \times \mathbf{F}_{\text{net}} \quad (\text{where } \mathbf{F}_{\text{net}} = \sum \mathbf{F}_i = \text{resultant force})$$

or $$\mathbf{N}_{\text{net}} = \mathbf{r} \times \mathbf{F}_{\text{net}}$$

Thus length of the arm, $$l = \frac{N_{\text{net}}}{F_{\text{net}}}$$

Here obviously $|\mathbf{F}_{\text{net}}| = 2F$ and it is directed toward right along AC. Take the origin at C. Then about C,

$$\mathbf{N} = \left(\sqrt{2}aF + \frac{a}{\sqrt{2}}F - \sqrt{2}aF\right), \text{ directed normally into the plane of figure}$$

(Here a = side of the square.)

Thus, $\mathbf{N} = F\dfrac{a}{\sqrt{2}}$, directed into the plane of the figure.

Hence, $$l = \frac{F(a/\sqrt{2})}{2F} = \frac{a}{2\sqrt{2}} = \frac{a}{2}\sin 45°$$

Thus the point of application of force is at the mid point of the side BC.

1.238 (a) Consider a strip of length dx at a perpendicular distance x from the axis about which we have to find the moment of inertia of the rod. The elemental mass of the rod equals

$$dm = \frac{m}{l}dx$$

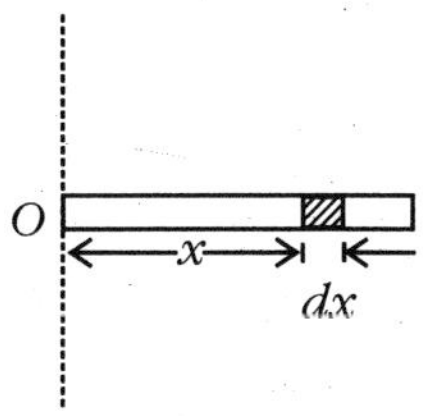

Moment of inertia of this element about the axis

$$dI = dmx^2 = \frac{m}{l}dx \cdot x^2$$

Thus, moment of inertia of the rod, as a whole about the given axis

$$I = \int_0^l \frac{m}{l}x^2\,dx = \frac{ml^2}{3}$$

(b) Let us imagine the plane of plate as x–y plane taking the origin at the intersection point of the sides of the plate (see figure).

Obviously, $$I_x = \int dm\ y^2$$

$$= \int_0^a \left(\frac{m}{ab}b\ dy\right)y^2$$

$$= \frac{ma^2}{3}$$

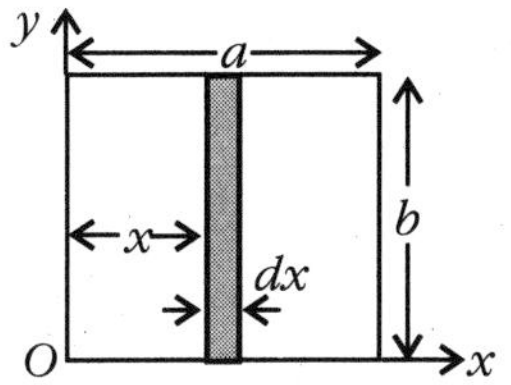

Similarly, $$I_y = \frac{mb^2}{3}$$

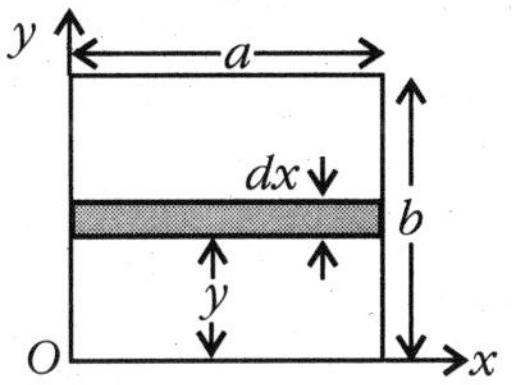

Hence from perpendicular axis theorem

$$I_z = I_x + I_y = \frac{m}{3}(a^2 + b^2)$$

which is the sought moment of inertia.

1.239 (a) Consider an elementary disk of thickness dz. Moment of inertia of this element about the z–axis, passing through its C.M. is

$$dI_z = \frac{(dm)R^2}{2} = \rho\ Sdz\frac{R^2}{2}$$

where ρ = density of the material of the plate and S = area of cross section of the plate. Thus, the sought moment of inertia

$$I_z = \frac{\rho SR^2}{2}\int_0^b dz = \frac{R^2}{2}\rho Sb$$

$$= \frac{\pi}{2}\rho bR^4 \quad (\text{as } S = \pi R^2)$$

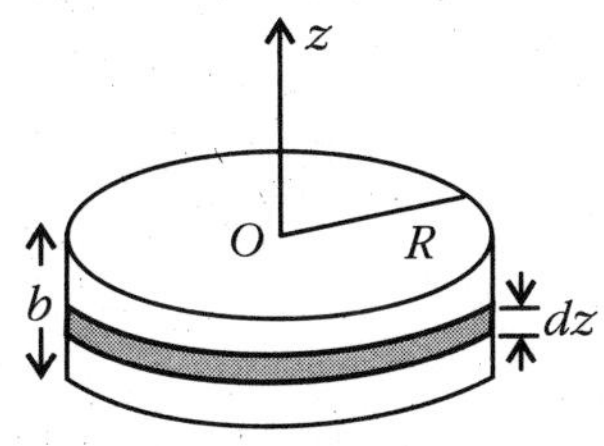

Putting all the values we get $I_z = 2.8$ g/m^2.

(b) Consider an element disk of radius r and thickness dz at a distance z from the point O. Then $r = z\tan\alpha$ and volume of the disk

$$= \pi z^2 \tan^2\alpha\ dz$$

Hence, its mass $$dm = \pi z^2 \tan\alpha \; dz \cdot \rho$$

where ρ = density of the cone $= m\Big/\frac{1}{3}\pi R^2 h.$

Moment of inertia of this element, about the axis OA,

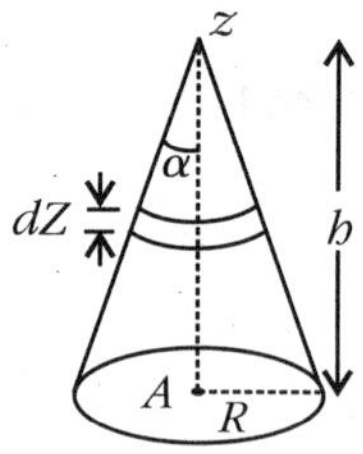

$$dI = dm\frac{r^2}{2}$$

$$= (\pi z^2 \tan^2 \alpha dz)\frac{z^2 \tan^2 \alpha}{2}$$

$$= \frac{\pi\rho}{2} z^4 \tan^4 \alpha dz$$

Thus the sought moment of inertia

$$I = \frac{\pi\rho}{2}\tan^4 \alpha \int_0^h z^4\, dz$$

$$= \frac{\pi\rho R^4 \cdot h^5}{10 h^4} \quad \left(\text{as} \tan\alpha = \frac{R}{h}\right)$$

Hence, $$I = \frac{3mR^2}{10} \quad \left(\text{putting}\; \rho = \frac{3m}{\pi R^2 h}\right)$$

1.240 (a) Let us consider a lamina of an arbitrary shape and indicate by 1, 2 and 3, three axes coinciding with x, y and z-axes and the plane of lamina as x–y plane.

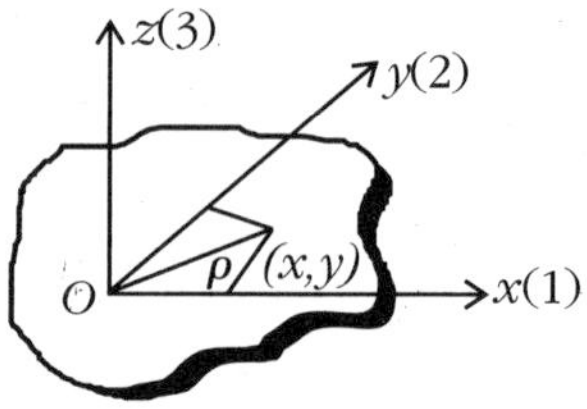

Now, moment of inertia of a point mass about x-axis,

$$dI_x = dmy^2$$

Thus moment of inertia of the lamina about this axis,

$$I_x = \int dmy^2$$

Similarly, $$I_y = \int dmx^2$$

and $$I_z = \int dm\rho^2$$

$$= \int dm(x^2 + y^2) \quad \text{as}\; \rho = \sqrt{x^2 + y^2}$$

Thus, $$I_z = I_x + I_y \quad \text{or} \quad I_3 = I_1 + I_2$$

(b) Let us take the plane of the disk as x–y plane and origin to the center of the disk (see figure). From the symmetry $I_x = I_y$. Let us consider a ring element of radius r and thickness dr, then the moment of inertia of the ring element about the z-axis is given by

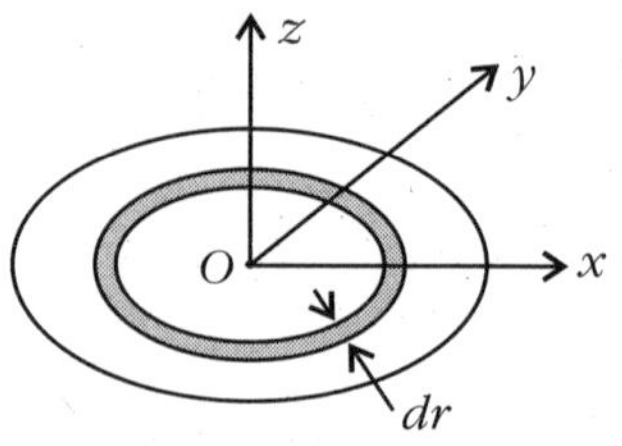

$$dI_z = dmr^2 = \frac{m}{\pi R^2}(2\pi r\, dr)r^2$$

Thus the moment of inertia of the disk about z-axis

$$I_z = \frac{2m}{R^2}\int_0^R r^3\, dr = \frac{mR^2}{2}$$

But we have $$I_z = I_x + I_y = 2I_x$$

Thus, $$I_x = \frac{I_z}{2} = \frac{mR^2}{4}$$

1.241 For simplicity let us use a mathematical trick. We consider the portion of the given disk as the superposition of two complete disks (without holes), one of positive density and radius R and other of negative density but of same magnitude and radius $R/2$.

As (area) $\propto$ (mass), the respective masses of the considered disks are $(4m/3)$ and $-m/3$ respectively, and these masses can be imagined to be situated at their respective centers (C.M.). Let us take point O as origin and point x-axis towards right. Obviously, the C.M. of the shaded position of given shape lies on the x-axis. Hence the C.M. (C) of the shaded portion is given by

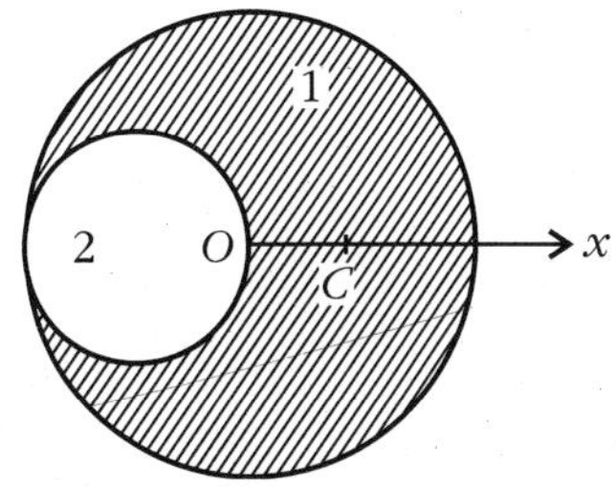

$$x_C = \frac{(-m/3)(-R/2) + (4m/3)0}{\left(-\frac{m}{3}\right) + \frac{4m}{3}} = \frac{R}{6}$$

Thus C.M. of the shape is at a distance $R/6$ from point O toward x-axis.

Using parallel axis theorem and bearing in mind that the moment of inertia of a complete homogenous disk of mass m_0 and radius r_0 equals $1/2m_0r_0^2$; the moment of inertia of the small disk of mass $(-m/3)$ and radius $R/2$ about the axis passing through point C and perpendicular to the plane of the disk is given by

$$I_{2C} = \frac{1}{2}\left(-\frac{m}{3}\right)\left(\frac{R}{2}\right)^2 + \left(-\frac{m}{3}\right)\left(\frac{R}{2} + \frac{R}{6}\right)^2$$
$$= -\frac{mR^2}{24} - \frac{4}{27}mR^2$$

Similarly, $$I_{1C} = \frac{1}{2}\left(\frac{4m}{3}\right)R^2 + \left(\frac{4m}{3}\right)\left(\frac{R}{6}\right)^2$$
$$= \frac{2}{3}mR^2 + \frac{mR^2}{27}$$

Thus, the sought moment of inertia,

$$I_C = I_{1C} + I_{2C} = \frac{37}{72}mR^2$$
$$= 0.15\ \text{kg}\cdot\text{m}^2$$

1.242 Moment of inertia of the shaded portion, about the axis passing through its center,

$$I = \frac{2}{5}\left(\frac{4}{3}\pi R^3\rho\right)R^2 - \frac{2}{5}\left(\frac{4}{3}\pi r^3\rho\right)r^2$$
$$= \frac{2}{5}\cdot\frac{4}{3}\pi\rho(R^5 - r^5)$$

Now, if $R = r + dr$, the shaded portion becomes a shell, which is the required shape to calculate the moment of inertia.

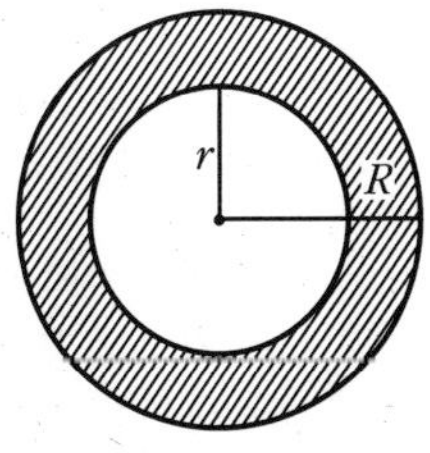

Now, $$I = \frac{2}{5}\cdot\frac{4}{3}\pi\rho\{(r + dr)^5 - r^5\}$$
$$= \frac{2}{5}\cdot\frac{4}{3}\pi\rho(r^5 + 5r^4\,dr + \cdots - r^5)$$

Neglecting higher terms, we get

$$= \frac{2}{3}(4\pi r^2\, dr\rho)r^2 = \frac{2}{3}mr^2$$

1.243 (a) Net force which is effective on the system (cylinder M + body m) is the weight of the body m in a uniform gravitational field, which is a constant. Thus, the initial acceleration of the body m is also constant.

From the conservation of mechanical energy of the said system in the uniform field of gravity at time $t = \Delta t$: $\Delta T + \Delta U_{gr} = 0$

or $$\frac{1}{2}mv^2 + \frac{1}{2}\frac{MR^2}{2}\omega^2 - mg\,\Delta h = 0$$

or $$\frac{1}{4}(2m+M)v^2 - mg\,\Delta h = 0 \quad (\text{as } v = \omega R \text{ at all times}) \tag{1}$$

But $$v^2 = 2w\Delta h$$

Hence using it in Eq. (1), we get

$$\frac{1}{4}(2m+M)2w\Delta h - mg\Delta h = 0 \quad \text{or} \quad w = \frac{2mg}{(2m+M)}$$

From the kinematic relationship, $\beta = \dfrac{w}{R} = \dfrac{2mg}{(2m+M)R}$

Thus the sought angular velocity of the cylinder

$$\omega(t) = \beta t = \frac{2mg}{(2m+M)R}t = \frac{gt}{(1+M/2m)R}$$

(b) Sought kinetic energy

$$T(t) = \frac{1}{2}mv^2 + \frac{1}{2}\frac{ml^2}{2}\omega^2$$

$$= \frac{1}{4}(2m+M)R^2\omega^2$$

$$= \frac{1}{2}mg^2\left(1+\frac{M}{2m}\right) \quad (\text{on substituting value of } \omega)$$

1.244 For equilibrium of the disk and axle

$$2T = mg \quad \text{or} \quad T = mg/2$$

As the disk unwinds, it has an angular acceleration β given by

$$2Tr = I\beta \quad \text{or} \quad \beta = \frac{2Tr}{I} = \frac{mgr}{I}$$

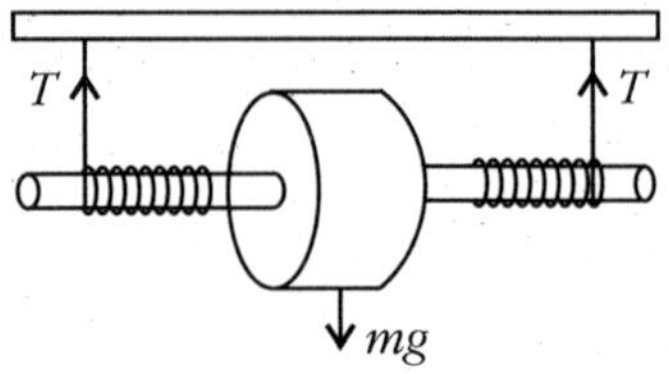

The corresponding linear acceleration is

$$r\beta = w = \frac{mgr^2}{I}$$

Since the disk remains stationary under the combined action of this acceleration and the acceleration ($-w$) of the bar which is transmitted to the axle, we must have $w = mgr^2/I$.

1.245 Let the rod be deviated through an angle φ from its initial position at an arbitrary instant of time, measured relative to the initial position in the positive direction. From the equation of increment of mechanical energy of the system

$$\Delta T = A_{ext}$$

or
$$\frac{1}{2} I\omega^2 = \int N_z\, d\varphi$$

or
$$\frac{1}{2}\frac{Ml^2}{3}\omega^2 = \int_0^{\varphi} Fl\cos\varphi\, d\varphi = Fl\sin\varphi$$

Thus,
$$\omega = \sqrt{\frac{6F\sin\varphi}{Ml}}$$

1.246 First of all, let us sketch a free diagram of each body. Since the cylinder is rotating and is massive, the tension will be different in both the sections of threads. From Newton's law in projection form for the bodies m_1 and m_2 and noting that $w_1 = w_2 = w = \beta R$ (as the thread is not slipping), we have ($m_1 > m_2$):

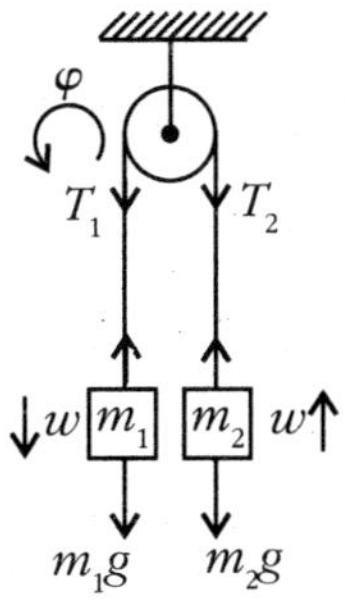

$$m_1 g - T_1 = m_1 w = m_1\beta R$$

and
$$T_2 - m_2 g = m_2 w \quad (1)$$

Now from the equation of rotational dynamics of a solid about stationary axis of rotation, i.e.,

$$N_z = I\beta_z \quad \text{(for the cylinder)}$$

or
$$(T_1 - T_2)R = I\beta = \frac{mR^2\beta}{2} \quad (2)$$

Simultaneous solution of the above equations yields:

$$\beta = \frac{(m_1 - m_2)g}{R(m_1 + m_2 + m/2)} \quad \text{and} \quad \frac{T_1}{T_2} = \frac{m_1(m + 4m_2)}{m_2(m + 4m_1)}$$

1.247 As the system ($m + m_1 + m_2$) is under constant forces, the acceleration of body m_1 and m_2 is constant. In addition to it the velocities and acceleration of bodies m_1 and m_2 are equal in magnitude (say v and w) because the length of the thread is constant.

From the equation of increment of mechanical energy, i.e., $\Delta T + \Delta U = A_{fr}$, at time t when block m_1 is distance h below from initial position corresponding to $t = 0$,

$$\frac{1}{2}(m_1 + m_2)v^2 + \frac{1}{2}\left(\frac{mR^2}{2}\right)\frac{v^2}{R^2} - m_2 gh = -km_1 gh \quad (1)$$

(as angular velocity $\omega = v/R$ for no slipping of thread).

But
$$v^2 = 2wh$$

So using it in Eq. (1), we get

$$w = \frac{2(m_2 - km_1)g}{m + 2(m_1 + m_2)} \quad (2)$$

Thus the work done by the friction force on m_1

$$A_{fr} = -km_1 gh = -km_1 g\left(\frac{1}{2}wt^2\right)$$
$$= -\frac{km_1(m_2 - km_1)g^2t^2}{m + 2(m_1 + m_2)} \quad \text{(using Eq. (2))}$$

1.248 In the problem, the rigid body is in translation equilibrium but there is an angular retardation. We first sketch the free body diagram of the cylinder. Obviously the friction forces, acting on the cylinder, are kinetic. From the condition of translational equilibrium for the cylinder,

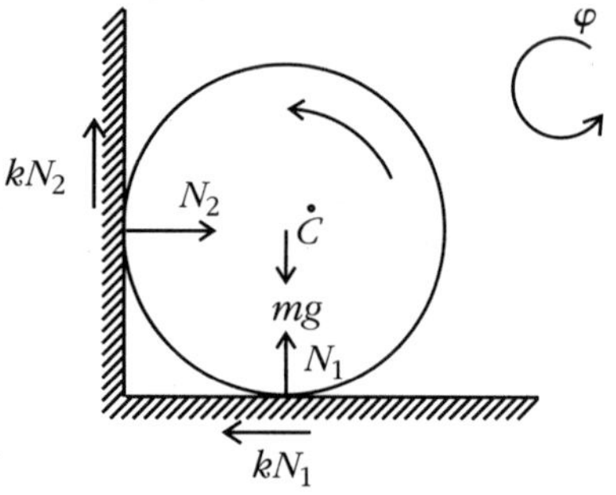

$$mg = N_1 + kN_2; \quad N_2 = kN_1$$

Hence,
$$N_1 = \frac{mg}{1+k^2}; \quad N_2 = k\frac{mg}{1+k^2}$$

For pure rotation of the cylinder about its rotation axis, $N_z = I\beta_z$

or
$$-kN_1R - kN_2R = \frac{mR^2}{2}\beta_z$$

or
$$-\frac{kmgR(1+k)}{1+k^2} = \frac{mR^2}{2}\beta_z$$

or
$$\beta_z = -\frac{2k(1+k)g}{(1+k^2)R}$$

Now, from the kinematical equation,

$$\omega^2 = \omega_0^2 + 2\beta_z\Delta\varphi$$

we have, $\Delta\varphi = \dfrac{\omega_0^2(1+k^2)R}{4k(1+k)g}$ (because $\omega = 0$)

Hence, the sought number of turns,

$$n = \frac{\Delta\varphi}{2\pi} = \frac{\omega_0^2(1+k^2)R}{8\pi k(1+k)g}$$

1.249 It is the moment of friction force which brings the disk to rest. The force of friction is applied to each section of the disk, and since these sections lie at different distances from the axis, the moments of the forces of friction differ from section to section.

To find N_z, where z is the axis of rotation of the disk let us partition the disk into thin rings (see figure). The force of friction acting on the considered element $dfr = k(2\pi r dr\sigma)g$, (where σ is the density of the disk).

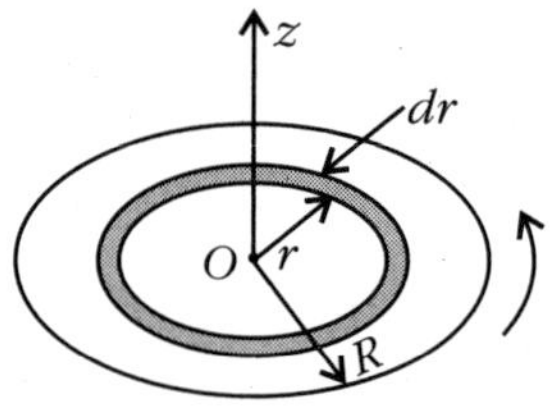

The moment of this force of friction is

$$dN_z = -rdfr = -2\pi k\sigma g r^2\, dr$$

Integrating with respect to r from zero to R, we get

$$N_z = -2\pi k\sigma g\int_0^R r^2\, dr = -\frac{2}{3}\pi k\sigma g R^3$$

For the rotation of the disk about the stationary axis z, from the equation $N_z = I\beta_z$

$$-\frac{2}{3}\pi k\sigma g R^3 = \frac{(\pi R^2\sigma)R^2}{2}\beta_z \quad \text{or} \quad \beta_z = -\frac{4kg}{3R}$$

Thus, from the angular kinematical equation $\omega_z = \omega_{0z} + \beta_z t$

$$0 = \omega_0 + \left(-\frac{4kg}{3R}\right)t \quad \text{or} \quad t = \frac{3R\omega_0}{4kg}$$

1.250 According to the question,

$$I\frac{d\omega}{dt} = -k\sqrt{\omega} \quad \text{or} \quad I\frac{d\omega}{\sqrt{\omega}} = -k\, dt$$

Integrating, $\sqrt{\omega} = -\frac{kt}{2I} + \sqrt{\omega_0}$ (1)

or $$\omega = \frac{k^2t^2}{4I^2} - \frac{\sqrt{\omega_0}kt}{I} + \omega_0 \text{ (noting that at } t = 0, \omega = \omega_0)$$

Let the flywheel stop at $t = t_0$, then from Eq. (1), $t_0 = 2I\sqrt{\omega_0}/k$

Hence sought average angular velocity

$$<\omega> = \frac{\int_0^{2I\sqrt{\omega_0}/k} (k^2t^2/4I^2 - \sqrt{\omega_0}kt/I + \omega_0)\, dt}{\int_0^{2I\sqrt{\omega_0}/k} dt} = \frac{\omega_0}{3}$$

1.251 Let us use the equation $dM_z/dt = N_z$, relative to the axis through O. (1)

For this purpose, let us find the angular momentum of the system M_z about the given rotation axis and the corresponding torque N_z. The angular momentum is

$$M_z = I\omega + mvR = \left(\frac{M}{2} + m\right) R^2\omega$$

where $I = \frac{M}{2}R^2$ and $v = \omega R$ (no cord slipping).

So, $$\frac{dM_z}{dt} = \left(\frac{MR^2}{2} + mR^2\right)\beta_z \quad (2)$$

The downward pull of gravity on the overhanging part is the only external force, which exerts a torque about the z-axis, passing through O and is given by,

$$N_z = \left(\frac{m}{l}\right)xgR$$

Hence, from the equation $\frac{dM_z}{dt} = N_z$

$$\left(\frac{MR^2}{2} + mR^2\right)\beta_z = \frac{m}{l}xgR$$

Thus, $$\beta_z = \frac{2mgx}{lR(M + 2m)} > 0$$

Note: *We may solve this problem using conservation of mechanical energy of the system (cylinder + thread) in the uniform field of gravity.*

1.252 (a) Let us indicate the forces acting on the sphere and their points of application. Choose positive direction of x and φ (rotation angle) along the incline in downward direction and in the sense of ω (for unidirectional rotation) respectively. Now from equations of dynamics of rigid body, i.e., $F_x = mw_{Cx}$ and $N_{Cz} = I_C\,\beta_z$ we get

$$mg\sin\alpha - fr = mw \quad (1)$$

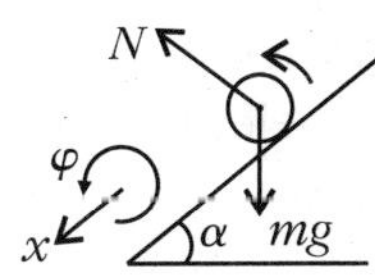

and $$frR = \frac{2}{5}mR^2\beta \quad (2)$$

But, $$fr \le kmg\cos\alpha \tag{3}$$

In addition, the absence of slipping provides the kinematical relationship between the accelerations as,

$$w = \beta R \tag{4}$$

The simultaneous solution of all the four equations yields

$$k\cos\alpha \ge \frac{2}{7}\sin\alpha \quad \text{or} \quad k \ge \frac{2}{7}\tan\alpha$$

(b) Solving Eqs. (1) and (2) [of part (a)], we get

$$w_C = \frac{5}{7} g\sin\alpha$$

As the sphere starts at $t = 0$ along positive x-axis, for pure rolling

$$v_C(t) = w_C t = \frac{5}{7} g\sin\alpha\, t \tag{5}$$

Hence, the sought kinetic energy is

$$T = \frac{1}{2}mv_C^2 + \frac{1}{2}\frac{2}{5}mR^2w^2 = \frac{7}{10}mv_C^2 \quad (\text{as } \omega = v_C/R)$$

$$= \frac{7}{10}m\left(\frac{5}{7}g\sin\alpha\, t\right)^2 = \frac{5}{14}mg^2\sin^2\alpha\, t^2$$

1.253 (a) Let us indicate the forces and their points of application for the cylinder. Choosing the positive direction for x and φ as shown in the figure, we write the equation of motion of the cylinder axis and the equation of moments in the C.M. frame relative to that axis, i.e., from equation $F_x = mw_C$ and $N_z = I_C\beta_z$ we get

$$mg - 2T = mw_C \quad \text{and} \quad 2TR = \frac{mR^2}{2}\beta$$

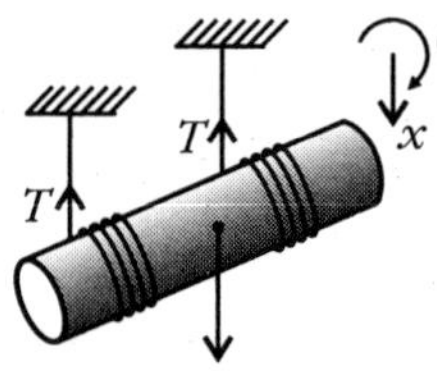

As there is no slipping of thread on the cylinder

$$w_C = \beta R$$

From these three equations

$$T = \frac{mg}{6} = 13\text{ N},\ \beta = \frac{2}{3}\frac{g}{R}$$

$$= 5\times 10^2\text{ rad/s}^2$$

(b) We have $$\beta = \frac{2}{3}\frac{g}{R}$$

So, $$w_C = \frac{2}{3}g > 0 \quad \text{or} \quad \text{in vector form } \mathbf{w}_C = \frac{2}{3}\mathbf{g}$$

$$P = \mathbf{F}\cdot\mathbf{v} = \mathbf{F}\cdot(\mathbf{w}_C t)$$

$$= m\mathbf{g}\cdot\left(\frac{2}{3}\mathbf{g}t\right) = \frac{2}{3}mg^2t$$

1.254 Let us depict the forces and their points of application corresponding to the cylinder attached with the elevator. Newton's second law for solid in vector form in the frame of elevator, gives

$$2\mathbf{T} + m\mathbf{g} + m(-\mathbf{w}_0) = m\mathbf{w}' \tag{1}$$

The equation of moment in the C.M. frame relative to the cylinder axis, i.e., from $N_z = I_C\beta_z$, is

$$2TR = \frac{mR^2}{2}\beta = \frac{mR^2}{2}\frac{w'}{R}$$

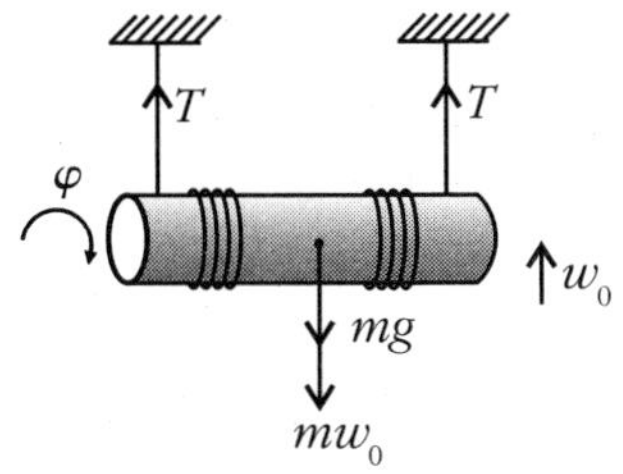

(as thread does not slip on the cylinder, $w' = \beta R$)

or $$T = \frac{mw'}{4}$$

From Eq. (1), $\mathbf{T}\uparrow\downarrow\mathbf{w}$, so in vector form,

$$\mathbf{T} = -\frac{m\mathbf{w}}{4} \quad (2)$$

Solving Eqs. (1) and (2), we get

$$\mathbf{w} = \frac{2}{3}(\mathbf{g} - \mathbf{w}_0)$$

and sought force $$\mathbf{F} = 2\mathbf{T} = \frac{1}{3}m(\mathbf{g} - \mathbf{w}_0)$$

1.255 Let us depict the forces and their points of application for the spool. Choosing the positive direction for x and φ as shown in the figure, we apply $F_x = mw_x$ and $N_{Cz} = I_C\beta_z$ and get $mg\sin\alpha - T = mw$, $Tr = I\beta$.

"Notice that if a point of a solid in plane motion is connected with a thread, the projection of velocity vector of the solid's point of contact along the length of the thread equals the velocity of the other end of the thread (if it is not slacked)".

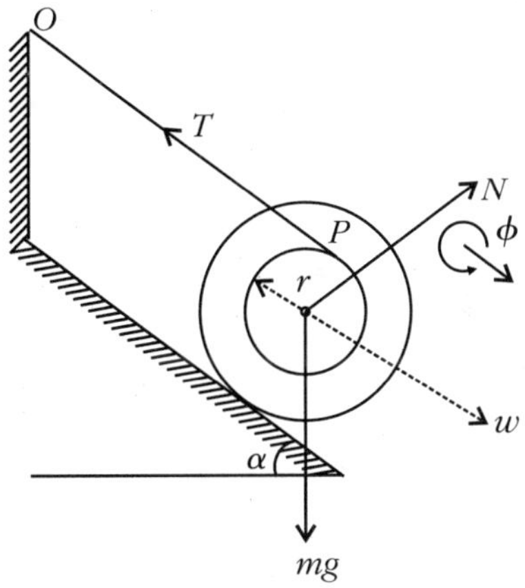

Thus in our problem, $v_P = v_0$ but $v_0 = 0$, hence point P is the instantaneous center of rotation of zero velocity for the spool. Therefore $v_C = \omega r$ and subsequently $w_C = \beta r$.

Solving the equations simultaneously, we get

$$w = \frac{g\sin\alpha}{1 + I/mr^2} = 1.6\,\text{m/s}^2$$

1.256 Let us sketch the force diagram for solid cylinder and apply Newton's second law in projection form along x- and y-axes (see figure). Then,

$$fr_1 + fr_2 = mw_C \quad (1)$$

and $$N_1 + N_2 - mg - F = 0$$

or $$N_1 + N_2 = mg + F \quad (2)$$

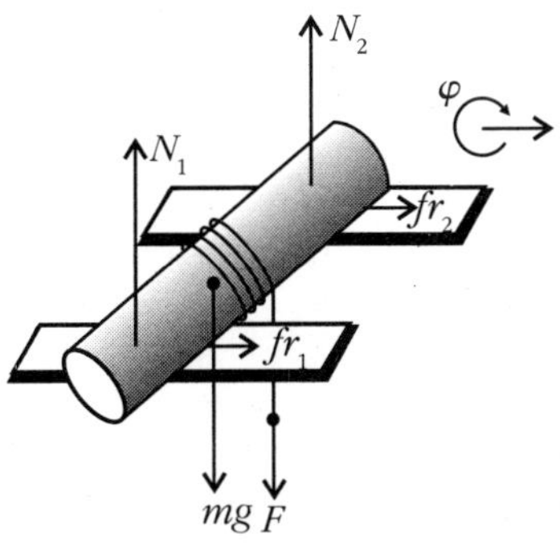

Now choosing positive direction of φ as shown in the figure and using $N_{Cz} = I_C\beta_z$, we get

$$FR - (fr_1 + fr_2)R = \frac{mR^2}{2}\beta = \frac{mR^2}{2}\frac{w_C}{R} \quad (3)$$

(as for pure rolling $w_C = \beta R$).

In addition,

$$fr_1 + fr_2 \le k(N_1 + N_2) \quad (4)$$

Solving the equations, we get

$$F \le \frac{3kmg}{2-3k}$$

or

$$F_{max} = \frac{3kmg}{2-3k}$$

and

$$w_{C(max)} = \frac{k(N_1 + N_2)}{m}$$

$$= \frac{k}{m}[mg + F_{max}] = \frac{k}{m}\left[mg + \frac{3kmg}{2-3k}\right] = \frac{2kg}{2-3k}$$

1.257 (a) Let us choose the positive direction of the rotation angle φ, such that w_{Cx} and β_z have identical signs (see figure). Equation of motion, $F_x = mw_{Cx}$ and $N_{Cz} = I_C\beta_z$ gives

$$F\cos\alpha - fr = mw_{Cx} \quad \text{and} \quad frR - Fr = I_C\beta_z = \gamma mR^2\beta_z$$

In the absence of the slipping of the spool, $w_{Cx} = \beta_z R$.

From the three equations we get

$$w_{Cx} = w_C = \frac{F[\cos\alpha - (r/R)]}{m(1+\gamma)}$$

(where $\cos\alpha > r/R$).

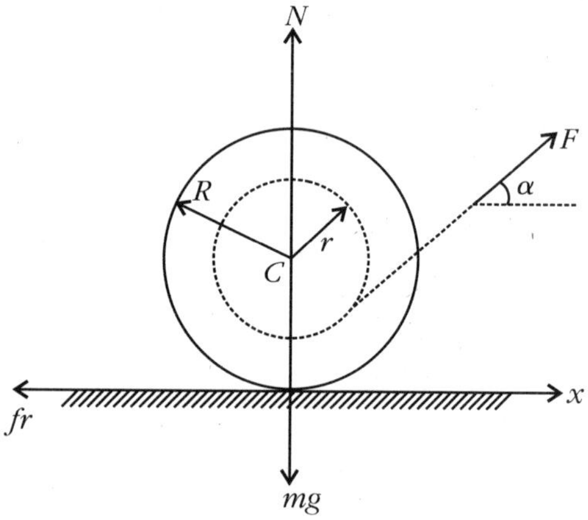

(b) As static friction (fr) does not work on the spool, from the equation of the increment of mechanical energy $A_{ext} = \Delta T$.

$$A_{ext} = \frac{1}{2}mv_C^2 + \frac{1}{2}\gamma mR^2\frac{v_C^2}{R^2} = \frac{1}{2}m(1+\gamma)v_C^2$$

$$= \frac{1}{2}m(1+\gamma)2w_C x = \frac{1}{2}m(1+\gamma)2w_C\left(\frac{1}{2}w_C t^2\right)$$

$$= \frac{F^2(\cos\alpha - r/R)^2 t^2}{2m(1+\gamma)}$$

Note: *At* $\cos\alpha = r/R$, *there is no rolling and for* $\cos\alpha < r/R$, $w_{cx} < 0$, *i.e., the spool will move towards negative x-axis and rotate in anticlockwise sense.*

1.258 For the cylinder, from equation $N_z = I\beta_z$ about its stationary axis of rotation,

$$2Tr = \frac{mr^2}{2}\beta \quad \text{or} \quad \beta = \frac{4T}{mr} \tag{1}$$

For the rotation of the lower cylinder, from equation $N_{Cz} = I_C\beta_z$,

$$2Tr = \frac{mr^2}{2}\beta'$$

or

$$\beta' = \frac{4T}{mr} = \beta$$

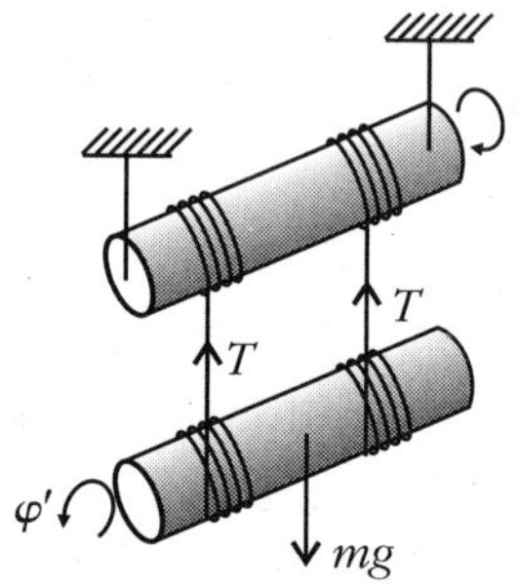

Now for the translational motion of lower cylinder, from equation $F_x = mw_{Cx}$,

$$mg - 2T = mw_C \tag{2}$$

As there is no slipping of threads on the cylinders,

$$w_C = \beta' r + \beta r = 2\beta r \quad (3)$$

Simultaneous solution of Eqs. (1), (2) and (3) yields

$$T = \frac{mg}{10}$$

1.259 Let us depict the forces acting on the pulley and weight A, and indicate positive direction for x and φ as shown in the figure. For the cylinder from the equation, $F_x = mw_x$ and $N_{Cz} = I_C\beta_z$, we get

$$Mg + T_A - 2T = Mw_C \quad (1)$$

and $$2TR + T_A(2R) = I\beta = \frac{Iw_C}{R} \quad (2)$$

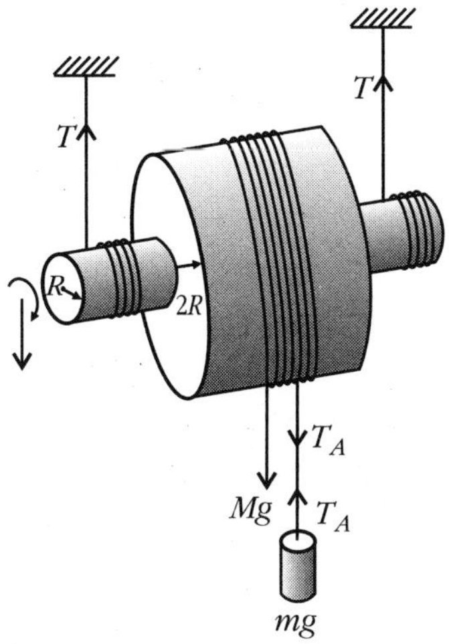

For the weight A from the equation

$$F_x = mw_x$$

We get, $$mg - T_A = mw_A \quad (3)$$

As there is no slipping of the threads on the pulleys,

$$w_A = w_C + 2\beta R = w_C + 2w_C = 3w_C \quad (4)$$

Simultaneous solution of above four equations gives

$$w_A = \frac{3(M + 3m)g}{M + 9m + I/R^2}$$

1.260 (a) For the translational motion of the system $(m_1 + m_2)$, from the equation: $F_x = mw_{Cx}$, we get

$$F = (m_1 + m_2)w_C \quad \text{or} \quad w_C = F/(m_1 + m_2) \quad (1)$$

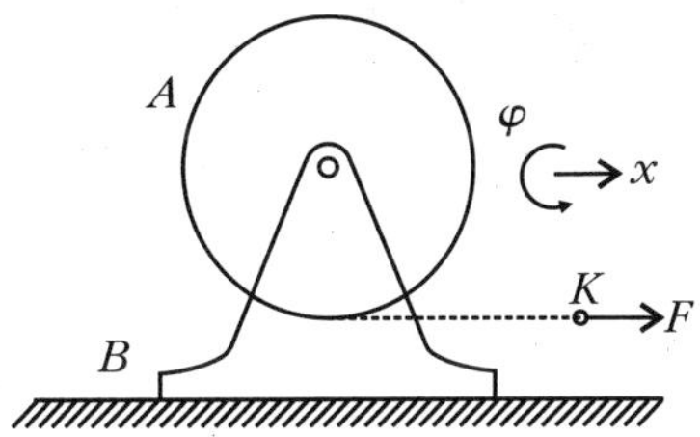

Now for the rotational motion of cylinder from the equation: $N_{Cz} = I_C\beta_z$, we get

$$Fr = \frac{m_1 r^2}{2}\beta \quad \text{or} \quad \beta r = \frac{2F}{m_1} \quad (2)$$

But, $$w_K = w_C + \beta r$$

So, $$w_K = \frac{F}{m_1 + m_2} + \frac{2F}{m_1} = \frac{F(3m_1 + 2m_2)}{m_1(m_1 + m_2)} \quad (3)$$

(b) From the equation of increment of mechanical energy: $\Delta T = A_{\text{ext}}$

Here, $\Delta T = T(t)$ so, $T(t) = A_{\text{ext}}$

As force F is constant and is directed along x-axis the sought work done is given by

$$A_{\text{ext}} = Fx$$

(where x is the displacement of the point of application of the force F during time interval t) or

$$F\left(\frac{1}{2}w_K t^2\right) = \frac{F^2 t^2(3m_1 + 2m_2)}{2m_1(m_1 + m_2)} = T(t) \quad \text{(using Eq. (3))}$$

Alternate:

$$T(t) = T_{\text{translation}}(t) + T_{\text{rotation}}(t)$$

$$= \frac{1}{2}(m_1 + m_2)\left(\frac{Ft}{(m_1 + m_2)}\right)^2 + \frac{1}{2}\frac{m_1 r^2}{2}\left(\frac{2Ft}{m_1 r}\right)^2 = \frac{F^2 t^2 (3m_1 + 2m_2)}{2m_1(m_1 + m_2)}$$

1.261 Choosing the positive direction for x and φ as shown in figure, let us write the equation of motion for the sphere $F_x = mw_{Cx}$ and $N_{Cz} = I_C \beta_z$

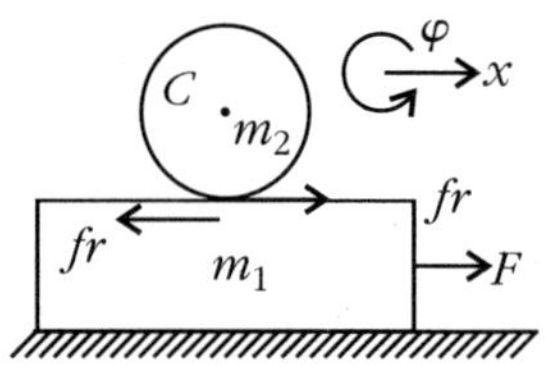

$$fr = m_2 w_2;$$

$$(fr)r = \frac{2}{5} m_2 r^2 \beta$$

(where w_2 is the acceleration of the C.M. of sphere).

For the plank, from equation $F_x = mw_x$, we have

$$F - fr = m_1 w_1$$

In addition, the condition for the absence of slipping of the sphere yields the kinematic relation between the accelerations

$$w_1 = w_2 + \beta r$$

Simultaneous solution of the four equations yields

$$w_1 = \frac{F}{m_1 + 2/7 m_2} \quad \text{and} \quad w_2 = \frac{2}{7} w_1$$

1.262 (a) Let us depict the forces acting on the cylinder and their point of applications for the cylinder and indicate positive direction of x and φ as shown in the figure. From the equations for the plane motion of a solid $F_x = mw_{Cx}$ and $N_{Cz} = I_C \beta_z$,

$$kmg = mw_{Cx} \quad \text{or} \quad w_{Cx} = kg \tag{1}$$

$$-kmg\,R = \frac{mR^2}{2}\beta_z \quad \text{or} \quad \beta_z = -2\frac{kg}{R} \tag{2}$$

Let the cylinder start pure rolling at $t = t_0$ after releasing on the horizontal floor at $t = 0$.

From the angular kinematical equation,

$$\omega_z = \omega_{oz} + \beta_z t,$$

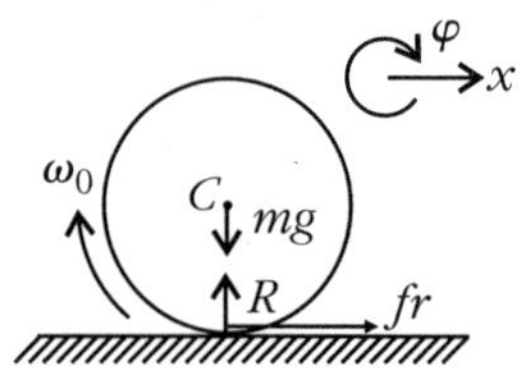

or $$\omega = \omega_0 - 2\frac{kg}{R}t \tag{3}$$

From the equation of the linear kinematics,

$$v_{Cx} = v_{oCx} + w_{Cx} t$$

or $$v_C = 0 + kgt_0 \tag{4}$$

But at the moment $t = t_0$, when pure rolling starts $v_C = \omega R$.

So, $$kgt_0 = \left(\omega_0 - 2\frac{kg}{R}t_0\right)R$$

Thus, $$t_0 = \frac{\omega_0 R}{3\,kg}$$

(b) As the cylinder picks up speed till it starts rolling, the point of contact has purely translatory movement equal to $1/2\ w_C t_0^2$ in the forward direction but there is also a backward movement of the point of contact of magnitude $(\omega_0 \tau_0 - 1/2\ \beta t_0^2)R$. Because of slipping the net displacement is backwards. The total work done is then,

$$A_{fr} = kmg\left[\frac{1}{2}w_C t_0^2 - \left(\omega_0 t_0 + \frac{1}{2}\beta t_0^2\right)R\right]$$

$$= kmg\left[\frac{1}{2}kgt_0^2 - \frac{1}{2}\left(-\frac{2kg}{R}\right)t_0^2 R - \omega_0 t_0 R\right]$$

$$= kmg\frac{\omega_0 R}{3kg}\left[\frac{\omega_0 R}{6} + \frac{\omega_0 R}{3} - \omega_0 R\right] = -\frac{m\omega_0^2 R^2}{6}$$

The same result can also be obtained by the work-energy theorem, $A_{fr} = \Delta T$.

1.263 Let us write the equation of motion for the center of the sphere at the moment of breaking-off

$$\frac{mv^2}{R+r} = mg\cos\theta$$

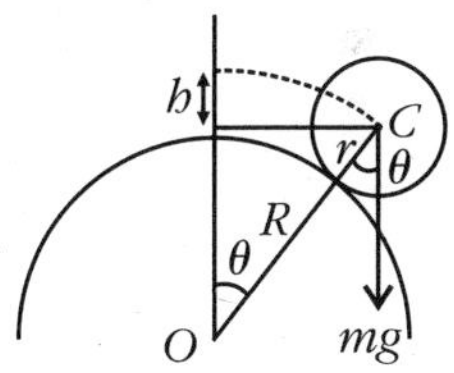

where v is the velocity of the center of the sphere at that moment, and θ is the corresponding angle (see figure). The velocity v can be found from the energy conservation law

$$mgh = \frac{1}{2}mv^2 + \frac{1}{2}I\omega^2$$

where I is the moment of inertia of the sphere relative to the axis passing through the sphere's center, i.e., $I = 2/5\ mr^2$. In addition,

$$v = \omega r \quad \text{and} \quad h = (R + r)(1 + \cos\theta)$$

From these four equations we obtain

$$\omega = \sqrt{10g(R+r)/17r^2}$$

1.264 Since the cylinder moves without sliding, the center of the cylinder rotates about the point O, while passing through the common edge of the planes. In other words, the point O becomes the foot of the instantaneous axis of rotation of the cylinder.

At any instant during this motion the velocity of the C.M. is v_1. When the angle (as shown in the figure) is β, we have

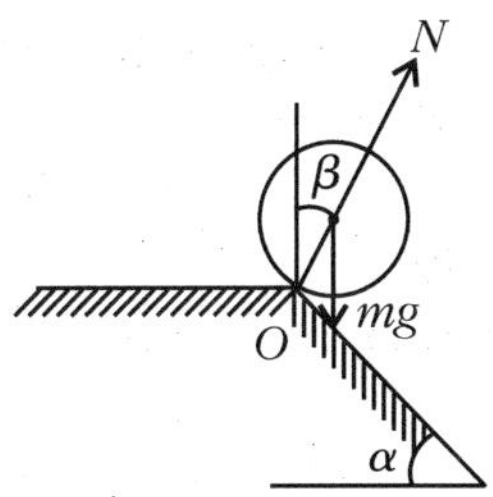

$$\frac{mv_1^2}{R} = mg\cos\beta - N$$

(where N is the normal reaction of the edge)

or $$v_1^2 = gR\cos\beta - \frac{NR}{m} \qquad (1)$$

From the energy conservation law,

$$\frac{1}{2}I_0\frac{v_1^2}{R^2} - \frac{1}{2}I_0\frac{v_0^2}{R^2} = mgR(1-\cos\beta)$$

But, $$I_0 = \frac{mR^2}{2} + mR^2 = \frac{3}{2}mR^2$$

(from the parallel axis theorem).

Thus, $$v_1^2 = v_0^2 + \frac{4}{3} gR(1 - \cos\beta)$$

From Eqs. (1) and (2)

$$v_0^2 = \frac{gR}{3}(7\cos\beta - 4) - \frac{NR}{m}$$

The angle β in this equation is clearly smaller than or equal to α, so putting $\beta = \alpha$, we get

$$v_0^2 - \frac{gR}{3}(7\cos\alpha - 4) - \frac{N_0 R}{M} \quad \text{(where } N_0 \text{ is the corresponding reaction.)}$$

Note that $N \geq N_0$. No jumping occurs during this turning if $N_0 > 0$. Hence, v_0 must be less than

$$v_{max} = \sqrt{\frac{gR}{3}(7\cos\alpha - 4)}$$
$$= 1.0 \text{ m/s (on substituting values)}$$

1.265 Clearly the tendency of bouncing of the hoop will be maximum when the small body A will be at the highest point of the hoop during its rolling motion. Let the velocity of C.M. of the hoop equals v at this position. The static friction does no work on the hoop, so from conservation of mechanical energy: $E_1 = E_2$

or $$0 + \frac{1}{2}mv_0^2 + \frac{1}{2}mR^2\left(\frac{v_0}{R}\right)^2 - mgR = \frac{1}{2}m(2v)^2 + \frac{1}{2}mv^2 + \frac{1}{2}mR^2\left(\frac{v}{R}\right)^2 + mgR$$

or $$3v^2 = v_0^2 - 2gR \tag{1}$$

From the equation $F_n = mw_n$ for body A at final position, as shown in Fig. (b):

$$mg + N' = m\omega^2 R = m\left(\frac{v}{R}\right)^2 R \tag{2}$$

As the hoop has no acceleration in vertical direction, so for the hoop,

$$N + N' = mg \tag{3}$$

From Eqs. (2) and (3),

$$N = 2mg - \frac{mv^2}{R} \tag{4}$$

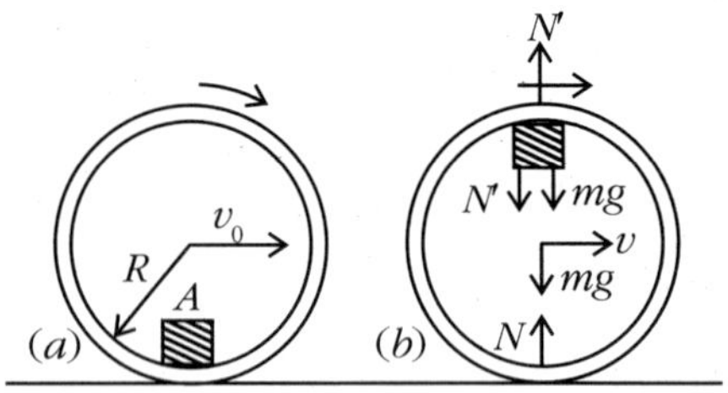

As the hoop does not bounce,

$$N \geq 0 \tag{5}$$

So, from Eqs. (1), (4) and (5),

$$\frac{8gR - v_0^2}{3R} \geq 0 \quad \text{or} \quad 8gR \geq v_0^2$$

Hence, $$v_0 \leq \sqrt{8gR}$$

1.266 Since the lower part of the belt is in contact with the rigid floor, velocity of this part becomes zero. The crawler moves with velocity v, hence, the velocity of upper part of the belt becomes $2v$ by the rolling condition and kinetic energy of upper part is given by

$$\frac{1}{2}\left(\frac{m}{2}\right)(2v)^2 = mv^2$$

which is also the sought kinetic energy, assuming that the length of the belt is much larger than the radius of the wheels.

1.267 The sphere has two types of motion, one is the rotation about its own axis and the other is motion in a circle of radius R. Hence the sought kinetic energy is,

$$T = \frac{1}{2} I_1 \omega_1^2 + \frac{1}{2} I_2 \omega_2^2 \quad (1)$$

where I_1 is the moment of inertia about its own axis, and I_2 is the moment of inertia about the vertical axis, passing through O.

But, $$I_1 = \frac{2}{5} mr^2 \text{ and } I_2 = \frac{2}{5} mr^2 + mR^2 \text{ (using parallel axis theorem)} \quad (2)$$

In addition

$$\omega_1 = \frac{v}{r} \quad \text{and} \quad \omega_2 = \frac{v}{R} \quad (3)$$

Using Eqs. (2) and (3) in Eq. (1), we get

$$T' = \frac{7}{10} mv^2 \left(1 + \frac{2r^2}{7R^2}\right)$$

1.268 For a point mass of mass dm, looked at from rotating frame C, the equation is

$$dm\,\mathbf{w}' = \mathbf{f} + dm\,\omega^2 \mathbf{r}' + 2dm(\mathbf{v}' \times \boldsymbol{\omega})$$

where $\mathbf{r}'$ = radius vector in the rotating frame with respect to rotation axis and $\mathbf{v}'$ = velocity in the same frame. The total centrifugal force is clearly

$$\mathbf{F}_{cf} = \sum dm\ \omega^2 \mathbf{r}' = m\omega^2 \mathbf{R}_C$$

where $\mathbf{R}_C$ is the radius vector of the C.M. of the body with respect to rotation axis. Also

$$\mathbf{F}_{cor} = 2m\mathbf{v}_C' \times \omega$$

where we have used the definitions,

$$m\mathbf{R}_C = \sum dm\,\mathbf{r}' \text{ and } m\mathbf{v}_C' = \sum dm\,\mathbf{v}'.$$

1.269 Consider a small element of length dx at a distance x from the point C, which is rotating in a circle of radius $r = x \sin\theta$.

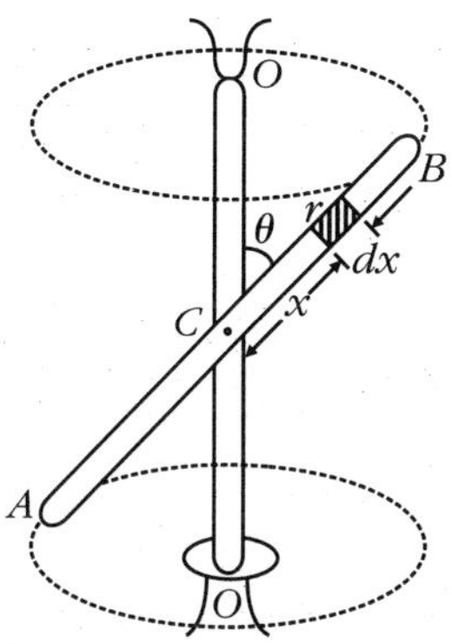

Now, mass of the element $= \left(\frac{m}{l}\right) dx$

So, centrifugal force acting on this element

$$= \left(\frac{m}{l}\right) dx\, \omega^2 x \sin\theta$$

and moment of this force about C is

$$|dN| = \left(\frac{m}{l}\right) dx\, \omega^2 x \sin\theta \cdot x\cos\theta$$

$$= \frac{m\omega^2}{2l} \sin 2\theta x^2\, dx$$

and hence, total moment

$$N = 2\int_0^{l/2} \frac{m\omega^2}{2l} \sin 2\theta x^2\, dx = \frac{1}{24} m\omega^2 l^2 \sin 2\theta$$

1.270 Let us consider the system in a frame rotating with the rod. In this frame, the rod is at rest and experiences not only the gravitational force mg and the reaction force $\mathbf{R}$, but also the centrifugal $\mathbf{F}_{cf}$. In the considered frame, from the condition of equilibrium, i.e., $N_{Oz} = 0$

or
$$N_{cf} = mg\frac{l}{2}\sin\theta \qquad (1)$$

where N_{cf} is the moment of centrifugal force about O. To calculate N_{cf}, let us consider an element of length dx, situated at a distance x from the point O. This element is subjected to a horizontal pseudo force $(m/l)\, dx\,\omega^2 x \sin\theta$. The moment of this pseudo force about the axis of rotation through the point O is

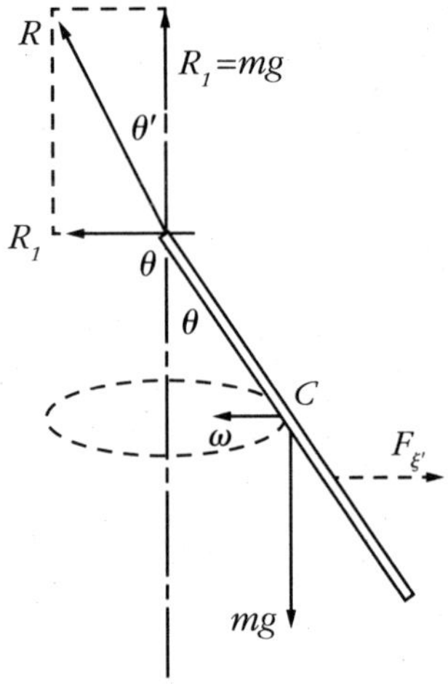

$$dN_{cf} = \left(\frac{m}{l}\right) dx\,\omega^2 x \sin\theta\, x\cos\theta$$

$$= \frac{m\omega^2}{l}\sin\theta\cos\theta\, x^2 dx$$

So,
$$N_{cf} = \int_0^l \frac{m\omega^2}{l}\sin\theta\cos\theta\, x^2\, dx \qquad (2)$$

$$= \frac{m\omega^2 l^2}{3}\sin\theta\cos\theta$$

It follows from Eqs. (1) and (2) that,

$$\cos\theta = \left(\frac{3g}{2\omega^2 l}\right) \quad \text{or} \quad \theta = \cos^{-1}\left(\frac{3g}{2\omega^2 l}\right) \qquad (3)$$

1.271 When the cube is given an initial velocity on the table in some direction (as shown in figure) it acquires an angular momentum about an axis on the table perpendicular to the initial velocity and (say) just below the center of gravity (C.G.). This angular momentum will disappear when the cube stops and this can only be due to a torque. Frictional forces cannot do this by themselves because they act in the plane containing the axis.

But if the forces of normal reaction act eccentrically (as shown in figure), their torque can bring about the vanishing of the angular momentum. We can calculate the distance Δx between the point of application of the normal reaction and the C.G. of the cube as follows. Take the moment about C.G. of all the forces. This must vanish because the cube does not turn on the table.

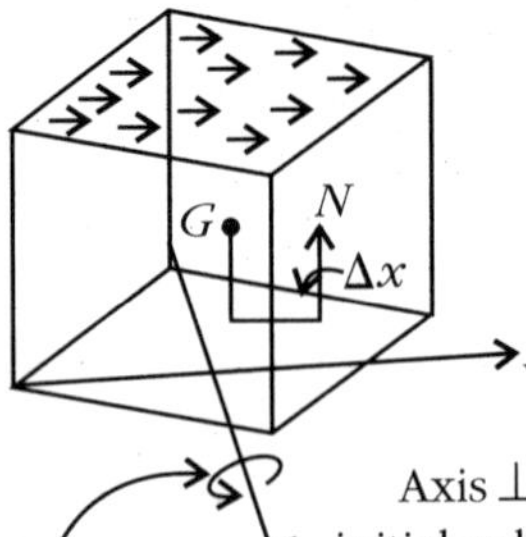

Then if the force of friction is fr

$$fr\frac{a}{2} = N\Delta x$$

But $N = mg$ and $fr = kmg$, so

$$\Delta x = \frac{ka}{2}$$

1.272 As the rod is smooth, in the process of motion of the given system, the kinetic energy and the angular momentum relative to rotation axis do not vary. Hence, it follows that

$$\frac{1}{2}\frac{Ml^2}{3}\omega_0^2 = \frac{1}{2}m(\omega^2 l^2 + v'^2) + \frac{1}{2}\frac{Ml^2}{3}\omega^2$$

(where ω is the final angular velocity of the rod)

and $$\frac{Ml^2}{3}\omega_0 = \frac{Ml^2}{3}\omega + ml^2\omega$$

From these equations we obtain

$$\omega = \frac{\omega_0}{(1+3m/M)}$$

and $$v' = \frac{\omega_0 l}{\sqrt{1+3m/M}}$$

1.273 As surface is smooth, for further motion the frame attached with the C.M. is inertial. Due to hitting of the ball, the angular impulse received by the rod about the C.M. is equal to $pl/2$. If ω is the angular velocity acquired by the rod, we have

$$\frac{ml^2}{12}\omega = \frac{pl}{2} \quad \text{or} \quad \omega = \frac{6p}{ml} \tag{1}$$

In the frame of C.M., the rod is rotating about an axis passing through its mid point with the angular velocity ω. Hence the force exerted by one half on the other = mass of one half × acceleration of C.M. of that part, in the frame of C.M.

$$= \frac{m}{2}\left(\omega^2 \frac{l}{4}\right) = m\frac{\omega^2 l}{8} = \frac{9p^2}{2ml} = 9 \text{ N}$$

1.274 (a) In the process of motion of the given system the kinetic energy and the angular momentum relative to rotation axis do not vary. Hence, it follows that

$$\frac{1}{2}mv^2 = \frac{1}{2}mv'^2 + \frac{1}{2}\left(\frac{Ml^2}{3}\right)\omega^2$$

and $$mv\frac{l}{2} = mv'\frac{l}{2} + \frac{Ml^2}{3}\omega$$

From these equations we obtain

$$v' = \left(\frac{3m-4m}{3m+4m}\right)v \quad \text{and} \quad \omega = \frac{4v}{l(1+4m/3M)}$$

As $\mathbf{v}'\uparrow\downarrow\mathbf{v}$, so in vector form

$$\mathbf{v}' = \left(\frac{3m-4M}{3m+4M}\right)\mathbf{v}$$

(b) Obviously the sought force provides the centripetal acceleration to the C.M. of the rod and is

$$F_n = mw_{cn}$$
$$= M\omega^2\frac{l}{2} = \frac{8Mv^2}{l(1+4M/3m)^2}$$

1.275 (a) About the axis of rotation of the rod, the angular momentum of the system is conserved. Thus if the velocity of the flying bullet is v,

$$mvl = \left(ml^2 + \frac{Ml^2}{3}\right)\omega$$

$$\omega = \frac{mv}{(m+(M/3))l} = \frac{3mv}{Ml} \quad (\text{as } m \ll M) \tag{1}$$

Now from the conservation of mechanical energy of the system (rod with bullet) in the uniform field of gravity,

$$\frac{1}{2}\left(ml^2 + \frac{Ml^2}{3}\right)\omega^2 = (M' + m)g\frac{l}{2}(1 - \cos\alpha) \tag{2}$$

[because C.M. of rod raises by the height $\frac{l}{2}(1 - \cos\alpha)$].

Solving Eqs. (1) and (2), we get

$$v = \left(\frac{M}{m}\right)\sqrt{\frac{2}{3}gl}\,\sin\frac{\alpha}{2} \quad \text{and} \quad \omega = \sqrt{\frac{6g}{l}}\,\sin\frac{\alpha}{2}$$

(b) Sought $$\Delta p = \left[m(\omega l) + M\left(\omega\frac{l}{2}\right)\right] - mv$$

where ωl is the velocity of the bullet and $\omega(l/2)$ equals the velocity of C.M. of the rod after the impact. Putting the value of v and ω, we get

$$\Delta p = \frac{1}{2}mv = M\sqrt{\frac{gl}{6}}\,\sin\frac{\alpha}{2}$$

This is caused by the reaction at hinge on the upper end.

(c) Let the rod starts swinging with angular velocity ω' in this case. Then, like in part (a),

$$mvx = \left(\frac{Ml^2}{3} + mx^2\right)\omega' \quad \text{or} \quad \omega' \approx \frac{3mvx}{Ml^2}$$

Final momentum is

$$P_f = mx\omega' + \int_0^l y\omega'\frac{M}{l}dy \approx \frac{M}{2}\omega' l \approx \frac{3}{2}mv\frac{x}{l}$$

So, $$\Delta p = p_f - p_i \approx mv\left(\frac{3x}{2l} - 1\right)$$

This vanishes for $x \approx 2/3l$.

1.276 (a) As force F on the body is radial so its angular momentum about the axis becomes zero and the angular momentum of system about the given axis is conserved. Thus

$$\frac{MR^2}{2}\omega_0 + m\omega_0 R^2 = \frac{MR^2}{2}\omega \quad \text{or} \quad \omega = \omega_0\left(1 + \frac{2m}{M}\right)$$

(b) From the equation of increment of the mechanical energy of the system

$$\Delta T = A_{ext}$$

$$\frac{1}{2}\frac{MR^2}{2}\omega^2 - \frac{1}{2}\left(\frac{MR^2}{2} + mR^2\right)\omega_0^2 = A_{ext}$$

Putting the value of ω from part (a) and solving, we get

$$A_{ext} = \frac{m\omega_0^2 R^2}{2}\left(1 + \frac{2m}{M}\right)$$

1.277 (a) Among the external forces are, the reaction force by the support at the center of the disk, weight of disk, and weight of the man, acting on the system (disk + man). First two forces pass through the center O of the disk, so no torque about O, that's why torque about the vertical axis of rotations OZ is also zero. The weight of the man is collinear to axis OZ, so it has no torque about this axis.

Let the man m_1 stand at point A (say) at the edge of the disk. Let Ox be a fixed line and such that at $t = 0$ the line joining O to A coincides with it. At time t, let $\angle AOx = \theta$ and $\angle AOP = \varphi'$ where P denotes the position of the man on the disk (relative to disk). The angle xOP increases at the rate $\dot{\varphi}' - \dot{\theta}$, therefore the velocity of the man at right angles to OP is $R(\dot{\varphi}' - \dot{\theta})$ and its angular momentum about the rotation axis Oz of the disk is $mR^2(\dot{\varphi}' - \dot{\theta})$.

The disk has angular momentum about the Oz axis which becomes $(m_2R^2/2)\,\dot{\theta}$ in the opposite sense to conserve the angular momentum of the (disk + man) system, which is zero.

Therefore,
$$\frac{m_2R^2}{2}\dot{\theta} - m_1R^2(\dot{\varphi}' - \dot{\theta}) = 0$$

$$m_1R^2\dot{\varphi}' = \left(\frac{m_2R^2}{2} + m_1R^2\right)\theta$$

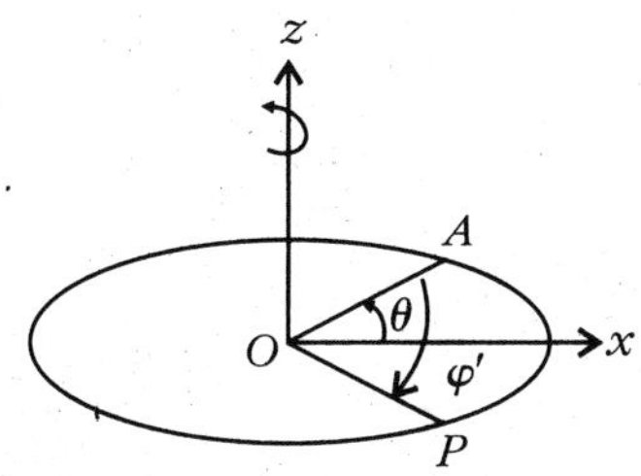

or
$$d\theta = \left[\frac{m_1}{m_1 + \left(\dfrac{m_2}{2}\right)}\right] d\varphi'$$

On integrating
$$\int_0^{\theta} d\theta = -\int_0^{\varphi'}\left(\frac{m_1}{m_1 + \dfrac{m_2}{2}}\right) d\varphi'$$

or
$$\theta = -\left(\frac{m_1}{m_1 + m_2/2}\right)\varphi' \qquad (1)$$

This gives the total angle of rotation of the disk.

(b) From Eq. (1)

$$\frac{d\theta}{dt} = -\left(\frac{m_1}{m_1 + m_2/2}\right)\frac{d\varphi'}{dt} = -\left(\frac{m_1}{m_1 + m_2/2}\right)\frac{v'(t)}{R}$$

Differentiating with respect to time

$$\frac{d^2\theta}{dt^2} = -\left(\frac{m_1}{m_1 + m_2/2}\right)\frac{1}{R}\frac{dv'(t)}{dt}$$

Thus, the sought force moment from the equation $N_z = I\beta_z$ is

$$N_z = \frac{m_2R^2}{2}\frac{d^2\theta}{dt^2} = -\frac{m_2R^2}{2}\left(\frac{m_1}{m_1 + m_2/2}\right)\frac{1}{R}\frac{dv'(t)}{dt}$$

Hence,
$$N_z = -\frac{m_1m_2R}{2m_1 + m_2}\frac{dv'(t)}{dt}$$

1.278 (a) From the law of conservation of angular momentum of the system relative to vertical axis z, it follows that

$$I_1\omega_{1z} + I_2\omega_{2z} = (I_1 + I_2)\omega_z$$

Hence,
$$\omega_z = (I_1\omega_{1z} + I_2\omega_{2z}/(I_1 + I_2) \qquad (1)$$

Note that for $\omega_z > 0$, the corresponding vector $\boldsymbol{\omega}$ coincides with the positive direction to the z-axis, and vice versa. As both discs rotate about the same vertical axis z, thus in vector form

$$\boldsymbol{\omega} = (I_1\boldsymbol{\omega}_1 + I_2\boldsymbol{\omega}_2)/(I_1 + I_2)$$

However, the problem makes sense only if $\boldsymbol{\omega}_1 \uparrow\uparrow \boldsymbol{\omega}_2$ or $\boldsymbol{\omega}_1 \uparrow\downarrow \boldsymbol{\omega}_2$.

(b) From the equation of increment of mechanical energy of a system: $A_{fr} = \Delta T$

$$A_{fr} = \frac{1}{2}(I_1 + I_2)\omega_z^2 - \left(\frac{1}{2}I_1\omega_{1z}^2 + \frac{1}{2}I_2\omega_{2z}^2\right)$$

Using Eq. (1)

$$A_{fr} = \frac{I_1 I_2}{2(I_1 + I_2)}(\omega_{1z} - \omega_{2z})^2$$

1.279 For the closed system (disk + rod), the angular momentum is conserved about any axis. Thus from the conservation of angular momentum of the system about the rotation axis of rod passing through its C.M. gives:

$$mv\frac{l}{2} = mv'\frac{l}{2} + \frac{\eta m l^2}{12}\omega \qquad (1)$$

(where v' is the final velocity of the disk and ω angular velocity of the rod).

For the closed system linear momentum is also conserved. Hence,

$$mv = mv' + \eta m v_C \qquad (2)$$

(where v_C is the velocity of C.M. of the rod)

From Eqs. (1) and (2) we get

$$v_C = \frac{l\omega}{3} \quad \text{and} \quad v - v' = \eta v_C \qquad (3)$$

Applying conservation of kinetic energy, as the collision is elastic

$$\frac{1}{2}mv^2 = \frac{1}{2}mv'^2 + \frac{1}{2}\eta m v_C^2 + \frac{1}{2}\frac{\eta m l^2}{12}\omega^2 \qquad (4)$$

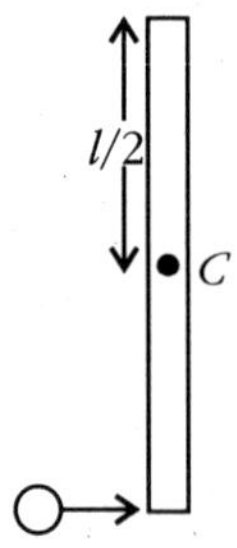

or

$$v^2 - v'^2 = 4\eta v_C^2 \qquad (5)$$

and hence from equation (3) and (5) $v + v' = 4v_C$

Then

$$v' = \frac{4-\eta}{4+\eta}v \quad \text{and} \quad \omega = \frac{12v}{(4+\eta)l}$$

Vectorially, noting that we have taken $\mathbf{v}'$ parallel to $\mathbf{v}$

$$\mathbf{v}' = \left(\frac{4-\eta}{4+\eta}\right)\mathbf{v}$$

So, $\mathbf{v}' = 0$ for $\eta = 4$ and $\mathbf{v}' \downarrow\uparrow \mathbf{v}$ for $\eta > 4$

Note: *Instead of Eq. (4) one can also write the equation of coefficient of restitution.*

1.280 See the diagram in the book (Fig. 1.72).

(a) When the shaft BB' is turned through 90°, the platform must start turning with angular velocity Ω so that the angular momentum remains constant. Here,

$$(I + I_0)\Omega = I_0\omega_0 \quad \text{or} \quad \Omega = \frac{I_0\omega_0}{I + I_0}$$

The work performed by the motor is therefore,

$$\frac{1}{2}(I+I_0)\Omega^2 = \frac{1}{2}\frac{I_0^2\omega_0^2}{I+I_0}$$

If the shaft is turned through 180°, angular velocity of the sphere changes sign. Thus from conservation of angular momentum,

$$I\Omega - I_0\omega_0 = I_0\omega_0$$

Here $-I_0\omega_0$ is the complete angular momentum of the sphere, i.e., we assume that the angular velocity of the sphere is just $-\omega_0$. Then

$$\Omega = 2I_0\frac{\omega_0}{I}$$

and the work done must be

$$\frac{1}{2}I\Omega^2 + \frac{1}{2}I_0\omega_0^2 - \frac{1}{2}I_0\omega_0^2 = \frac{2I_0^2\omega_0^2}{I}$$

(b) In the first part of (a), the angular momentum vector of the sphere is precessing with angular velocity Ω. Thus torque needed is

$$I_0\omega_0\Omega = \frac{I_0^2\omega_0^2}{I+I_0}$$

1.281 The total centrifugal force can be calculated by,

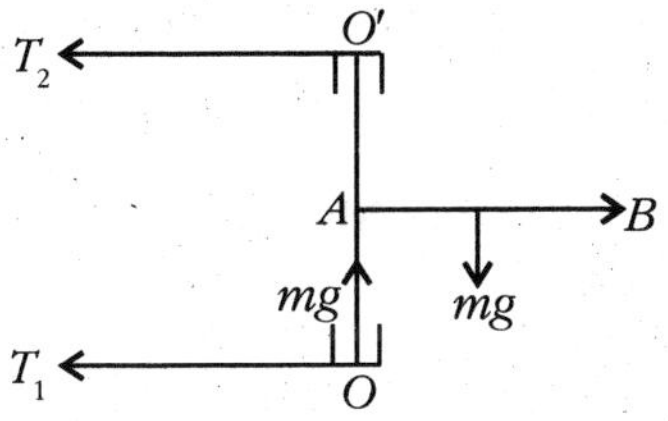

$$\int_0^{l_0}\frac{m}{l_0}\omega^2 x\,dx = \frac{1}{2}ml_0\omega^2$$

Then for equilibrium,

$$(T_2 - T_1)\frac{l}{2} = mg\frac{l_0}{2}$$

and $$T_2 + T_1 = \frac{1}{2}ml_0\omega^2$$

Thus T_1 vanishes, when

$$\omega^2 = \frac{2g}{l},\quad \omega = \sqrt{\frac{2g}{l}} = 6 \text{ rad/s}$$

Then $$T_2 = mg\frac{l_0}{l} = 25 \text{ N}$$

1.282 See the diagram in the problem book (Fig. 1.71).

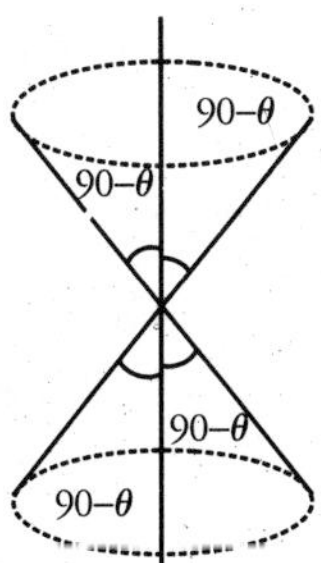

(a) The angular velocity ω about OO' can be resolved into a component parallel to the rod and a component $\omega \sin\theta$ perpendicular to the rod through C. The component parallel to the rod does not contribute to the angular momentum.

$$M = I\omega\sin\theta = \frac{1}{12}ml^2\,\omega\sin\theta$$

Also, $$M_z = M \sin\theta = \frac{1}{12} ml^2 \,\omega \sin^2\theta$$

This can be obtained directly also.

(b) The modulus of **M** does not change but the modulus of the change of **M** is

$$|\Delta\mathbf{M}| = 2M \sin(90° - \theta) = \frac{1}{12} ml^2\omega \sin 2\theta$$

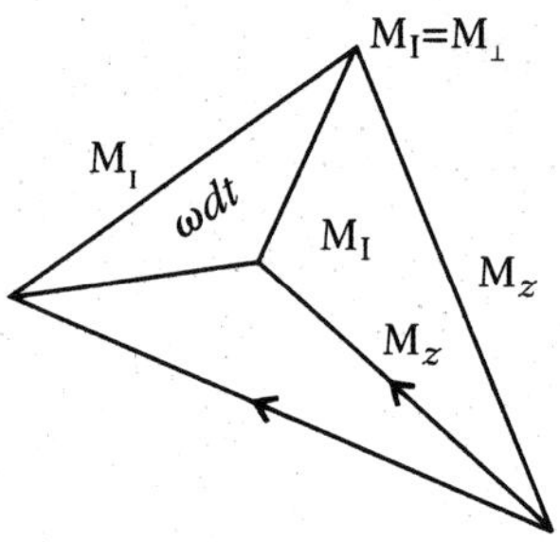

(c) Here, $$M_\perp = M \cos\theta = I\omega \sin\theta \cos\theta$$

Now, $$\left|\frac{d\mathbf{M}}{dt}\right| = I\omega \sin\theta \cos\theta \frac{\omega dt}{dt} = \frac{1}{24} ml^2\omega^2 \sin 2\theta$$

as **M** precesses with angular velocity ω.

1.283 In fact, the angular momentum vector **M** of a precessing top relative to the supporting point O may be represented as the sum of the angular momentum $\mathbf{M}_\omega$ associated with the top, spinning about its axis and some additional angular momentum $\mathbf{M}'$ caused by the top precession about the vertical axis, i.e.

$$\mathbf{M} = \mathbf{M}_\omega + \mathbf{M}'$$

Since the top axis coincides with one of the principal axes of inertia, $\mathbf{M}_\omega = I\omega$, where I is the moment of inertia of a top with respect to that axis. Moreover, it is clear that as the precession gets slower, the corresponding angular momentum $\mathbf{M}'$ diminishes. If $\omega >> \omega'$ (angular velocity of precession), then in all practical cases $M_\omega >> M'$, and therefore the resultant angular momentum **M** essentially coincides with $\mathbf{M}_\omega$ both in magnitude and direction. We can thus assume that

$\mathbf{M} = I\omega$, and is directed along the symmetry axis

Here **M** only changes its direction, so $d\mathbf{M} \perp \mathbf{M}$. (See Figure)

It has two components, the part $M_{||} = I\omega \cos\theta$ is constant and the part $M_\perp = I\omega \sin\theta$ precesses, then

$$\left|\frac{d\mathbf{M}}{dt}\right| = \left|\frac{d\mathbf{M}_\perp}{dt}\right| = I\omega \sin\theta \;\omega' = mgl \sin\theta$$

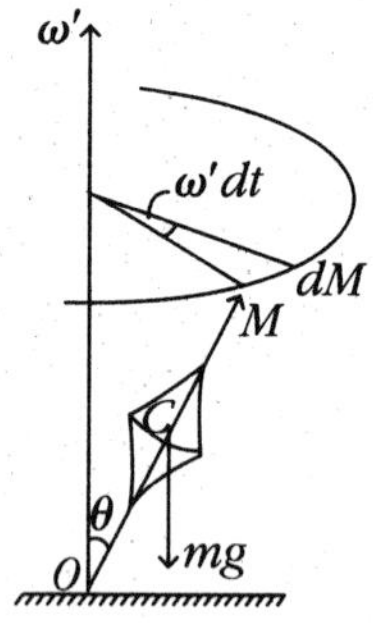

or, $$\omega' = \text{precession frequency} = \frac{mgl}{I\omega} = 0.7 \text{ rad/s}$$

This force is the centripetal force due to precession. It acts inward and has the magnitude

$$|\mathbf{F}| = \left|\sum m_i \omega'^2 \rho_i\right| = m\omega'^2 l \sin\theta = 12 \text{ mN}$$

ρ_i is the distance of the ith element from the axis. This is the force that the table will exert on the top. See the diagram in the answer sheet.

1.284 See the diagram in the book (Fig. 1.73).

The moment of inertia of the disc about its symmetry axis is $1/2\ mR^2$. If the angular velocity of the disk is ω, then the angular momentum is $1/2\ mR^2\omega$. The precession frequency being $2\pi n$, we have

$$\left|\frac{d\mathbf{M}}{dt}\right| = \frac{1}{2} mR^2\omega \times 2\pi n$$

This must equal $m(g + w)l$, the effective gravitational torques (g being replaced by $g + w$ in the elevator). Thus,

$$\omega = \frac{(g+w)l}{\pi R^2 n} = 300 \text{ rad/s}$$

1.285 The effective g is $\sqrt{g^2 + w^2}$ inclined at angle $\tan^{-1} w/g$ with the vertical. Then with reference to the new "vertical" we proceed as in solution of problem 1.283. Thus,

$$\omega' = \frac{ml\sqrt{g^2 + w^2}}{I\omega} = 0.8 \text{ rad/s}$$

The vector $\boldsymbol{\omega}'$ forms an angle $\theta = \tan^{-1}\frac{\omega}{g} = 6°$ with the normal vertical.

1.286 The moment of inertia of the sphere is $2/5\ mR^2$ and hence the value of angular momentum is $2/5\ mR^2\omega$. Since it precesses at speed ω' the torque required is

$$\frac{2}{5} mR^2 \omega\omega' = F'l$$

So, $$F' = \frac{2}{5} mR^2 \omega\omega'/l = 300 \text{ N (the force } F' \text{ must be vertical)}$$

1.287 The moment of inertia is $1/2\ mr^2$ and angular momentum is $1/2\ mr^2\omega$. The axle oscillates about a horizontal axis making an instantaneous angle

$$\varphi = \varphi_m \sin\frac{2\pi t}{T}$$

This means that there is a variable precession with a rate of precession $d\varphi/dt$. The maximum value of this is $2\pi\varphi_m/T$. When angle between the axle and the axis is at its maximum value, a torque $I\omega\Omega$ acts on it which is equal to

$$\frac{1}{2} mr^2 \omega \frac{2\pi\varphi_m}{T} = \frac{\pi mr^2 \omega\varphi_m}{T}$$

The corresponding gyroscopic force will be

$$\frac{\pi mr^2 \omega\varphi_m}{lT} = 90 \text{ N}$$

1.288 The revolutions per minute of the flywheel being n, the angular momentum of the flywheel is $I \times 2\pi n$. The rate of precession is v/R.

Thus, $$N = 2\pi INv/R = 5.97 \text{ kNm}$$

1.289 As in the previous problem, a couple $2\pi Inv/R$ must come in play. This can be done if a force, $2\pi Inv/Rl$ acts on the rails in opposite directions in addition to the centrifugal and other forces. The force on the outer rail is increased and that on the inner rail decreased. The additional force in this case has the magnitude 1.4 kN.

1.6 Elastic Deformations of a Solid Body

1.290 Variation of length with temperature is given by

$$l_t = l_0(1+\alpha\Delta t) \quad \text{or} \quad \frac{\Delta l}{l_0} = \alpha\Delta t = \varepsilon \qquad (1)$$

But, $$\varepsilon = \frac{\sigma}{E}$$

Thus, $\sigma = \alpha\Delta t E$, which is the sought stress or pressure.

Putting the value of α and E from Appendix and taking $\Delta t = 100°C$, we get

$$\sigma = 2.2\times10^3 \text{ atm}$$

1.291 (a) Consider a transverse section of the tube and concentrate on an element which subtends angle $\Delta\varphi$ at the center. The forces acting on a portion of length Δl on the element are:

(1) Tensile forces side ways of magnitude $\sigma\Delta r\Delta l$.

The resultant of these is $2\sigma\Delta r\Delta l\sin\Delta\varphi/2 \approx \sigma\Delta r\Delta l\Delta\varphi$ radially towards the centre.

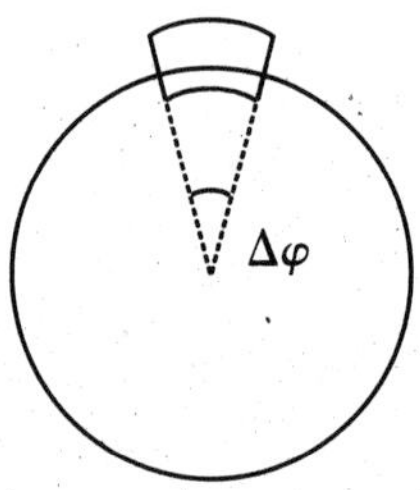

(2) The force due to fluid pressure = $pr\Delta\varphi\Delta l$.

Since these balance, we get $p_{max} \approx \sigma_m \Delta r/r$

where σ_m is the maximum tensile force.

Putting the values, we get $p_{max} = 19.7$ atm.

(b) Consider an element of area $dS = \pi(r\Delta\varphi/r)^2$ about z-axis chosen arbitrarily. There are tangential tensile forces all around the ring of the cap. Their resultant is

$$\sigma\left[2\pi\left(r\frac{\Delta\varphi}{2}\right)\Delta r\right]\sin\frac{\Delta\varphi}{2}$$

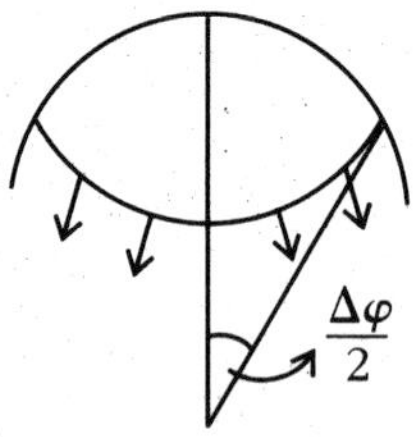

Hence, in the limit,

$$p_m\pi = \left(\frac{r\Delta\varphi}{2}\right)^2 = \sigma_m\pi\left(\frac{r\Delta\varphi}{2}\right)\Delta r\Delta\varphi$$

or $$p_m = \left(\frac{2\sigma_m\Delta r}{r}\right) = 39.5 \text{ atm}$$

1.292 Let us consider an element of rod at a distance ρ from its rotation axis (see figure). From Newton's second law in projection form directed towards the rotation axis,

$$-dT = (dm)\omega^2\rho = \frac{m}{l}\omega^2\rho\,d\rho$$

On integrating $$-T = \frac{m\omega^2}{l}\frac{l^2}{2} + C \text{ (constant)}$$

But at $$\rho = \pm\frac{l}{2} \quad \text{or} \quad \text{free end, } T = 0$$

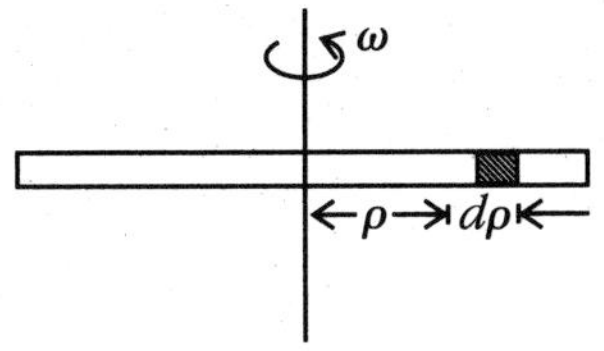

Thus, $$0 = \frac{m\omega^2}{2l}\frac{l^2}{4} + C \quad \text{or} \quad C = -\frac{m\omega^2 l}{8}$$

Hence $$T = \frac{m\omega^2}{2}\left(\frac{1}{2} - \frac{\rho^2}{l}\right)$$

Thus $$T_{max} = \frac{m\omega^2 l}{8} \quad \text{(at mid point)}$$

Condition required for the problem is

$$T_{max} = S\sigma_m$$

So, $$\frac{m\omega^2 l}{8} = S\sigma_m \quad \text{or} \quad \omega = \frac{2}{l}\sqrt{\frac{2\sigma_m}{\rho}}$$

Hence the sought number of rps

$$n = \frac{\omega}{2\pi} = \frac{1}{\pi l}\sqrt{\frac{2\sigma_m}{\rho}} \quad \text{(using the table } n = 0.8 \times 10^3 \text{ rps)}$$

1.293 Let us consider an element of the ring (see figure). From Newton's law $F_n = mw_n$ for this element, we get

$$Td\theta = \left(\frac{m}{2\pi}d\theta\right)\omega^2 r \quad \text{(see solution of problem 1.93 or 1.92)}$$

So, $$T = \frac{m}{2\pi}\omega^2 r$$

Condition for the problem is

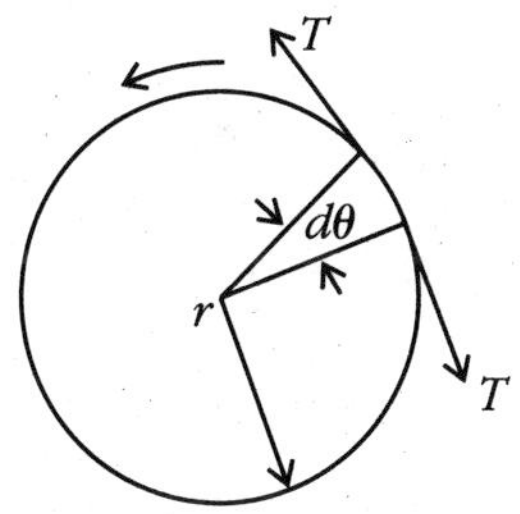

$$\frac{T}{\pi r^2} \le \sigma_m \quad \text{or} \quad \frac{m\omega^2 r}{2\pi^2 r^2} \le \sigma_m$$

or $$\omega^2_{max} = \frac{2\pi^2\sigma_m r}{\pi r^2(2\pi r\rho)} = \frac{\sigma_m}{\rho r^2}$$

Thus, sought number of rps is

$$n = \frac{\omega_{max}}{2\pi} = \frac{1}{2\pi r}\sqrt{\frac{\sigma_m}{\rho}}$$

Using the table of appendices $n = 23$ rps.

1.294 Let the point O descend by the distance x (see figure). From the condition of equilibrium of point O,

$$2T\sin\theta = mg \quad \text{or} \quad T = \frac{mg}{2\sin\theta} = \frac{mg}{2x}\sqrt{(l/2)^2 + x^2} \tag{1}$$

Now, $$\frac{T}{\pi(d/2)^2} = \sigma = \varepsilon E \quad \text{or} \quad T = \varepsilon E\pi\frac{d^2}{4} \tag{2}$$

(Here σ is stress and ε is strain.)

In addition to this,

$$\varepsilon = \frac{\sqrt{(l/2)^2 + x^2} - l/2}{l/2} = \sqrt{1 + \left(\frac{2x}{l}\right)^2} - 1 \tag{3}$$

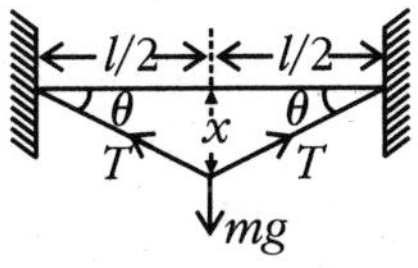

From Eqs. (1), (2) and (3)

$$x - \frac{x}{\sqrt{1+(2x/l)^2}} = \frac{mgl}{\pi E d^2}$$

or
$$x = l\left(\frac{mg}{2\pi E d^2}\right)^{1/3} = 2.5 \text{ cm}$$

1.295 Let us consider an element of the rod at a distance x from the free end (see figure). For the considered element $T - T'$ are internal restoring forces which produce elongation and dT provide the acceleration to the element. For the element, from Newton's law

$$dT = (dm)w = \left(\frac{m}{l}dx\right)\frac{F_0}{m} = \frac{F_0}{l}dx$$

As free end has zero tension, on integrating the above expression,

$$\int_0^T dT = \frac{F_0}{l}\int_0^x dx \quad \text{or} \quad T = \frac{F_0}{l}x$$

Elongation in the considered element of length dx is given by

$$\partial\xi = \frac{\sigma(x)}{E}dx = \frac{T}{SE}dx = \frac{F_0 x\,dx}{SEl}$$

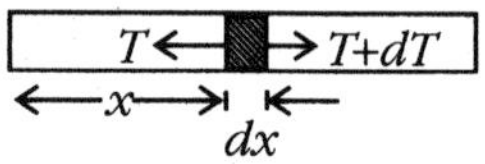

Thus, total elongation
$$\xi = \frac{F_0}{SEl}\int_0^l x\,dx = \frac{F_0 l}{2SE}$$

Hence, the sought strain is

$$\sigma = \frac{\xi}{l} = \frac{F_0}{2SE}$$

1.296 Let us consider an element of the rod at a distance r from its rotation axis. As the element rotates in a horizontal circle of radius r, we have from Newton's second law in projection form directed towards the axis of rotation

$$T - (T + dT) = (dm)\omega^2 r$$

or
$$-dT = \left(\frac{m}{l}dr\right)\omega^2 r = \frac{m}{l}\omega^2 r\,dr$$

At the free end tension becomes zero. Integrating the above expression, we get

$$-\int_T^0 dT = \frac{m}{l}\omega^2\int_r^l r\,dr$$

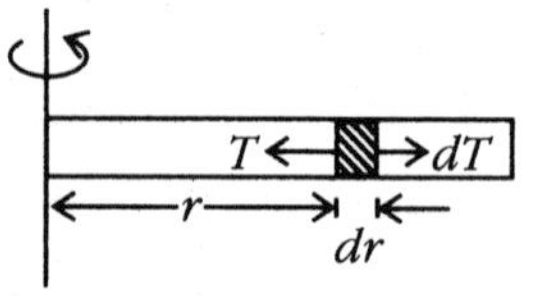

Thus,
$$T = \frac{m\omega^2}{l}\left(\frac{l^2 - r^2}{2}\right) = \frac{m\omega^2 l}{2}\left(1 - \frac{r^2}{l^2}\right)$$

Elongation in elemental length dr is given by

$$\partial\xi = \frac{\sigma(r)}{E}dr = \frac{T}{SE}dr$$

(where S is the cross sectional area of the rod and T is the tension in the rod at the considered element)

or
$$\partial\xi = \frac{m\omega^2 r}{2SE}\left(1 - \frac{r^2}{l^2}\right)dr$$

Thus, the sought elongation

$$\xi = \int d\xi = \frac{m\omega^2 l}{2SE}\int_0^l \left(1-\frac{r^2}{l^2}\right)dr$$

or

$$\xi = \frac{m\omega^2 l}{2SE}\frac{2l}{3} = \frac{(Sl\rho)}{3SE}\omega^2 l^3$$

$$= \frac{1}{3}\frac{\rho\omega^2 l^3}{E} \quad \text{(where } \rho \text{ is the density of the copper)}$$

1.297 Volume of a solid cylinder

$$V = \pi r^2 l$$

So,

$$\frac{\Delta V}{V} = \frac{\pi 2r\Delta r l}{\pi r^2 l} + \frac{\pi r^2 \Delta l}{\pi r^2 l} = \frac{2\Delta r}{r} + \frac{\Delta l}{l} \tag{1}$$

But longitudinal $\Delta l/l$ and accompanying lateral strain $\Delta r/r$ are related as

$$\frac{\Delta r}{r} = -\mu\frac{\Delta l}{l} \tag{2}$$

Using Eq. (2) in Eq. (1), we get

$$\frac{\Delta V}{V} = \frac{\Delta l}{l}(1-2\mu) \tag{3}$$

But

$$\frac{\Delta l}{l} = \frac{-F/\pi r^2}{E}$$

(Because the increment in the length of cylinder Δl is negative.)

So,

$$\frac{\Delta V}{V} = \frac{-F}{\pi r^2 E}(1-2\mu) \quad \text{(where } \mu \text{ is the Poisson's ratio for copper)}$$

Thus,

$$\Delta V = \frac{-Fl}{E}(1-2\mu)$$

$$= -1.6 \text{ mm}^3$$

(Negative sign means that the volume of the cylinder has decreased.)

1.298 (a) As free end has zero tension, thus the tension in the rod at a vertical distance y from its lower end

$$T = \frac{m}{l}gy \tag{1}$$

Let ∂l be the elongation of the element of length dy, then

$$\partial l = \frac{\sigma(y)}{E}dy$$

$$= \frac{T}{SE}dy = \frac{mgydy}{SlE} = \frac{\rho gydy}{E} \quad \text{(where } \rho \text{ is the density of copper)}$$

Thus, the sought elongation is

$$\Delta l = \int \partial l = \rho g\int_0^l \frac{ydy}{E} = \frac{1}{2}\rho gl^2/E \tag{2}$$

(b) If the longitudinal (tensile) strain is $\varepsilon = \Delta l/l$, the accompanying lateral (compressive) strain is given by

$$\varepsilon' = \frac{\Delta r}{r} = -\mu\varepsilon \tag{3}$$

Then, since $V = \pi r^2 l$, we have

$$\frac{\Delta V}{V} = \frac{2\Delta r}{r} + \frac{\Delta l}{l}$$

$$= (1 - 2\mu)\frac{\Delta l}{l} \quad \text{(using Eq. (3))}$$

(where $\Delta l/l$ is given in part (a) and μ is the Poisson ratio for copper.)

1.299 (a) Since the hydrostatic pressure is uniform, the stress on each face of the bar is the same (see figure (a)). The change in length of the bar can be thought of as the sum of changes in length that would occur in the three independent cases shown in the figures (b), (c) and (d).

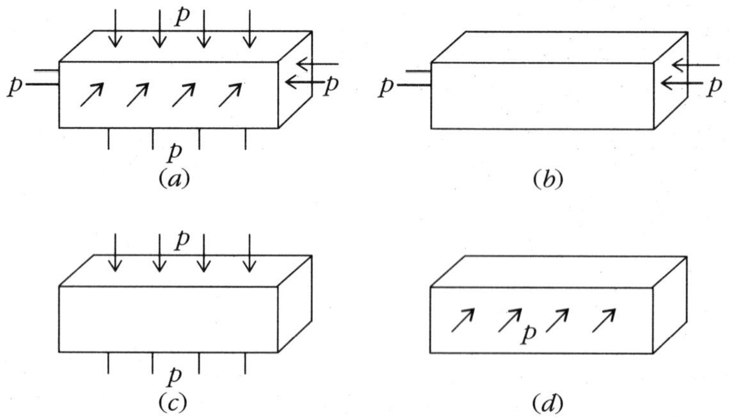

If we push on the ends of the block with a pressure p, the compressional strain is p/E, and it is negative,

$$\frac{\Delta l_1}{l} = -\frac{p}{E}$$

If we push on the two sides of the block with pressure p, the compressional strain is again p/Y, but now we want the lengthwise strain. We can get that from the sideways strain multiplied by $\sim\mu$. The sideways strain is

$$\frac{\Delta w}{w} = -\frac{p}{E}$$

So,
$$\frac{\Delta l_2}{l} = +\mu\frac{p}{E}$$

If the we push on the top of the block, the compressional strain is once more p/E, and the corresponding strain in the sideways direction is again $-\mu p/\text{E}$. We get

$$\frac{\Delta l_3}{l} = +\mu\frac{p}{E}$$

Combining the results of the three problems i.e., taking $\Delta l = \Delta l_1 + \Delta l_2 + \Delta l_3$, we get

$$\frac{\Delta l}{l} = -\frac{p}{E}(1 - 2\mu) \tag{1}$$

The problem is, of course, symmetrical in all three directions; it follows that

$$\frac{\Delta w}{w} = \frac{\Delta h}{h} = -\frac{p}{E}(1-2\mu) \tag{2}$$

The change in the volume under hydrostatic pressure is also of some interest. Since $V = lwh$, we can write, for small displacements.

$$\frac{\Delta V}{V} = \frac{\Delta l}{l} + \frac{\Delta w}{w} + \frac{\Delta h}{h}$$

Using Eqs. (1) and (2), we have

$$\frac{\Delta V}{V} = -3\frac{p}{E}(1-2\mu) \tag{3}$$

(b) But volume strain

$$\frac{\Delta V}{V} = -\frac{p}{K} \quad \text{(where } K \text{ is called the bulk modulus)} \tag{4}$$

From Eqs. (3) and (4), we have

$$K = \frac{E}{3(1-2\mu)}$$

Hence the compressibility

$$\beta = \frac{1}{K} = \frac{3(1-2\mu)}{E}$$

1.300 A beam clamped at one end and supporting an applied load at the free end is called a cantilever. The theory of cantilevers is discussed in the advanced text book on mechanics. The key result is that elastic forces in the beam generate a couple, whose moment, called the moment of resistances, balances the external bending moment due to weight of the beam, load etc. The moment of resistance, also called internal bending moment (I.B.M) is given by

$$\text{I.B.M.} = EI/R$$

Here R is the radius of curvature of the beam at the representative point (x, y). I is called the geometrical moment of inertia of the cross section relative to the axis passing through the natural layer which remains unstretched Fig. (a).

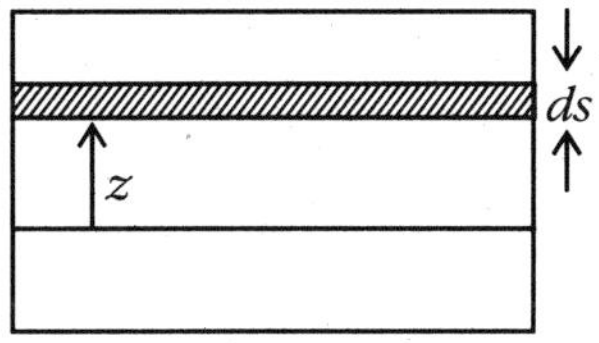

(a)

$$I = \int z^2\, ds \tag{1}$$

The section of the beam beyond P exerts the bending moment $N(x)$ and we have,

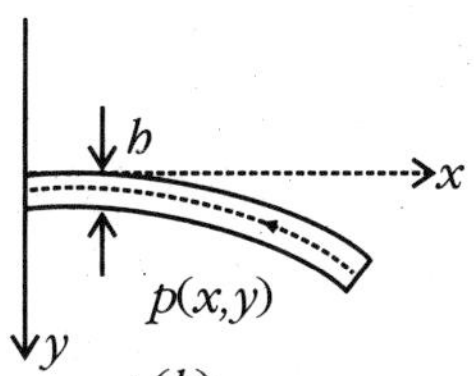

(b)

$$\frac{El}{R} = N(x) \tag{2}$$

If there is no load other than that due to the weight of the beam, then

$$N(x) = \frac{1}{2}\rho g(l-x)^2\, b$$

(where ρ = density of steel).

Hence, at $x = 0$

$$\left(\frac{1}{R}\right)_0 = \frac{\rho g l^2 b}{2El}$$

Here, b = width of the beam perpendicular to paper.

Also, $$I = \int_{-b/2}^{b/2} z^2\, bdz = \frac{bh^3}{12}$$

Hence, $$\left(\frac{1}{R}\right)_0 = \frac{6\rho g l^2}{Eh^2}$$

$$R = \frac{1}{6}\frac{Eh^2}{\rho g l^2} = 0.121 \text{ km}$$

1.301 We use the equation given above and use the result that when y is small

$$\frac{1}{R} = \frac{d^2y}{dx^2} \quad \text{thus,} \quad \frac{d^2y}{dx^2} = \frac{N(x)}{EI}$$

(a) Here $N(x) = N_0$ is a constant. Then integration gives

$$\frac{dy}{dx} = \frac{N_0 x}{EI} + C_1$$

But, $$\left(\frac{dy}{dx}\right) = 0, \quad \text{for } x = 0, \quad \text{so} \quad C_1 = 0.$$

Integrating again, $$y = \frac{N_0 x^2}{2EI}$$

where we have used $y = 0$ for $x = 0$ to set the constant of integration at zero. This is the equation of a parabola. The sag of the free end is

$$\lambda = y(x = l) = \frac{N_0 l^2}{2EI}$$

(b) In this case $N(x) = F(l - x)$ because the load F at the extremity is balanced by a similar force at F directed upward and they constitute a couple. Then

$$\frac{d^2y}{dx^2} = \frac{F(l-x)}{EI}$$

Integrating, $$\frac{dy}{dx} = \frac{F(lx - x^2/2)}{EI} + C_1$$

As before $C_1 = 0$. Integrating again, using $y = 0$ for $x = 0$,

$$y = \frac{F(lx^2/2 - x^3/6)}{EI} \quad \left(\text{here } \lambda = \frac{Fl^3}{3EI}\right)$$

Here for a square cross section

$$I = \int_{-a/2}^{a/2} z^2\, a\, dz = \frac{a^4}{12}$$

1.302 One can think of it as analogous to the previous case but with a beam of length $l/2$ loaded upward by a force $F/2$.

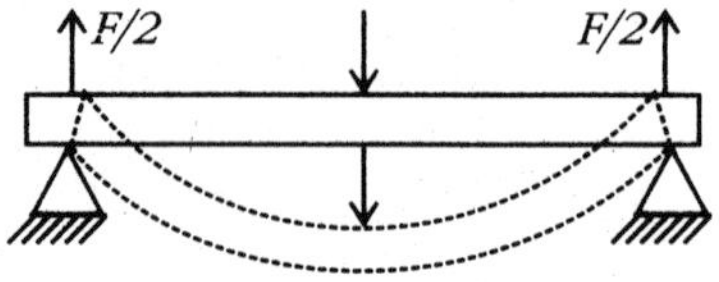

Thus, $$\lambda = \frac{Fl^3}{48EI}$$

on using the last result of the previous problem.

1.303 (a) In this case $N(x) = \frac{1}{2}\rho g b(l - x)^2$ (where b = width of the girder).

Also $$I = \frac{bh^3}{12}$$

Then, $$\frac{Ebh^2}{12}\frac{d^2y}{dx^2} = \frac{\rho g b}{2}(l^2 - 2lx + x^2)$$

Integrating, $$\frac{dy}{dx} = \frac{6\rho g}{Eh^2}\left(l^2x - lx^2 + \frac{x^2}{3}\right)$$

Using $$\frac{dy}{dx} = 0 \quad \text{for } x = 0$$

Integrating again, $$y = \frac{6\rho g}{Eh^2}\left(\frac{l^2x^2}{2} - \frac{lx^2}{3} + \frac{x^4}{12}\right)$$

Thus, $$\lambda = \frac{6\rho g l^2}{Eh^2}\left(\frac{1}{2} - \frac{1}{3} + \frac{1}{12}\right) = \frac{6\rho g l^4}{Eh^2}\frac{3}{2} = \frac{3\rho g l^4}{2Eh^2}$$

(b) As before, $EI\frac{d^2y}{dx^2} = N(x)$ where $N(x)$ is the bending moment due to section PB.

Thus, bending moment is clearly

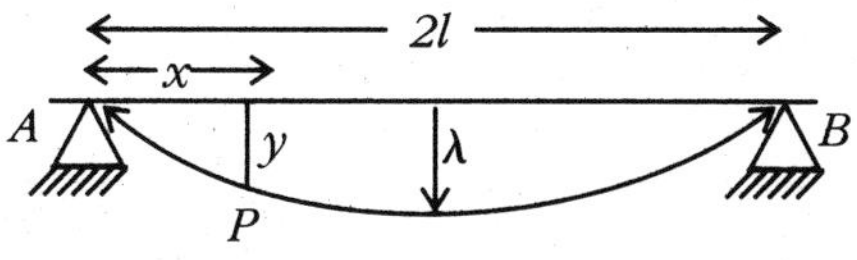

$$N = \int_x^{2l} w\,d\xi(\xi - x) - wl(2l - x)$$

$$= w\left(2l^2 - 2xl + \frac{x^2}{2}\right) - wl(2l - x)$$

$$= w\left(\frac{x^2}{2} - xl\right)$$ (here $w = \rho\, gb\, h$ is weight of the beam per unit length)

Now integrating, $$EI\frac{dy}{dx} = w\left(\frac{x^3}{6} - \frac{x^2 l}{2}\right) + c_0$$

or since $$\frac{dy}{dx} = 0 \quad \text{for } x = l,\ c_0 = \frac{wl^3}{3}$$

Integrating again, $$EIy = w\left(\frac{x^4}{24} - \frac{x^3 l}{6}\right) + \frac{wl^3 x}{3} + c_1$$

As $y = 0$ for $x = 0$, $c_1 = 0$. From this we find

$$\lambda = y(x = l) = \frac{5wl^4/24}{EI} = \frac{5\rho g l^4}{2Eh^2}$$

1.304 The deflection of the plate can be noticed by going to a co-rotating frame. In this frame each element of the plate experiences a pseudo force proportional to its mass. These forces have a moment which constitutes the bending moment of the problem. To calculate this moment we note that the acceleration of an element at a distance ξ from the axis is $a = \xi\beta$ and the moment of the forces exerted by the section between x and l is

$$N = \rho l h \beta \int_l^x \xi^2 d\xi = \frac{1}{3}\rho l h \beta(l^3 - x^3)$$

From the fundamental equation

$$El\frac{d^2y}{dx^2} = \frac{1}{3}\rho lb\beta(l^3 - x^3)$$

The moment of inertia $$I = \int_{-b/2}^{+b/2} z^2 l dz = \frac{lb^3}{12}$$

Note that the neutral surface (i.e., the surface which contains lines which are neither stretched nor compressed) is a vertical plane here and z is perpendicular to it.

$$\frac{d^2y}{dx^2} = \frac{4\rho\beta}{Eb^2}(l^3 - x^3)$$

Integrating $$\frac{dy}{dx} = \frac{4\rho\beta}{Eb^2}\left(l^3x - \frac{x^4}{4}\right) + c_1$$

Since $\frac{dy}{dx} = 0$, for $x = 0$, $c_1 = 0$. Integrating again,

$$y = \frac{4\rho\beta}{Eb^2}\left(\frac{l^3x^2}{2} - \frac{x^5}{20}\right) + c_2$$

$$C_2 = 0 \quad (\text{because } y = 0 \text{ for } x = 0)$$

Thus, $$\lambda = y(x = l) = \frac{9\rho\beta l^5}{5Eb^2}$$

1.305 (a) Consider a hollow cylinder of length l, outer radius $r + \Delta r$, inner radius r, fixed at one end and twisted at the other by means of a couple of moment N. The angular displacement φ, at a distance l from the fixed end, is proportional to both l and N. Consider an element of length dx at the twisted end. It is moved by an angle φ as shown. A vertical section is also shown and the twisting of the parallelopipe of length l and area $\Delta r dx$ under the action of the twisting couple can be discussed by elementary means. If f is the tangential force generated, then shearing stress is $f/\Delta r\, dx$ and this must equal

$$G\theta = G\frac{r\varphi}{l}, \quad \text{since } \theta = \frac{r\varphi}{l}$$

Hence, $$f = G\Delta r\, dx\frac{r\varphi}{l}$$

The force f has moment fr about the axis and so the total moment is

$$N = G\Delta r\frac{\varphi}{l}r^2\int dx = \frac{2\pi r^3\Delta r\varphi}{l}G$$

(b) For a solid cylinder we must integrate over r. Thus

$$N = \int_0^r \frac{2\pi r^3\, dr\varphi G}{l} = \frac{\pi r^4 G\varphi}{2l}$$

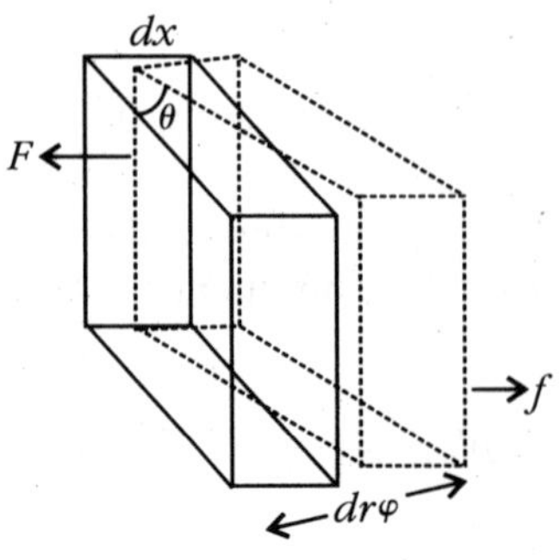

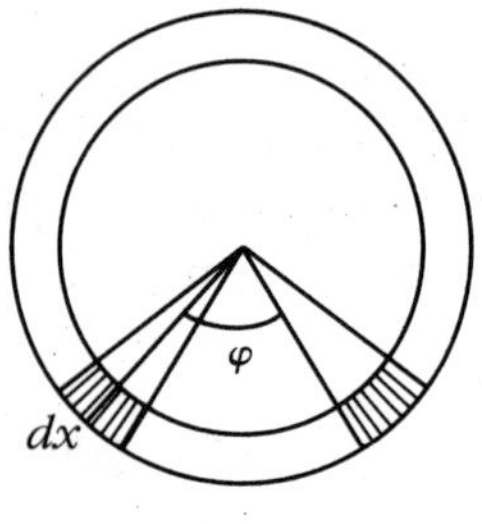

1.306 Clearly
$$N = \int_{d_1/2}^{d_2/2} \frac{2\pi r^3\, dr \varphi G}{l} = \frac{\pi}{32 l} G\varphi\left(d_2^4 - d_1^4\right)$$

Using
$$G = 81\text{ GPa} = 8.1\times 10^{10}\text{ Nm}^{-2}$$
$$d_2 = 5\times 10^{-2}\text{ m},\ d_1 = 3\times 10^{-2}\text{ m}$$
$$\varphi = 2.0° = \frac{\pi}{90}\text{ radians},\ l = 3\text{ m}$$
$$N = \frac{\pi\times 8.1\times\pi}{32\times 3\times 90}(625-81)\times 10^2\text{ Nm}$$
$$= 0.5033\times 10^3\text{ Nm} \approx 0.5\text{ kNm}$$

1.307 The maximum power that can be transmitted by means of a shaft rotating about its axis is clearly $N\omega$ where N is moment of the couple producing the maximum permissible torsion, φ. Thus,
$$P = \frac{\pi r^4 G\omega}{2l}\cdot\omega = 16.9\text{ kW}$$

1.308 Consider an elementary ring of width dr at a distance r from the axis. The part outside exerts couple $N + \frac{dN}{dr}dr$ on this ring while the part inside exerts a couple N in the opposite direction. We have for equilibrium
$$\frac{dN}{dr}dr = -dI\beta$$
where dI is the moment of inertia of the elementary ring, β is the angular acceleration and minus sign is needed because the couple $N(r)$ decreases, with distance, vanishing at the outer radius, $N(r_2) = 0$. Now
$$dI = \frac{m}{\pi\left(r_2^2 - r_1^2\right)} 2\pi r\, dr\, r^2$$

Thus,
$$dN = -\frac{2m\beta}{\pi\left(r_2^2 - r_1^2\right)} r^3\, dr$$

On integration,
$$N = \frac{1}{2}\frac{m\beta}{\left(r_2^2 - r_1^2\right)}\left(r_2^4 - r_1^4\right)$$

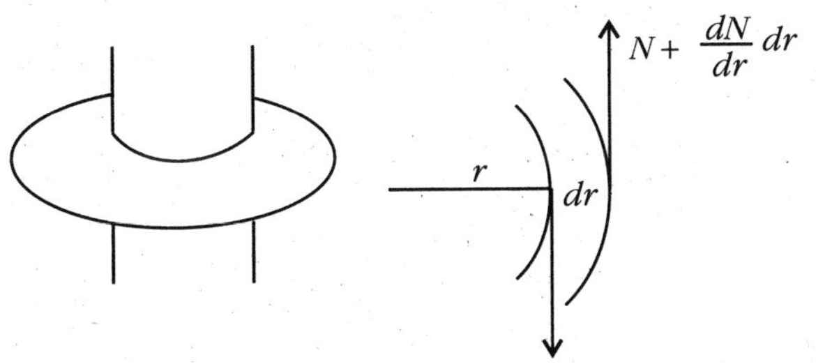

1.309 We assume that the deformation is wholly due to external load, neglecting the effect of the weight of the rod (see next problem). Then a well known formula says, elastic energy per unit volume
$$= \frac{1}{2}\text{stress}\times\text{strain} = \frac{1}{2}\sigma\varepsilon$$

This gives $\frac{1}{2}\frac{m}{\rho}E\varepsilon^2 \approx 0.04$ kJ for the total deformation energy.

1.310 When a rod is deformed by its own weight the stress increases as one moves up, the stretching force being the weight of the portion below the element considered.

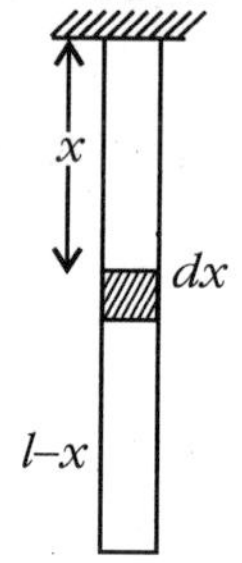

The stress on the element dx is

$$\rho \pi r^2 (l-x)\, g/\pi r^2 = \rho g (l-x)$$

The extension of the element is

$$\Delta dx = d\Delta x = \rho g (l-x)\; dx/E$$

Integrating we get the extension of the whole rod as

$$\Delta l = \frac{1}{2}\frac{\rho g l^2}{E}$$

The elastic energy of the element is

$$\frac{1}{2}\rho g(l-x)\frac{\rho g(l-x)}{E}\pi r^2 dx$$

Integrating

$$\Delta U = \frac{1}{6}\frac{\pi r^2 \rho^2 g^2 l^3}{E} = \frac{2}{3}\pi r^2 l E\left(\frac{\Delta l}{l}\right)^2$$

1.311 The work done to make a loop out of a steel band appears as the elastic energy of the loop and may be calculated from the same.

If the length of the band is l, the radius of the loop $R = l/2\pi$. Now consider an element $ABCD$ of the loop. The elastic energy of this element can be calculated by the same sort of arguments as used to derive the formula for internal bending moment. Consider a fiber at a distance z from the neutral surface PQ. This fiber experiences a force p and undergoes an extension ds where $ds = Zd\varphi$, while $PQ = s = Rd\varphi$. Thus, strain $ds/s = Z/R$.

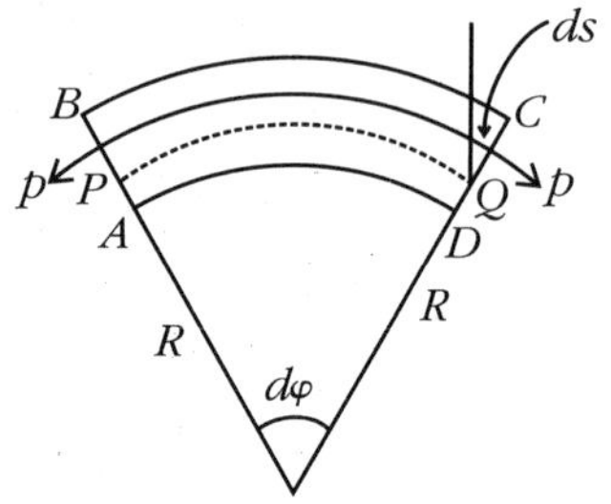

If α is the cross sectional area of the fiber, the elastic energy associated with it is

$$\frac{1}{2}E\left(\frac{Z}{R}\right)^2 Rd\varphi\alpha$$

Summing over all the fibers we get

$$\frac{Ed\varphi}{2R}\sum \alpha Z^2 = \frac{Ed\varphi}{2RI}$$

For the whole loop this gives, using $\int d\varphi = 2\pi$

$$\frac{EI\pi}{R} = \frac{2EI\pi^2}{l} \quad \left(\text{Putting } R = \frac{l}{2\pi}\right)$$

Now,

$$I = \int_{-\delta/2}^{\delta/2} z^2 b\, dZ = \frac{b\delta^3}{12}$$

So the energy is

$$\frac{1}{6}\frac{\pi^2 E b\delta^3}{l} = 0.08 \text{ kJ}$$

Alternate:

Suppose that the steel band was made into a hoop of radius R, then length of the hoop $l = 2\pi R$.

Consider an infinitesimally thin section of radius ρ and thickness $d\rho$ in the hoop. The length of this section of the hoop is $2\pi\rho$. Hence the longitudinal strain corresponding to this section is

$$\varepsilon = \frac{2\pi\rho - 2\pi R}{2\pi R} = \frac{\rho}{R} - 1$$

So elastic energy density is

$$u = \frac{1}{2}E\varepsilon^2 = \frac{1}{2}E\left(\frac{\rho}{R} - 1\right)^2$$

Thus the sought work done is nothing but elastic energy stored.

$$U = \int u dV = \int_{R-\frac{\delta}{2}}^{R+\frac{\delta}{2}} \frac{1}{2}E\left(\frac{\rho}{R} - 1\right)^2 2\pi\rho d\rho\, h = \frac{\pi^2 E h \delta^3}{l} \quad \text{(on integrating)}$$

1.312 A rod should be treated as a solid cylinder. Let the upper end of the cylindrical rod be fixed and let a couple be applied to the lower end in a plane perpendicular to its length (with its axis coinciding with that of the cylinder) twisting it through an angle θ (radians). This is an example of "pure shear". In equilibrium position the twisting couple is equal and opposite to the restoring couple. Let us calculate the value of this couple.

Imagine the cylinder to consist of a large number of co-axial, hollow cylinders, and consider one such hollow cylinder of radius ρ, and radial thickness $d\rho$ Fig. (a). Each radius of the lower end is turned through the same angle θ, but the displacement is the greatest at the rim, decreasing as the centre is approached, where it is reduced to zero.

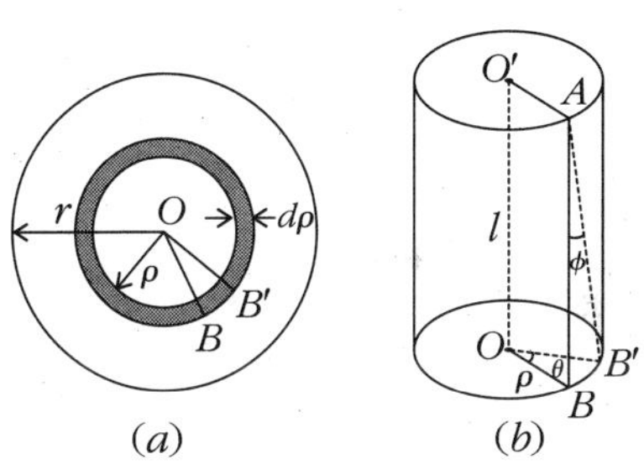

(a) (b)

Let AB Fig. (b) be a line, parallel to the axis, before the cylinder is twisted. On twisting, since the point B shifts to B', the line AB takes up the position AB'. The angle through which this hollow cylinder is sheared is, therefore, $BAB' = \phi$, say. Then clearly,

$$BB' = l\phi \quad \text{(see Figs. (a) and (b))}$$

Also $BB' = \rho\theta$, so, angle of shear or shear strain

$$\phi = \frac{\rho\theta}{l}$$

Obviously, ϕ will have the maximum value where ρ is the greatest, i.e., the maximum strain is on the outermost part of the cylinder, and the least, on the innermost. In other words, the shearing stress is not uniform all through.

Thus, although the angle of shear is the same for any one hollow cylinder, it is different for different cylinders, being the greatest for the outermost and the least for the innermost one.

Since $$G = \frac{\text{shearing stress}}{\text{strain or angle of shear}} = \frac{\sigma}{\phi}$$

We have $$\sigma = G\phi = \frac{G\rho\theta}{l}$$

Now, face area of this hollow cylinder $= 2\pi\rho\, d\rho$.

Therefore total shearing force on this area $= 2\pi\rho\, d\rho \times \dfrac{G\rho\theta}{l} = 2\pi G\dfrac{\theta}{l}\rho^2\, d\rho$

Therefore, moment of this force about the axis OO' Fig. (b) of the cylinder is equal to

$$\left(\frac{2\pi G\theta\rho^2 d\rho}{l}\right)\rho = \frac{2\pi G\theta\rho^3 d\rho}{l}$$

Integrating this expression between the limits, $\rho = 0$ and $\rho = r$, we have total twisting couple on the cylinder,

$$N(\theta) = \int_0^r 2\pi G\frac{\theta}{l}\rho^3 d\rho$$

$$= \frac{2\pi G\theta}{l}\int_0^r \rho^3 d\rho = \frac{2\pi G\theta}{l}\left[\frac{\rho^4}{4}\right]_0^r = \frac{\pi G\theta r^4}{2l} = C\theta$$

where $C = \pi G r^4/2l$ is twisting couple per unit twist of the cylinder or torsional rigidity of the cylinder. Thus, when the rod is twisted through an angle θ, a couple

$$N(\theta) = \frac{\pi r^4 G}{2l}\theta$$

appears to resist this. Work done in twisting the rod by an angle φ or elastic energy stored is then

$$U = \int_0^\varphi N(\theta)\, d(\theta) = \frac{\pi r^4 G}{4l}\varphi^2 = 7\text{ J} \quad \text{(on substituting values)}$$

1.313 The energy between radii r and $r + dr$ is given by differentiation of

$$U = \frac{\pi r^4 G}{4l}\varphi^2 \quad \text{(see solution of problem 1.312)}$$

So, $$dU = \frac{\pi r^3 dr}{l}G\varphi^2$$

Its density is $$u = \frac{dU}{dV} = \frac{\pi r^3 dr}{2\pi r dr\, l}\frac{G\varphi^2}{l} = \frac{1}{2}\frac{G\varphi^2 r^2}{l^2}$$

Alternate:

The sought energy density

$$u = \frac{1}{2}G\phi^2 \quad \text{(where } \phi \text{ is angle of shear or shear strain)}$$

But, $$\phi = \left(r\frac{\varphi}{l}\right) \quad \text{(see solution of problem 1.312)}$$

So, $$u = \frac{1}{2}\frac{G\varphi^2 r^2}{l^2}$$

1.314 The energy density is as usual 1/2 stress × strain. Stress is the pressure $\rho g h$. Strain is $\beta \times \rho g h$ by definition of β. Thus

$$u = \frac{1}{2}\beta(\rho g h)^2 = 23.5\text{ kJ/m}^3 \quad \text{(on substituting values)}$$

1.7 Hydrodynamics

1.315 Between points 1 and 2, fluid particles are in nearly circular motion and therefore have centripetal acceleration. The force for this acceleration, like for any other situation in an ideal fluid, can only come from the pressure variation along the line joining 1 and 2. This requires that pressure at 1 should be greater than the pressure at 2 so that the fluid particles can have required acceleration. If there is no turbulence, the motion can be taken as irrotational. Then by considering

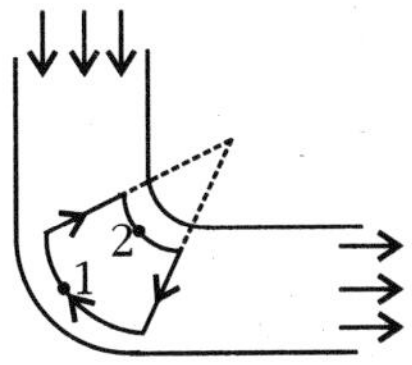

$$\oint \mathbf{v} \cdot d\mathbf{l} = 0$$

along the circuit shown, we infer that $v_2 < v_1$.

The portion of the circuit near 1 and 2 are streamlines while the other two arms are at right angle to streamlines.

(By electrostatic analogy, the density of streamlines is proportional to the velocity.)

1.316 From the conservation of mass

$$v_1 S_1 = v_2 S_2 \tag{1}$$

But $S_1 < S_2$ as shown in the figure of the problem book, therefore

$$v_1 > v_2$$

As every streamline is horizontal between 1 and 2, Bernoulli's theorem becomes

$$p + \frac{1}{2}\rho v^2 = \text{constant}$$

which gives $$p_1 < p_2 \quad \text{as} \quad v_1 > v_2$$

As the difference in height of the water column is Δh, therefore

$$p_2 - p_1 = \rho g \Delta h \tag{2}$$

From Bernoulli's theorem between points 1 and 2 of a streamline

$$p_1 + \frac{1}{2}\rho v_1^2 = p_2 + \frac{1}{2}\rho v_2^2$$

or $$p_2 - p_1 = \frac{1}{2}\rho(v_1^2 - v_2^2)$$

or $$\rho g \Delta h = \frac{1}{2}\rho(v_1^2 - v_2^2) \quad \text{(using Eq. 2)} \tag{3}$$

Using Eq. (1) in (3), we get

$$v_1 = S_2\sqrt{\frac{2g\Delta h}{S_2^2 - S_1^2}}$$

Hence the sought volume of water flowing per second

$$Q = v_1 S_2 = S_1 S_2\sqrt{\frac{2g\Delta h}{S_2^2 - S_1^2}}$$

1.317 Applying Bernoulli's theorem for the point A and B,

$$p_A = p_B + \frac{1}{2}\rho v^2 \quad \text{as} \quad v_A = 0$$

or

$$\frac{1}{2}\rho v^2 = p_A - p_B = \Delta h \rho_0 g$$

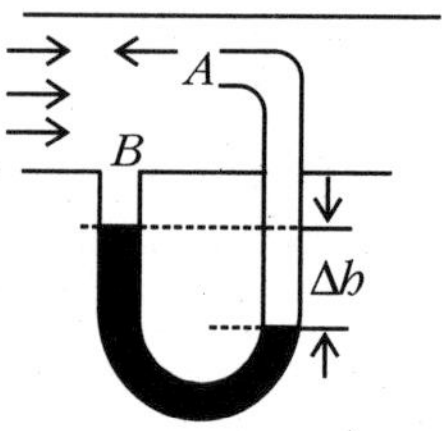

So,

$$v = \sqrt{\frac{2\Delta h \rho_0 g}{\rho}}$$

Thus, rate of flow of gas,

$$Q = Sv = S\sqrt{\frac{2\Delta h \rho_0 g}{\rho}}$$

The gas flows over the tube past it at B. But at A the gas becomes stationary as the gas will move into the tube which already contains gas.

In applying Bernoulli's theorem we should remember that

$$\frac{p}{\rho} + \frac{1}{2}v^2 + gz$$

is constant along a streamline. In the present case, we are really applying Bernoulli's theorem somewhat indirectly. The streamline at A is not the streamline at B. Nevertheless the result is correct. To be convinced of this, we need only apply Bernoulli's theorem to the streamline that goes through A by comparing the situation at A with that above B on the same level. In steady conditions, this agrees with the result derived because there cannot be a transverse pressure differential.

1.318 Since, the density of water is greater than that of kerosene oil, it will collect at the bottom. Now, pressure due to water level equals $h_1\rho_1 g$ and pressure due to kerosene oil level equals $h_2\rho_2 g$. So, net pressure becomes $h_1\rho_1 g + h_2\rho_2 g$.

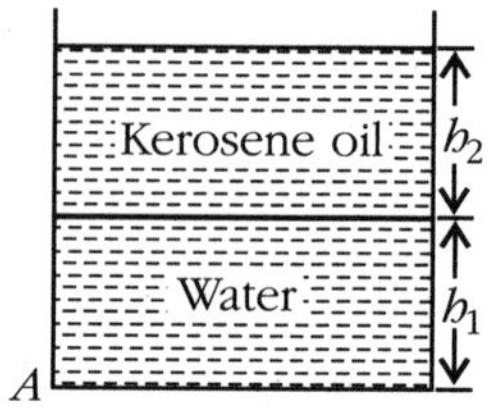

From Bernoulli's theorem, this pressure energy will be converted into kinetic energy while flowing through the hole A.

i.e.,

$$h_1\rho_1 g + h_2\rho_2 g = \frac{1}{2}\rho_1 v^2$$

Hence,

$$v = \sqrt{2\left(h_1 + h_2\frac{\rho_2}{\rho_1}\right)g} = 3 \text{ m/s}$$

1.319 Let H be the total height of the water column and the hole is made at a height h from the bottom.

Then from Bernoulli's theorem

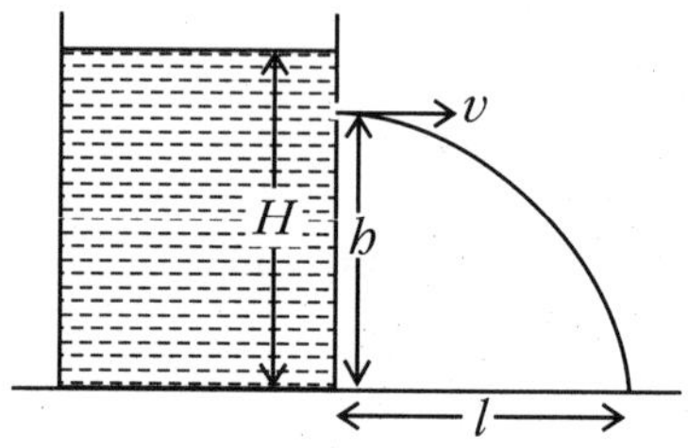

$$\frac{1}{2}\rho v^2 = (H - h)\rho g$$

or

$$v = \sqrt{(H - h)2g}$$

which is directed horizontally.

For the horizontal range,

$$l = vt$$

$$= \sqrt{2g(H - h)} \cdot \sqrt{\frac{2h}{g}}$$

$$= 2\sqrt{(Hh - h^2)}$$

Now, for maximum l

$$\frac{d(Hh - h^2)}{dh} = 0$$

which yields $$h = \frac{H}{2} = 25 \text{ cm}$$

1.320 Let the velocity of the water jet near the orifice be v'. Then applying Bernoulli's theorem,

$$\frac{1}{2}\rho v^2 = h_0 \rho g + \frac{1}{2}\rho v'^2$$

or $$v' = \sqrt{v^2 - 2gh_0} \tag{1}$$

Here the pressure term on both sides is the same and is equal to the atmospheric pressure.

Now, if water rises up to a height h, then at this height, whole of its kinetic energy will be converted into potential energy. So,

$$\frac{1}{2}\rho v'^2 = \rho g h \quad \text{or} \quad h = \frac{v'^2}{2g}$$

$$= \frac{v^2}{2g} - h_0 = 20 \text{ cm} \quad \text{(using Eq. 1)}$$

1.321 Water flows through the small clearance into the orifice. Let d be the clearance, then from the equation of continuity

$$(2\pi R_1 d)v_1 = (2\pi r d)v = (2\pi R_2 d)v_2$$

or $$v_1 R_1 = vr = v_2 R_2 \tag{1}$$

where v_1, v_2 and v are respectively the inward radial velocities of the fluid at 1, 2 and 3. Now by Bernouli's theorem just before 2 and just after it in the clearance

$$p_0 + h\rho g = p_2 + \frac{1}{2}\rho v_2^2 \tag{2}$$

Applying the same theorem at 3 and 1, we find that this also equals

$$p + \frac{1}{2}\rho v^2 = p_0 + \frac{1}{2}\rho v_1^2 \tag{3}$$

(since the pressure in the orifice is p_0).

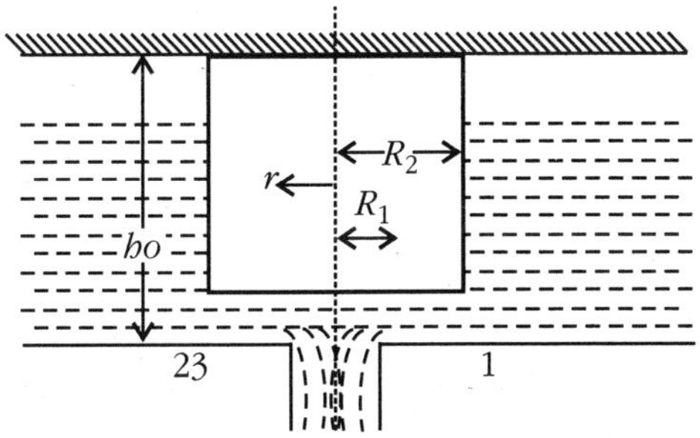

From Eqs. (2) and (3) we also have

$$v_1 = \sqrt{2gh} \tag{4}$$

and $$p = p_0 + \frac{1}{2}\rho v_1^2\left(1 - \left(\frac{v}{v_1}\right)^2\right)$$

$$= p_0 + h\rho g\left(1 - \left(\frac{R_1}{r}\right)^2\right) \quad \text{(using Eqs. 1 and 4)}$$

1.322 As constant force is acting on the piston, so inside water pressure p has to be constant.

As area of orifice is very small in comparison to the piston, so from Bernoulli's Eq. from just inside and outside points of the orifice we get

$$p = p_0 + \frac{1}{2}\rho v_{\text{efflux}}^2$$

or, $$(p - p_0) = \frac{1}{2}\rho v_{\text{efflux}}^2 \quad (1)$$

From Eq. (1) for constant p, we have v_{efflux} = constant also.

So, $$v_{\text{efflux}} = \frac{V}{st} \quad (2)$$

As $(p - p_0)$ is constatnt, so sought work done

$$W = (p - p_0)V$$

$$= \frac{1}{2}\rho v_{\text{efflux}}^2 V, \quad \text{using equation (1)}$$

$$= \frac{1}{2}\rho\frac{V^3}{s^2t^2}, \quad \text{using equation (2)}$$

Note: *After equation (2), one can apply work energy theorem to get the sought work done.*

1.323 Water jet coming out from the orifice of area s at an arbitrary instant of time when the height of water column in the cylindrical vessel is H.

$$v_{\text{efflux}} = \sqrt{2gH}$$

As the rate of out-flux should be equal to the rate with which volume of the water in the vessel decreases

$$-S\frac{dH}{dt} = s\sqrt{2gH}$$

Hence, $$t = -\frac{S}{s\sqrt{2g}}\int_h^0 \frac{dH}{\sqrt{H}} = \frac{S}{s}\sqrt{\frac{2h}{g}}$$

1.324 In a rotating frame (with constant angular velocity), the Eulerian equation is

$$-\nabla p + \rho\mathbf{g} + 2\rho(\mathbf{v}' \times \boldsymbol{\omega}) + \rho\omega^2\mathbf{r} = \rho\frac{d\mathbf{v}'}{dt}$$

In the frame of rotating tube the liquid in the "column" is practically static because the orifice is sufficiently small. Thus the Eulerian equation in projection form along $\mathbf{r}$ (which is the position vector of an arbitrary liquid element of length dr relative to the rotation axis) reduces to

$$\frac{-dp}{dr} + \rho\omega^2 r = 0$$

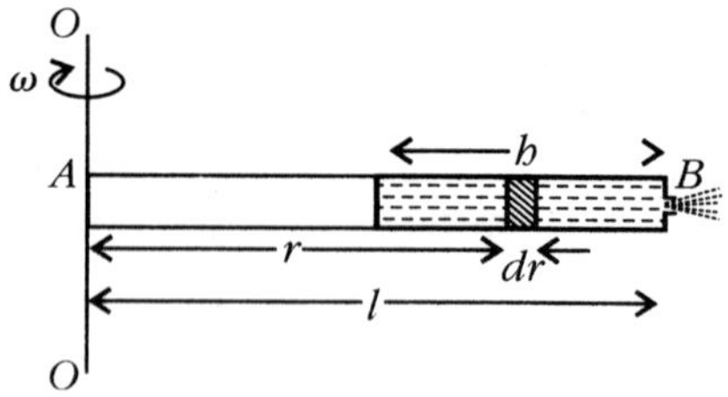

or $$dp = \rho w^2 r\, dr$$

So, $$\int_{p_0}^{p} dp = \rho\omega^2 \int_{(l-h)}^{r} r\, dr$$

Thus, $$p(r) = p_0 + \frac{\rho\omega^2}{2}[r^2 - (l - h)^2] \quad (1)$$

Hence, the pressure at end B just before the orifice is

$$p(l) = p_0 + \frac{\rho\omega^2}{2}(2lh - h^2) \quad (2)$$

Then applying Bernoulli's theorem at the orifice for the points just inside and outside of the end B

$$p_0 + \frac{1}{2}\rho\omega^2(2lh - h^2) = p_0 + \frac{1}{2}\rho v^2 \quad \text{(where } v \text{ is the sought velocity)}$$

So, $$v = wh\sqrt{\frac{2l}{h} - 1}$$

1.325 The Euler's equation is

$$\rho\frac{d\mathbf{v}}{dt} = \mathbf{f} - \nabla p = -\nabla(p + \rho gz) \quad \text{(where } z \text{ is vertically upwards)}$$

Now,
$$\frac{d\mathbf{v}}{dt} = \frac{\partial \mathbf{v}}{\partial t} + (\mathbf{v}\cdot\nabla)\mathbf{v} \tag{1}$$

But
$$(\mathbf{v}\cdot\nabla)\mathbf{v} = \nabla\left(\frac{1}{2}v^2\right) - \mathbf{v}\times \text{Curl } \mathbf{v} \tag{2}$$

If we consider the steady (i.e., $\partial \mathbf{v}/\partial t = 0$) flow of an incompressible fluid then ρ = constant and as the motion is irrotational, Curl $\mathbf{v} = 0$

So, from Eqs. (1) and (2)

$$\rho\nabla\left(\frac{1}{2}v^2\right) = -\nabla(p + \rho gz)$$

$$\nabla\left(p + \frac{1}{2}\rho v^2 + \rho gz\right) = 0$$

or
$$p + \frac{1}{2}\rho v^2 + \rho gz = \text{constant}$$

1.326 Net reaction force acting by outgoing and incoming liquids, from and into a container

$$\mathbf{F} = \Sigma -\rho_i(\mathbf{v}_i\cdot\mathbf{S}_i)\mathbf{v}_i,$$

Where $\mathbf{v}_i$ is velocity of outgoing (incoming) liquid of density ρ_i relative to the container at the location having the small opening of area $\mathbf{S}_i$.

Let the velocity of water flowing through A be v_A and that through B be v_B,

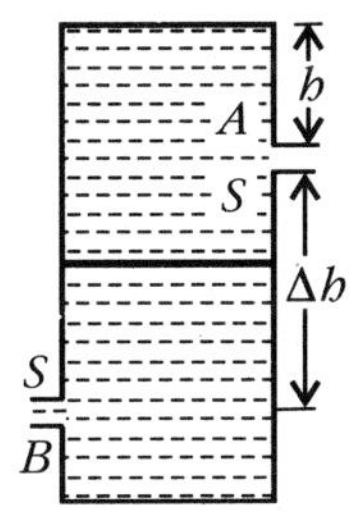

Hence, the sought net force,

$$F = \rho S(v_B^2 - v_A^2) \quad \text{towards right as } \mathbf{F}_A \uparrow\downarrow \mathbf{F}_B \tag{1}$$

Applying Bernoulli's theorem to the liquid flowing out of A we get

$$p_0 + \rho gh = p_0 + \frac{1}{2}\rho v_A^2$$

or,
$$v_A^2 = 2gh \tag{2}$$

And similarly at B

$$p_0 + \rho g(h + \Delta h) = p_0 + \frac{1}{2}\rho v_B^2$$

or,
$$v_B^2 = 2g(h + \Delta h) \tag{3}$$

Hence from equations (2) and (3)

$$v_B^2 - v_A^2 = 2g\,\Delta h \tag{4}$$

Using equation (4) in equation (1)

$$F = 2\rho gS\Delta h = 0.50 \text{ N}$$

1.327 Consider an element of height dy at a distance y from the top. The velocity of the fluid coming out of the element is

$$v = \sqrt{2gy}$$

The force of reaction dF due to this is $dF = \rho dAv^2 = \rho(bdy)2gy$, as in the previous problem.

Integrating
$$F = \rho gb \int_{h-1}^{h} 2y\,dy$$
$$= \rho gb[h^2 - (h-l)^2] = \rho gbl(2h - l)$$
$$= 5 \text{ N} \quad \text{(on substituting values)}$$

(The slit runs from a depth $h - l$ to a depth h from the top.)

1.328 Let the velocity of water flowing through the tube at a certain instant of time be u, then $u = Q/\pi r^2$, where Q is the rate of flow of water and πr^2 is the cross section area of the tube.

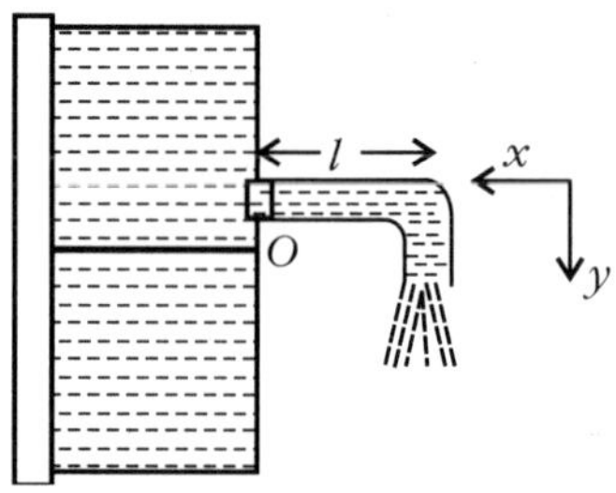

From impulse momentum theorem, for the stream of water striking the tube corner, in x-direction in the time interval dt,

$$F_x\, dt = -\rho Q u\, dt \quad \text{or} \quad F_x = -\rho Q u$$

Similarly,
$$F_y = \rho Q u$$

Therefore, the force exerted on the water stream by the tube,

$$\mathbf{F} = -\rho Q u \mathbf{i} + \rho Q u \mathbf{j}$$

According to third law, the reaction force on the tube's wall by the stream equals $(-\mathbf{F})$

$$\rho Q u \mathbf{i} - \rho Q u \mathbf{j}$$

Hence, the sought moment of force about O becomes

$$\mathbf{N} = -l\mathbf{i} \times (\rho Q u \mathbf{i} - \rho Q u \mathbf{j}) = \rho Q u l \mathbf{k} = \frac{\rho Q^2}{\pi r^2} l \mathbf{k}$$

and
$$|\mathbf{N}| = \frac{\rho Q^2 l}{\pi r^2} = 0.70 \text{ Nm}$$

1.329 Let us take an arbitrary cross section of radius r of the narrowing tube. If p is the inside pressure at the location of taken ring element of radius r and width dl, and p_0 is outside atmosphere pressure, from symmetry of the problem, net force exerted on the taken ring element due to the inside and outside pressure difference.

$$= dF \sin\theta = (p - p_0) 2\pi r dl \sin\theta$$
$$= (p - p_0) 2\pi r dr \quad (\text{because } dl \sin\theta = dr)$$

Hence the sought net force on the tube

$$F = F_x = \int dF \sin\theta = \int (p - p_0)\, 2\pi r\, dr \qquad (1)$$

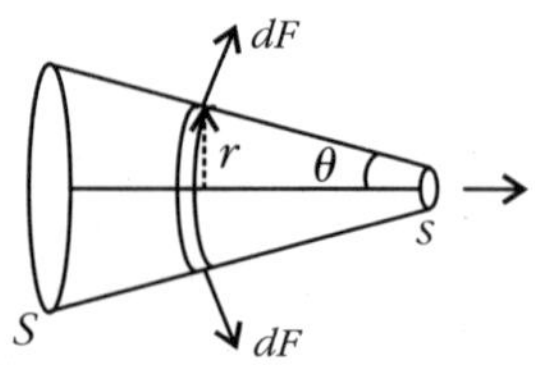

From Bernoulli's equation

$$p - p_0 = \frac{1}{2}\rho(v_{\text{efflux}}^2 - v^2),$$

where
$$v_{\text{efflux}} = \sqrt{2gh} \qquad (2)$$

and v is the speed of water at the location of taken ring element.

From conservation of mass

$$\rho v \pi r^2 = \rho v_{\text{efflux}} \pi r_1^2$$

or $$v = v_{\text{efflux}} \frac{r_1^2}{r^2} \qquad (3)$$

(where r_1 is the radius at open end).

Using Eq. (3) in Eq. (2), we get

$$p - p_0 = \frac{1}{2}\rho\left(v_{\text{efflux}}^2 - v_{\text{efflux}}^2 \frac{r_1^4}{r^4}\right)$$

$$= \frac{1}{2}\rho v_{\text{efflux}}^2\left(1 - \frac{r_1^4}{r^4}\right) = \rho g h\left(1 - \frac{r_1^4}{r^4}\right) \qquad (4)$$

Using Eq. (4) in Eq. (1), we get

$$F = \int_{r_1}^{r_2} \rho g h\left(1 - \frac{r_1^4}{r^4}\right) 2\pi r dr \qquad (5)$$

After integration and using $\pi r_2^2 = S$ and $\pi r_1^2 = s$, we get

$$F = \frac{\rho g h(S-s)^2}{S} = 6 \text{ N} \quad \text{(on substituting the values)}$$

Note: *If we try to calculate F from the momentum change of the liquid flowing out we will be wrong even as regards the sign of the force. There is of course the effect of pressure at S and s but quantitative derivation of F from Newton's law is difficult.*

1.330 The Euler's equation is

$$\rho \frac{d\mathbf{v}}{dt} = \mathbf{f} - \nabla p$$

in the space fixed frame where $\mathbf{f} = -\rho g \mathbf{k}$ downward. We assume incompressible fluid so ρ is constant.

Then $\mathbf{f} = -\nabla(\rho g z)$ where z is the height vertically upwards from some fixed origin. We go to the rotating frame where the equation becomes

$$\rho \frac{d\mathbf{v}}{dt} = -\nabla(p + \rho g z) + \rho\omega^2 \mathbf{r} + 2\rho(\mathbf{v} \times \boldsymbol{\omega})$$

The additional terms on the right are the well known coriolis and centrifugal forces. In the frame rotating with the liquid $\mathbf{v} = 0$, so

$$\nabla\left(p + \rho g z - \frac{1}{2}\rho\omega^2 r^2\right) = 0$$

or $$p + \rho g z - \frac{1}{2}\rho\omega^2 r^2 = \text{constant}$$

On the free surface p = constant, thus

$$z = \frac{\omega^2}{2g} r^2 + \text{constant}$$

If we choose the origin at point $r = 0$ (i.e., the axis) of the free surface then "constant" = 0 and

$$z = \frac{\omega^2}{2g} r^2 \quad \text{(the parabolic of revolution)}$$

At the bottom z = constant.

So, $$p = \frac{1}{2}\rho\omega^2 r^2 + \text{constant}$$

If $p = p_0$ on the axis at the bottom, then

$$p = p_0 + \frac{1}{2}\rho\omega^2 r^2$$

1.331 When the disk rotates the fluid in contact with it co-rotates, but the fluid in contact with the walls of the cavity does not rotate. A velocity gradient is then set up leading to viscous forces.

At a distance r from the axis, the linear velocity is ωr so there is a velocity gradient $\omega r/h$ both in the upper and lower clearance. The corresponding force on the element whose radial width is dr is

$$\eta 2\pi r dr \frac{\omega r}{h} \quad \left(\text{from the formula } F = \eta A \frac{dv}{dx}\right)$$

The torque due to this force is

$$\eta 2\pi r dr \frac{\omega r}{h} r$$

and the net torque considering both the upper and lower clearance is

$$2\int_0^R \eta 2\pi r^3 dr \frac{\omega}{h} = \pi R^4 \eta \frac{\omega}{h}$$

So, power developed is

$$P = \frac{\pi R^4 \omega^2 \eta}{h} = 9.05 \text{ W} \quad \text{(on substituting values)}$$

(As instructed, end effects, i.e., rotation of fluid in the clearance $r \geq R$, has been neglected.)

1.332 Let us consider a coaxial cylinder of radius r and thickness dr, then force of friction or viscous force on this elemental layer is given by

$$F = 2\pi r l \eta \frac{dv}{dr}$$

This force must be constant from layer to layer so that steady motion may be possible.

So, $$\frac{F dr}{r} = 2\pi l \eta \, dv \qquad (1)$$

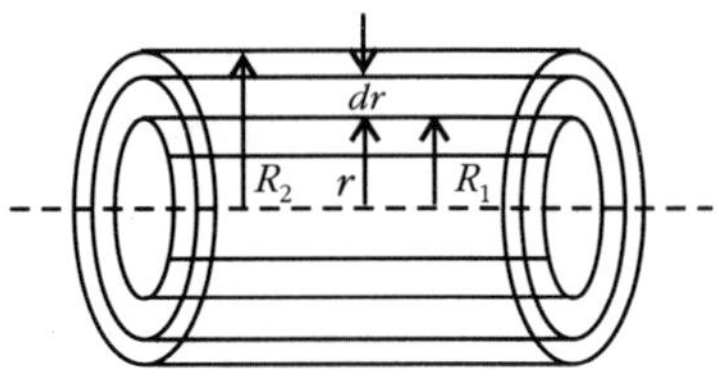

Integrating, $$F\int_{R_2}^{r} \frac{dr}{r} = 2\pi l \eta \int_0^v dv$$

or $$F \ln\left(\frac{r}{R_2}\right) = 2\pi l \eta v \qquad (2)$$

Putting $r = R_1$, and $v = v_0$ we get,

$$F \ln\frac{R_1}{R_2} = 2\pi l \eta v_0 \qquad (3)$$

Dividing Eqs. (2) by (3), we get

$$v = v_0 \frac{\ln r/R_2}{\ln R_1/R_2}$$

Note: *The force F is supplied by the agency which tries to carry the inner cylinder with velocity* v_0.

1.333 (a) Let us consider an elemental cylinder of radius r and thickness dr then from Newton's formula

$$F = 2\pi r l \eta r \frac{d\omega}{dr} = 2\pi l \eta r^2 \frac{d\omega}{dr} \qquad (1)$$

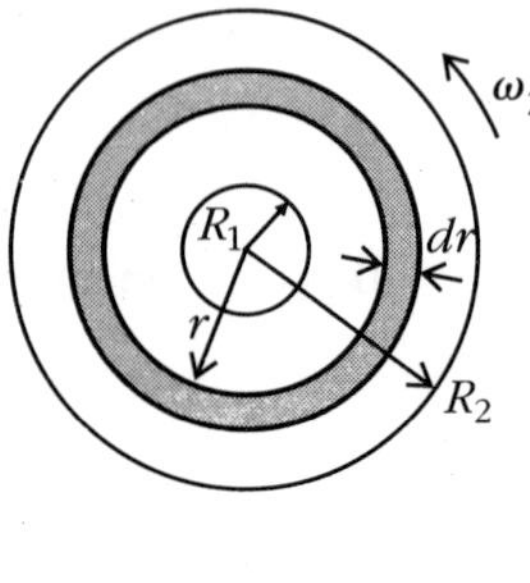

and moment of this force acting on the element,

$$N = 2\pi r^2 l \eta \frac{d\omega}{dr} r = 2\pi r^3 l \eta \frac{d\omega}{dr}$$

or $$2\pi l \eta d\omega = N \frac{dr}{r^3} \qquad (2)$$

As in the previous problem N is constant when conditions are steady

Integrating, $$2\pi l\eta\int_0^{\omega} d\omega = N\int_{R_1}^{r}\frac{dr}{r^3}$$

or $$2\pi l\eta\omega = \frac{N}{2}\left[\frac{1}{R_1^2}-\frac{1}{r^2}\right] \qquad (3)$$

Putting $r = R_2$, $\omega = \omega_2$, we get

$$2\pi l\eta\omega_2 = \frac{N}{2}\left[\frac{1}{R_1^2}-\frac{1}{R_2^2}\right] \qquad (4)$$

From Eqs. (3) and (4)

$$\omega = \omega_2\frac{R_1^2R_2^2}{R_2^2-R_1^2}\left[\frac{1}{R_1^2}-\frac{1}{r^2}\right]$$

(b) From Eq. (4) for $r = R_2$,

Sought torque per unit length $$= 4\pi\eta\omega_2\frac{R_1^2R_2^2}{R_2^2-R_1^2}$$

1.334 (a) Let dV be the volume flowing per second through the cylindrical shell of thickness dr then,

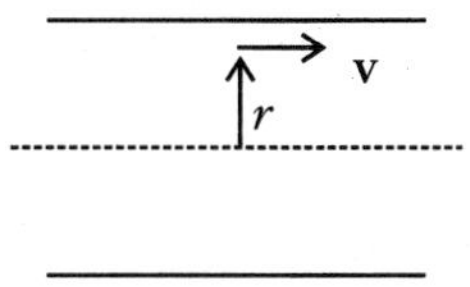

$$dV = (2\pi r dr)v_0\left(1-\frac{r^2}{R^2}\right) = 2\pi v_0\left(r-\frac{r^3}{R^2}\right)dr$$

The total volume is

$$V = 2\pi v_0\int_0^R\left(r-\frac{r^3}{R^2}\right)dr = 2\pi v_0\frac{R^2}{4} = \frac{\pi}{2}R^2v_0$$

(b) Let dE be the kinetic energy within the above cylindrical shell. Then,

$$dT = \frac{1}{2}(dm)v^2 = \frac{1}{2}\left[\rho(2\pi r dr)l\right]v^2$$

$$= \frac{1}{2}(2\pi l\rho)r dr v_0^2\left(1-\frac{r^2}{R^2}\right)^2 = \pi l\rho v_0^2\left[r-\frac{2r^3}{R^2}+\frac{r^5}{R^4}\right]dr$$

Hence, total energy of the fluid,

$$T = \pi l\rho v_0^2\int_0^R\left(r-\frac{2r^3}{R^2}+\frac{r^5}{R^4}\right)dr = \frac{\pi R^2\rho l v_0^2}{6}$$

(c) Here frictional force is the shearing force on the tube, exerted by the fluid, which equals $-\eta S dv/dr$.

Given, $$v = v_0\left(1-\frac{r^2}{R^2}\right)$$

So,
$$\frac{dv}{dr} = -2v_0 \frac{r}{R^2}$$

At $r = R$,
$$\frac{dv}{dr} = -\frac{2v_0}{R}$$

Then, viscous force is given by $F = -\eta(2\pi Rl)\left(\frac{dv}{dr}\right)_{r=R}$

$$= -2\pi R\eta l\left(-\frac{2v_0}{R}\right) = 4\pi\eta v_0 l$$

(d) Taking a cylindrical shell of thickness dr and radius r, viscous force,

$$F = -\eta(2\pi rl)\frac{dv}{dr}$$

Let Δp be the pressure difference, then net force on the element is

$$\Delta p\pi r^2 + 2\pi\eta lr\frac{dv}{dr}$$

But, since the flow is steady, $F_{net} = 0$

or
$$\Delta p = \frac{-2\pi l\eta r\, dv/dr}{\pi r^2} = \frac{-2\pi l\eta r(-2v_0 r/R^2)}{\pi r^2} = \frac{4\eta v_0 l}{R^2}$$

1.335 The loss of pressure head in traveling a distance l is seen from the middle section to be $h_2 - h_1 = 10$ cm. Since $h_2 - h_1 = h_1$ in our problem and $h_3 - h_2 = 15$ cm $= 5 + h_2 - h_1$, we see that pressure head of 5 cm remains incompensated and must be converted into kinetic energy of the liquid flowing out. Thus,

$$\frac{\rho v^2}{2} = \rho g\Delta h \quad (\text{where } \Delta h = (h_3 - h_2))$$

Thus,
$$v = \sqrt{2g\Delta h} \cong 1 \text{ m/s}$$

1.336 We know that, Reynold's number (R_e) is defined as, $R_e = \rho vl/\eta$, where v is the velocity, l is the characteristic length, and η the coefficient of viscosity. In the case of circular cross-section the characteristic length is the diameter of cross-section d, and v is taken as average velocity of flow of liquid.

Now, R_{e_1} (Reynold's number at x_1 from the pipe end) $= \dfrac{\rho d_1 v_1}{m\eta}$, where v_1 is the velocity at distance x_1.

Similarly,
$$R_{e_2} = \frac{\rho d_2 v_2}{\eta}$$

So,
$$\frac{R_{e_1}}{R_{e_2}} = \frac{d_1 v_1}{d_2 v_2}$$

From equation of continuity,

$$A_1 v_1 = A_2 v_2$$

or
$$\pi r_1^2 v_1 = \pi r_2^2 v_2 \quad \text{or} \quad d_1 v_1 r_1 = d_2 v_2 r_2$$

$$\frac{d_1 v_1}{d_2 v_2} = \frac{r_2}{r_1} = \frac{r_0 e^{-\alpha x_2}}{r_0 e^{-\alpha x_1}} = e^{-\alpha\Delta x} \quad (\text{as } x_2 - x_1 = \Delta x)$$

Thus,
$$\frac{R_{e_2}}{R_{e_1}} = e^{\alpha\Delta x} = 5$$

1.337 We know that Reynold's number for turbulent flow is greater than that for laminar flow.

Now, $(R_e)_l = \dfrac{\rho v d}{\eta} = \dfrac{2\rho_1 v_1 r_1}{\eta_1}$ and $(R_e)_t = \dfrac{2\rho_2 v_2 r_2}{\eta}$

But, $(R_e)_t \geq R$

So, $v_{2_{\min}} = \dfrac{\rho_1 v_1 r_1 \eta_2}{\rho_2 r_2 \eta_1} = 5\ \mu\text{m/s}$ (on substituting values)

1.338 We have
$$R = \frac{v\rho_0 d}{\eta}$$

and v is given by
$$6\pi\eta r v = \frac{4\pi}{3} r^2(\rho - \rho_0)g$$

(where ρ = density of lead, ρ_0 = density of glycerine)

$$v = \frac{2}{9\eta}(\rho - \rho_0)gr^2 = \frac{1}{18\eta}(\rho - \rho_0)gd^2$$

Thus, $\dfrac{1}{2} = \dfrac{1}{18\eta^2}(\rho - \rho_0)g\rho_0 d^3$

and $d = [9\eta^2/\rho_0(\rho - \rho_0)g]^{1/3} = 5.2$ mm (on substituting values)

1.339
$$m\frac{dv}{dt} = mg - 6\pi\eta r v$$

or $\dfrac{dv}{dt} + \dfrac{6\pi\eta r}{m} v = g$

or $\dfrac{dv}{dt} + kv = g \quad \left(\text{where } k = \dfrac{6\pi\eta r}{m}\right)$

or $e^{kt}\dfrac{dv}{dt} + ke^{kt}v = ge^{kt}$ or $\dfrac{d}{dt}e^{kt}v = ge^{kt}$

or $ve^{kt} = \dfrac{g}{k}e^{kt} + C$ or $v = \dfrac{g}{k} + Ce^{-kt}$ (where C is constant)

Since $v = 0$ for $t = 0$, so $0 = \dfrac{g}{k} + C$

or, $C = -\dfrac{g}{k}$

Thus, $v = \dfrac{g}{k}(1 - e^{-kt})$

The steady state velocity is g/k.

v differs from g/k by n, where
$$-e^{-kt} = n$$

or $t = \dfrac{1}{k}\ln n$

Thus, $\dfrac{1}{k} = -\dfrac{(4\pi/3)r^3 P}{6\pi\eta r} = -\dfrac{4r^2\rho}{18\eta} = -\dfrac{d^2\rho}{18\eta}$

(We have neglected buoyancy in olive oil.)

$$t = \frac{-\rho d^2}{18\eta}\ln n$$
$$= 0.20\ \text{s} \quad \text{(on substituting values)}$$

1.8 Relativistic Mechanics

1.340 From the formula for length contraction

$$\left(l_0 - l_0\sqrt{1-\frac{v^2}{c^2}}\right) = \eta l_0$$

So,
$$1-\frac{v^2}{c^2} = (1-\eta)^2 \quad \text{or} \quad v = c\sqrt{\eta(2-\eta)}\,.$$

1.341 (a) In the frame in which the triangle is at rest, the space coordinates of the vertices are (000), $\left(a\frac{\sqrt{3}}{2}, +\frac{a}{2}, 0\right)\left(a\frac{\sqrt{3}}{2}, -\frac{a}{2}, 0\right)$, all measured at the same time t. In the moving frame the corresponding coordinates at time t' are

$$A : (vt', 0, 0),\ B : \left(\frac{a}{2}\sqrt{3}\sqrt{1-\beta^2} + vt', \frac{a}{2}, 0\right) \quad \text{and} \quad C : \left(\frac{a}{2}\sqrt{3}\sqrt{1-\beta^2} + vt', -\frac{a}{2}, 0\right)$$

The perimeter P is then

$$P = a + 2a\left(\frac{3}{4}(1-\beta^2) + \frac{1}{4}\right)^{1/2} = a\left[1 + \sqrt{4-3\beta^2}\right]$$

(b) The coordinates in the rest frame are shown at time t. The coordinates in the moving frame are

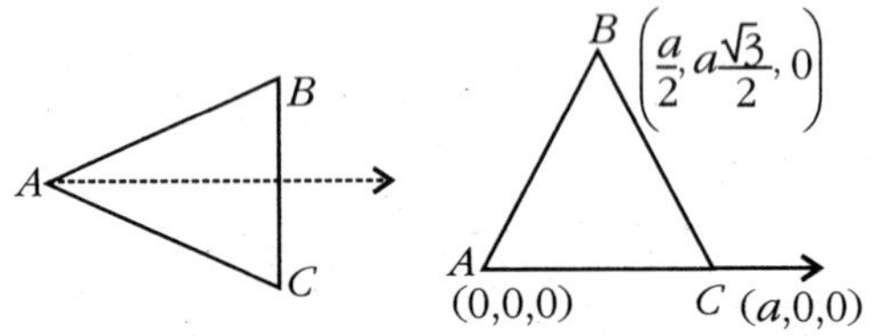

$$A : (vt', 0, 0),\ B : \left(\frac{a}{2}\sqrt{1-\beta^2} + vt', a\frac{\sqrt{3}}{2}, 0\right),\ C : \left(a\sqrt{1-\beta^2} + vt', 0, 0\right)$$

The perimeter P is then

$$P = a\sqrt{1-\beta^2} + \frac{a}{2}[1-\beta^2+3]^{1/2} \times 2$$

$$= a\left(\sqrt{1-\beta^2} + \sqrt{4-\beta^2}\right) \quad \left(\text{where } \beta = \frac{v}{c}\right)$$

1.342 In the rest frame, the coordinates of the ends of the rod are in terms of proper length l_0

$$A : (0, 0, 0),\ B : (l_0 \cos\theta_0, l_0 \sin\theta_0, 0)$$

at time t. In the laboratory frame the coordinates at time t' are

$$A : (vt', 0, 0),\ B : (l_0 \cos\theta_0\sqrt{1-\beta^2} + vt', l_0 \sin\theta_0, 0)$$

Therefore, we can write

$$l\cos\theta = l_0\cos\theta_0\sqrt{1-\beta^2} \quad \text{and} \quad l\sin\theta = l_0\sin\theta_0$$

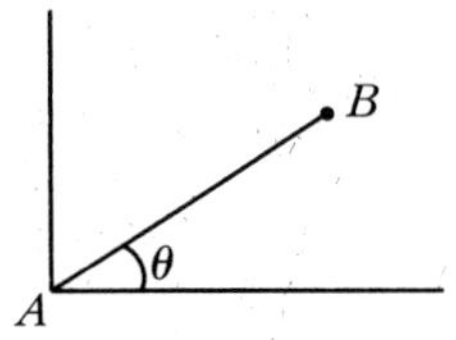

Hence,
$$l_0^2 = (l^2)\left(\frac{\cos^2\theta + (1-\beta^2)\sin^2\theta}{1-\beta^2}\right)$$

or $$l_0 = l\sqrt{\frac{1-\beta^2\sin^2\theta}{1-\beta^2}} \quad \left(\text{where } \beta = \frac{v}{c}\right)$$

$$= 1.08 \text{ m} \quad \text{(on substituting values)}$$

1.343 In the frame K in which the cone is at rest, the coordinates of A are (0, 0, 0) and of B are $(h, h\tan\theta, 0)$. In the frame K', which is moving with velocity v along the axis of the cone, the coordinates of A and B at time t' are

$$A:(-vt', 0, 0),\ B:\left(h\sqrt{1-\beta^2} - vt',\ h\tan\theta,\ 0\right)$$

Thus the taper angle in the frame K' is

$$\tan\theta' = \frac{\tan\theta}{\sqrt{1-\beta^2}} = \left(\frac{y'_B - y'_A}{x'_B - x'_A}\right) \quad \text{or} \quad \theta' = 59° \quad \text{(on substituting values)}$$

The lateral surface area is

$$S = \pi h'^2 \sec\theta' \tan\theta'$$

$$= \pi h^2(1-\beta^2)\frac{\tan\theta}{\sqrt{1-\beta^2}}\sqrt{1+\frac{\tan^2\theta}{1-\beta^2}}$$

$$= S_0\sqrt{1-\beta^2\cos^2\theta}$$

$$= 3.3 \text{ m}^2 \quad \text{(on substituting values)}$$

Here, $S_0 = \pi h^2 \sec\theta \tan\theta$ is the lateral surface area in the rest frame and

$$h' = h\sqrt{1-\beta^2} \quad \left(\text{where } \beta = \frac{v}{c}\right)$$

1.344 Because of time dilation, a moving clock reads less time. We write,

$$t - \Delta t = t\sqrt{1-\beta^2} \quad \left(\text{where } \beta = \frac{v}{c}\right)$$

Thus, $$1 - \frac{2\Delta t}{t} + \left(\frac{\Delta t}{t}\right)^2 = 1 - \beta^2$$

or $$v = c\sqrt{\frac{\Delta t}{t}\left(2 - \frac{\Delta t}{t}\right)}$$

$$= 0.6 \times 10^8 \text{ m/s} \quad \text{(on substituting values)}$$

1.345 In the frame K, the length l of the rod is related to the time of flight Δt by

$$l = v\Delta t$$

In the reference frame fixed to the rod (frame K') the proper length l_0 of the rod is given by

$$l_0 = v\Delta t'$$

But, $$l_0 = \frac{l}{\sqrt{1-\beta^2}} = \frac{v\Delta t}{\sqrt{1-\beta^2}} \quad \left(\text{where } \beta = \frac{v}{c}\right)$$

Thus, $$v\Delta t' = \frac{v\Delta t}{\sqrt{1-\beta^2}}$$

So, $\quad 1-\beta^2 = \left(\frac{\Delta t}{\Delta t'}\right)^2 \quad$ or $\quad v = c\sqrt{1-\left(\frac{\Delta t}{\Delta t'}\right)^2}$

and $\quad l_0 = c\sqrt{(\Delta t')^2 - (\Delta t)^2} = c\Delta t'\sqrt{1-\left(\frac{\Delta t}{\Delta t'}\right)} = 4.5 \text{ m} \quad$ (on substituting values)

1.346 The distance travelled in the laboratory frame of reference is $v\Delta t$ where v is the velocity of the particle. But by time dilation

$$\Delta t = \frac{\Delta t_0}{\sqrt{1-v^2/c^2}}$$

So,

$$v = c\sqrt{1-(\Delta t_0/\Delta t)^2}$$

Thus, the distance traversed is

$$s = c\Delta t\sqrt{1-(\Delta t_0/\Delta t)^2}$$

$$= 5 \text{ m} \quad \text{(on substituting values)}$$

1.347 (a) If τ_0 is the proper lifetime of the muon, the life time in the moving frame is

$$\frac{\tau_0}{\sqrt{1-v^2/c^2}} \quad \text{and hence} \quad l = \frac{v\tau_0}{\sqrt{1-v^2/c^2}}$$

Thus,

$$\tau_0 = \frac{l}{v}\sqrt{1-v^2/c^2}$$

$$= 1.4 \text{ μs} \quad \text{(on substituting values)}$$

(b) The distance travelled "from the muon's standpoint" is simply $v\tau_0 = l\sqrt{1-v^2/c^2}$. The words "from the muon's stand point" are not part of any standard terminology.

1.348 In the frame K in which the particles are at rest, their positions are A and B whose coordinates may be taken as,

$$A : (0, 0, 0),\ B = (l_0, 0, 0)$$

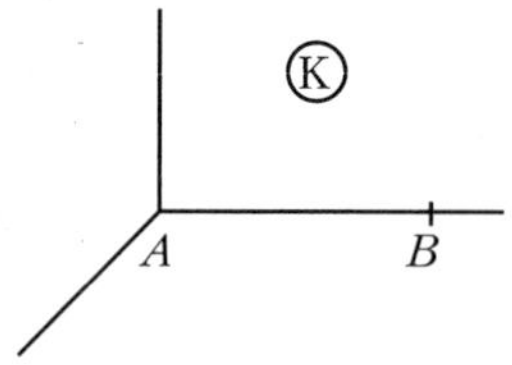

In the frame K' with respect to which K is moving with a velocity v, the coordinates of A and B at time t' in the moving frame are

$$A = (vt', 0, 0),\ B = (l_0\sqrt{1-\beta^2} + vt', 0, 0) \quad \left(\text{where } \beta = \frac{v}{c}\right)$$

Suppose B hits a stationary target in K' after time t'_B while A hits it after time $t_B + \Delta t$. Then,

$$l_0\sqrt{1-\beta^2} + vt'_B = v(t'_B + \Delta t)$$

So,

$$l_0 = \frac{v\Delta t}{\sqrt{1-v^2/c^2}}$$

$$= 17 \text{ m} \quad \text{(on substituting values)}$$

1.349 In the reference frame fixed to the ruler the rod is moving with a velocity v and suffers Lorentz contraction. If l_0 is the proper length of the rod, its measured length will be

$$\Delta x_1 = l_0\sqrt{1-\beta^2} \quad \left(\text{where } \beta = \frac{v}{c}\right)$$

In the reference frame fixed to the rod, the ruler suffers Lorentz contraction and we must have

$$\Delta x_2\sqrt{1-\beta^2} = l_0$$

Thus, $$l_0 = \sqrt{\Delta x_1 \Delta x_2} = 6.0 \text{ m} \quad \text{(on substituting values)}$$

and $$1-\beta^2 = \frac{\Delta x_1}{\Delta x_2}$$

or $$v = c\sqrt{1-\frac{\Delta x_1}{\Delta x_2}} = 2.2\times 10^8 \text{ m/s} \quad \text{(on substituting values)}$$

1.350 The coordinates of the ends of the rods in the frame fixed to the left rod are shown in the figure. The points B and D coincide when

$$l_0 = c_1 - vt_0 \quad \text{or} \quad t_0 = \frac{c_1 - l_0}{v}$$

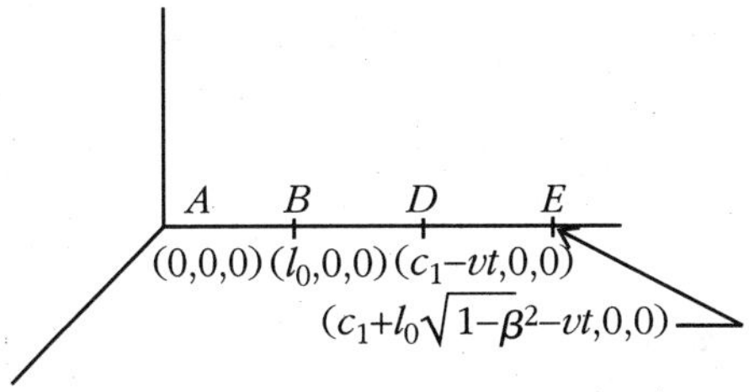

The points A and E coincide when

$$0 = c_1 + l_0\sqrt{1-\beta^2} - vt_1$$

$$t_1 = \frac{c_1 + l_0\sqrt{1-\beta^2}}{v}$$

Thus, $$\Delta t = t_1 - t_0 = \frac{l_0}{v}\left(1+\sqrt{1-\beta^2}\right)$$

or $$\left(\frac{v\Delta t}{l_0} - 1\right)^2 = 1-\beta^2 = 1 - \frac{v^2}{c^2}$$

From this $$v = \frac{2c^2\Delta t/l_0}{1 + c^2\Delta t^2/l_0^2} = \frac{2l_0/\Delta t}{1+(l_0/c\Delta t)^2}$$

1.351 In K_0, the rest frame of the particles, the events corresponding to the decay of the particles are

$$A : (0,\ 0,\ 0,\ 0) \quad \text{and} \quad B : (0,\ l_0,\ 0,\ 0)$$

In the reference frame K, the corresponding coordinates are by Lorentz transformation

$$A : (0,\ 0,\ 0,\ 0),\ B : \left(\frac{vl_0}{c^2\sqrt{1-\beta^2}},\ \frac{l_0}{\sqrt{1-\beta^2}},\ 0,\ 0\right)$$

Now, $l_0\sqrt{1-\beta^2} = l$, by Lorentz Fitzgerald contraction formula.

Thus the time lag of the decay time of B is

$$\Delta t = \frac{vl_0}{c^2\sqrt{1-\beta^2}} = \frac{vl}{c^2(1-\beta^2)} = \frac{vl}{c^2 - v^2}$$

$$= 20 \text{ μs} \quad \text{(on substituting values)}$$

B decays later. (B is the forward particle in the direction of motion.)

1.352 (a) In the reference frame K with respect to which the rod is moving with velocity v, the coordinates of A and B are

$$A : t,\ x_A + v(t - t_A),\ 0,\ 0$$

$$B : t,\ x_B + v(t - t_B),\ 0,\ 0$$

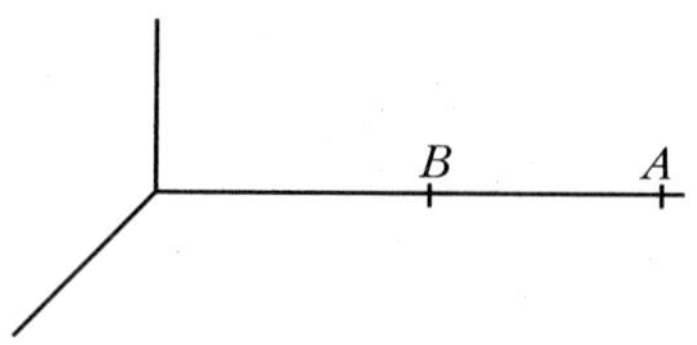

Thus, $$l = x_A - x_B - v(t_A - t_B) = l_0\sqrt{1-\beta^2}$$

So, $$l_0 = \frac{x_A - x_B - v(t_A - t_B)}{\sqrt{1-v^2/c^2}}$$

(b) $\pm l_0 - v(t_A - t_B) = l = l_0\sqrt{1-v^2/c^2}$

(Since $x_A - t_B$ can be either $+l_0$ or $-l_0$.)

Thus, $$v(t_A - t_B) = (\pm 1 - \sqrt{1-v^2/c^2})l_0$$

i.e., $$t_A - t_B = \frac{l_0}{v}\left(1 - \sqrt{1-v^2/c^2}\right)$$

or $$t_B - t_A = \frac{l_0}{v}\left(1 + \sqrt{1-v^2/c^2}\right)$$

1.353 At the instant the picture is taken the coordinates of A, B, A', B' in the rest frame of AB are

$A : (0, 0, 0, 0)$

$B : (0, l_0, 0, 0)$

$B' : (0, 0, 0, 0)$

$A' : (0, \text{or} - l_0\sqrt{1-v^2/c^2}, 0, 0)$

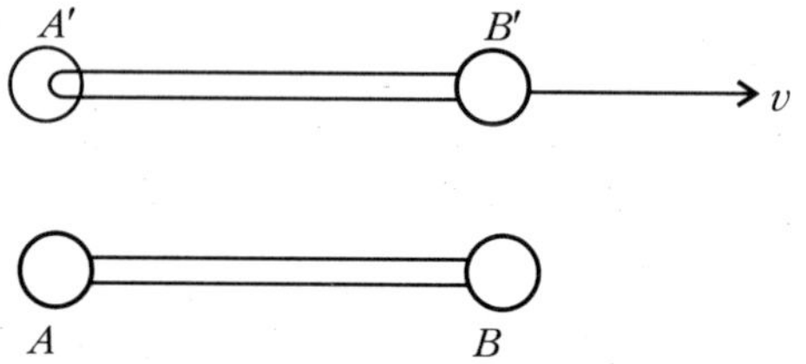

(a) In this frame the coordinates of B' at other times are $B' : (t, vt, 0, 0)$. So, B' is opposite to B at time $t(B) = l_0/v$. In the frame in which B', A' is at rest, the time corresponding to this is given by Lorentz transformation.

$$t(B') = \frac{1}{\sqrt{1-v^2/c^2}}\left(\frac{l_0}{v} - \frac{vl_0}{c^2}\right) = \frac{l_0}{v}\sqrt{1-v^2/c^2}$$

(b) Similarly in the rest frame of A, B, the coordinates of A at other times are

$$A' : \left(t, -l_0\sqrt{1-v^2/c^2} + vt, 0, 0\right)$$

A' is opposite to A at time $t(A) = \frac{l_0}{v}\sqrt{1-v^2/c^2}$

The corresponding time in the frame in which A', B' are at rest is

$$t(A') = \gamma t(A) = \frac{l_0}{v}$$

1.354 By Lorentz transformation $$t' = \frac{1}{\sqrt{1-v^2/c^2}}\left(t - \frac{vx}{c^2}\right)$$

So at time $t = 0$, $$t' = -\frac{vx}{c^2}\frac{1}{\sqrt{1-v^2/c^2}}$$

If $x > 0, t' < 0$ and $x < 0, t' > 0$, so we get the diagram given below "in terms of the K-clocks".

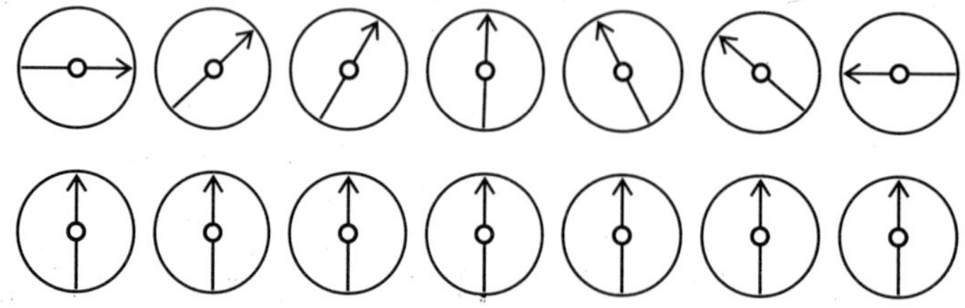

The situation in terms of the K' clock is reversed.

1.355 Suppose $x(t)$ is the locus of points in the frame K at which the readings of the clocks of both reference system are permanently identical, then by Lorentz transformation

$$t' = \frac{1}{\sqrt{1-V^2/c^2}}\left(t - \frac{Vx(t)}{c^2}\right) = t$$

On differentiating $\dot{x}(t) = \frac{c^2}{V}(1-\sqrt{1-V^2/c^2})$

$$= \frac{c}{\beta}(1-\sqrt{1-\beta^2}) \quad \left(\text{where } \beta = \frac{V}{c}\right)$$

Let $\beta = \tan h\theta$, $0 \le \theta < \infty$. Then,

$$\dot{x}(t) = \frac{c}{\tan h\theta}\left(1-\sqrt{1-\tan h^2\theta}\right) = c\frac{\cos h\theta}{\sin h\theta}\left(1-\frac{1}{\cos h\theta}\right)$$

$$= c\frac{\cos h\theta - 1}{\sin h\theta} = c\sqrt{\frac{\cos h\theta - 1}{\cos h\theta + 1}} = c\tan h\frac{\theta}{2} \le V$$

($\tan h\theta$ is a monotonically increasing function of θ.)

1.356 We can take the coordinates of the two events to be

$$A:(0,0,0,0); \quad B:(\Delta t, a, 0, 0)$$

For B to be the effect and A to be cause we must have $\Delta t > \frac{|a|}{c}$.

In the moving frame, the coordinates of A and B become

$$A:(0,0,0,0),\ B:\left[\gamma\left(\Delta t - \frac{aV}{c^2}\right), \gamma(a - V\Delta t), 0, 0\right] \quad \left(\text{where } \gamma = \frac{1}{\sqrt{1-(V^2/c^2)}}\right)$$

Since,

$$(\Delta t')^2 - \frac{a'^2}{c^2} = \gamma^2\left[\left(\Delta t - \frac{aV}{c^2}\right)^2 - \frac{1}{c^2}(a - V\Delta t)^2\right] = (\Delta t)^2 - \frac{a^2}{c^2} > 0$$

we must have $\Delta t' > \frac{|a'|}{c}$.

1.357 (a) The four-dimensional interval between A and B (assuming $\Delta y = \Delta z = 0$) is:

$$5^2 - 3^2 = 16 \text{ units}$$

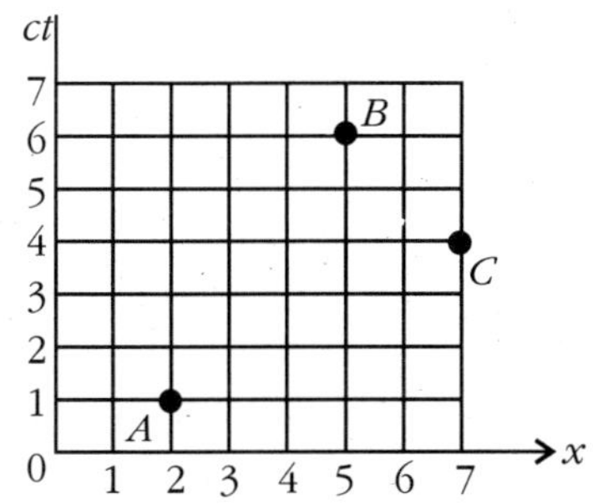

Therefore, the time interval between these two events in the reference frame in which the events occurred at the same place is

$$c(t'_B - t'_A) = \sqrt{16} = 4 \text{ m}$$

or

$$t'_B - t'_A = \frac{4}{c} = \frac{4}{3}\times 10^{-8}\text{ s}$$

(b) The four dimensional interval between A and C (assuming $\Delta y = \Delta z = 0$) is:

$$3^2 - 5^2 = -16$$

So, the distance between the two events in the frame in which they are simultaneous is 4 units = 4 m.

1.358 By the velocity addition formula

$$v'_x = \frac{v_x - V}{1 - V v_x / c^2}$$

$$v'_y = \frac{v_y \sqrt{1 - V^2/c^2}}{1 - v_x V / c^2}$$

and
$$v' = \sqrt{v'^2_x + v'^2_y} = \frac{\sqrt{(v_x - V)^2 + v_y^2 (1 - V^2/c^2)}}{1 - v_x V / c^2}.$$

1.359 (a) By definition, the velocity of approach is

$$v_{\text{approach}} = \frac{dx_1}{dt} - \frac{dx_2}{dt} = v_1 - (-v_2) = v_1 + v_2 = 1.25c$$

in the reference frame K.

(b) The relative velocity is obtained by the transformation law

$$v_r = \frac{v_1 - (-v_2)}{1 - \dfrac{v_1(-v_2)}{c^2}} = \frac{v_1 + v_2}{1 + \dfrac{v_1 v_2}{c^2}}$$
$$= 0.19c \quad \text{(where } c \text{ is the velocity of light)}$$

1.360 The velocity of one of the rods in the reference frame fixed to the other rod is

$$V = \frac{v + v}{1 + v^2/c^2} = \frac{2v}{1 + \beta^2}$$

The length of the moving rod in this frame is

$$l = l_0 \sqrt{1 - \frac{4v^2/c^2}{(1 + \beta^2)^2}} = l_0 \frac{1 - \beta^2}{1 + \beta^2}$$

1.361 The approach velocity is defined by

$$\mathbf{V}_{\text{approach}} = \frac{d\mathbf{r}_1}{dt} - \frac{d\mathbf{r}_2}{dt} = \mathbf{v}_1 - \mathbf{v}_2$$

in the laboratory frame.

So,
$$V_{\text{approach}} = \sqrt{v_1^2 + v_2^2}$$

On the other hand, the relative velocity can be obtained by using the velocity addition formula and has the components

$$\left[-v_1, v_2 \sqrt{1 - \left(\frac{v_1^2}{c^2} \right)} \right] \quad \text{so,} \quad V_r = \sqrt{v_1^2 + v_2^2 - \frac{v_1^2 v_2^2}{c^2}}$$

1.362 The components of the velocity of the unstable particle in the frame K are

$$\left(V, v' \sqrt{1 - \frac{V^2}{c^2}}, 0 \right)$$

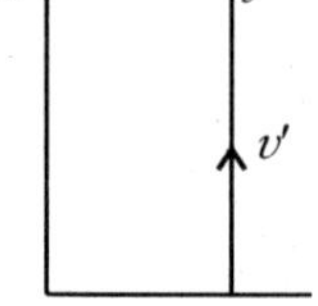

So the velocity relative to K is

$$\sqrt{V^2 + v'^2 - \frac{v'^2 V^2}{c^2}}$$

The lifetime in this frame dilates to

$$\Delta t_0 \Big/ \sqrt{1 - \frac{V^2}{c^2} - \frac{v'^2}{c^2} + \frac{v'^2 V^2}{c^4}}$$

and the distance traversed is

$$s = \Delta t_0 \frac{\sqrt{V^2 + v'^2 - (v'^2V^2)/c^2}}{\sqrt{1 - V^2/c^2}\sqrt{1 - v'^2/c^2}}$$

1.363 In the frame K' the components of the velocity of the particle are

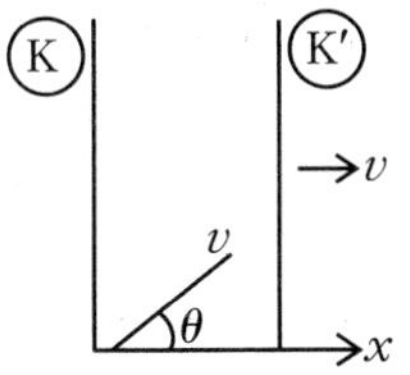

$$v'_x = \frac{v\cos\theta - V}{1 - \dfrac{vV\cos\theta}{c^2}}$$

$$v'_y = \frac{v\sin\theta\sqrt{1 - V^2/c^2}}{1 - \dfrac{vV\cos\theta}{c^2}}$$

Hence, $$\tan\theta' = \frac{v'_y}{v'_x} = \frac{v\sin\theta}{v\cos\theta - V}\sqrt{1 - V^2/c^2}$$

1.364 In K' the coordinates of A and B are

$$A : (t', 0, -v't', 0);\ B : (t', l, -v't', 0)$$

After performing Lorentz transformation to the frame K, we get

$$A : t = \gamma t';\quad B : t = \gamma\left(t' + \frac{Vl}{c^2}\right)$$

$$x = \gamma Vt';\quad x = \gamma(l + Vt')$$

$$y = v't';\quad y = -v't'$$

$$z = 0;\quad z = 0$$

By translating $t' \to t' - \dfrac{Vl}{c^2}$, we can write the coordinates of B as $B : t = \gamma t'$.

$$x = \gamma l\left(1 - \frac{V^2}{c^2}\right) + Vt'\gamma = l\sqrt{1 - \frac{V^2}{c^2}} + Vt'\gamma$$

$$y = -v'\left(t' - \frac{Vl}{c^2}\right), z = 0$$

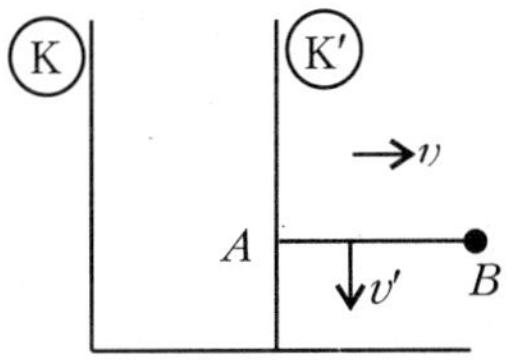

Thus, $$\Delta x = l\sqrt{1 - \left(\frac{V^2}{c^2}\right)}, \Delta y = \frac{v'Vl}{c^2}$$

Hence, $$\tan\theta' = \frac{v'V}{c^2\sqrt{1 - (V/c)^2}}$$

1.365

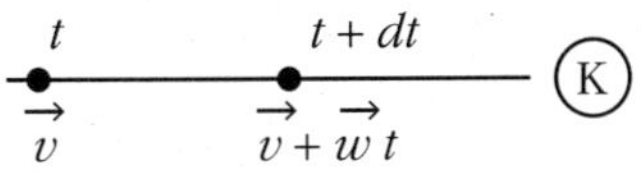

(a) In K the velocities at time t and $t + dt$ are respectively v and $v + w\,dt$ along x-axis which is parallel to the vector $\mathbf{V}$. In the frame K' moving with velocity $\mathbf{V}$ with respect to K, the velocities are, respectively,

$$\frac{v - V}{1 - vV/c^2} \quad \text{and} \quad \frac{v + wdt - V}{1 - (v + wdt)V/c^2}$$

The latter velocity is written as

$$\frac{v-V}{1-vV/c^2}+\frac{wdt}{1-vV/c^2}+\frac{v-V}{(1-vV/c^2)}\frac{wV}{c^2}dt=\frac{v-V}{1-vV/c^2}+\frac{wdt(1-V^2/c^2)}{(1-vV/c^2)^2}$$

Also by Lorentz transformation

$$dt'=\frac{dt-Vdx/c^2}{\sqrt{1-V^2/c^2}}=dt\frac{1-vV/c^2}{\sqrt{1-V^2/c^2}}$$

Thus the acceleration in the K' frame is

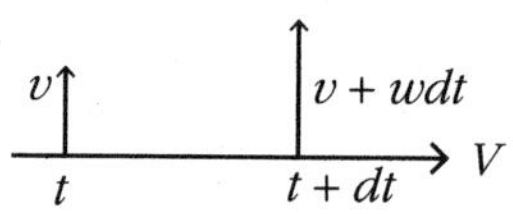

$$w'=\frac{dv'}{dt'}=\frac{w}{(1-vV/c^2)^3}\left(1-\frac{V^2}{c^2}\right)^{3/2}$$

(b) In the K frame, the velocities of the particle at the time t and $t+dt$ are, respectively $(0, v, 0)$ and $(0, v+wdt, 0)$, where $\mathbf{V}$ is along x-axis.

In the K' frame the velocities are $(-V, v\sqrt{1-V^2/c^2}, 0)$

and $(-V, (v+wdt)\sqrt{1-V^2/c^2}, 0)$, respectively.

Thus the acceleration

$$w'=\frac{wdt\sqrt{(1-V^2/c^2)}}{dt'}=w\left(1-\frac{V^2}{c^2}\right)\quad \text{along the } y\text{-axis}$$

$$\left(\text{We have used } dt'=\frac{dt}{\sqrt{1-V^2/c^2}}.\right)$$

1.366 In the instantaneous rest frame $v=V$ and

$$w'=\frac{w}{(1-v^2/c^2)^{3/2}}\quad \text{(from solution of problem 1.365 (a))}$$

So,
$$\frac{dv}{(1-v^2/c^2)^{3/2}}=w'\,dt$$

w' is constant by assumption. Thus, integration gives

$$v=\frac{w't}{\sqrt{1+(w't/c)^2}}$$

Integrating once again
$$x=\frac{c^2}{w'}\left(\sqrt{1+\left(\frac{w't}{c}\right)^2}-1\right)$$

$$=0.91 \text{ light year}$$

The percentage difference of rocket velocity from velocity of light is given by

$$\frac{c-v}{c}=\frac{1}{2}\left(\frac{c}{w't}\right)^2=0.47\%$$

1.367 The boost time τ_0 in the reference frame fixed to the rocket is related to the time τ elapsed on the Earth by

$$\tau_0 = \int_0^\tau \sqrt{1-\frac{v^2}{c^2}}\,dt = \int_0^\tau \left[1-\frac{(w't/c)^2}{1+(w't/c)}\right]^{1/2} dt$$

$$= \int_0^\tau \frac{dt}{1+(w't/c)} = \frac{c}{w'}\int_0^{\frac{w'\tau}{c}} \frac{d\xi}{\sqrt{1+\xi^2}} = \frac{c}{w'}\ln\left[\frac{w'\tau}{c}+\sqrt{1+\left(\frac{w'\tau}{c}\right)^2}\right]$$

$= 3.5$ months (on substituting values)

1.368

$$m = \frac{m_0}{\sqrt{1-\beta^2}}$$

For $\beta \approx 1$, $\quad \dfrac{m}{m_0} \approx \dfrac{1}{\sqrt{2(1-\beta)}} = \dfrac{1}{\sqrt{2\eta}}$

≈ 70 (on substituting values)

1.369 We define the density ρ in the frame K in such a way that $\rho\,dxdydz$ is the rest mass dm_0 of the element. That is $\rho\,dxdydz = \rho_0 dx_0 dy_0 dz_0$, where ρ_0 is the proper density, dx_0, dy_0, dz_0 are the dimensions of the element in the rest frame K_0. Now,

$$dy = dy_0,\; dz = dz_0,\; dx = dx_0\sqrt{1-v^2/c^2}$$

if the frame K is moving with velocity v relative to the frame K_0. Thus

$$\rho = \frac{\rho_0}{\sqrt{1-v^2/c^2}}$$

Defining η by $\quad \rho = \rho_0(1+\eta)$

We get $\quad 1+\eta = \dfrac{1}{\sqrt{1-v^2/c^2}}$ or $\dfrac{v^2}{c^2} = 1-\dfrac{1}{(1+\eta)^2} = \dfrac{\eta(2+\eta)}{1(1+\eta)^2}$

or $\quad v = c\sqrt{\dfrac{\eta(2+\eta)}{(1+\eta)^2}} = \dfrac{c\sqrt{\eta(2+\eta)}}{1+\eta}$

$= 0.6c$ (where c is the velocity of light)

1.370 We have $\quad \dfrac{m_0 v}{\sqrt{1-v^2/c^2}} = p$ or $\dfrac{m_0}{\sqrt{1-v^2/c^2}} = \sqrt{m_0^2+\dfrac{p^2}{c^2}}$

or $\quad 1-\dfrac{v^2}{c^2} = \dfrac{m_0^2c^2}{m_0^2c^2+p^2} = 1-\dfrac{p^2}{p^2+m_0^2c^2}$

or $\quad v = \dfrac{cp}{\sqrt{p^2+m_0^2c^2}} = \dfrac{c}{\sqrt{1+(m_0c/p)^2}}$

So, $\quad \dfrac{c-v}{c} = \left[1-\left(1+\left(\dfrac{m_0c}{p}\right)^2\right)^{-1/2}\right]\times 100\% \approx 0.44\%$ (on substituting values)

1.371 By definition of η,

$$\frac{m_0 v}{\sqrt{1-v^2/c^2}} = \eta m_0 v \quad \text{or} \quad 1-\frac{v^2}{c^2} = \frac{1}{\eta^2}$$

or
$$v = c\sqrt{1-\frac{1}{\eta^2}} = \frac{c}{\eta}\sqrt{\eta^2-1} = \frac{1}{2}c\sqrt{3} \quad \text{(on substituting for } \eta\text{)}$$

1.372 The work done is equal to change in kinetic energy which is different in the two cases. Classically, i.e., in non-relativistic mechanics, the change in kinetic energy is

$$\frac{1}{2}m_0c^2((0.8)^2-(0.6)^2) = \frac{1}{2}m_0c^2 0.28 = 0.14\, m_0c^2$$

Relativistically, it is

$$\frac{m_0c^2}{\sqrt{1-(0.8)^2}} - \frac{m_0c^2}{\sqrt{1-(0.6)^2}} = \frac{m_0c^2}{0.6} - \frac{m_0c^2}{0.8} = m_0c^2(1.666-1.250)$$

$$= 0.416\, m_0c^2 \approx 0.42\, m_0c^2$$

1.373
$$\frac{m_0c^2}{\sqrt{1-\frac{v^2}{c^2}}} = 2m_0c^2$$

or
$$\sqrt{1-\frac{v^2}{c^2}} = \frac{1}{2} \quad \text{or} \quad 1-\frac{v^2}{c^2} = \frac{1}{4}$$

or
$$\frac{v}{c} = \frac{\sqrt{3}}{2}, \text{ i.e,} \quad v = c\frac{\sqrt{3}}{2} = 2.6\times10^8 \text{ m/s}$$

1.374 Relativistically,

$$\frac{T}{m_0c^2} = \left(\frac{1}{\sqrt{1-\beta^2}}-1\right) = \frac{1}{2}\beta^2 + \frac{3}{8}\beta^2$$

So,
$$\beta_{rel}^2 \cong \frac{2T}{m_0c^2} - \frac{3}{4}(\beta_{rel}^2)^2 \approx \frac{2T}{m_0c^2} - \frac{3}{4}\left(\frac{2T}{m_0c^2}\right)^2$$

Thus,
$$\beta_{rel} = \left[\frac{2T}{m_0c^2} - 3\frac{T^2}{m_0^2c^4}\right]^{1/2} = \sqrt{\frac{2T}{m_0c^2}}\left(1-\frac{3}{4}\frac{T}{m_0c^2}\right)$$

But classically,
$$\beta_{cl} = \sqrt{\frac{2T}{m_0c^2}}$$

so,
$$\frac{\beta_{rel}-\beta_{cl}}{\beta_{cl}} = \frac{3}{4}\frac{T}{m_0c^2} = \varepsilon$$

Hence, if $\frac{T}{m_0c^2} < \frac{4}{3}\varepsilon \approx 0.013$, then the velocity β is given by the classical formula with an error less than ε.

1.375 From the formula

$$E = \frac{m_0c^2}{\sqrt{1-v^2/c^2}}, \quad p = \frac{m_0v}{\sqrt{1-v^2/c^2}}$$

we find
$$E^2 = c^2p^2 + m_0^2c^4 \quad \text{or} \quad (m_0c^2+T)^2 = c^2p^2 + m_0^2c^4$$

or
$$T(2m_0c^2+T) = c^2p^2, \text{ i.e., } p = \frac{1}{c}\sqrt{T(2m_0c^2+T)} = 1.09 \text{ GeV/c}$$

1.376 Let the total force exerted by the beam on the target surface be F and the power liberated there be P. Then, using the result of the previous problem, we see

$$F = Np = \frac{N}{c}\sqrt{T(T+2m_0c^2)} = \frac{I}{ec}\sqrt{T(T+2m_0c^2)}$$

since, $I = Ne$, N being the number of particles striking the target per second. Also,

$$p = N\left(\frac{m_0c^2}{\sqrt{1-v^2/c^2}} - m_0c^2\right) = \frac{I}{e}T$$

These will be, respectively, equal to the pressure and power developed per unit area of the target if I is current density.

1.377 In the frame fixed to the sphere, the momentum transferred to the elastically scattered particle is

$$\frac{2mv}{\sqrt{1-v^2/c^2}}$$

The density of the moving element is $n\frac{1}{\sqrt{1-v^2/c^2}}$ (from solution of problem 1.369) and the momentum transferred per unit time per unit area is the pressure, given by

$$p = \frac{2mv}{\sqrt{1-v^2/c^2}}n\frac{1}{\sqrt{1-v^2/c^2}}v$$

In the frame fixed to the gas, when the sphere hits a stationary particle, the latter recoils with a velocity

$$= \frac{v+v}{1+v^2/c^2} = \frac{2v}{1+v^2/c^2}$$

The momentum transferred is $$\frac{\dfrac{m\cdot 2v}{1+v^2/c^2}}{\sqrt{1-\dfrac{4v^2/c^2}{(1-v^2/c^2)^2}}} = \frac{2mv}{1-v^2/c^2}$$

and the pressure is $$\frac{2mv}{1-v^2/c^2}nv = \frac{2mnv^2}{1-v^2/c^2}$$

1.378 The equation of motion is

$$\frac{d}{dt}\left(\frac{m_0v}{\sqrt{1-v^2/c^2}}\right) = F$$

Integrating, $$\frac{v/c}{\sqrt{1-v^2/c^2}} = \frac{\beta}{\sqrt{1-\beta^2}} = \frac{Ft}{m_0c} \quad \text{(using } v = 0 \text{ for } t = 0\text{)}$$

$$\frac{\beta^2}{1-\beta^2} = \left(\frac{Ft}{m_0c}\right)^2 \quad \text{or} \quad \beta^2 = \frac{(Ft)^2}{(Ft)^2+(m_0c)^2} \quad \text{or} \quad v = \frac{Fct}{\sqrt{(m_0c)^2+(Ft)^2}}$$

or $$x = \int\frac{Fct\,dt}{\sqrt{F^2t^2+m_0^2c^2}} = \frac{c}{F}\int\frac{\xi\,d\xi}{\sqrt{\xi^2+(m_0c)^2}} = \frac{c}{F}\sqrt{F^2t^2+m_0^2c^2} + \text{constant}$$

or using $x = 0$ at $t = 0$, we get, $x = \sqrt{c^2t^2+\left(\dfrac{m_0c^2}{F}\right)^2} - \dfrac{m_0c^2}{F}$

1.379 Since, $x = \sqrt{a^2 + c^2t^2}$

so,
$$\dot{x} = v = \frac{c^2t}{a^2 + c^2t^2}$$

or
$$\frac{v}{\sqrt{1 - v^2/c^2}} = \frac{c^2t}{a}$$

Thus,
$$\frac{d}{dt}\left(\frac{m_0 v}{\sqrt{1 - v^2/c^2}}\right) = \frac{m_0 c^2}{a} = F$$

1.380 $\mathbf{F} = \frac{d}{dt}\left(\frac{m_0 \mathbf{v}}{\sqrt{1 - v^2/c^2}}\right) = m_0 \frac{\dot{\mathbf{v}}}{\sqrt{1 - v^2/c^2}} + m_0 \frac{\vec{v}}{c^2}\mathbf{v} \cdot \dot{\mathbf{v}} \frac{1}{(1 - v^2/c^2)^{3/2}}$

Thus,
$$\mathbf{F}_{\perp} = m_0 \mathbf{w}\sqrt{1 - \beta^2},\ \mathbf{w} = \dot{\mathbf{v}},\ \mathbf{w} \perp \mathbf{v}$$

$$\mathbf{F}_{\parallel} = m_0 \frac{\mathbf{w}}{(1 - \beta^2)^{3/2}},\ \mathbf{w} = \dot{\mathbf{v}},\ \mathbf{w} \parallel \mathbf{v}$$

1.381 By definition,

$$E = m_0 \frac{c^2}{\sqrt{1 - v_x^2/c^2}} = \frac{m_0 c^3\, dt}{ds},\ p_x = m_0 \frac{v_x}{\sqrt{1 - v_x^2/c^2}} = \frac{c m_0\, dx}{ds}$$

where $ds^2 = c^2\, dt^2 - dx^2$ is the invariant interval $(dy = dz = 0)$.

Thus,
$$p_x' = c m_0 \frac{dx'}{ds} = c m_0 \gamma \frac{(dx - V dt)}{ds} = \frac{p_x - VE/c^2}{\sqrt{1 - V^2/c^2}}$$

$$E' = m_0 c^3 \frac{dt'}{ds} = c^3 m_0 \gamma \frac{(dt - V dx/c^2)}{ds} = \frac{E - V p_x}{\sqrt{1 - \frac{V^2}{c^2}}}$$

1.382 For a photon moving in the x direction

$$\varepsilon = c p_x$$

and
$$p_y = p_z = 0$$

In the moving frame,
$$\varepsilon' = \frac{1}{\sqrt{1 - \beta^2}}\left(\varepsilon - V\frac{\varepsilon}{c}\right) = \varepsilon\sqrt{\frac{1 - V/c}{1 + V/c}}$$

Note that
$$\varepsilon' = \frac{\varepsilon}{2} \text{ if, } \frac{1}{4} = \frac{1 - \beta}{1 + \beta} \text{ or } \beta = \frac{3}{5},\ V = \frac{3c}{5}$$

1.383 As before

$$E = m_0 c^3 \frac{dt}{ds}$$

$$p_x = m_0 c \frac{dx}{ds}$$

Similarly $$p_y = m_0 c\frac{dy}{ds}, \quad p_z = m_0 c\frac{dz}{ds}$$

Then $$E^2 = c^2 p^2 = E^2 - c^2\left(p_x^2 + p_y^2 + p_z^2\right)$$

$$= m_0^2 c^4 \frac{(c^2 dt^2 - dx^2 - dy^2 - dz^2)}{ds^2} = m_0^2 c^4 \text{ is invariant}$$

1.384 (b) In the C.M. frame, the total momentum is zero, Thus

$$\frac{V}{c} = \frac{cp_{1x}}{E_1 + E_2} = \frac{\sqrt{T(T + 2m_0c^2)}}{T + 2m_0c^2} = \sqrt{\frac{T}{T + 2m_0c^2}}$$

$$V = c\sqrt{\frac{T}{T + 2m_0c^2}} = 2.12 \times 10^8 \text{ m/s}$$

where we have used the result of problem 1.375.

(a) Then

$$\frac{1}{\sqrt{1 - V^2/c^2}} = \frac{1}{\sqrt{1 - \dfrac{T}{T + 2m_0c^2}}} = \sqrt{\frac{T + 2m_0c^2}{2m_0c^2}}$$

Total energy in the C.M. frame is

$$\frac{2m_0c^2}{\sqrt{1 - V^2/c^2}} = 2m_0c^2\sqrt{\frac{T + 2m_0c^2}{2m_0c^2}} = \sqrt{2m_0c^2(T + 2m_0c^2)} = \tilde{T} + 2m_0c^2$$

So, $$\tilde{T} = 2m_0c^2\left(\sqrt{1 + \frac{T}{2m_0c^2}} - 1\right) = 777 \text{ MeV}$$

Also $$2\sqrt{c^2\tilde{p}^2 + mc^4} = \sqrt{2m_0^2c^2(T + 2m_0c^2)} \Rightarrow 4c^2\tilde{p}^2 = 2m_0c^2T$$

or $$\tilde{p} = \sqrt{\frac{1}{2}m_0T} = 940 \text{ MeV/c}$$

1.385 $M_0c^2 = \sqrt{E^2 - c^2p^2}$

$$\sqrt{(2m_0c^2 + T)^2 - T(2m_0c^2 + T)} = \sqrt{2m_0c^2(2m_0c^2 + T)} = c\sqrt{2m_0(2m_0c^2 + T)}$$

Also $$cp = \sqrt{T(T + 2m_0c^2)}, \quad v = \frac{c^2p}{E} = c\sqrt{\frac{T}{T + 2m_0c^2}}$$

1.386 Let T' = kinetic energy of a proton striking another stationary particle of the same rest mass. Then, combined kinetic energy in the C.M. frame

$$= 2m_0c^2\left(\sqrt{1 + \frac{T'}{2m_0c^2}} - 1\right) = 2T, \quad \left(\frac{T}{m_0c^2} + 1\right)^2 = 1 + \frac{T'}{2m_0c^2}$$

$$\frac{T'}{2m_0c^2} = \frac{T(2m_0c^2 + T)}{m_0^2c^4} \quad \text{or} \quad T' = \frac{2T(T + 2m_0c^2)}{m_0c^2}$$

1.387 We have

$$E_1 + E_2 + E_3 = m_0c^2, \ \mathbf{p}_1 + \mathbf{p}_2 + \mathbf{p}_3 = 0$$

Hence,
$$(m_0c^2 - E_1)^2 - c^2|\vec{\mathbf{p}}|^2 = (E_2 + E_3)^2 - (\vec{\mathbf{p}}_2 + \vec{\mathbf{p}}_3)^2c^2$$

$$\text{L.H.S.} = m_0^2c^4 - 2E_1m_0c^2 + m_1^2c^4 = (m_0^2 + m_1^2)c^4 - 2m_0c^2E_1$$

The R.H.S. is an invariant. We can evaluate it in any frame. Choose the C.M. frame of the particles 2 and 3.

In this frame, R.H.S. $= (E_2' - E_3')^2 \geq (m_2 + m_3)^2c^4$

Thus,
$$(m_0^2 + m_1^2)c^4 - 2m_0c^2E_1 \geq (m_2 + m_3)^2c^4$$

or
$$2m_0c^2E_1 \leq \{m_0^2 + m_1^2 - (m_2 + m_3)^2\}c^4 \quad \text{or} \quad E_1 \leq \frac{m_0^2 + m_1^2 - (m_2 + m_3)^2}{2m_0}c^2$$

1.388 The velocity of ejected gases is u relative to the rocket. In an Earth centered frame it is

$$\frac{v - u}{1 - vu/c^2}$$

in the direction of the rocket. The momentum conservation equation then reads

$$(m + dm)(v + dv) + \frac{v - u}{1 - uv/c^2}(-dm) = mv$$

or
$$mdv - \left(\frac{v - u}{1 - uv/c^2} - v\right)dm = 0$$

Here, $-dm$ is the mass of the ejected gases. So,

$$mdv - \frac{-u + uv^2/c^2}{1 - uv/c^2}dm = 0 \quad \text{or} \quad mdv + u\left(1 - \frac{v^2}{c^2}\right)dm = 0$$

(neglecting $1 - uv/c^2$, since u is non-relativistic).

Integrating, using $\beta = \frac{v}{c}$, we get

$$\int \frac{d\beta}{1 - \beta^2} + \frac{u}{c}\int \frac{dm}{m} = 0$$

or
$$\ln \frac{1 + \beta}{1 - \beta} + \frac{u}{c} \ln m = \text{constant}$$

The constant $= \frac{u}{c} \ln m_0$, since $\beta = 0$ initially.

Thus,
$$\frac{1 - \beta}{1 + \beta} = \left(\frac{m}{m_0}\right)^{u/c} \quad \text{or} \quad \beta = \frac{1 - (m/m_0)^{u/c}}{1 + (m/m_0)^{u/c}}$$

THERMODYNAMICS AND MOLECULAR PHYSICS 2

2.1 Equation of the Gas State. Processes

2.1 Let m_1 and m_2 be the masses of the gas in the vessel before and after the gas is released. Hence mass of the gas released,

$$\Delta m = m_1 - m_2$$

Now from ideal gas equation

$$p_1 V = m_1 \frac{R}{M} T_0 \quad \text{and} \quad p_2 V = m_2 \frac{R}{M} T_0$$

as V and T are same before and after the release of the gas.

So, $$(p_1 - p_2)V = (m_1 - m_2)\frac{R}{M}T_0 = \Delta m \frac{R}{M} T_0$$

or $$\Delta m = \frac{(p_1 - p_2)VM}{RT_0} = \frac{\Delta p VM}{RT_0} \tag{1}$$

We also know $$p = \rho \frac{R}{M} T \quad \text{so,} \quad \frac{M}{RT_0} = \frac{\rho}{p_0} \tag{2}$$

(where p_0 = standard atmospheric pressure and T_0 = 273 K).

From Eqs. (1) and (2), we get

$$\Delta m = \rho V \frac{\Delta p}{p_0} = 1.3 \times 30 \times \frac{0.78}{1} = 30 \text{ g}$$

2.2 Let V is volume of each vessel. For the vessel which contained the ideal gas,

$$p_1 V = \nu_1 R T_1$$

When heated, some gas, passes into the evacuated vessel till pressure difference becomes Δp. Let p_1' and p_2' be the pressure on the two sides of the valve. Then

$$p_1' V = \nu_1' R T_2$$

and $$p_2' V = \nu_2' R T_2 = (\nu_1 - \nu_1') R T_2$$

Putting the values of ν_1 and ν_2 from the first two equations

$$p_2'V = \left(\frac{p_1V}{RT_1} - \frac{p_1'V}{RT_2}\right)RT_2 \quad \text{or} \quad p_2' = \left(\frac{p_1}{T_1} - \frac{p_1'}{T_2}\right)T_2$$

But,
$$p_1' - p_2' = \Delta p$$

So,
$$p_2' = \left(\frac{p_1}{T_1} - \frac{p_2' + \Delta p}{T_2}\right)T_2$$

$$= \frac{p_1T_2}{T_1} - p_2' - \Delta p$$

or
$$p_2' = \frac{1}{2}\left(\frac{p_1T_2}{T_1} - \Delta p\right) = 0.08 \text{ atm}$$

2.3 Let the mixture contain ν_1 and ν_2 moles of H_2 and He, respectively. If molecular weights of H_2 and He are M_1 and M_2, then their respective masses in the mixture are

$$m_1 = \nu_1M_1 \quad \text{and} \quad m_2 = \nu_2M_2$$

Therefore, for the total mass of the mixture we get,

$$m = m_1 + m_2 \quad \text{or} \quad m = \nu_1M_1 + \nu_2M_2 \tag{1}$$

Also, if ν is the total number of moles of the mixture in the vessels, then we know,

$$\nu = \nu_1 + \nu_2 \tag{2}$$

Solving Eqs. (1) and (2) for ν_1 and ν_2, we get

$$\nu_1 = \frac{(\nu M_2 - m)}{M_2 - M_1}, \; \nu_2 = \frac{m - \nu M_1}{M_2 - M_1}$$

Therefore, we get

$$m_1 = M_1\frac{(\nu M_2 - m)}{M_2 - M_1} \quad \text{and} \quad m_2 = M_2\frac{(m - \nu M_1)}{M_2 - M_1}$$

or
$$\frac{m_1}{m_2} = \frac{M_1(\nu M_2 - m)}{M_2(m - \nu M_1)}$$

One can also express the above result in terms of the effective molecular weight M of the mixture, defined as

$$M = \frac{m}{\nu} = m\frac{RT}{pV}$$

Thus,
$$\frac{m_1}{m_2} = \frac{M_1}{M_2}\cdot\frac{M_2 - M}{M - M_1} = \frac{1 - M/M_2}{M/M_1 - 1}$$

Using the data and table, we get

$$M = 3.0\,\text{g} \text{ and } \frac{m_1}{m_2} = 0.50$$

2.4 We know, for the mixture, N_2 and CO_2 (being regarded as ideal gases, their mixture too behaves like an ideal gas), $p_0V = \nu RT$, where, ν is the total number of moles of the gases (mixture) present, p_0 is the pressure and V is the volume of the vessel.

If ν_1 and ν_2 are number of moles of N_2 and CO_2, respectively present in the mixture, then $\nu = \nu_1 + \nu_2$.

Now number of moles of N_2 and CO_2 is, by definition, given by,

$$\nu_1 = \frac{m_1}{M_1} \quad \text{and} \quad \nu_2 = \frac{m_2}{M_2}$$

where, m_1 is the mass of N_2 (molecular weight = M_1) in the mixture and m_2 is the mass of CO_2 (molecular weight = M_2) in the mixture.

Therefore density of the mixture is given by

$$\rho = \frac{m_1 + m_2}{V} = \frac{m_1 + m_2}{(\nu RT/p_0)}$$
$$= \frac{p_0}{RT}\frac{m_1 + m_2}{\nu_1 + \nu_2} = \frac{p_0\,(m_1 + m_2)M_1M_2}{RT(m_1M_2 + m_2M_1)}$$
$$= 1.5 \text{ kg/m}^3 \text{ (on substituting values)}$$

2.5 (a) The mixture contains ν_1, ν_2 and ν_3 moles of O_2, N_2 and CO_2, respectively. Then the total number of moles of the mixture

$$\nu = \nu_1 + \nu_2 + \nu_3$$

We know, ideal gas equation for the mixture

$$pV = \nu RT \quad \text{or} \quad p = \frac{\nu RT}{V}$$

or

$$p = \frac{(\nu_1 + \nu_2 + \nu_3)RT}{V} = 1.968 \text{ atm (on substituting values)}$$

(b) Mass of oxygen (O_2) present in the mixture: $m_1 = \nu_1 M_1$

Mass of nitrogen (N_2) present in the mixture: $m_2 = \nu_2 M_2$

Mass of carbon dioxide (CO_2) present in the mixture: $m_3 = \nu_3 M_3$

So, mass of the mixture:

$$m = m_1 + m_2 + m_3 = \nu_1M_1 + \nu_2M_2 + \nu_3M_3$$

Molecular mass of the mixture

$$M = \frac{\text{Mass of the mixture}}{\text{Total number of moles}}$$
$$= \frac{\nu_1M_1 + \nu_2M_2 + \nu_3M_3}{\nu_1 + \nu_2 + \nu_3} = 36.7 \text{ g/mol (on substituting values)}$$

2.6 Let p_1 and p_2 be the pressure in the upper and lower part of the cylinder, respectively at temperature T_0. At the equilibrium position for the piston:

$$p_1 S + mg = p_2 S$$

or $$p_1 + \frac{mg}{S} = p_2 \text{ (where } m \text{ is the mass of the piston)}$$

But, $$p_1 = \frac{RT}{\eta V_0} \text{ (where } V_0 \text{ is the initial volume of the lower part)}$$

So, $$\frac{RT_0}{\eta V_0} + \frac{mg}{S} = \frac{RT}{V_0} \Rightarrow \frac{mg}{S} = \frac{RT_0}{V_0}\left(1 - \frac{1}{\eta}\right) \quad (1)$$

Let T' be the sought temperature and at this temperature the volume of the lower part becomes V', then according to the problem the volume of the upper part becomes $\eta' V$.

Hence, $$\frac{mg}{S} = \frac{RT'}{V'}\left(1 - \frac{1}{\eta'}\right) \quad (2)$$

From Eqs. (1) and (2),

$$\frac{RT}{V_0}\left(1 - \frac{1}{\eta}\right) = \frac{RT'}{V'}\left(1 - \frac{1}{\eta'}\right) \text{ or } T' = \frac{T_0(1 - 1/\eta)V'}{V_0(1 - 1/\eta')} \quad (3)$$

As, the total volume must be constant, so

$$V_0(1+\eta) = V'(1+\eta') \quad \text{or} \quad V' = \frac{V_0(1+\eta)}{1+\eta'}$$

Putting the value of V'in Eq. (3), we get

$$T = \frac{T_0(1 - 1/\eta)V_0 \dfrac{(1+\eta)}{(1+\eta')}}{V_0(1 - 1/\eta')}$$

$$= \frac{T_0(\eta^2 - 1)\eta'}{(\eta'^2 - 1)\eta} = 0.42 \text{ kK}$$

2.7 Let ρ_1 be the density after the first stroke. The mass remains constant.

So, $$V\rho = (V + \Delta V)\rho_1 \Rightarrow \rho_1 = \frac{V\rho}{(V + \Delta V)}$$

Similarly, if ρ_2 is the density after second stroke

$$V\rho_1 = (V + \Delta V)\rho_2$$

or $$\rho_2 = \left(\frac{V}{V + \Delta V}\right)\rho_1 = \left(\frac{V}{V + \Delta V}\right)^2 \rho_0$$

In this way after nth stroke,

$$\rho_n = \left(\frac{V}{V + \Delta V}\right)^n \rho_0$$

Since pressure $\propto$ density,

$$p_n = \left(\frac{V}{V + \Delta V}\right)^n p_0 \text{ (because temperature is constant)}$$

It is required by p_n/p_0 to be $1/\eta$,

So, $$\frac{1}{\eta} = \left(\frac{V}{V + \Delta V}\right)^n \quad \text{or} \quad \eta = \left(\frac{V + \Delta V}{V}\right)^n$$

Hence, $$n = \frac{\ln \eta}{\ln (1 + \Delta V/V)}$$

2.8 From the ideal gas equation $$p = \frac{m}{M}\frac{RT}{V}$$

$$\frac{dp}{dt} = \frac{RT}{MV}\frac{dm}{dt} \tag{1}$$

In each stroke, volume v of the gas is ejected, where v is given by

$$v = \frac{V}{m_N}[m_{N-1} - m_N]$$

In case of continuous ejection, if m_{N-1} corresponds to mass of gas in the vessel at time t, then m_N is the mass at time $t + \Delta t$, where Δt, is the time in which volume v of the gas has come out.

The rate of evacuation is therefore $v/\Delta t$, i.e.,

$$C = \frac{v}{\Delta t} = -\frac{V}{m(t + \Delta t)} \cdot \frac{m(t + \Delta t) - m(t)}{\Delta t}$$

In the limit $\Delta t \to 0$, we get

$$C = \frac{V}{m}\frac{dm}{dt} \tag{2}$$

From Eqs. (1) and (2)

$$\frac{dp}{dt} = -\frac{C}{V}\frac{mRT}{MV} = -\frac{C}{V}p \quad \text{or} \quad \frac{dp}{p} = -\frac{C}{V}dt$$

Integrating we get, $$\int_p^{p_0} \frac{dp}{p} = -\frac{C}{V}\int_t^0 dt$$

or $$\ln\frac{p}{p_0} = -\frac{C}{V}t$$

Thus, $$p = p_0 e^{-Ct/V}$$

2.9 Let ρ be the instantaneous density, then instantaneous mass $= V\rho$. In a short interval dt the volume is increased by Cdt.

So, $V\rho = (V + Cdt)(\rho + d\rho)$ (because mass remains constant in a short interval dt).

So,
$$\frac{d\rho}{\rho} = -\frac{C}{V}dt$$

Since pressure $\propto$ density,
$$\frac{dp}{p} = -\frac{C}{V}dt$$

or
$$\int_{p_1}^{p_2} -\frac{dp}{p} = \frac{C}{V}t$$

or
$$t = \frac{V}{C}\ln\frac{p_1}{p_2} = \frac{V}{C}\ln\frac{1}{\eta} = 1.0 \text{ min}$$

2.10 The physical system consists of one mole of gas confined in the smooth vertical tube. Let m_1 and m_2 be the masses of upper and lower pistons and S_1 and S_2 be their respective areas.

For the lower piston

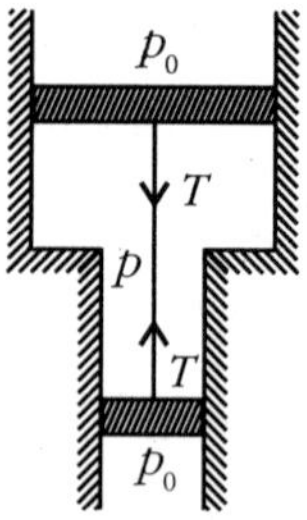

$$pS_2 + m_2 g = p_0 S_2 + T$$

or
$$T = (p - p_0)S_2 + m_2 g \qquad (1)$$

Similarly for the upper piston
$$p_0 S_1 + T + m_1 g = pS_1$$

or
$$T = (p - p_0)S_1 - m_1 g \qquad (2)$$

From Eqs. (1) and (2)
$$(p - p_0)(S_1 - S_2) = (m_1 + m_2)g$$

or
$$(p - p_0)\Delta S = mg$$

So,
$$p = \frac{mg}{\Delta S} + p_0 = \text{constant}$$

From the gas law, $pV = \nu RT$ or $p\Delta V = \nu R\Delta T$ (because p is constant).

So,
$$\left(p_0 + \frac{mg}{\Delta S}\right)\Delta Sl = R\Delta T$$

Hence,
$$\Delta T = \frac{1}{R}(p_0\Delta S + mg)l$$
$$= 0.9 \text{ K}$$

2.11 (a) Given that $p = p_0 - \alpha V^2 = p_0 - \alpha(RT/p)^2$ (as $V = RT/p$ for one mole of gas).

Thus, $$T = \frac{1}{R\sqrt{\alpha}}\, p\sqrt{p_0 - p} = \frac{1}{R\sqrt{\alpha}}\sqrt{p_0 p^2 - p^3} \qquad (1)$$

For T_{max}, $$\frac{d}{dp}(p_0 p^2 - p^3) = 0$$

which yields, $$p = 2/3\, p_0$$

Hence, using this value of p in Eq. (1), we get

$$T_{max} = \frac{1}{R\sqrt{\alpha}} \cdot \frac{2}{3} p_0 \sqrt{p_0 - \frac{2}{3} p_0} = \frac{2}{3}\left(\frac{p_0}{R}\right)\sqrt{\frac{p_0}{3\alpha}}$$

(b) Given that $$p = p_0 e^{-\beta V} = p_0 e^{-\beta KT/p}$$

So, $$\frac{\beta RT}{p} = \ln\frac{p_0}{p} \quad \text{and} \quad T = \frac{p}{\beta R}\ln\frac{p_0}{p} \qquad (2)$$

For T_{max}, the condition is $dT/dp = 0$, which yields $p = p_0/e$.

Hence using this value of p in Eq. (2), we get

$$T_{max} = \frac{p_0}{e\beta R}$$

2.12 Given that $T = T_0 + \alpha V^2 = T_0 + \alpha(R^2T^2/p^2)$ (as $V = RT/p$ for one mole of gas).

So, $$p = \sqrt{\alpha}\, RT(T - T_0)^{-1/2} \qquad (1)$$

For P_{min}, $$\frac{dp}{dT} = 0, \text{ which gives} \qquad (2)$$

$$T = 2T_0$$

From Eqs. (1) and (2), we get

$$p_{min} = \sqrt{\alpha}\, R2T_0(2T_0 - T_0)^{-1/2} = 2R\sqrt{\alpha T_0}$$

2.13 Consider a thin layer at a height h and thickness dh. Let p and $dp + p$ be the pressure on the two sides of the layer. The mass of the layer is $Sdh\rho$. Equating vertical downward force to the upward force acting on the layer. We get

$$Sdh\rho g + (p + dp)S = pS$$

So, $$\frac{dp}{dh} = -\rho g \qquad (1)$$

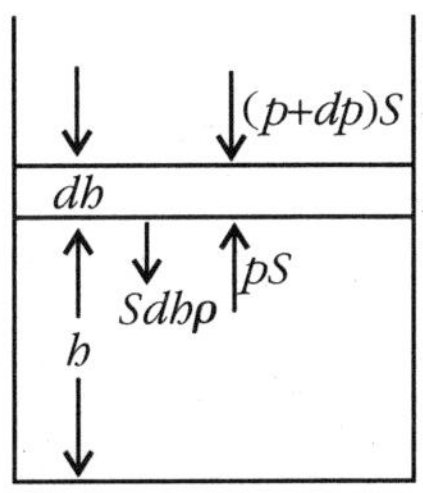

But, $$p = \frac{\rho}{M}RT$$

So, $$dp = \frac{\rho R}{M} dT$$

or $$-\frac{\rho R}{M} dT = \rho g dh \quad \text{(using Eq. 1)}$$

So, $$\frac{dT}{dh} = -\frac{gM}{R} = -34\,\text{K/km}$$

That means, temperature of air drops by 34 K at a height of 1 km above bottom.

2.14 From the previous problem we have,

$$\frac{dp}{dh} = -\rho g \tag{1}$$

But, from $p = C\rho^n$ (where C is a constant), we get

$$dp/d\rho = Cn\rho^{n-1} \tag{2}$$

We have from gas law,

$$p = \rho \frac{R}{M} T \quad \text{or} \quad C\rho^n = \rho \frac{R}{M} \cdot T$$

So, $$T = \frac{M}{R} C\rho^{n-1} \quad \text{(using Eq. 2)}$$

Thus, $$\frac{dT}{d\rho} = \frac{M}{R} \cdot C(n-1)\rho^{n-2} \tag{3}$$

But, $$\frac{dT}{dh} = \frac{dT}{d\rho} \cdot \frac{d\rho}{dp} \cdot \frac{dp}{dh}$$

Substituting from Eqs. (1), (2) and (3) we get

$$\frac{dT}{dh} = \frac{M}{R} C(n-1)\rho^{n-2} \frac{1}{Cn\rho^{n-1}} (-\rho g) = \frac{-Mg(n-1)}{nR}$$

2.15 We have, $dp = -\rho g dh$ and from gas law $\rho = (M/RT)p$.

Thus, $$\frac{dp}{p} = \frac{Mg}{RT} dh$$

Integrating, we get

$$\int_{p_0}^{p} \frac{dp}{p} = -\frac{Mg}{RT} \int_0^h dh$$

or $$\ln \frac{p}{p_0} = -\frac{Mg}{RT} h$$ (where p_0 is the pressure at the surface of the Earth)

So, $$p = p_0 e^{-Mgh/RT}$$

Under standard conditions, $p_0 = 1$ atm, $T = 273\,\text{K}$

Pressure at a height of 5 km $= 1 \times e^{-28\times 9.81\times 5000/8.314\times 273} = 0.5$ atm

Pressure in a mine at a depth of 5 km $= 1 \times e^{-28\times 9.81\times(-5000)/8.314\times 273} = 2$ atm

2.16 We have $dp = -\rho g dh$ and from gas law $p = (\rho/M)RT$.

Thus, $$dp = \frac{d\rho}{M}RT \quad \text{(at constant temperature)}$$

So, $$\frac{d\rho}{\rho} = -\frac{gM}{RT}dh$$

Integrating within limits

$$\int_{\rho_0}^{\rho}\frac{d\rho}{\rho} = -\int_0^h \frac{gM}{RT}dh$$

or $$\ln\frac{\rho}{\rho_0} = -\frac{gM}{RT}h$$

So, $$\rho = \rho_0 e^{-Mgh/RT} \quad \text{and} \quad h = -\frac{RT}{Mg}\ln\frac{\rho}{\rho_0}$$

(a) Given that $T = 273$ K and $\rho_0/\rho = e$.

Thus, $$h = -\frac{RT}{Mg}\ln e^{-1}$$
$$= \frac{RT}{Mg} = 8\,\text{km}.$$

(b) Given that $T = 273$ K and $(\rho_0 - \rho)/\rho_0 = 0.01$ or $\rho/\rho_0 = 0.99$.

Thus, $$h = -\frac{RT}{Mg}\ln\frac{\rho}{\rho_0} = 0.09\ \text{km (on substituting values)}$$

2.17 From the Barometric formula, we have

$$p = p_0 e^{-Mgh/RT}$$

and from gas law $$\rho = \frac{pM}{RT}$$

So, at constant temperature from these two equations

$$\rho = \frac{Mp_0}{RT}e^{-Mgh/RT} = \rho_0 e^{-Mgh/R} \quad (1)$$

Eq. (1) shows that density varies with height in the same manner as pressure. Let us consider the mass element of the gas contained in the column.

$$dm = \rho(Sdh) = \frac{Mp_0}{RT}e^{-Mgh/RT}Sdh$$

Hence the sought mass,

$$m = \frac{Mp_0S}{RT}\int_0^h e^{-Mgh/RT}dh = \frac{p_0S}{g}(1 - e^{-Mgh/RT})$$

2.18 As the gravitational field is constant the centre of gravity and the centre of mass (C.M.) are same. The location of C.M.

$$h = \frac{\int_0^\infty h\,dm}{\int_0^\infty dm} = \frac{\int_0^\infty h\rho\,dh}{\int_0^\infty \rho\,dh}$$

But from Barometric formula and gas law $\rho = \rho_0 e^{-Mgh/RT}$

So,
$$h = \frac{\int_0^\infty h(e^{-Mgh/RT})dh}{\int_0^\infty (e^{-Mgh/RT})dh} = \frac{RT}{Mg}$$

2.19 (a) We know that the variation of pressure with height of a fluid is given by

$$dp = -\rho g dh$$

But from gas law
$$p = \frac{\rho}{M}RT \quad \text{or} \quad \rho = \frac{pM}{RT}$$

From these two equations

$$dp = -\frac{pMg}{RT}dh \tag{1}$$

or
$$\frac{dp}{p} = \frac{-Mgdh}{RT_0(1-ah)}$$

Integrating we get,
$$\int_{p_0}^{p}\frac{dp}{p} = \frac{-Mg}{RT_0}\int_0^h \frac{dh}{(1-ah)}$$

or
$$\ln\frac{p}{p_0} = \ln(1-ah)^{Mg/aRT_0}$$

Hence,
$$p = p_0(1-ah)^{Mg/aRT_0} \quad \left(\text{obviously } h < \frac{1}{a}\right).$$

(b) Proceed up to Eq. (1) of part (a), and then put $T = T_0(1 + ah)$ and proceed further in the same fashion to get

$$p = \frac{p_0}{(1+ah)^{Mg/aRT_0}}$$

2.20 Let us consider the mass element of the gas (thin layer) in the cylinder at a distance r from its open end as shown in the figure.

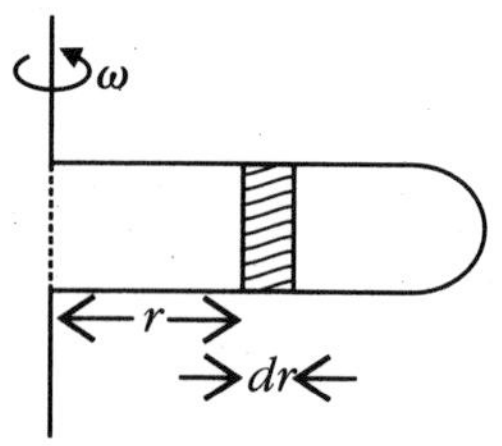

Using Newton's second law for the element

$$F_n = mw_n$$

$$(p + dp)S - pS = (\rho S\,dr)\,\omega^2 r$$

or $$dp = \rho\omega^2 rdr = \frac{pM}{RT}\omega^2 rdr$$

So, $$\frac{dp}{p} = \frac{M\omega^2}{RT} rdr$$

or $$\int_{p_0}^{p} \frac{dp}{p} = \frac{M\omega^2}{RT}\int_0^r rdr$$

Thus, $$\ln\frac{p}{p_0} = \frac{M\omega^2}{2RT}r^2 \quad \text{or} \quad p = p_0 e^{M\omega^2 r^2/2RT}$$

2.21 For an ideal gas law

$$p = \frac{\rho}{M}RT$$

On substitution, we get $$p = 0.082 \times 300 \times \frac{500}{44}\text{ atm} = 279.5\text{ atm}$$

From Van der Waal gas equation

$$\left(p + \frac{\nu^2 a}{V^2}\right)(V - \nu b) = \nu RT \quad (\text{where } V = \nu V_M)$$

or $$p = \frac{\nu RT}{V - \nu b} - \frac{a\nu^2}{V^2} = \frac{mRT/M}{V - mb/M} - \frac{am^2}{V^2 M^2}$$

$$= \frac{\rho RT}{M - \rho b} - \frac{a\rho^2}{M^2} = 79.2\text{ atm}$$

2.22 (a) $p = \left[\frac{RT}{V_M - b} - \frac{a}{V_M^2}\right](1+\eta) = \frac{RT}{V_M}$, (where V_M is the molar volume)

(The pressure is less for a Van der Waal gas than for an ideal gas.)

or $$\frac{a(1+\eta)}{V_M^2} = RT\left[\frac{-1}{V_M} + \frac{-1+\eta}{V_M - b}\right] = RT\frac{\eta V_M + b}{V_M(V_M - b)}$$

or $$T = \frac{a(1+\eta)(V_M - b)}{RV_M(\eta V_M + b)}$$

$$= \frac{1.35 \times 1.1 \times (1 - 0.039)}{0.082 \times (0.139)} \approx 125\text{K}$$

(b) The corresponding pressure is

$$p = \frac{RT}{V_M - b} - \frac{a}{V_M^2} = \frac{a(1+\eta)}{V_M(\eta V_M + b)} - \frac{a}{V_M^2}$$

$$= \frac{a}{V_M^2}\frac{(V_M + \eta V_M - \eta V_M - b)}{(\eta V_M + b)} = \frac{a}{V_M^2}\frac{(V_M - b)}{(V_M + b)}$$

$$= \frac{1.35}{1} \times \frac{0.961}{0.139} \approx 9.3\text{ atm}$$

2.23 Since, $$p_1 = RT_1\frac{1}{V-b} - \frac{a}{V^2} \quad \text{and} \quad p_2 = RT_2\frac{1}{V-b} - \frac{a}{V^2}$$

So, $$p_2 - p_1 = \frac{R(T_2 - T_1)}{V-b}$$

or $$V - b = \frac{R(T_2 - T_1)}{p_2 - p_1} \Rightarrow b = V - \frac{R(T_2 - T_1)}{p_2 - p_1}$$

Also, $$p_1 = T_1\frac{p_2 - p_1}{T_2 - T_2} - \frac{a}{V^2}$$

$\Rightarrow$ $$\frac{a}{V^2} = \frac{T_1(p_2 - p_1)}{T_2 - T_1} - p_1 = \frac{T_1p_2 - p_1T_2}{T_2 - T_1}$$

or $$a = V^2\frac{T_1p_2 - p_1T_2}{T_2 - T_1}$$

Using $T_1 = 300$ K, $p_1 = 90$ atm, $T_2 = 350$ K, $p_2 = 110$ atm, $V = 0.250$, we get $a = 1.87$ atm l^2/mol^2, and $b = 0.045$ l/mol.

2.24 $$p = \frac{RT}{V-b} - \frac{a}{V^2} - V\left(\frac{\partial p}{\partial V}\right)_T = \frac{RTV}{(V-b)^2} - \frac{2a}{V^2}$$

or $$\kappa = \frac{-1}{V}\left(\frac{\partial V}{\partial p}\right)_T$$

$$= \left[\frac{RTV^3 - 2a(V-b)^2}{V^2(V-b)^2}\right]^{-1} = \frac{V^2(V-b)^2}{[RTV^3 - 2a(V-b)^2]}$$

2.25 For an ideal gas, $\kappa_0 = V/RT$.

Now, $$\kappa = \frac{(V-b)^2}{RTV}\left\{1 - \frac{2a(V-b)^2}{RTV^3}\right\}^{-1} = \kappa_0\left(1 - \frac{b}{V}\right)^2\left\{1 - \frac{2a}{RTV}\left(1 - \frac{b}{V}\right)^2\right\}^{-1}$$

$$= \kappa_0\left\{1 - \frac{2b}{V} + \frac{2a}{RTV}\right\} \text{ (to leading order in } a, b)$$

Now, $$\kappa > \kappa_0 \quad \text{if} \quad \frac{2a}{RTV} > \frac{2b}{V} \quad \text{or} \quad T < \frac{a}{bR}$$

If a and b do not vary much with temperature, then the effect at high temperature is clearly determined by b and its effect is repulsive so compressibility is less.

2.2 The First Law of Thermodynamics. Heat Capacity

2.26 Internal energy of air, treating it as an ideal gas is given by

$$U = \frac{m}{M}C_V T = \frac{m}{M}\frac{R}{\gamma-1}T = \frac{pV}{\gamma-1} \quad (1)$$

(using $(C_V = R/(\gamma - 1)$, since $C_p - C_V = R$ and $C_p/C_V = \gamma)$).

Thus at constant pressure U = constant, because the volume of the room is a constant. Putting the value of $p = p_{atm}$ and V in Eq. (1), we get U = 10 M J.

2.27 From energy conservation

$$U_i + \frac{1}{2}(\nu M)v^2 = U_f$$

or $$\Delta U = (1/2)\nu M v^2 \quad (1)$$

Using $$U = \frac{\nu RT}{\gamma-1} \quad \text{(from previous problem)} \quad (2)$$

We get $$\Delta U = \frac{\nu R}{\gamma-1}\Delta T$$

Hence from Eqs. (1) and (2),

$$\Delta T = \frac{Mv^2(\gamma-1)}{2R}$$

2.28 On opening the valve, the air will flow from the vessel at higher pressure to the vessel at lower pressure till both vessels have the same air pressure. If this air pressure is p, the total volume of the air in the two vessels will be $(V_1 + V_2)$. Also if ν_1 and ν_2 be the number of moles of air initially in the two vessels, we have

$$p_1V_1 = \nu_1RT_1 \quad \text{and} \quad p_2V_2 = \nu_2RT_2 \quad (1)$$

After the air is mixed, the total number of moles are $(v_1 + v_2)$ and the mixture is at temperature T.

Hence, $$p(V_1 + V_2) = (\nu_1 + \nu_2)\, RT \quad (2)$$

Let us look at the two portions of air as one single system. Since this system is contained in a thermally insulated vessel, no heat exchange is involved in the process. That is, total heat transfer for the combined system $Q = 0$.

Moreover, this combined system does not perform mechanical work either. The walls of the containers are rigid and there are no pistons, etc to the pushed, looking at the total system, we know $A = 0$.

Hence, internal energy of the combined system does not change in the process. Initially, energy of the combined system is equal to the sum of internal energies of the two portions of air:

$$U_i = U_1 + U_2 = \frac{\nu_1RT_1}{\gamma-1} + \frac{\nu_2RT_2}{\gamma-1} \quad (3)$$

Final internal energy of $(\nu_1 + \nu_2)$ moles of air at temperature T is given by

$$U_f = \frac{(\nu_1+\nu_2)RT}{\gamma-1} \quad (4)$$

Therefore, $U_i = U_f$ implies

$$T = \frac{\nu_1T_1 + \nu_2T_2}{\nu_1+\nu_2} = \frac{p_1V_1 + p_2V_2}{(p_1V_1/T_1)+(p_2V_2/T_2)} = T_1T_2\frac{p_1V_1 + p_2V_2}{p_1V_1T_2 + p_2V_2T_1}$$

From Eq. (2) therefore, final pressure is given by

$$p = \frac{\nu_1 + \nu_2}{V_1 + V_2} RT = \frac{R}{V_1 + V_2}(\nu_1 T_1 + \nu_2 T_2) = \frac{p_1 V_1 + p_2 V_2}{V_1 + V_2}$$

This process is an example of free adiabatic expansion of ideal gas.

2.29 By the first law of thermodynamics, $Q = \Delta U + A$.

Here $A = 0$, as the volume remains constant.

So,
$$Q = \Delta U = \frac{\nu R}{\gamma - 1}\Delta T$$

From gas law, $p_0 V = \nu R T_0$

So,
$$\Delta U = -\frac{p_0 V \Delta T}{T_0(\gamma - 1)} = -0.25 \text{ kJ}$$

Hence amount of heat lost $= -\Delta U = 0.25$ kJ

2.30 By the first law of thermodynamics, $Q = \Delta U + A$.

But,
$$\Delta U = \frac{p\Delta V}{\gamma - 1} = \frac{A}{\gamma - 1} \quad \text{(as } p \text{ is constant)}$$

So,
$$Q = \frac{A}{\gamma - 1} + A = \frac{\gamma A}{\gamma - 1} = \frac{1.4}{1.4 - 1} \times 2 = 7\text{J}$$

2.31 Under an isobaric process,

$$A = p\Delta V = R\Delta T \text{ (as } \nu = 1) = 0.6 \text{ kJ.}$$

From the first law of thermodynamics,

$$\Delta U = Q - A = Q - R\Delta T = 1 \text{ kJ}$$

Again increment in internal energy, $\Delta U = \frac{R\Delta T}{\gamma - 1}$ (for $\nu = 1$)

Thus,
$$Q - R\Delta T = \frac{R\Delta T}{\gamma - 1} \quad \text{or} \quad \gamma = \frac{Q}{Q - R\Delta T} = 1.6$$

2.32 Given $\nu = 2$ moles of the gas. In the first phase, under isochoric process, $A_1 = 0$, therefore from gas law if pressure is reduced n times, the temperature, i.e., new temperature becomes T_0/n.

Now from first law of thermodynamics,

$$Q_1 = \Delta U_1 = \frac{\nu R \Delta T}{\gamma - 1}$$

$$= \frac{\nu R}{\gamma - 1}\left(\frac{T_0}{n} - T_0\right) = \frac{\nu R T_0 (1 - n)}{n(\gamma - 1)}$$

During the second phase (under isobaric process), $A_2 = p\Delta V = \nu R \Delta T$.

Thus from first law of thermodynamics,

$$Q_2 = \Delta U_2 + A_2 = \frac{\nu R \Delta T}{\gamma - 1} + \nu R \Delta T$$

$$= \frac{\nu R\left(T_0 - \frac{T_0}{n}\right)\gamma}{\gamma - 1} = \frac{\nu R T_0 (n - 1)\gamma}{n(\gamma - 1)}$$

Hence the total amount of heat absorbed,

$$Q = Q_1 + Q_2 = \frac{\nu R T_0 (1-n)}{n(\gamma-1)} + \frac{\nu R T_0 (n-1)\gamma}{n(\gamma-1)}$$

$$= \frac{\nu R T_0 (n-1)}{n(\gamma-1)}(-1+\gamma) = \nu R T_0\left(1-\frac{1}{n}\right) = 2.5 \text{ kJ} \quad \text{(on substituting values)}$$

2.33 Number of moles in the mixture $\nu = \nu_1 + \nu_2$

At a certain temperature, $U = U_1 + U_2$ or $\nu C_V = \nu_1 C_{V_1} + \nu_2 C_{V_2}$

Thus,
$$C_V = \frac{\nu_1 C_{V_1} + \nu_2 C_{V_2}}{\nu} = \frac{\left(\nu_1 \dfrac{R}{\gamma_1 - 1} + \nu_2 \dfrac{R}{\gamma_2 - 1}\right)}{\nu}$$

Similarly,
$$C_p = \frac{\nu_1 C_{p_1} + \nu_2 C_{p_2}}{\nu}$$

$$= \frac{\nu_1 \gamma_1 C_{V_1} + \nu_2 \gamma_2 C_{V_2}}{\nu} = \frac{\left(\nu_1 \dfrac{\gamma_1 R}{\gamma_1 - 1} + \nu_2 \dfrac{\gamma_2 R}{\gamma_2 - 1}\right)}{\nu}$$

Thus,
$$\gamma = \frac{C_p}{C_V} = \frac{\nu_1 \dfrac{\gamma_1}{\gamma_1 - 1} R + \nu_2 \dfrac{\gamma_2}{\gamma_2 - 1} R}{\nu_1 \dfrac{R}{\gamma_1 - 1} + \nu_2 \dfrac{R}{\gamma_2 - 1}}$$

$$= \frac{\nu_1 \gamma_1 (\gamma_2 - 1) + \nu_2 \gamma_2 (\gamma_1 - 1)}{\nu_1 (\gamma_2 - 1) + \nu_2 (\gamma_1 - 1)}$$

$$= 1.33 \text{ (on substituting values)}$$

2.34 From the previous problem,

$$C_V = \frac{\nu_1 \dfrac{R}{\gamma_1 - 1} + \nu_2 \dfrac{R}{\gamma_2 - 1}}{\nu_1 + \nu_2} = 15.2 \text{ J/mol K (on substituting values)}$$

and
$$C_p = \frac{\nu_1 \dfrac{\gamma_1 R}{\gamma_1 - 1} + \nu_2 \dfrac{\gamma_2 R}{\gamma_2 - 1}}{\nu_1 + \nu_2} = 23.85 \text{ J/mol K (on substituting values)}$$

Now molar mass of the mixture $(M) = \dfrac{\text{Total mass}}{\text{Total number of moles}} = \dfrac{20+7}{(1/2)+(1/4)} = 36$

Hence,
$$c_V = \frac{C_V}{M} = 0.42 \text{ J/gK} \quad \text{and} \quad c_p = \frac{C_p}{M} = 0.66 \text{ J/gk}$$

2.35 Let S be the area of the piston and F be the force exerted by the external agent.

Then, $F + pS = p_0 S$ (see figure) at an arbitrary instant of time. Here p is the pressure at the instant the volume is V. (Initially the pressure inside is p_0). Work done by the agent,

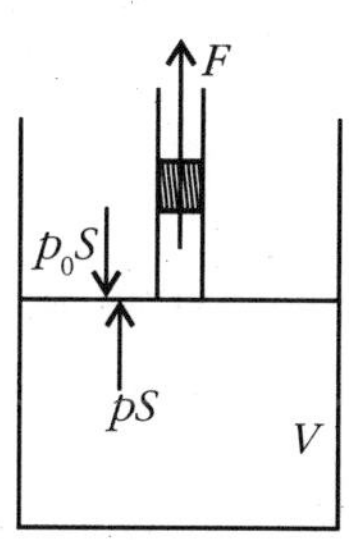

$$A = \int_{V_0}^{\eta V_0} F\,dx$$

$$= \int_{V_0}^{\eta V_0} (p_0 - p) S \cdot dx = \int_{V_0}^{\eta V_0} (p_0 - p)\,dV$$

$$= p_0(\eta-1)V_0 - \int_{V_0}^{\eta V_0} pdV = p_0(\eta-1)V_0 - \int_{V_0}^{\eta V_0} \nu RT\cdot\frac{dV}{V}$$

$$= (\eta-1)p_0V_0 - \nu RT\ \ln\ \eta = (\eta-1)\nu RT - \nu RT\ \ln\ \eta$$

$$= \nu RT(\eta-1-\ln\eta) = RT(\eta-1-\ln\eta)\ (\text{for } \nu = 1 \text{ mole})$$

2.36 Let the agent move the piston to the right by x. In equilibrium position,

$$p_1S + F_{\text{agent}} = p_2S \quad \text{or} \quad F_{\text{agent}} = (p_2 - p_1)S$$

Work done by the agent in an infinitesimal change dx is

$$F_{\text{agent}}\cdot dx = (p_2 - p_1)Sdx = (p_2 - p_1)dV$$

By applying pV = constant, for the two parts,

$$p_1(V_0 + Sx) = p_0V_0 \quad \text{and} \quad p_2(V_0 - Sx) = p_0V_0$$

So, $$p_2 - p_1 = \frac{p_0V_0 2Sx}{V_0^2} - S^2x^2 = \frac{2p_0V_0V}{V_0^2 - V^2}\ (\text{where } Sx = V)$$

When the volume of the left end is η times the volume of the right end,

$$(V_0 - V) = \eta(V_0 - V) \quad \text{or} \quad V = \frac{\eta-1}{\eta+1}V_0$$

$$A = \int_0^V (p_2 - p_1)dV = \int_0^V \frac{2p_0V_0V}{V_0^2 - V^2}dV = -p_0V_0\ [\ln\ (V_0^2 - V^2)]_0^V$$

$$= -p_0V_0\ [\ln(V_0^2 - V^2) - \ln V_0^2]$$

$$= -p_0V_0\left[\ln\left\{V_0^2 - \left(\frac{\eta-1}{\eta+1}\right)^2 V_0^2\right\} - \ln V_0^2\right]$$

$$= -p_0V_0\left(\ln\frac{4\eta}{(\eta+1)^2}\right) = p_0V_0\ln\frac{(\eta+1)^2}{4\eta}$$

2.37 In the isothermal process, heat transfer to the gas is given by

$$Q_1 = \nu RT_0\ln\frac{V_2}{V_1} = \nu RT_0\ln\eta\left(\text{for } \eta = \frac{V_2}{V_1} = \frac{p_1}{p_2}\right)$$

In the isochoric process, $A = 0$. Thus heat transfer to the gas is given by

$$Q_2 = \Delta U = \nu C_V\Delta T = \frac{\nu R}{\gamma-1}\Delta T\left(\text{for } C_V = \frac{R}{\gamma-1}\right)$$

But, $$\frac{p_2}{p_1} = \frac{T_0}{T} \Rightarrow T = T_0\frac{p_1}{p_2} = \eta T_0\left(\text{for } \eta = \frac{p_1}{p_2}\right)$$

or $$\Delta T = \eta T_0 - T_0 = (\eta-1)T_0$$

So, $$Q_2 = \frac{\nu R}{\gamma-1}\cdot(\eta-1)T_0$$

Thus, net heat transfer to the gas is

$$Q = \nu RT_0\ln\eta + \frac{\nu R}{\gamma-1}\cdot(\eta-1)T_0$$

or $$\frac{Q}{\nu RT_0} = \ln\eta + \frac{\eta-1}{\gamma-1} \Rightarrow \frac{Q}{\nu RT_0} - \ln\eta = \frac{\eta-1}{\gamma-1}$$

or $$\gamma = 1 + \frac{\eta-1}{\frac{Q}{\nu RT_0} - \ln\eta} = 1 + \frac{6-1}{\left(\frac{80\times10^3}{3\times8.314\times273}\right) - \ln 6} = 1.4$$

2.38 (a) From ideal gas law, $p = (\nu R/V)T = kT$ (where $k = \nu R/V$).

For isochoric process, V = constant, thus $p = kT$, represents a straight line passing through the origin and its slope is equal to k.

For isobaric process p = constant, thus on $p - T$ curve, it is a horizontal straight line parallel to T-axis, if T is along horizontal (or x-axis).

For isothermal process, T = constant, thus on $p - T$ curve, it represents a vertical straight line if T is taken along horizontal (or x-axis).

For an adiabatic process $T^\gamma p^{1-\gamma}$ = constant.

On differentiating, we get

$$(1-\gamma)p^{-\gamma}dpT^\gamma + \gamma p^{1-\gamma}\cdot T^{\gamma-1}\cdot dT = 0$$

So, $$\frac{dp}{dT} = \left(\frac{\gamma}{1-\gamma}\right)\left(\frac{p^{1-\gamma}}{p^{-\gamma}}\right)\left(\frac{T^{\gamma-1}}{T^\gamma}\right) = \left(\frac{\gamma}{\gamma-1}\right)\frac{p}{T}$$

The approximate plots of isochoric, isobaric, isothermal, and adiabatic processes are drawn in the answer sheet.

(b) As p is not considered variable, we have from ideal gas law

$$V = \frac{\nu R}{p}T = k'T\left(\text{where } k' = \frac{\nu R}{p}\right)$$

On $V - T$ co-ordinate system let us, take T along x-axis.

For an isochoric process, V = constant, thus k' = constant and $V = k'T$ obviously represents a straight line passing through the origin of the coordinate system and k' is its slope.

For isothermal process T = constant. Thus on the stated coordinate system it represents a straight line parallel to the V-axis.

For an adiabatic process, $TV^{\gamma-1}$ = constant.

On differentiating, we get

$$(\gamma-1)V^{\gamma-2}\,dVT + V^{\gamma-1}dT = 0$$

So, $$\frac{dV}{dT} = -\left(\frac{1}{\gamma-1}\right)\cdot\frac{V}{T}$$

The approximate plots of isochoric, isobaric, isothermal and adiabatic processes are drawn in the answer sheet.

2.39 (a) According to $T-p$ relation in adiabatic process, $T^\gamma = kp^{\gamma-1}$ (where k = constant).

Therefore, $(T_2/T_1)^\gamma = (p_2/p_1)^{\gamma-1}$

So, $T^\gamma/T_0^\gamma = \eta^{\gamma-1}$(for $\eta = p_2/p_1$)

Hence, $T = T_0\cdot\eta^{\gamma-1/\gamma} = 290\times10^{(1.4-1)/1.4} = 0.56$ kK

(b) Using the solution of part (a), sought work done

$$A = \frac{\nu R \Delta T}{\gamma - 1} = \frac{\nu R T_0}{\gamma - 1}\left(\eta^{(\gamma-1)/\gamma} - 1\right) = 5.61 \text{ kJ (on substituting values)}$$

2.40 Let (p_0, V_0, T_0) be the initial state of the gas.

We know that work done by the gas in adiabatic process is

$$A_{\text{adia}} = \frac{-\nu R \Delta T}{\gamma - 1}$$

But from the equation $TV^{\gamma-1}$ = constant, we get $\Delta T = T_0\,(\eta^{\gamma-1} - 1)$.

Thus,
$$A_{\text{adia}} = \frac{-\nu R T_0 (\eta^{\gamma-1} - 1)}{\gamma - 1}$$

On the other hand, work done by the gas in isothermal process is

$$A_{\text{iso}} = \nu R T_0 \ln(1/\eta) = -\nu R T_0 \ln \eta$$

Thus,
$$\frac{A_{\text{adia}}}{A_{\text{iso}}} = \frac{\eta^{\gamma-1} - 1}{(\gamma - 1)\ln \eta} = \frac{5^{0.4} - 1}{0.4 \times \ln 5} = 1.4$$

2.41 Since here the piston is conducting and it is moved slowly, the temperature on the two sides increases and is maintained at the same value.

Elementary work done by the agent = work done in compression – work done in expansion, i.e., $dA = p_2 dV - p_1 dV = (p_2 - p_1)dV$, where p_1 and p_2 are pressures at any instant of the gas on expansion and compression side, respectively.

From the gas law $p_1(V_0 + Sx) = \nu RT$ and $p_2\,(V_0 - Sx) = \nu RT$, for each section (x is the displacement of the piston towards section 2).

So,
$$p_2 - p_1 = \nu RT \frac{2sx}{V_0^2 - S^2x^2} = \nu RT \cdot \frac{2V}{V_0^2 - V^2} \quad (\text{as } Sx = V)$$

So,
$$dA = \nu RT \frac{2V}{V_0^2 - V^2} dV$$

Also, from the first law of thermodynamics

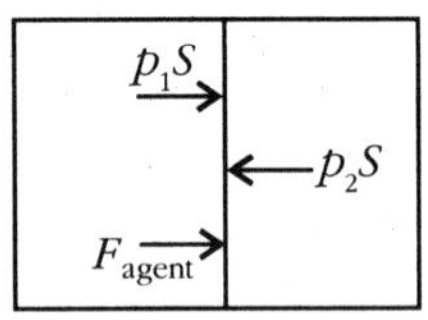

$$dA = -dU = -2\nu \frac{R}{\gamma - 1} dT \quad (\text{as } dQ = 0)$$

So, work done on the gas $= -dA = 2\nu \dfrac{R}{\gamma - 1} dT$

Thus,
$$2\nu \frac{R}{\gamma - 1} dT = \nu RT \frac{2V \cdot dV}{V_0^2 - V^2}$$

or
$$\frac{dT}{T} = \gamma - 1 \frac{VdV}{V_0^2 - V^2}$$

When the left end is η times the volume of the right end,

$$(V_0 + V) = \eta(V_0 - V) \quad \text{or} \quad V = \frac{\eta - 1}{\eta + 1} V_0$$

On integrating,
$$\int_{T_0}^{T} \frac{dT}{T} = (\gamma - 1)\int_0^V \frac{VdV}{V_0^2 - V^2}$$

or $$\ln\frac{T}{T_0} = (\gamma-1)\left[-\frac{1}{2}\ln(V_0^2 - V^2)\right]_0^V$$

$$= -\frac{\gamma-1}{2}[\ln(V_0^2 - V^2) - \ln V_0^2]$$

$$= \frac{\gamma-1}{2}\left[\ln V_0^2 - \ln V_0^2\left\{1-\left(\frac{\eta-1}{\eta+1}\right)^2\right\}\right] = \frac{\gamma-1}{2}\ln\frac{(\eta+1)^2}{4\eta}$$

Hence, $$T = T_0\left(\frac{(\eta+1)^2}{4\eta}\right)^{\frac{\gamma-1}{2}}$$

2.42 From energy conservation as in the derivation of Bernouli's theorem it reads

$$\frac{p}{\rho} + \frac{1}{2}\mathbf{v}^2 + gz + u + Q_d = \text{constant} \tag{1}$$

In Eq. (1), u is the internal energy per unit mass and in this case is the thermal energy per unit mass of the gas. As the gas vessel is thermally insulated $Q_d = 0$. Just inside the vessel,

$$u = \frac{C_V T}{M} = \frac{RT}{M(\gamma-1)} \quad \text{also} \quad \frac{p}{\rho} = \frac{RT}{M}$$

Inside the vessel $\mathbf{v} = 0$ and just outside $p = 0$, and $u = 0$. In general gz is not very significant for gases. Thus applying Eq. (1) just inside and outside the hole, we get

$$\frac{1}{2}v^2 = \frac{p}{\rho} + u$$

$$= \frac{RT}{M} + \frac{RT}{M(\gamma-1)} = \frac{\gamma RT}{M(\gamma-1)}$$

Hence, $$v^2 = \frac{2\gamma RT}{M(\gamma-1)}$$

or $$v = \sqrt{\frac{2\gamma RT}{M(\gamma-1)}} = 3.22 \text{ km/s}$$

Alternate:

From Eulerian equation

$$-\nabla p + \rho\mathbf{g} = \rho\mathbf{a}$$

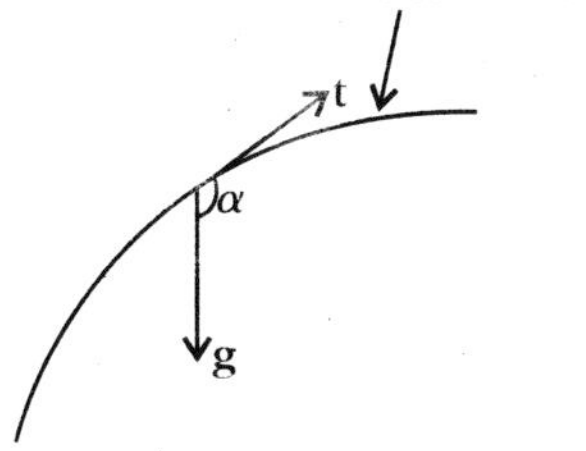

In projection form along a stream line for steady flow, it becomes,

$$\frac{-\partial p}{\partial s} + g\cos\alpha = \rho v\frac{\partial v}{\partial s}$$

In this problem,

$$\frac{-dp}{ds} = \rho v\frac{dv}{ds} \quad \text{(only because } \alpha \text{ becomes } 90°\text{)}$$

Experiments show, when the particles move sufficiently fast it can be assumed that the process is adiabatic. Then the connection between the pressure and the density is given by the relation,

$$\frac{p}{\rho^\gamma} = \frac{p_i}{\rho_i^\gamma} = \text{constant}$$

where γ is the adiabatic exponent dependent on the nature of the gas and ρ_i is the density of the gas inside the vessel. The adiabatic relation is a consequence of the fact that during the expansion of a gas there is no heat exchange with the surrounding, that is there is no loss or gain of heat.

On finding the density as a function of the pressure and substituting, we obtain the equation

$$-p^{-1/\gamma}\frac{dp}{ds} = \frac{\rho_i}{p_i^{1/\gamma}}v\frac{dv}{ds}$$

This equation should be integrated along the stream tube. Since the pressure in the vessel is p_i and the pressure outside the vessel is p_0, we must integrate from p_i to p_0 for the pressure and from zero to $v = v_0$ (the outlet velocity) for the velocity.

Thus,
$$-\int_{p_i}^{p_0} p^{-1/\gamma}dp = \frac{\rho_i}{p_i^{1/\gamma}}\int_0^{v_0} v\,dv$$

On performing the integration we find the outlet velocity,

$$v = \sqrt{\frac{2\gamma}{\gamma-1}\frac{p_i}{\rho_i}\left[1-\left(\frac{p_0}{p_i^{1/\gamma}}\right)^{\frac{\gamma-1}{\gamma}}\right]}$$

According the problem $p_0 = 0$ because outside is vacuum.

Hence,
$$v = \sqrt{\frac{2\gamma}{\gamma-1}\frac{p_i}{\rho_i}} = \sqrt{\frac{2\gamma}{\gamma-1}\frac{RT}{M}} = 3.3 \text{ km/s (on substituting values)}$$

Note: *The velocity here is the velocity of hydrodynamic flow of the gas into vacuum. This requires that the diameter of the hole is not too small (D > mean free path l). In the opposite case (D << l), the flow is called effusion. Then the above result does not apply and kinetic theory methods are needed.*

2.43 The differential work done by the gas,

$$dA = pdV = \frac{\nu RT^2}{a}\left(-\frac{a}{T^2}\right)dT = -\nu RdT \text{ (as } pV = \nu RT \text{ and } V = a/T)$$

So,
$$A = -\int_T^{T+\Delta T} \nu RdT = -\nu R\Delta T$$

From the first law of thermodynamics

$$Q = \Delta U + A = \frac{\nu R}{\gamma-1}\Delta T - \nu R\Delta T$$

$$= \nu R\Delta T\cdot\frac{2-\gamma}{\gamma-1} = R\Delta T\cdot\frac{2-\gamma}{\gamma-1} \text{ (for } \nu = 1 \text{ mole)}$$

2.44 According to the problem: $A \propto U$ or $dA = aU$ (where a is proportionality constant).

So,
$$pdV = \frac{a\nu RdT}{\gamma-1} \tag{1}$$

From ideal gas law, $pV = \nu RT$, which on differentiating gives

$$pdV + Vdp = \nu RdT \tag{2}$$

Thus, from Eqs. (1) and (2)

$$pdV = \frac{a}{\gamma-1}(pdV + Vdp)$$

or
$$pdV\left(\frac{a}{\gamma-1}-1\right)+\frac{a}{\gamma-1}Vdp=0$$

or
$$pdV(k-1)+kVdp=0 \text{ (where } k=\frac{a}{\gamma-1}=\text{ another constant)}$$

or
$$pdV\frac{k-1}{k}+Vdp=0$$

or
$$npdV+Vdp=0\ \left(\text{where } \frac{k-1}{k}=n=\text{ratio}\right)$$

Dividing both the sides by pV, we get
$$n\frac{dV}{V}+\frac{dp}{p}=0$$

On integrating we get,
$$n\ln V+\ln p=\ln C \quad \text{(where } C \text{ is constant)}$$

or
$$\ln(pV^n)=\ln C \quad \text{or} \quad pV^n=C$$

2.45 In the polytropic process work done by the gas is given by,
$$A=\frac{\nu R[T_i-T_f]}{n-1}$$

(where T_i and T_f are initial and final temperatures of the gas in an adiabatic process).

Also,
$$\Delta U=\frac{\nu R}{\gamma-1}(T_f-T_i)$$

By the first law of thermodynamics,
$$Q=\Delta U+A$$
$$=\frac{\nu R}{\gamma-1}(T_f-T_i)+\frac{\nu R}{n-1}(T_i-T_f)$$
$$=(T_f-T_i)\nu R\left[\frac{1}{\gamma-1}-\frac{1}{n-1}\right]=\frac{\nu R[n-\gamma]}{(n-1)(\gamma-1)}\Delta T$$

According to definition of molar heat capacity when number of moles $\nu=1$ and $\Delta T=1$, then Q = molar heat capacity.

Here,
$$C_n=\frac{R(n-\gamma)}{(n-1)(\gamma-1)}<0 \text{ for } 1<n<\gamma$$

2.46 Let the process be polytropic, according to the law pV^n = constant.

Thus,
$$p_fV_f^n=p_iV_i^n$$

or
$$(p_i/p_f)=\beta$$

So,
$$\alpha^n=\beta \quad \text{or} \quad \ln\beta=n\ln\alpha \quad \text{or} \quad n=\frac{\ln\beta}{\ln\alpha}$$

In the polytropic process molar heat capacity is given by
$$C_n=\frac{R(n-\gamma)}{(n-1)(\gamma-1)}=\frac{R}{\gamma-1}-\frac{R}{n-1}$$
$$=\frac{R}{\gamma-1}-\frac{R\ln\alpha}{\ln\beta\ln\alpha}\ \left(\text{where } n=\frac{\ln\beta}{\ln\alpha}\right)$$

So, $$C_n = \frac{8.314}{1.66-1} - \frac{8.314 \ln 4}{\ln 8 - \ln 4} = -42 \text{ J/mol K}$$

2.47 (a) Increment of internal energy for ΔT becomes

$$\Delta U = \frac{\nu R \Delta T}{\gamma - 1} = \frac{R\Delta T}{\gamma - 1} = -324 \text{ J (as } \nu = 1 \text{ mole)}$$

From first law of thermodynamics

$$Q = \Delta U + A = \frac{R\Delta T}{\gamma - 1} - \frac{R\Delta T}{n-1} = 0.11 \text{ kJ}$$

(b) Sought work done,

$$A_n = \int p dV = \int_{V_i}^{V_f} \frac{k}{V^n} dV \text{ (where } pV^n = k = p_i V_i^n = p_f V_f^n)$$

$$= \frac{k}{1-n}(V_f^{1-n} - V_i^{1-n}) = \frac{\left(p_f V_f^n V_f^{1-n} - p_i V_i^n V_i^{1-n}\right)}{1-n}$$

$$= \frac{p_f V_f - p_i V_i}{1-n} = \frac{\nu R(T_f - T_i)}{1-n}$$

$$= \frac{\nu R \Delta T}{n-1} = \frac{R\Delta T}{n-1} = 0.43 \text{ kJ (as } \nu = 1 \text{ mole)}$$

2.48 Law of the process is, $p = \alpha V$ or $pV^{-1} = \alpha$, so the process is polytropic with index $n = -1$. As $p = \alpha V$ so, $p_i = \alpha V_0$ and $p_f = \alpha\eta V_0$.

(a) Increment of the internal energy is given by

$$\Delta U = \frac{\nu R}{\gamma - 1}[T_f - T_i] = \frac{p_f V_f - p_i V_i}{\gamma - 1}$$

Putting $p_i = \alpha V_0$ and $p_f = \alpha \eta V_0$, we get

$$\Delta U = \frac{\alpha V_0^2(\eta^2 - 1)}{\gamma - 1}$$

(b) Work done by the gas is given by

$$A = \frac{p_i V_i - p_f V_f}{n-1} = \frac{\alpha V_0^2 - \alpha\eta V_0 \cdot \eta V_0}{-1-1}$$

$$= \frac{\alpha V_0^2(1-\eta^2)}{-2} = \frac{1}{2}\alpha V_0^2(\eta^2 - 1)$$

(c) Molar heat capacity is given by

$$C_n = \frac{R(n-\gamma)}{(n-1)(\gamma-1)} = \frac{R(-1-\gamma)}{(-1-1)(\gamma-1)} = \frac{R}{2} \cdot \frac{\gamma+1}{\gamma-1}$$

2.49 (a) $\Delta U = \dfrac{\upsilon R}{\gamma - 1}\Delta T$ and $Q = \nu C_n \Delta T$, where C_n is the molar heat capacity in the process. It is given that $Q = -\Delta U$.

So, $$C_n \Delta T = -\frac{R}{\gamma - 1}\Delta T \quad \text{or} \quad C_n = -\frac{R}{\gamma - 1}$$

(b) By the first law of thermodynamics,

$$dQ = dU + dA$$

or
$$2dQ = dA \quad (\text{as } dQ = -dU)$$
$$2\nu C_n dT = pdV \quad \text{or} \quad \frac{2R\nu}{\gamma-1}dT + pdV = 0$$

So,
$$\frac{2R\nu}{\gamma-1}dT + \frac{\mathbf{v}RT}{V}dV = 0 \quad \text{or} \quad \frac{2dT}{(\gamma-1)T} + \frac{dV}{V} = 0$$

or
$$\frac{dT}{T} + \frac{\gamma-1}{2}\frac{dV}{V} = 0 \Rightarrow TV^{(\gamma-1)/2} = \text{ constant}$$

(c) We know $C_n = \dfrac{(n-\gamma)R}{(n-1)(\gamma-1)}$

But from part (a), we have $C_n = -R/(\gamma - 1)$.

Thus,
$$-\frac{R}{\gamma-1} = \frac{(n-\gamma)R}{(n-1)(\gamma-1)}$$

which yields
$$n = \frac{1+\gamma}{2}$$

From part (b), we know $TV^{(\gamma-1)/2}$ = constant.

So,
$$\frac{T_0}{T} = \left(\frac{V}{V_0}\right)^{(\gamma-1)/2} = \eta^{(\gamma-1)/2} \text{ (where } T \text{ is the final temperature)}$$

Work done by the gas for one mole is given by
$$A = \frac{R(T_0 - T)}{n-1} = \frac{2RT_0}{\gamma-1}[1 - \eta^{(1-\gamma/2)}]$$

2.50 Given $p = aT^\alpha$ (for one mole of gas).

So,
$$pT^{-\alpha} = a \quad \text{or} \quad p\left(\frac{pV}{R}\right)^{-\alpha} = a$$

$\Rightarrow$
$$p^{1-\alpha}V^{-\alpha} = aR^{-\alpha} \quad \text{or} \quad pV^{\alpha/(\alpha-1)} = \text{constant}$$

Here polytropic exponent $n = \alpha/(\alpha-1)$.

(a) In the polytropic process for one mole of gas,
$$A = \frac{R\Delta T}{1-n} = \frac{R\Delta T}{1-[\alpha/(\alpha-1)]} = R\Delta T(1-\alpha)$$

(b) Molar heat capacity is given by
$$C = \frac{R}{\gamma-1} - \frac{R}{n-1} = \frac{R}{\gamma-1} - \frac{R}{[\alpha/(\alpha-1)]-1}$$
$$= \frac{R}{\gamma-1} + R(\alpha-1)$$

2.51 Given $U = aV^\alpha$

or
$$\mathbf{v}C_V T = aV^\alpha \quad \text{or} \quad \mathbf{v}C_V\frac{pV}{\mathbf{v}R} = aV^\alpha$$

$\Rightarrow$
$$aV^\alpha \cdot \frac{R}{C_V} \cdot \frac{1}{p_V} = 1 \quad \text{or} \quad V^{\alpha-1} \cdot p^{-1} = \frac{C_V}{Ra}$$

or
$$pV^{1-\alpha} = Ra/C_V = \text{constant} = a\ (\gamma - 1)\ [\text{as } C_V = R/(\gamma - 1)]$$

So polytropic index $n = 1 - \alpha$.

(a) Work done by the gas is given by

$$A = \frac{-\nu R\Delta T}{n-1} \quad \text{and} \quad \Delta U = \frac{\nu R\Delta T}{\gamma - 1}$$

Hence, $$A = \frac{-\Delta U(\gamma-1)}{n-1} = \frac{\Delta U(\gamma-1)}{\alpha} \quad (\text{as } n = 1-\alpha)$$

By the first law of thermodynamics

$$Q = \Delta U + A$$

$$= \Delta U + \frac{\Delta U(\gamma-1)}{\alpha} = \Delta U\left[1 + \frac{\gamma-1}{\alpha}\right]$$

(b) Molar heat capacity is given by

$$C = \frac{R}{\gamma-1} - \frac{R}{n-1} = \frac{R}{\gamma-1} - \frac{R}{1-\alpha-1}$$

$$= \frac{R}{\gamma-1} + \frac{R}{\alpha} \quad (\text{as } n = 1-\alpha)$$

2.52 (a) By the first law of thermodynamics

$$dQ = dU + dA = \nu C_V dT + pdV$$

Molar specific heat according to definition is given by

$$C = \frac{dQ}{\nu dT} = \frac{\nu C_V dT + pdV}{\nu dT}$$

$$= \frac{\nu C_V dT + (\nu RT/V)dV}{\nu dT} = C_V + \frac{RT}{V}\frac{dV}{dT}$$

We have $$T = T_0 e^{\alpha V}$$

On differentiating, $dT = \alpha T_0 e^{\alpha V} \cdot dV$

So, $$\frac{dV}{dT} = \frac{1}{\alpha T_0 e^{\alpha V}}$$

Hence, $$C = C_V + \frac{RT}{V} \cdot \frac{1}{\alpha T_0 e^{\alpha V}} = C_V + \frac{RT_0 e^{\alpha V}}{\alpha V T_0 e^{\alpha V}} = C_V + \frac{R}{\alpha V}$$

(b) For the process, $p = p_0 e^{\alpha V}$.

So, $$p = RT/V = p_0 e^{\alpha V}$$

or $$T = (p_0/R)e^{\alpha V} \cdot V$$

So, $$C = C_V + \frac{RT}{V}\frac{dV}{dT} = C_V + p_0 e^{\alpha V} \cdot \frac{R}{p_0 e^{\alpha V}(1+\alpha V)} = C_V + \frac{R}{1+\alpha V}$$

2.53 (a) Using solution of Problem 2.52,

$$C = C_V + \frac{RT}{V}\frac{dV}{dT} = C_V + \frac{pdV}{dT} \quad (\text{for one mole of gas})$$

We have, $$p = p_0 + \frac{\alpha}{V} \quad \text{or} \quad \frac{RT}{V} = p_0 + \frac{\alpha}{V} \Rightarrow RT = p_0 V + \alpha$$

Therefore, $$RdT = p_0 dV$$

So, $dV/dT = R/p_0$

Hence, $$C = C_V + \left(p_0 + \frac{\alpha}{V}\right)\cdot\frac{R}{p_0} = \frac{R}{\gamma-1} + \left(1 + \frac{\alpha}{p_0V}\right)$$

$$R = \frac{R\gamma}{\gamma-1} + \frac{\alpha R}{p_0V}$$

(b) Work done is given by

$$A = \int_{V_1}^{V_2}\left(p_0 + \frac{\alpha}{V}\right)dV = p_0(V_2 - V_1) + \alpha\ln\frac{V_2}{V_1}$$

$$\Delta U = C_V(T_2 - T_1) = C_V\left(\frac{p_2V_2}{R} - \frac{p_1V_1}{R}\right) \text{ (for one mole)}$$

$$= \frac{R}{(\gamma-1)R}(p_2V_2 - p_1V_1)$$

$$= \frac{1}{\gamma-1}\left[(p_0 + \alpha V_2)V_2 - \left(p_0 + \frac{\alpha}{V_1}\right)V_1\right] = \frac{p_0(V_2 - V_1)}{\gamma-1}$$

By the first law of thermodynamics,

$$Q = \Delta U + A$$

$$= \frac{p_0(V_2 - V_1)}{(\gamma-1)} + p_0(V_2 - V_1) + \alpha\ \ln\frac{V_2}{V_1}$$

$$= \frac{\gamma p_0(V_2 - V_1)}{(\gamma-1)} + \alpha\ \ln\frac{V_2}{V_1}$$

2.54 (a) From solution of Problem 2.52, heat capacity is given by

$$C = C_V + \frac{RT}{V}\frac{dV}{dT}$$

We have $$T = T_0 + \alpha V \Rightarrow V = \frac{T}{\alpha} - \frac{T_0}{\alpha}$$

On differentiating, we get

$$\frac{dV}{dT} = \frac{1}{\alpha}$$

Hence, $$C = C_V + \frac{RT}{V}\cdot\frac{1}{\alpha} = \frac{R}{\gamma-1} + \frac{R(T_0 + \alpha V)}{V}\cdot\frac{1}{\alpha}$$

$$= \frac{R}{\gamma-1} + R\left(\frac{T_0}{\alpha V} + 1\right) = \frac{\gamma R}{\gamma-1} + \frac{RT_0}{\alpha V} = C_V + \frac{RT}{\alpha V} = C_p + \frac{RT_0}{\alpha V}$$

(b) Given that $T = T_0 + \alpha V$ and as $T = pV/R$ for one mole of an ideal gas, so

$$p = \frac{R}{V}(T_0 + \alpha V) = \frac{RT_0}{V} + \alpha R$$

Now, $$A = \int_{V_1}^{V_2} p\,dV = \int_{V_1}^{V_2}\left(\frac{RT_0}{V} + \alpha R\right)dV \text{ (for one mole)}$$

$$= RT_0\ln\frac{V_2}{V_1} + \alpha(V_2 - V_1)R$$

Also, $$\Delta U = C_V(T_2 - T_1)$$
$$= C_V[T_0 + \alpha V_2 - (T_0 + \alpha V_1)] = \alpha C_V(V_2 - V_1)$$

By the first law of thermodynamics

$$Q = \Delta U + A$$
$$= \frac{\alpha R}{\gamma - 1}(V_2 - V_1) + RT_0 \ln\frac{V_2}{V_1} + \alpha R(V_2 - V_1)$$
$$= \alpha R(V_2 - V_1)\left[1 + \frac{1}{\gamma - 1}\right] + RT_0 \ln\frac{V_2}{V_1}$$
$$= \alpha C_p(V_2 - V_1) + RT_0 \ln\frac{V_2}{V_1}$$

2.55 Heat capacity is given by

$$C = C_V + \frac{RT}{V}\frac{dV}{dT}$$

(a) Given that

$$C = C_V + \alpha T$$

So, $$C_V + \alpha T = C_V + \frac{RT}{V}\frac{dV}{dT} \quad \text{or} \quad \frac{\alpha}{R}dT = \frac{dV}{V}$$

Integrating both sides, we get

$$\frac{\alpha}{R}T = \ln V + \ln C_0 = \ln VC_0 \text{ (where } C_0 \text{ is a constant)}$$

or $$V \cdot C_0 = e^{-\alpha T/R}$$

Hence, $$V \cdot e^{\alpha T/R} = \frac{1}{C_0} = \text{constant}$$

(b) According to the problem $C = C_V + \beta V$.

But, $$C = C_V + \frac{RT}{V}\frac{dV}{dT}$$

So, $$\frac{RT}{V}\frac{dV}{dT} = \beta V$$

or $$\frac{R}{\beta}V^{-2}\, dV = \frac{dT}{T}$$

Integrating both sides, we get

$$\frac{R}{\beta}V^{-1} = \ln T + \ln C_0 = T \cdot C_0 \text{ (where } C_0 \text{ is a constant)}$$

So, $$\ln T \cdot C_0 = -\frac{R}{\beta V}$$

or $$T \cdot C_0 = e^{-R/\beta V}$$

Hence, $$Te^{R/\beta V} = \frac{1}{C_0} = \text{constant}$$

(c) Given that $$C = C_V + ap \quad \text{and} \quad C = C_V + \frac{RT}{V}\frac{dV}{dT}$$

So, $$C_V + ap = C_V + \frac{RT}{V}\frac{dV}{dT} \Rightarrow ap = \frac{RT}{V}\frac{dV}{dT}$$

or $$a\frac{RT}{V} = \frac{RT}{V}\frac{dV}{dT} \quad \left(\text{as } p = \frac{RT}{V} \text{ for one mole of gas}\right)$$

or $$dV/dT = a \Rightarrow dV = adT \quad \text{or} \quad dT = dV/a$$

So, $$T = \frac{V}{a} + \text{constant} \quad \text{or} \quad V - aT = \text{constant}$$

2.56 (a) By the first law of thermodynamics

$$A = Q - \Delta U$$

or $$A = CdT - C_V dT = (C - C_V)\, dT \text{ (for one mole)}$$

Given that $C = \alpha/T$

So, $$A = \int_{T_0}^{\eta T_0} \left(\frac{\alpha}{T} - C_V\right) dT = \alpha \ln \frac{\eta T_0}{T_0} - C_V(\eta T_0 - T_0)$$

$$= \alpha \ln \eta - C_V T_0 (\eta - 1)$$

$$= \alpha \ln \eta - \frac{RT_0}{\gamma - 1}(\eta - 1)$$

(b) Given that $$C = +\frac{dQ}{dT} = \frac{RT}{V}\frac{dV}{dT} + C_V$$

and $$C = \alpha/T$$

So, $$C_V + \frac{RT}{V}\frac{dV}{dT} = \frac{\alpha}{T}$$

or $$\frac{R}{\gamma - 1}\frac{1}{RT} + \frac{dV}{V} = \frac{\alpha}{RT^2}dT$$

or $$\frac{dV}{V} = \frac{\alpha}{RT^2}dT - \frac{1}{\gamma - 1}\cdot\frac{dT}{T}$$

or $$(\gamma - 1)\frac{dV}{V} = \frac{\alpha(\gamma - 1)}{RT^2}dT - \frac{dT}{T}$$

Integrating both sides, we get

$$(\gamma - 1)\ln V = -\frac{\alpha(\gamma - 1)}{RT} - \ln T + \ln K$$

or $$\ln\left[V^{\gamma-1}\cdot\frac{T}{K}\right] = \frac{-\alpha(\gamma - 1)}{RT}$$

$$\ln\left[V^{\gamma-1}\cdot\frac{pV}{RK}\right] = \frac{-\alpha(\gamma - 1)}{pV}$$

or $$pV^{\gamma}/RK = e^{-a(\gamma-1)/pV}$$

or $$pV^{\gamma}e^{a(\gamma-1)/pV} = RK = \text{constant}$$

2.57 The work done is

$$A = \int_{V_1}^{V_2} p\,dV = \int_{V_1}^{V_2}\left(\frac{RT}{V-b} - \frac{a}{V^2}\right)dV$$

$$= RT\ln\frac{V_2-b}{V_1-b} + a\left(\frac{1}{V_2} - \frac{1}{V_1}\right)$$

2.58 (a) The increment in the internal energy is

$$\Delta U = \int_{V_1}^{V_2}\left(\frac{\partial U}{\partial V}\right)_T dV$$

But from the second law of thermodynamics

$$\left(\frac{\partial U}{\partial V}\right)_T = T\left(\frac{\partial s}{\partial V}\right)_T - p = T\left(\frac{\partial p}{\partial T}\right)_V - p$$

On the other hand

$$p = \frac{RT}{V-b} - \frac{a}{V^2}$$

or $$T\left(\frac{\partial p}{\partial T}\right)V = \frac{RT}{V-b} \quad \text{and} \quad \left(\frac{\partial U}{\partial V}\right)_T = \frac{a}{V^2}$$

So, $$\Delta U = a\left(\frac{1}{V_1} - \frac{1}{V_2}\right) = 0.11 \text{ kJ (on substituting values)}$$

(b) From the first law of thermodynamics

$$Q = A + \Delta U = RT\ln\frac{V_2-b}{V_1-b} = 3.8 \text{ kJ (on substituting values)}$$

2.59 (a) From the first law, for an adiabatic process, $dQ = dU + pdV = 0$.

From the previous problem

$$dU = \left(\frac{\partial U}{\partial T}\right)_V dT + \left(\frac{\partial U}{\partial V}\right)_T dV = C_V dT + \frac{a}{V^2}dV$$

So, $$0 = C_V dT + \frac{RTdV}{V-b}$$

This equation can be integrated if we assume that C_V and b are constant, then

$$\frac{R}{C_V}\frac{dV}{V-b} + \frac{dT}{T} = 0$$

or $$\ln T + \frac{R}{C_V}\ln(V-b) = \text{constant} \quad \text{or} \quad T(V-b)^{R/C_V} = \text{constant}$$

(b) We use
$$dU = C_V dT + \frac{a}{V^2} dV$$

Now,
$$dQ = C_V dT + \frac{RT}{V-b} dV$$

So along constant p,
$$C_p = C_V + \frac{RT}{V-b}\left(\frac{\partial V}{\partial T}\right)_p$$

Thus,
$$C_p - C_V = \frac{RT}{V-b}\left(\frac{\partial V}{\partial T}\right)_p$$

But,
$$p = \frac{RT}{V-b} \frac{a}{V^2}$$

On differentiating,
$$0 = \left(-\frac{RT}{(V-b)^2} + \frac{2a}{V^2}\right)\left(\frac{\partial V}{\partial T}\right)_p + \frac{R}{V-b}$$

or
$$T\left(\frac{\partial V}{\partial T}\right)_p = \frac{RT/V-b}{\dfrac{RT}{(V-b)^2} - \dfrac{2a}{V^3}} = \frac{V-b}{1 - \dfrac{2a(V-b)^2}{RTV^3}}$$

and
$$C_p - C_V = \frac{R}{1 - \dfrac{2a(V-b)^2}{RTV^3}}$$

2.60 From the first law
$$Q = U_f - U_i + A = 0 \text{ (as the vessels are thermally insulated)}$$
As this is free expansion, $A = 0$, so, $U_f = U_i$.

But,
$$U = \nu C_V T - \frac{a\nu^2}{V}$$

So,
$$C_V(T_f - T_i) = \left(\frac{a}{V_1 + V_2} - \frac{a}{V_1}\right)\nu = \frac{-aV_2\nu}{V_1(V_1 + V_2)}$$

or
$$\Delta T = \frac{-a(\gamma - 1)V_2\nu}{RV_1(V_1 + V_2)}$$

On substitution we get, $\Delta T = -3$ K.

2.61 From the first law
$$Q = U_f - U_i + A = U_f - U_i \text{ (as } A = 0 \text{ in free expansion).}$$

So at constant temperature,
$$Q = \frac{-a\nu^2}{V_2} - \left(-\frac{a\nu^2}{V_1}\right) = a\nu^2 \frac{V_2 - V_1}{V_1 \cdot V_2}$$
$$= 0.33 \text{ kJ (on substituting values)}$$

2.3 Kinetic Theory of Gases. Boltzmann's Law and Maxwell's Distribution

2.62 From the formula $p = nkT$

$$n = \frac{p}{kT} = \frac{4\times10^{-15}\times1.01\times10^{5}}{1.38\times10^{-23}\times300}\ \text{m}^{-3}$$
$$\approx 1\times10^{11}\,\text{per m}^3 = 10^5\ \text{cm}^{-3}$$

Mean distance between molecules

$$(10^{-5}\text{cm}^3)^{1/3} = 10^{1/3}\times10^{-2}\ \text{cm} \approx 0.2\ \text{mm}$$

2.63 After dissociation each N_2 molecule becomes two N-atoms and so contributes, 2×3 degrees of freedom. Thus the number of moles becomes

$$\frac{m}{M}(1+\eta)$$

and
$$p = \frac{mRT}{MV}(1+\eta) = 1.9\ \text{atm (on substituting values)}$$

(Here M is the molecular weight in grams of N_2.)

2.64 Let n_1 = number density of He atoms, n_2 = number density of N_2 molecules.

Then,
$$\rho = n_1m_1 + n_2m_2 \tag{1}$$

(where m_1 and m_2 are masses of helium and nitrogen molecules.)

Also,
$$p = (n_1+n_2)kT \tag{2}$$

From Eqs. (1) and (2), we get

$$n_1 = \frac{\left(\frac{p}{kT} - \frac{\rho}{m_2}\right)}{\left(1-\frac{m_1}{m_2}\right)} = 1.6\times10^{-19}\ \text{cm}^{-3}$$

2.65 Number of nitrogen molecules coming to strike the differential area ds of the wall per unit time are given by

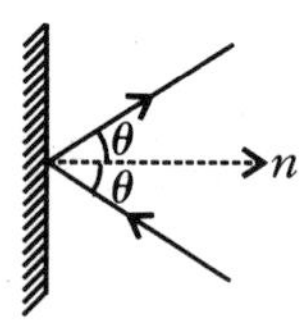

$$\frac{dN}{dt} = n|(\mathbf{v}\cdot d\mathbf{S})| = (nvdS\cos\theta)$$

where $n = N/V$ = number density or concentration of N_2 molecules in the beam.

The change in linear momentum of each of N_2 molecule is

$$\Delta\mathbf{p}_{\text{molecule}} = 2mv\cos\theta\,\mathbf{n}\ \text{(see figure)}$$

So $\Delta\mathbf{p}$ for all the molecules striking the wall per unit time, i.e., the force exerted by the wall on incoming N_2 molecules is

$$(nvds\cos\theta)\Delta\mathbf{p}_{\text{molecule}} = 2nv^2\cos^2\theta dS\ \mathbf{n}$$

Therefore from Newton's law, the force exerted by N_2 molecules on the wall is

$$2nmv^2\cos^2\theta\, dS(-\mathbf{n})$$

Thus the sought pressure i.e., normal force per unit differential is

$$p = \frac{|d\vec{F}_{\text{normal}}|}{ds} = 2nmv^2\cos^2\theta$$

2.66 From the formula

$$v = \sqrt{\frac{\gamma p}{\rho}} \quad \text{or} \quad \gamma = \frac{\rho v^2}{p}$$

If i = number of degrees of freedom of the gas, then

$$C_p = C_V + RT \quad \text{and} \quad C_V = \frac{i}{2}RT$$

$$\gamma = \frac{C_p}{C_V} = 1 + \frac{2}{i} \quad \text{or} \quad i = \frac{2}{\gamma - 1}$$

$$= \frac{2}{(\rho v^2/p) - 1} = 5 \text{ (on substituting values)}$$

2.67 $v_{\text{sound}} = \sqrt{\frac{\gamma p}{\rho}} = \sqrt{\frac{\gamma RT}{M}}$ and $v_{\text{rms}} = \sqrt{\frac{3kT}{m}} = \sqrt{\frac{3RT}{M}}$

So,
$$\frac{v_{\text{sound}}}{v_{\text{rms}}} = \sqrt{\frac{\gamma}{3}} = \sqrt{\frac{i+2}{3i}}$$

(a) For monoatomic gases $i = 3$, therefore

$$\frac{v_{\text{sound}}}{v_{\text{rms}}} = \sqrt{\frac{5}{9}} = 0.75$$

(b) For rigid diatomic molecules $i = 5$, therefore

$$\frac{v_{\text{sound}}}{v_{\text{rms}}} = \sqrt{\frac{7}{15}} = 0.68$$

2.68 For a general non-collinear, non-planar molecule, mean energy

$$= \frac{3}{2}kT \text{ (translational)} + \frac{3}{2}kT \text{ (rotational)} + (3N - 6)kT \text{ (vibrational)}$$

$$= (3N - 3)kT \text{ (per molecule)}$$

For linear molecules, mean energy

$$= \frac{3}{2}kT \text{ (translational)} + kT \text{ (rotational)} + (3N - 5)kT \text{ (vibrational)}$$

$$= \left(3N - \frac{5}{2}\right)kT \text{ (per molecule)}$$

Translational energy corresponds to fraction $\frac{1}{2(N-1)}$ and $\frac{1}{2N - 5/3}$ in non-linear and linear molecules, respectively.

2.69 (a) A diatomic molecule has 3 translational, 2 rotational and one vibrational degrees of freedom. The corresponding energy per mole is

$$\frac{3}{2}RT \text{ (for translational)} + 2 \times \frac{1}{2}RT \text{ (for rotational)} + 1 \times RT \text{ (for vibrational)}$$

$$= \frac{7}{2}RT$$

Thus,
$$C_V = \frac{7}{2}R \quad \text{and} \quad \gamma = \frac{C_p}{C_V} = \frac{9}{7}$$

(b) For linear N-atomic molecules, energy per mole

$$= \left(3N - \frac{5}{2}\right)RT \text{ (as before)}$$

So, $$C_V = \left(3N - \frac{5}{2}\right)R \quad \text{and} \quad \gamma = \frac{6N-3}{6N-5}$$

(c) For non-collinear N-atomic molecules

$$C_V = 3(N-1)\text{ R} \quad \text{(as before from Problem 2.68)}$$

and $$\gamma = \frac{3N-2}{3N-3} = \frac{N-2/3}{N-1}$$

2.70 In the isobaric process, work done is $A = pdV = RdT$ (per mole).

On the other hand heat transferred $Q = C_p dT$.

Now $C_p = (3N-2)R$ for non-collinear molecules and $C_p = [3N-(3/2)]R$ for linear molecules.

Thus, $$\frac{A}{Q} = \begin{cases} \dfrac{1}{3N-2} & \text{non-collinear} \\[2ex] \dfrac{1}{3N-(3/2)} & \text{linear} \end{cases}$$

For monoatomic gases, $C_p = 5/2$ and $A/Q = 2/5$.

2.71 Given specific heats c_p, c_V (per unit mass), then

$$M(c_p - c_V) = R$$

or $$M = \frac{R}{c_p - c_V} = 32 \text{ g/mol} \quad \text{(on substituting values)}$$

Also, $$\gamma = \frac{c_p}{c_V} = \frac{2}{i} + 1$$

or $$i = \frac{2}{(c_p/c_V)-1} = \frac{2c_V}{c_p - c_V} = 5 \quad \text{(on substituting values)}$$

2.72 (a) Given, $$C_p = 29 \text{ J/K mol} = \frac{29}{8.3}R$$

$$C_V = \frac{20.7}{8.3}R, \quad \gamma = \frac{29}{20.7} = 1.4 = \frac{7}{5}$$

$$i = 2\left(\frac{C_p}{R} - 1\right) = 5$$

(b) In the process, pT = constant and (T^2/V) = constant.

So, $$2\frac{dT}{T} - \frac{dV}{V} = 0$$

Thus, $$CdT = c_V dT + pdV = c_V dT + \frac{RT}{V}dV = c_V dT + \frac{2RT}{T}dT$$

or $$C = C_V + 2R = \left(\frac{29}{8.3}\right)R \Rightarrow C_V = \frac{12.4}{8.3}R = \frac{3}{2}R$$

Hence, $$i = 2\left[\frac{C}{R} + \frac{1}{(n-1)}\right] = 3 \quad \text{(monoatomic)}$$

(where $n = 1/2$ is the polytropic index).

2.73 We know that,

$$\frac{1}{R}C_V = \frac{3}{2}\nu_1 + \frac{5}{2}\nu_2$$

(since a monoatomic gas has $C_V = (3/2)R$ and a diatomic gas has $C_V = (5/2)R$)

and
$$\frac{1}{R}C_p = \frac{3}{2}\nu_1 + \frac{5}{2}\nu_2 + \nu_1 + \nu_2$$

Hence,
$$\gamma = \frac{C_p}{C_V} = \frac{5\nu_1 + 7\nu_2}{3\nu_1 + 5\nu_2}$$

2.74 The internal energy of the molecules is

$$U = \frac{1}{2}mN < |\vec{u} - \vec{v}|^2 > = \frac{1}{2}mN <u^2 - v^2>$$

(where $\vec{u}$ = velocity of the vessel, N = number of molecules, each of mass m).

When the vessel is stopped, internal energy becomes $(1/2)mN <u^2>$.

So, there is an increase in internal energy of $\Delta U = (1/2)mNv^2$. This will lead to a rise in temperature of

$$\Delta T = \frac{(1/2)mNv^2}{(i/2)R}$$

$$= \frac{mNv^2}{iR}$$

Since there is no flow of heat, this change of temperature will lead to an excess pressure, given by

$$\Delta p = \frac{R\Delta T}{V} = \frac{mNv^2}{iV}$$

and finally
$$\frac{\Delta p}{p} = \frac{Mv^2}{iRT} = 2.2\%$$

(where M = molecular weight of N_2, i = number of degrees of freedom of $N_2 = 5$).

2.75 (a) From the equipartition theorem

$$\bar{\varepsilon} = \frac{3}{2}kT = 6\times10^{-21}\,\text{J} \quad \text{and} \quad v_{\text{rms}} = \sqrt{\frac{3kT}{m}} = \sqrt{\frac{3RT}{M}} = 0.47\ \text{km/s}$$

(b) In equilibrium, the mean kinetic energy of the droplet will be equal to that of a molecule.

So,
$$\frac{1}{2}\frac{\pi}{6}d^3\rho v_{\text{rms}}^2 = \frac{3}{2}kT \quad \text{or} \quad v_{\text{rms}} = 3\sqrt{\frac{2kT}{\pi d^3\rho}} = 0.15\ \text{m/s}$$

2.76 Given that, $i = 5$, $C_V = 5/2R$, $\gamma = 7/5$.

So,
$$v'_{\text{rms}} = \sqrt{\frac{3RT'}{M}} = \frac{1}{\eta}v_{\text{rms}} = \frac{1}{\eta}\sqrt{\frac{3RT}{M}} \quad \text{or} \quad T' = \frac{1}{\eta^2}T$$

Now, in an adiabatic process

$$TV^{\gamma-1} = TV^{2/i} = \text{constant} \quad \text{or} \quad VT^{i/2} = \text{constant}$$

So,
$$V'\left(\frac{1}{\eta^2}T\right)^{i/2} = VT^{i/2} \quad \text{or} \quad V'\eta^{-i} = V \quad \text{or} \quad V' = \eta^i V$$

The gas must be expanded η^i times, i.e., 7.6 times.

2.77 Here, $$C_V = \frac{5}{2}\frac{m}{M}R \text{ (as } i = 5)$$

(where m = mass of the gas, M = molecular weight of the gas).

If v_{rms} increases η times, the temperature will have increased η^2 times. This will require (neglecting expansion of the vessels), a heat flow of amount

$$\frac{5}{2}\frac{m}{M}R(\eta^2 - 1)T = 10 \text{ kJ}$$

2.78 The root mean square angular velocity is given by

$$\frac{1}{2}I\omega^2 = 2 \times \frac{1}{2}kT \text{ (for 2 degrees of rotations)}$$

or $$\omega = \sqrt{\frac{2kT}{I}} = 6.3 \times 10^{12} \text{ rad/s}$$

2.79 Under compression, the temperature will rise, as

$$TV^{\gamma-1} = \text{constant} \quad \text{or} \quad TV^{2/i} = \text{constant}$$

So, $$T'(\eta^{-1}V_0)^{2/i} = T_0 V_0^{2/i} \quad \text{or} \quad T' = \eta^{+2/i}T_0$$

So mean kinetic energy of rotation per molecule in the compressed state is given by

$$kT' = kT_0\eta^{2/i} = 0.72 \times 10^{-20} \text{ J}$$

2.80 Number of collisions is given by

$$\frac{1}{4}n\langle v\rangle = \nu$$

Now, $$\frac{\nu'}{\nu} = \frac{n'\langle v'\rangle}{n\langle v\rangle} = \frac{1}{\eta}\sqrt{\frac{T'}{T}}$$

(When the gas is expanded η times, n decreases by a factor η.)

Also $$T'(\eta V)^{2/i} = TV^{2/i} \quad \text{or} \quad T' = \eta^{-2/i}T$$

So, $$\frac{\nu'}{\nu} = \frac{1}{\eta}\eta^{-1/i} = \eta^{-(i+1)/i}$$

i.e., collisions decrease by a factor $\eta^{(i+1)/i}$ (where $i = 5$).

2.81 In a polytropic process pV^n = constant where n is called the polytropic index. For this process

$$pV^n = \text{constant} \quad \text{or} \quad TV^{n-1} = \text{constant}$$

So, $$\frac{dT}{T} + (n-1)\frac{dV}{V} = 0$$

Then, $$dQ = CdT = dU + pdV = C_V dT + pdV$$

$$= \frac{i}{2}RdT + \frac{RT}{V}dV = \frac{i}{2}RdT - \frac{1}{n-1}RdT = \left(\frac{i}{2} - \frac{1}{n-1}\right)RdT$$

Now, $$C = R \text{ so, } \frac{i}{2} - \frac{1}{n-1} = 1$$

or $$\frac{1}{n-1} = \frac{i}{2} - 1 = \frac{i-2}{2} \quad \text{or} \quad n = \frac{i}{i-2}$$

Now, $$\frac{\nu'}{\nu} = \frac{n'<v'>}{n\ <v>} = \frac{1}{\eta}\sqrt{\frac{T'}{T}} = \frac{1}{\eta}\left(\frac{V}{V'}\right)^{\frac{n-1}{2}}$$

$$= \frac{1}{\eta}\left(\frac{1}{\eta}\right)^{1/(i-2)} = \left(\frac{1}{\eta}\right)^{(i-1)/(i-2)}$$

$$= (\eta)^{-(i-1)/(i-2)} \text{ times} = \frac{1}{2.52} \text{ times}$$

2.82 If α is the polytropic index, then

$$pV^{\alpha} = \text{constant}, \quad TV^{\alpha-1} = \text{constant}$$

Now, $$\frac{\nu'}{\nu} = \frac{n'<v'>}{n<v>} = \frac{V}{V'}\sqrt{\frac{T'}{T}} = \frac{VT^{-1/2}}{V'T'^{-1/2}} = 1$$

Hence, $$\frac{1}{\alpha-1} = -\frac{1}{2} \quad \text{or} \quad \alpha = -1$$

Then, $$C = \frac{iR}{2} + \frac{R}{2} = 3R$$

2.83 $$v_p = \sqrt{\frac{2kT}{m}} = \sqrt{\frac{2RT}{M}} = \sqrt{\frac{2p}{\rho}} = 0.45 \text{ km/s}$$

$$v_{avg} = \sqrt{\frac{8p}{\pi\rho}} = 0.51 \text{ km/s} \quad \text{and} \quad v_{rms} = \sqrt{\frac{3p}{\rho}} = 0.55 \text{ km/s}$$

2.84 (a) The formula is

$$df(u) = \frac{4}{\sqrt{\pi}}u^2e^{-u^2}du \quad (\text{where } u = v/v_p)$$

Now, $$\text{Prob}\left(\left|\frac{v-v_p}{v_p}\right| < \delta\eta\right) = \int_{1-\delta\eta}^{1+\delta\eta} df(u)$$

$$= \frac{4}{\sqrt{\pi}}e^{-1} \times 2\,\delta\eta = \frac{8}{\sqrt{\pi}e}\delta\eta = 0.0166 = 1.66\%$$

(b) $$\text{Prob}\left(\left|\frac{v-v_{rms}}{v_{rms}}\right| < \delta\eta\right) = \text{Prob}\left(\left|\frac{v}{v_p} - \frac{v_{rms}}{v_p}\right| < \delta\eta\frac{v_{rms}}{v_p}\right)$$

$$= \text{Prob}\left(\left|u - \sqrt{\frac{3}{2}}\right| < \sqrt{\frac{3}{2}}\delta\eta\right)$$

$$= \int_{\sqrt{\frac{3}{2}}-\sqrt{\frac{3}{2}}\delta\eta}^{\sqrt{\frac{3}{2}}+\sqrt{\frac{3}{2}}\delta\eta} \frac{4}{\sqrt{\pi}}u^2e^{-u^2}du$$

$$= \frac{4}{\sqrt{\pi}} \times \frac{3}{2}e^{-3/2} \times 2\sqrt{\frac{3}{2}}\delta\eta$$

$$= \frac{12\sqrt{3}}{\sqrt{2\pi}}e^{-3/2}\delta\eta = 0.0185 = 1.85\%$$

2.85 (a) $v_{\text{rms}} - v_p = \left(\sqrt{3}-\sqrt{2}\right)\sqrt{\dfrac{kT}{m}} = \Delta v$

$$T = \frac{m}{k}\left(\frac{\Delta v}{\sqrt{3}-\sqrt{2}}\right)^2 = 384 \text{ K}$$

(b) Clearly v is the most probable speed at this temperature.

So, $$\sqrt{\frac{2kT}{m}} = v \quad \text{or} \quad T = \frac{mv^2}{2k} = 342 \text{ K}$$

2.86 (a) We have,

$$\frac{v_1^2}{v_p^2}e^{-v_1^2/v_p^2} = \frac{v_2^2}{v_p^2}e^{-v_2^2/v_p^2} \Rightarrow \left(\frac{v_1}{v_2}\right)^2 = e^{v_1^2-v_2^2/v_p^2} \quad \text{or} \quad v_p^2 = \frac{2kT}{m} = \frac{v_1^2-v_2^2}{(\text{In } v_1^2/v_2^2)}$$

So, $$T = \frac{m(v_1^2-v_2^2)}{2k \ln v_1^2/v_2^2} = 330 \text{ K}$$

(b) $F(v) = \dfrac{4}{\sqrt{\pi}}\dfrac{v^2}{v_p^2}e^{-v^2/v_p^2} \times \dfrac{1}{v_p}\left(\dfrac{1}{v_p}\text{comes from } F(v)dv = df(u), du = \dfrac{dv}{v_p}\right)$

Thus, $$\frac{v^2}{v_{p_1}^3}e^{-v^2/v_{p_1}^2} = \frac{v^2}{v_{p_2}^3}e^{-v^2/v_{p_2}^2}, \text{ Now } v_{p_1}^2 = \frac{2kT_0}{m} \text{ and } v_{p_2}^2 = \frac{2kT_0\eta}{m}$$

Now, $$e^{\frac{-mv^2}{2kT_0}\left(1-\frac{1}{\eta}\right)} = \frac{1}{\eta^{3/2}} \quad \text{or} \quad \frac{mv^2}{2kT_0}\left(1-\frac{1}{\eta}\right) = \frac{3}{2}\text{In}\,\eta$$

Thus, $$v = \sqrt{\frac{3kT_0}{m}}\sqrt{\frac{\ln\eta}{1-1/\eta}}$$

2.87 $v_{p_{\text{N}}} = \sqrt{\dfrac{2kT}{m_{\text{N}}}} = \sqrt{\dfrac{2RT}{M_{\text{N}}}}, v_{p_{\text{O}}} = \sqrt{\dfrac{2RT}{M_{\text{O}}}}$

$$v_{p_{\text{N}}} - v_{p_{\text{O}}} = \Delta v = \sqrt{\frac{2RT}{M_{\text{N}}}}\left(1-\sqrt{\frac{M_{\text{N}}}{M_{\text{O}}}}\right)$$

$$T = \frac{M_N(\Delta v)^2}{2R\left(1-\sqrt{\frac{M_{\text{N}}}{M_{\text{O}}}}\right)^2} = \frac{m_N(\Delta v)^2}{2k\left(1-\sqrt{\frac{m_{\text{N}}}{m_{\text{O}}}}\right)^2} = 363 \text{ K}$$

2.88 $$\frac{v^2}{v_{p_{\text{H}}}^3}e^{-v^2/v_{p_{\text{H}}}^2} = \frac{v^2}{v_{p_{\text{He}}}^3}e^{-v^2/v_{p_{\text{He}}}^2} \text{ or } e^{v^2\left(\frac{m_{\text{He}}}{2kT}-\frac{m_{\text{H}}}{2kT}\right)} = \left(\frac{m_{\text{He}}}{m_{\text{H}}}\right)^{3/2}$$

$$v^2 = 3kT\frac{\ln m_{\text{He}}/m_{\text{H}}}{m_{\text{He}}-m_{\text{H}}}$$

On substitution, we get, $v = 1.60$ km/s.

2.89 $$dN(v) = \frac{N4}{\sqrt{\pi}}\frac{v^2dv}{v_p^3}e^{-v^2/v_p^2}$$

For a given range v to $v + dv$, $(dn(v))$ is maximum when

$$\frac{\delta}{\delta v_p}\frac{dN(v)}{Nv^2dv} = 0 = \left(-3v_p^{-4} + \frac{2v^2}{v_p^6}\right)e^{-v^2/v_p^2}$$

or
$$v^2 = \frac{3}{2}v_p^2 = \frac{3kT}{m}$$

Thus,
$$T = \frac{1}{3}\frac{mv^2}{k}$$

2.90 Given
$$d^3v = 2\pi v_\perp dv_\perp dv_x$$

Thus,
$$dN(v) = N\left(\frac{m}{2\pi kT}\right)^{3/2} e^{-\frac{m}{2kT}(v_x^2+v_\perp^2)} 2\pi v_\perp dv_\perp dv_x$$

2.91
$$<v_x> = 0 \text{ (by symmetry)}$$

$$<|v_x|> = \frac{\int_{-\infty}^{\infty} |v_x| e^{-mv_x^2/2kT} dv_x}{\int_{-\infty}^{\infty} e^{-mv_x^2/2kT} dv_x} = \frac{\int_0^{\infty} v_x e^{-mv_x^2/2kT} dv_x}{\int_0^{\infty} e^{-mv_x^2/2kT} dv_x}$$

$$= \frac{\sqrt{\frac{2kT}{m}}\int_0^{\infty} ue^{-u^2} du}{\int_0^{\infty} e^{-u^2} du} = \frac{\sqrt{\frac{2kT}{m}}\int_0^{\infty} \frac{1}{2}e^{-x} dx}{\int_0^{\infty} e^{-x}\frac{dx}{2\sqrt{x}}}$$

$$= \frac{\sqrt{\frac{2kT}{m}}\Gamma(1)}{\Gamma\left(\frac{1}{2}\right)} = \sqrt{\frac{2kT}{m\pi}}$$

2.92
$$< v_x^2 > = \frac{\int_0^{\infty} v_x^2 e^{-mv_x^2/2kT} dv_x}{\int_0^{\infty} e^{-mv_x^2/2kT} dv_x} = \frac{\frac{2kT}{m}\int_0^{\infty} xe^{-x}\frac{dx}{2\sqrt{x}}}{\int_0^{\infty} e^{-x}\frac{dx}{2\sqrt{x}}} = \frac{\frac{2kT}{m}\Gamma\left(\frac{3}{2}\right)}{\Gamma\left(\frac{1}{2}\right)} = \frac{kT}{m}$$

2.93 Here vdA is equal to number of molecules hitting an area dA of the wall per second

i.e.,
$$udV = \int_0^{\infty} dN(v_x)v_x dA$$

or
$$v = \int_0^{\infty} n\left(\frac{m}{2\pi kT}\right)^{1/2} e^{-mv_x^2/2kT} v_x dv_x$$

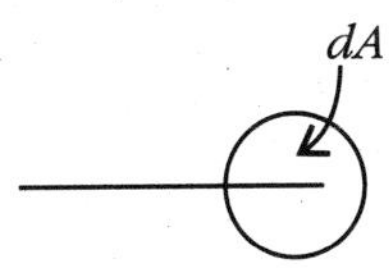

$$= \int_0^{\infty} \frac{n}{\sqrt{\pi}}\left(\frac{2kT}{m}\right)^{1/2} e^{-u^2} udu$$

$$= \frac{1}{2}n\sqrt{\frac{2kT}{m\pi}} = n\sqrt{\frac{kT}{2m\pi}} = \frac{1}{4}n<v> \quad \left(\text{where } <v> = \sqrt{\frac{8kT}{m\pi}}\right)$$

2.94 Let the number of molecules per unit volume with x component of velocity in the range v_x to $v_x + dv_x$, be

$$dn(v_x) = n\left(\frac{m}{2\pi kT}\right)^{1/2} e^{-mv_x^2/2kT} dv_x$$

Then, $$p = \int_0^\infty 2mv_x \cdot v_x dn\,(v_x)$$

$$= \int_0^\infty 2mv_x^2 n\left(\frac{m}{2\pi kT}\right)^{1/2} e^{-mv_x^2/2kT} dv_x$$

$$= 2mn\frac{1}{\sqrt{\pi}}\frac{2kT}{m}\int_0^\infty u^2 e^{-u^2} du$$

$$= \frac{4}{\sqrt{\pi}} nkT \int_0^\infty xe^{-x} \frac{dx}{2\sqrt{x}} = nkT$$

2.95 $$<1/v> = \int_0^\infty \left(\frac{m}{2\pi kT}\right)^{3/2} e^{-mv^2/2kT} 4\pi v^2 dv \frac{1}{v}$$

$$= \left(\frac{m}{2\pi kT}\right)^{3/2} 4\pi \frac{1}{2}\frac{2kT}{m}\int_0^\infty e^{-x} dx$$

$$= 2\left(\frac{m}{2\pi kT}\right)^{1/2} = \left(\frac{2m}{\pi kT}\right)^{1/2} = \left(\frac{16}{\pi^2}\frac{mn}{8kT}\right)^{1/2} = \frac{4}{\pi <v>}$$

2.96 $$dN(v) = N\left(\frac{m}{2\pi kT}\right)^{3/2} e^{-mv^2/kT} 4\pi v^2 dv = dN(\varepsilon) = \frac{dN(\varepsilon)}{d\varepsilon} d\varepsilon$$

or $$\frac{dN(\varepsilon)}{d\varepsilon} = N\left(\frac{m}{2\pi kT}\right)^{3/2} e^{-mv^2/2RT} 4\pi v^2 \frac{dv}{d\varepsilon}$$

Now, $$\varepsilon = \frac{1}{2}mv^2 \quad \text{so} \quad \frac{dv}{d\varepsilon} = \frac{1}{mv}$$

or $$\frac{dN(\varepsilon)}{d\varepsilon} = N\left(\frac{m}{2\pi kT}\right)^{3/2} e^{-\varepsilon/kT} 4\pi \sqrt{\frac{2\varepsilon}{m}} \frac{1}{m}$$

$$= N\frac{2}{\sqrt{\pi}}(kT)^{-3/2} e^{-\varepsilon/kT} \varepsilon^{1/2}$$

i.e., $$dN(\varepsilon) = N\frac{2}{\sqrt{\pi}}(kT)^{-3/2} e^{-\varepsilon/kT} \varepsilon^{1/2} d\varepsilon$$

The most probable kinetic energy is given by

$$\frac{d}{d\varepsilon}\frac{dN(\varepsilon)}{d\varepsilon} = 0$$

or $$\frac{1}{2}\varepsilon^{-1/2} e^{-\varepsilon/kT} - \frac{\varepsilon^{1/2}}{kT} e^{\varepsilon/\mathrm{kT}} = 0 \quad \text{or} \quad \varepsilon = \frac{1}{2}kT = \varepsilon_p$$

The corresponding velocity is $v = \sqrt{\dfrac{kT}{m}} \neq v_p$

2.97 The mean kinetic energy is

$$<\varepsilon> = \frac{\int_0^\infty \varepsilon^{3/2} e^{-\varepsilon/kT} d\varepsilon}{\int_0^\infty \varepsilon^{1/2} e^{-\varepsilon/kT} d\varepsilon} = kT\frac{\Gamma(5/2)}{\Gamma(3/2)} = \frac{3}{2}kT$$

Thus,

$$\frac{\delta N}{N} = \int_{\frac{3}{2}kT(1-\delta\eta)}^{\frac{3}{2}(1+\delta\eta)kT} \frac{2}{\sqrt{\pi}}(kT)^{-3/2} e^{-\varepsilon/kT} \varepsilon^{1/2} d\varepsilon$$

$$= \frac{2}{\sqrt{\pi}} e^{-3/2} \left(\frac{3}{2}\right)^{3/2} 2\delta\eta = 3\sqrt{\frac{6}{\pi}} e^{-3/2}\delta\eta$$

If $\delta\eta = 1\%$, the fraction of molecules is 0.9%.

2.98

$$\frac{\Delta N}{N} = \frac{2}{\sqrt{\pi}}(kT)^{-3/2} \int_{\varepsilon_0}^{\infty} \sqrt{\varepsilon}\, e^{-\varepsilon/kT} d\varepsilon$$

$$= \frac{2}{\sqrt{\pi}}(kT)^{-3/2} \sqrt{\varepsilon_0} \int_{\varepsilon_0}^{\infty} e^{-\varepsilon/kT} d\varepsilon \quad (\varepsilon_0 >> kT)$$

$$= \frac{2}{\sqrt{\pi}}(kT)^{-3/2} \sqrt{\varepsilon_0}\, kT\, e^{-\varepsilon_0/kT} = 2\sqrt{\frac{\varepsilon_0}{\pi kT}} e^{-\varepsilon_0/kT}$$

(In evaluating the integral, we have taken out $\sqrt{\varepsilon}$ as $\sqrt{\varepsilon_0}$ since the integral is dominated by the lower limit.)

2.99 (a) $F(v) = Av^3 e^{-mv^2/2kT}$

For the most probable value of the velocity

$$\frac{dF(v)}{dv} = 0 \quad \text{or} \quad 3Av^2 e^{-mv^2/2kT} - Av^3 \frac{2mv}{2kT} e^{-mv^2/2kT} = 0$$

So,

$$v_p = \sqrt{\frac{3kT}{m}}$$

This should be compared with the value $v_p = \sqrt{\frac{2kT}{m}}$ for the Maxwellian distribution.

(b) In terms of energy, $\varepsilon = \frac{1}{2}mv^2$

$$F(\varepsilon) = Av^3 e^{-mv^2/2kT} \frac{dv}{d\varepsilon}$$

$$= A\left(\frac{2e}{m}\right)^{3/2} e^{-\varepsilon/kT} \frac{1}{\sqrt{2m\varepsilon}} = A\frac{2\varepsilon}{m^2} e^{-\varepsilon/kT}$$

From this the probable energy comes out as follows:

$$F'(\varepsilon) = 0 \Rightarrow \frac{2A}{m^2}\left(e^{-\varepsilon/kT} - \frac{\varepsilon}{kT} e^{-\varepsilon/kT}\right) = 0 \quad \text{or} \quad \varepsilon_p = kT$$

2.100 The number of molecules reaching a unit area of wall at angle between θ and $\theta + d\theta$ to its normal per unit time is

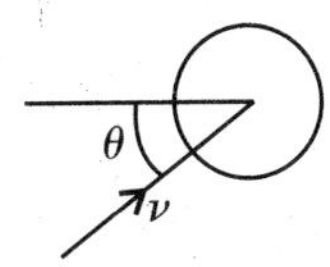

$$d\nu = \int_{v=0}^{v=\infty} dn(v) \frac{d\Omega}{4\pi} v\cos\theta$$

$$= \int_0^{\infty} n\left(\frac{m}{2\pi kT}\right)^{3/2} e^{-mv^2/2kT} v^3\, dv \sin\theta\cos\theta\, d\theta \times 2\pi$$

$$= n\left(\frac{2kT}{m\pi}\right)^{1/2} \int_0^{\infty} e^{-x} x dx \sin\theta\cos\theta\, d\theta = n\left(\frac{2kT}{m\pi}\right)^{1/2} \sin\theta\cos\theta\, d\theta$$

2.101 The number of molecules reaching per unit area of the wall with velocities in the interval v to $v + dv$ per unit time is

$$dv = \int_{\theta=0}^{\theta=\pi/2} dn(v)\frac{d\Omega}{4\pi} v\cos\theta$$

$$= \int_{\theta=0}^{\theta=\pi/2} n\left(\frac{m}{2\pi kT}\right)^{3/2} e^{-mv^2/2kT} v^3 dv \sin\theta\cos\theta\, d\theta \times 2\pi$$

$$= n\pi\left(\frac{m}{2\pi kT}\right)^{3/2} e^{-mv^2/2kT} v^3 dv$$

2.102 If the force exerted is **F**, then by the law of variation of concentration with height

$$n(Z) = n_0 e^{-FZ/kT}$$

So,
$$\eta = e^{F\Delta h/kT} \quad \text{or} \quad F = \frac{kT\ln\eta}{\Delta h} = 9\times10^{-20}\ \text{N}$$

2.103 Here,
$$F = \frac{\pi}{6}d^3\Delta\rho g = \frac{RT\ln\eta}{N_a h} \quad \text{or} \quad N_a = \frac{6RT\ln\eta}{\pi d^3 g\Delta\rho h}$$

Given
$T = 290$ K, $\eta = 2$, $h = 4\times10^{-5}$ m, $d = 4\times10^{-7}$ m, $g = 9.8$ m/s^2,

$\Delta\rho = 0.2\times10^3$ kg/m^3 and $R = 8.31$ J/K.

Hence,
$$N_a = \frac{6\times8.31\times290\times\ln2}{\pi\times64\times9.8\times200\times4}\times10^{26} = 6.36\times10^{23}\ \text{mol}^{-1}$$

2.104
$$\eta = \frac{\text{concentration of } H_2}{\text{concentration of } N_2} = \eta_0\frac{e^{-M_{N_2}gh/RT}}{e^{-M_{H_2}gh/RT}} = \eta_0 e^{(-M_{N_2}+M_{H_2})gh/RT}$$

$$\frac{\eta}{\eta_0} = e^{(-M_{N_2}+M_{H_2})gh/RT} = 1.39$$

2.105 We know that

$$n_1(h) = n_1 e^{-m_1 gh/kt} \quad \text{and} \quad n_2(h) = n_2 e^{-m_2 gh/kt}$$

They are equal at a height h, where $n_1/n_2 = e^{gh(m_1-m_2)/kT}$

or
$$h = \frac{kT}{g}\frac{\ln n_1 - \ln n_2}{m_1 - m_2}$$

2.106 At a temperature T the concentration $n(z)$ varies with height according to

$$n(z) = n_0 e^{-mgz/kT}$$

This means that particles per unit area of the base of cylinder are

$$\int_0^\infty n(z)dz$$

$$= \int_0^\infty n_0 e^{-mgz/kT} dz = \frac{n_0 kT}{mg}$$

Clearly this cannot change. Thus $n_0 kT = p_0$ = pressure at the bottom of the cylinder, must not change with change of temperature.

2.107 $$<U> = \frac{\int_0^\infty mgze^{-mgz/kt}\,dz}{\int_0^\infty e^{-mgz/kT}\,dz} = kT\frac{\int_0^\infty xe^{-x}\,dx}{\int_0^\infty e^{-x}\,dx} = kT\frac{\Gamma(2)}{\Gamma(1)} = kT$$

When there are many kinds of molecules, this formula holds for each kind and the average energy is given by

$$<U> = \frac{\Sigma f_i kT}{\Sigma f_i} = kT$$

where $f_i \propto$ fractional concentration of each kind of molecule at the ground level.

2.108 The constant acceleration is equivalent to a pseudo force wherein a concentration gradient is set up. Then

$$e^{-M_A wl/RT} = 1 - \eta$$

or $$w = -\frac{RT\ln(1-\eta)}{M_A l} \approx \frac{\eta RT}{M_A l} \approx 70 \text{ g}$$

2.109 In a centrifuge rotating with angular velocity ω about an axis, there is a centrifugal acceleration w_r^2, where r is the radial distance from the axis. In a fluid if there are suspended colloidal particles they experience an additional force. If m is the mass of each particle then its volume is m/ρ and the excess force on this particle is $m/\rho(\rho - \rho_0)\omega^2 r$ outward, corresponding to a potential energy of $-(m/2\rho)(\rho - \rho_0)\omega^2 r^2$.

This gives rise to a concentration variation, given by

$$n(r) = n_0 \exp\left(\frac{m}{2\rho kT}(\rho - \rho_0)\omega^2 r^2\right)$$

Thus, $$\frac{n(r_2)}{n(r_1)} = \eta_0 = \exp\left(\frac{M}{2\rho RT}(\rho - \rho_0)\omega^2(r_2^2 - r_1^2)\right)$$

(where $\frac{m}{k} = \frac{M}{R}$ and $M = N_A m$ is the molecular weight).

Thus, $$M = \frac{2\rho RT\ln\eta}{(\rho - \rho_0)\omega^2(r_2^2 - r_1^2)}$$

2.110 The potential energy associated with each molecule is $-(1/2)m\omega^2 r^2$ and there is a concentration variation, given by

$$n(r) = n_0 \exp\left(\frac{m\omega^2 r^2}{2kT}\right) = n_0 \exp\left(\frac{M\omega^2 r^2}{2RT}\right)$$

Thus, $$\eta = \exp\left(\frac{M\omega^2 l^2}{2RT}\right) \quad \text{or} \quad \omega = \sqrt{\frac{2RT}{Ml^2}\ln\eta}$$

Using $M = 12 + 32 = 44$ g, $l = 100$ cm, $R = 8.31 \times 10^7$ erg/K, $T = 300$ K, we get,

$$\omega = 280 \text{ rad/s}$$

2.111 Here, the concentration variation is

$$n(r) = n_0 \exp\left(-\frac{ar^2}{kT}\right)$$

(a) The number of molecules located at the distance between r and $r + dr$ is

$$4\pi r^2 dr\, n(r) = 4\pi n_0 \exp\left(-\frac{ar^2}{kT}\right) r^2 dr$$

(b) r_p is given by $\frac{d}{dr} r^2\, n(r) = 0$ or $2r - \frac{2ar^3}{kT} = 0$ or $r_p = \sqrt{\frac{kT}{a}}$

(c) The fraction of molecules lying between r and $r + dr$ is

$$\frac{dN}{N} = \frac{4\pi r^2 dr n_0 \exp(-ar^2/kT)}{\int_0^\infty 4\pi r^2 dr n_0 \exp(ar^2/kT)}$$

$$\int_0^\infty 4\pi r^2 dr \exp\left(-\frac{ar^2}{kT}\right) = \left(\frac{kT}{a}\right)^{3/2} 4\pi \int_0^\infty x \frac{dx}{2\sqrt{x}} \exp(-x)$$

$$= \left(\frac{kT}{a}\right)^{3/2} 2\pi \Gamma\left(\frac{3}{2}\right) = \left(\frac{\pi kT}{a}\right)^{3/2}$$

Thus, $$\frac{dN}{N} = \left(\frac{a}{\pi kT}\right)^{3/2} 4\pi r^2 dr \exp\left(\frac{-ar^2}{kT}\right)$$

(d) $$dN = N\left(\frac{a}{\pi kT}\right)^{3/2} 4\pi r^2 dr \exp\left(\frac{-ar^2}{kT}\right)$$

So, $$n(r) = N\left(\frac{a}{\pi kT}\right)^{1/2} \exp\left(\frac{-ar^2}{kT}\right)$$

When T decreases η times, $n(0) = n_0$ will increase $\eta^{3/2}$ times.

2.112 (a) Let $$U = ar^2 \quad \text{or} \quad r = \sqrt{\frac{U}{a}}$$

So, $$dr = \sqrt{\frac{1}{a}} \frac{dU}{2\sqrt{U}} = \frac{dU}{2\sqrt{aU}}$$

So, $$dN = n_0 4\pi \frac{U}{a} \frac{dU}{2\sqrt{aU}} \exp\left(\frac{U}{-kT}\right)$$

$$= 2\pi n_0 a^{-3/2} U^{1/2} \exp\left(\frac{U}{-kT}\right) dU$$

(b) The most probable value of U is given by

$$\frac{d}{dU}\left(\frac{dN}{dU}\right) = 0\left(\frac{1}{2\sqrt{U}} - \frac{U^{1/2}}{kT}\right) \exp\left(\frac{-U}{kT}\right) \quad \text{or} \quad U_p = \frac{1}{2} kT$$

From 2.111 (b), the potential energy at the most probable distance is kT.

2.4 The Second Law of Thermodynamics. Entropy

2.113 The efficiency is given by

$$\eta = \frac{T_1 - T_2}{T_1} \text{ (for } T_1 > T_2\text{)}$$

Now, in the two cases the efficiencies are

$$\eta_b = \frac{T_1 + \Delta T - T_2}{T_1 + \Delta T} \text{ (}T_1\text{ increased)}$$

$$\eta_l = \frac{T_1 - T_2 + \Delta T}{T_1} \text{ (}T_2\text{ decreased)}$$

Thus, $$\eta_b < \eta_l$$

2.114 (a) For H_2, $\gamma = 7/5$. From the figure

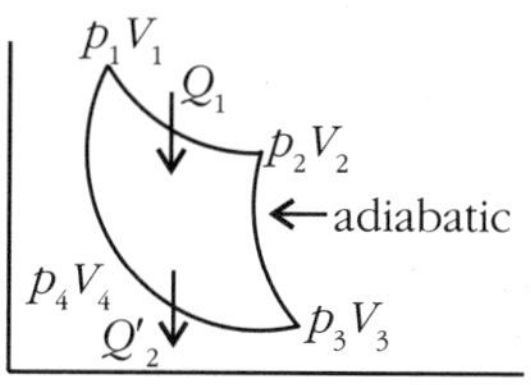

$$p_1V_1 = p_2V_2, p_3V_3 = p_4V_4$$

and $$p_2V_2^\gamma = p_3V_3^\gamma,\ p_1V_1^\gamma = p_4V_4^\gamma$$

Define n by $$V_3 = n\ V_2.$$

Then $$p_3 = p_2\ n^{-\gamma}$$

So, $$p_4V_4 = p_3V_3 = p_2V_2n^{1-\gamma} = p_1V_1n^{1-\gamma}$$

since, $$p_4V_4^\gamma = p_1V_1^\gamma \text{ so, } V_4^{1-\gamma} = V_1^{1-\gamma}\ n^{1-\gamma} \text{ or } V_4 = nV_1$$

Also, $$Q_1 = p_2V_2 \ln\frac{V_2}{V_1}, Q_2' = p_3V_3 \ln\frac{V_3}{V_4} = p_2V_2n^{1-\gamma} \ln\frac{V_3}{V_4}$$

Finally $$\eta = 1 - \frac{Q_2'}{Q_1} = 1 - n^{1-\gamma} = 0.242$$

(b) Define n by $$p_3 = p_2/n$$

Then, $$p_2V_2^\gamma = \frac{p_2}{n}V_3\gamma \quad \text{or} \quad V_3 = n^{1/\gamma}V_2$$

So we get the formulae here by substituting n by $n^{1/\gamma}$ in the previous case.

Then, $$\eta = 1 - n^{(1/\gamma)-1} = 1 - n^{-(2/7)} = 0.18$$

2.115 Used as a refrigerator, the refrigerating efficiency of a heat engine is given by

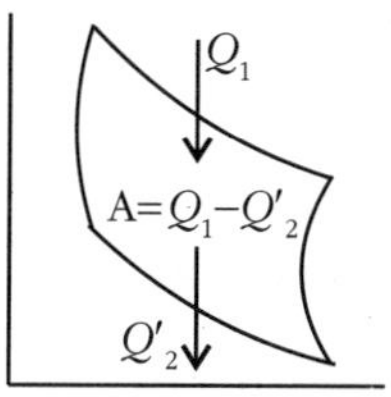

$$\varepsilon = \frac{Q_2'}{A} = \frac{Q_2'}{Q_1 - Q_2'} = \frac{Q_2'/Q_1}{1 - \frac{Q_2'}{Q_1}} = \frac{1-\eta}{\eta} = 9$$

(where η is the efficiency of the heat engine).

2.116 Given $V_2 = nV_1,\ V_4 = nV_3$

Q_1 = Heat taken at the upper temperature

$= RT_1 \ln n + RT_2 \ln n = R(T_1 + T_2) \ln n$

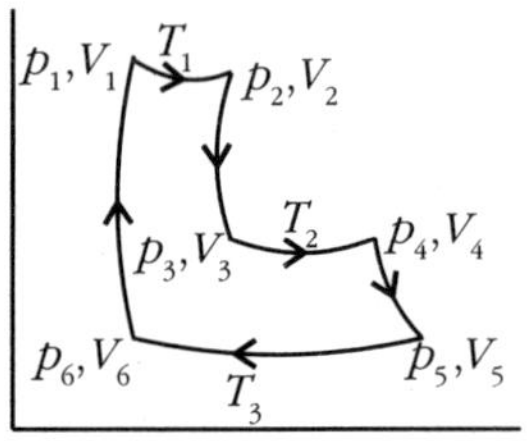

Now,
$$T_1V_2^{\gamma-1} = T_2V_3^{\gamma-1}$$

or
$$V_3 = \left(\frac{T_1}{T_2}\right)^{1/\gamma-1} V_2$$

Similarly,
$$V_5 = \left(\frac{T_2}{T_3}\right)^{1/\gamma-1} V_4,$$

$$V_6 = \left(\frac{T_1}{T_3}\right)^{1/\gamma-1} V_1$$

Thus, heat ejected at the lower temperature

$$Q_2 = -RT_3 \ln \frac{V_6}{V_5}$$

$$= -RT_3 \ln \left(\frac{T_1}{T_2}\right)^{1/\gamma-1} \frac{V_1}{V_4} = -RT_3 \ln \left(\frac{T_1}{T_2}\right)^{1/\gamma-1} \frac{V_2}{n^2V_3}$$

$$= -RT_3 \ln \left(\frac{T_1}{T_2}\right)^{1/\gamma-1} \frac{1}{n^2}\left(\frac{T_1}{T_2}\right)^{-1/\gamma-1} = 2RT_3 \ln n$$

Thus,
$$\eta = 1 - \frac{2T_3}{T_1 + T_2}$$

2.117 Since,
$$Q_2' = C_V(T_2 - T_3) = \frac{C_V}{R} V_2(p_2 - p_3)$$

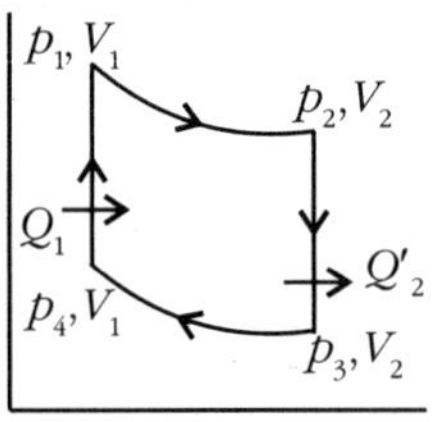

and
$$Q_1 = \frac{C_V}{R} V_1(p_1 - p_4)$$

So,
$$\eta = 1 - \frac{V_2(p_2 - p_3)}{V_1(p_1 - p_4)}$$

On the other hand,

$$p_1V_1^{\gamma} = P_2V_2^{\gamma},\ p_3V_2^{\gamma} = p_4V_1^{\gamma} \text{ also, } V_2 = nV_1$$

Thus,
$$p_1 = p_2n^{\gamma} \quad \text{and} \quad p_4 = p_3n^{\gamma}$$

So,
$$\eta = 1 - n^{1-\gamma} \text{ (where } \gamma = 7/5 \text{ for } N_2)$$

Therefore,
$$\eta = 0.602 = 60.2\%$$

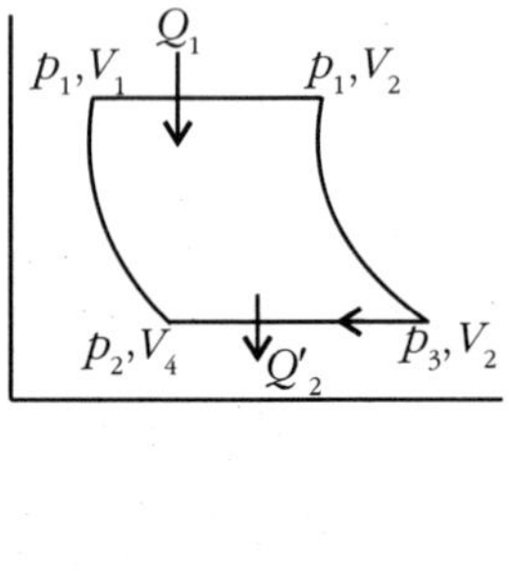

2.118 Since,
$$Q_1 = \frac{C_p}{R} p_1(V_2 - V_1)$$

and
$$Q_2' = \frac{C_p}{R} p_2(V_3 - V_4)$$

So,
$$\eta = 1 - \frac{p_2(V_3 - V_4)}{p_1(V_2 - V_1)}$$

Now,
$$p_1 = np_2 \quad \text{or} \quad V_3 = n^{1/\gamma}V_2$$
$$p_2V_4^{\gamma} = p_1V_1^{\gamma} \quad \text{or} \quad V_4 = n^{1/\gamma}V_1$$

So,
$$\eta = 1 - \frac{1}{n} \cdot n^{1/\gamma} = 1 - n^{(1/\gamma)-1}$$

2.119 Since the absolute temperature of the gas rises n times both in the isochoric heating and the isobaric expansion, we have

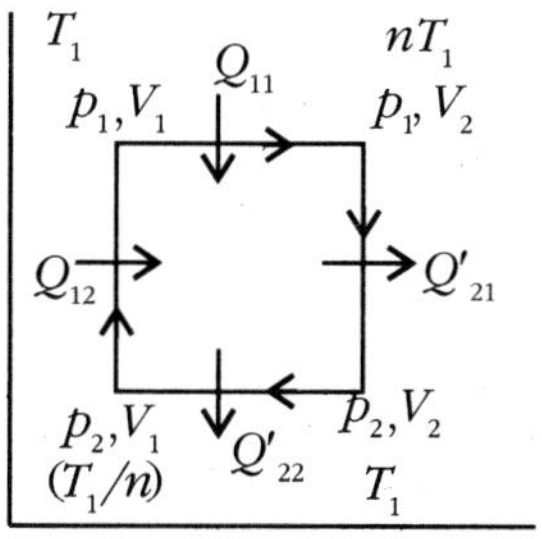

$$p_1 = np_2 \quad \text{and} \quad V_2 = nV_1$$

Heat taken is
$$Q_1 = Q_{11} + Q_{12}$$

where,
$$Q_{11} = C_P(n-1)T_1$$

and
$$Q_{12} = C_V T_1\left(1 - \frac{1}{n}\right)$$

Heat rejected is
$$Q_2' = Q_{21}' + Q_{22}'$$

where,
$$Q_{21}' = C_V T_1(\text{n}-1), Q_{22}' = C_p T_1\left(1 - \frac{1}{n}\right)$$

Thus,
$$\eta = 1 - \frac{Q_2'}{Q_1} = 1 - \frac{C_V(n-1) + C_p(1-1/n)}{C_p(n-1) + C_V(1-1/n)}$$

$$= \left(1 - \frac{\eta - 1 + \gamma(1-1/n)}{\gamma(n-1) + (1-1/n)}\right) = 1 - \frac{1 + \dfrac{\gamma}{n}}{\gamma + \dfrac{1}{n}} = 1 - \frac{n+\gamma}{1+n\gamma}$$

2.120 (a) Here,
$$p_2 = np_1, p_1V_1 = p_0V_0,$$

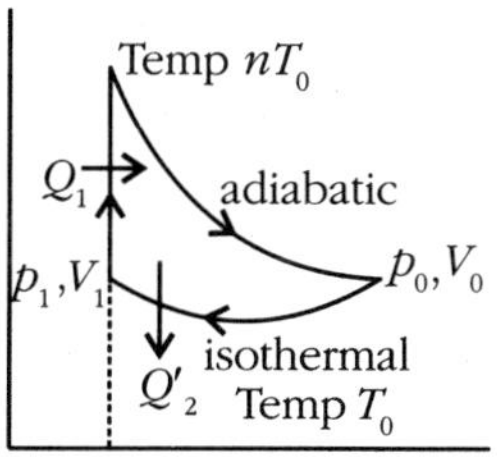

and
$$np_1V_1^{\gamma} = p_0V_0^{\gamma}$$

Also,
$$Q_2' = RT_0 \ln\frac{V_0}{V_1}, Q_1 = C_V T_0(n-1)$$

But,
$$nV_1^{\gamma-1} = V_0^{\gamma-1} \quad \text{or} \quad V_1 = V_0 n^{-1/\gamma-1}$$

$$Q_2' = RT_0 \ln n^{1/\gamma-1} = \frac{RT_0}{\gamma-1}\ln n$$

Thus,
$$\eta = 1 - \frac{\ln n}{n-1}\left(\text{on using } C_V = \frac{R}{\gamma-1}\right)$$

(b) Here,
$$V_2 = nV_1, \; p_1V_1 = p_0V_0$$

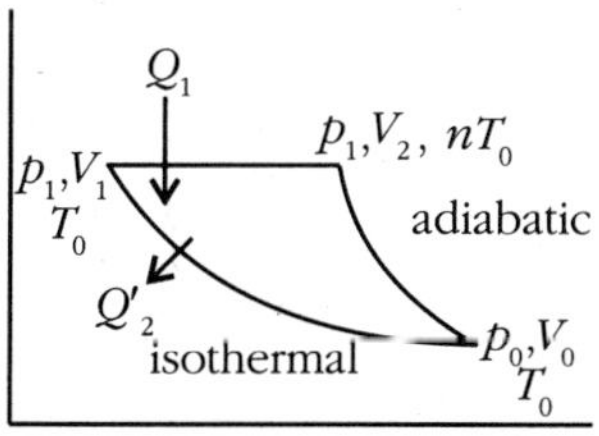

and
$$p_1(n\,V_1)^{\gamma} = p_0V_0^{\gamma}$$

i.e.,
$$n^{\gamma}V_1^{\gamma-1} = V_0^{\gamma-1} \quad \text{or} \quad V_1 = n^{-\gamma/\gamma-1}V_0$$

Also, $$Q_1 = C_p T_0 (n-1),$$

$$Q_2' = RT_0 \ln \frac{V_0}{V_1}$$

or $$Q_2' = RT_0 \ln n^{\gamma/(\gamma-1)} = \frac{R\gamma}{\gamma-1} T_0 \ln n = C_p T_0 \ln n$$

Thus, $$\eta = 1 - \frac{\ln n}{n-1}$$

2.121 Here the isothermal process proceeds at the maximum temperature instead of at the minimum temperature of the cycle as in Problem 2.120.

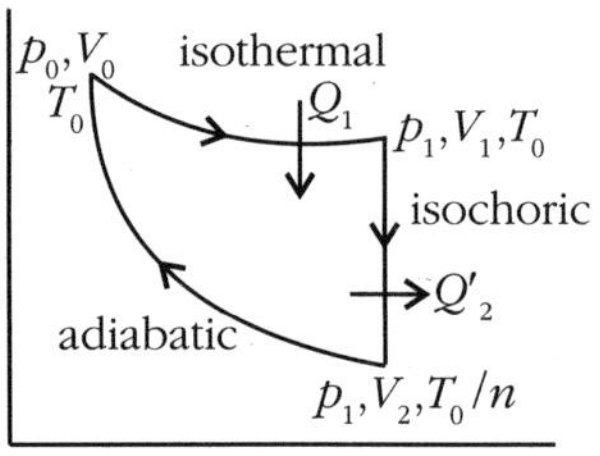

(*a*)

(a) Here $$p_1V_1 = p_0V_0,\ p_2 = \frac{p_1}{n}$$

and $$p_2V_1^\gamma = p_0V_0^\gamma \text{ or } p_1V_1^\gamma = np_0V_0^\gamma$$

i.e., $$V_1^{\gamma-1} = nV_0^{\gamma-1} \text{ or } V_1 = n^{1/\gamma-1}V_0$$

$$Q_2' = C_V T_0\left(1 - \frac{1}{n}\right),\ Q_1 = RT_0 \ln \frac{V_1}{V_0} = \frac{RT_0}{\gamma-1}\ln n = C_V T_0 \ln n$$

Thus, $$\eta = 1 - \frac{Q_2'}{Q_1} = 1 - \frac{n-1}{n\ln n}$$

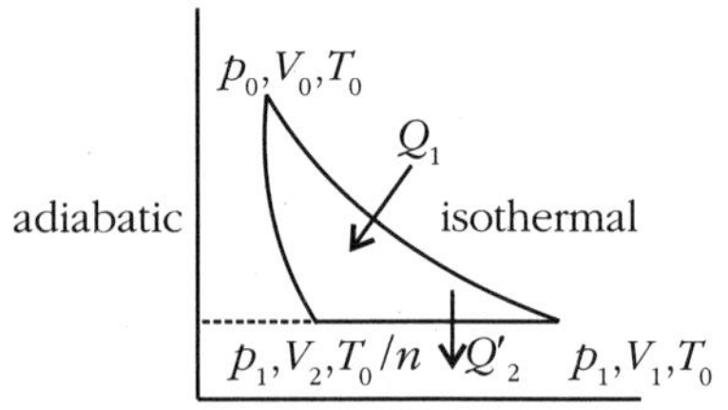

(*b*)

(b) Here, $$V_2 = \frac{V_1}{n},\ p_0V_0 = p_1V_1$$

$$p_0V_0^\gamma = p_1V_2^\gamma = p_1n^{-\gamma}V_1^\gamma \text{ or } V_0^{\gamma-1} = n^{-\gamma}V_1^{\gamma-1} \text{ or } V_1 = n^{(\gamma/\gamma-1)}V_0$$

$$Q_2' = C_p T_0\left(1 - \frac{1}{n}\right), Q_1 = RT_0 \ln \frac{V_1}{V_0} = \frac{R\gamma}{\gamma-1} T_0 \ln n = C_p T_0 \ln n$$

Thus, $$\eta = 1 - \frac{n-1}{n\ln n}$$

2.122 The section from (p_1, V_1, T_0) to $(p_2, V_2, T_0/n)$ is a polytropic process of index α. We shall assume that the corresponding specific heat C is positive.

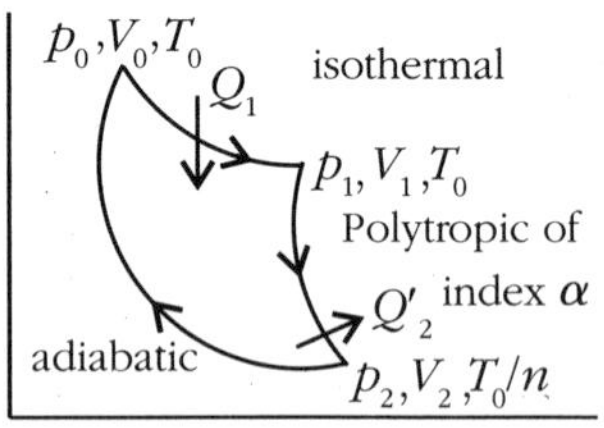

Here, $$dQ = CdT = C_VdT + pdV$$

Now, $$pV^\alpha = \text{constant or } TV^{\alpha-1} = \text{constant}$$

So, $$pdV = \frac{RT}{V}dV = -\frac{R}{\alpha-1}dT$$

Then, $$C = C_V - \frac{R}{\alpha-1} = R\left(\frac{1}{\gamma-1} - \frac{1}{\alpha-1}\right)$$

We have $$p_1V_1 = RT_0,\ p_2V_2 = \frac{RT_0}{n} = \frac{p_1V_1}{n}$$

$$p_0V_0 = p_1V_1 = np_2V_2,\ p_0V_0^\gamma = p_2V_2^\gamma$$

$$p_1V_1^\alpha = p_2V_2^\alpha \text{ or } V_0^{\gamma-1} = \frac{1}{n}V_2^{\gamma-1} \text{ or } V_2 = V_0n^{1/(\gamma-1)}$$

$$V_1^{\alpha-1} = \frac{1}{n}V_2^{\alpha-1} \text{ or } V_1 = n^{-1/(\alpha-1)}\ V_2 = n^{1/(\gamma-1)-1/(\alpha-1)}\ V_0$$

Now, $$Q_2' = CT_0\left(1-\frac{1}{n}\right), Q_1 = RT_0 \ln\frac{V_1}{V_0} = RT_0\left(\frac{1}{\gamma-1}-\frac{1}{\alpha-1}\right)\ln n = CT_0 \ln n$$

Thus, $$\eta = 1-\frac{n-1}{n\ln n}$$

2.123 (a) Here, $$Q_2' = C_p\left(T_1-\frac{T_1}{n}\right) = C_pT_1\left(1-\frac{1}{n}\right), Q_1 = C_V\left(T_0-\frac{T_1}{n}\right)$$

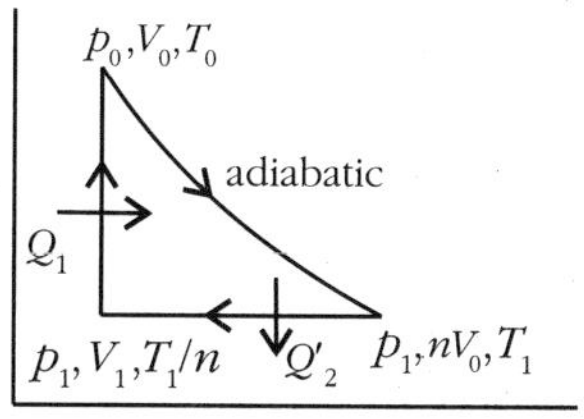

(a)

Along the adiabatic line

$$T_0V_0^{\gamma-1} = T_1(nV_0)^{\gamma-1} \quad \text{or} \quad T_0 = T_1 n^{\gamma-1}$$

So, $$Q_1 = C_V\frac{T_1}{n}(n^\gamma - 1)$$

Thus, $$\eta = 1-\frac{\gamma(n-1)}{n^\gamma - 1}$$

(b) Here $$Q_2' = C_V(nT_1 - T_0), Q_1 = C_p \cdot T_1(n-1)$$

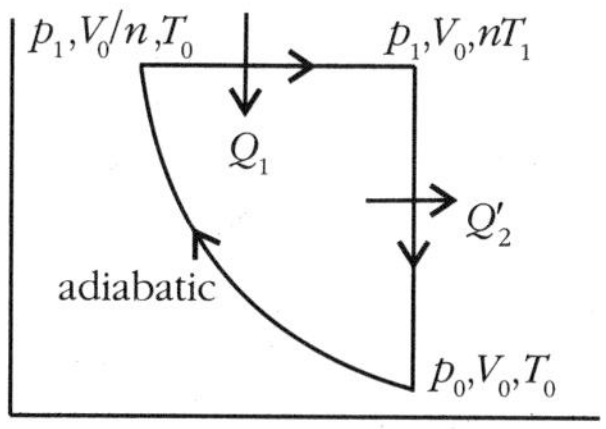

(b)

Along the adiabatic line $TV^{\gamma-1}$ = constant.

So, $$T_0V_0^{\gamma-1} = T_1\left(\frac{V_0}{n}\right)^{\gamma-1} \quad \text{or} \quad T_1 = n^{\gamma-1}\, T_0$$

Thus, $$\eta = 1-\frac{n^\gamma - 1}{\gamma n^{\gamma-1}(n-1)}$$

2.124 (a) Here, $$Q_2' = C_pT_0\left(1-\frac{1}{n}\right), Q_1'' = RT_0 \ln n$$

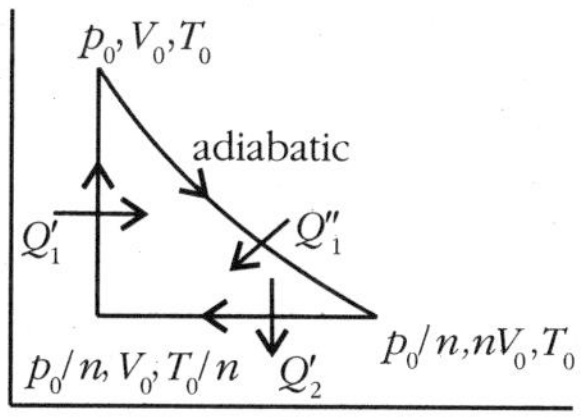

(a)

and $$Q_1' = C_VT_0\left(1-\frac{1}{n}\right)$$

Now, $$Q_1 = Q_1' + Q_1''$$

So, $$\eta = 1-\frac{Q_2'}{Q_1} = 1-\frac{C_p(1-1/n)}{C_V(1-1/n)+R\ln n}$$

$$= 1-\frac{\gamma}{1+\dfrac{R}{C_V}\dfrac{n\ln n}{n-1}} = 1-\frac{\gamma(n-1)}{n-1+(\gamma-1)n\ln n}$$

(b) Here, $$Q_1 = C_pT_0(n-1), Q_2'' = C_VT_0(n-1),$$

and $$Q_2''' = RT_0 \ln n$$

Now, $$Q_2' = Q_2'' + Q_2'''$$

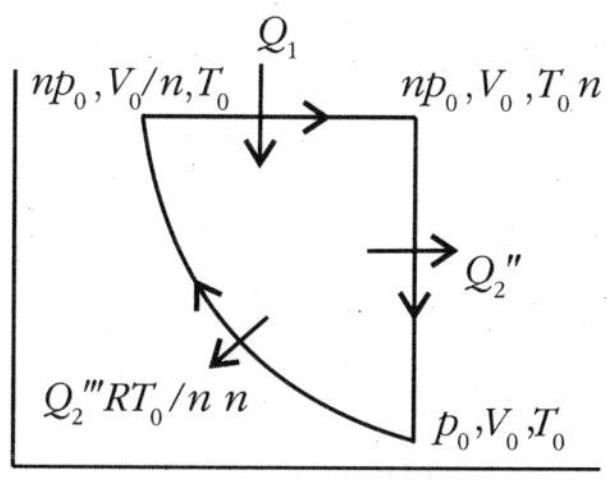

(b)

So, $$\eta = 1-\frac{Q_2'}{Q_1} = 1-\frac{n-1+(\gamma-1)\ln n}{\gamma(n-1)}$$

2.125 We have

$$Q_1' = \tau R T_0 \ln \nu, Q_2'' = C_V T_0(\tau - 1)$$

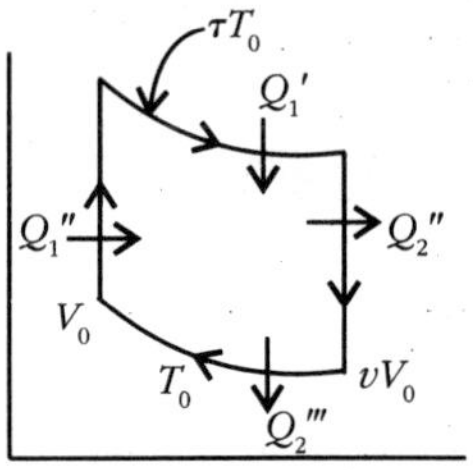

and $$Q_2''' = RT_0 \ln \nu, Q_1'' = C_V T_0(\tau - 1)$$

as well as $$Q_1 = Q_1' + Q_1''$$

and $$Q_2' = Q_2'' + Q_2'''$$

So,
$$\eta = 1 + \frac{Q_2'}{Q_1}$$

$$= 1 - \frac{C_V(\tau - 1) + R\ln \nu}{C_V(\tau - 1) + \tau R \ln \nu}$$

$$= 1 - \frac{\frac{\tau - 1}{\gamma - 1} + \ln \nu}{\frac{\tau - 1}{\gamma - 1} + \tau \ln \nu} = \frac{(\tau - 1)\ln \nu}{\tau \ln \nu + \frac{\tau - 1}{\gamma - 1}}$$

2.126 Here, $$Q_1' = C_p T_0(\tau - 1), Q_1'' = \tau R T_0 \ln n$$

and $$Q_2'' = C_p T_0(\tau - 1), Q_2''' = \tau R T_0 \ln n$$

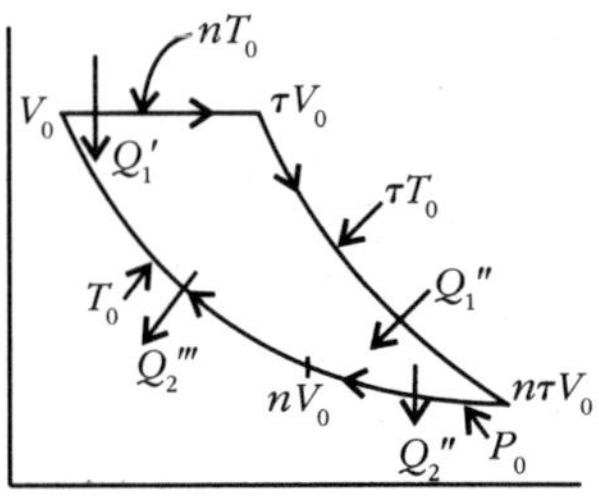

In addition, we have

$$Q_1 = Q_1' + Q_1''$$

and $$Q_2' = Q_2'' + Q_2'''$$

So,
$$\eta = 1 - \frac{Q_2'}{Q_1} = 1 - \frac{C_p(\tau - 1) + R\ln n}{C_p(\tau - 1) + \tau R \ln n}$$

$$= 1 - \frac{\tau - 1 + \left(1 - \frac{1}{\gamma}\right)\ln n}{\tau - 1 + \left(1 - \frac{1}{\gamma}\right)\tau \ln n} = \frac{(\tau - 1)\ln n}{\tau \ln n + \frac{\gamma(\tau - 1)}{\gamma - 1}}$$

2.127 Because of the linearity of the section BC, whose equation is

$$\frac{p}{p_0} = \frac{\nu V}{V_0} \times (\equiv p = \alpha V)$$

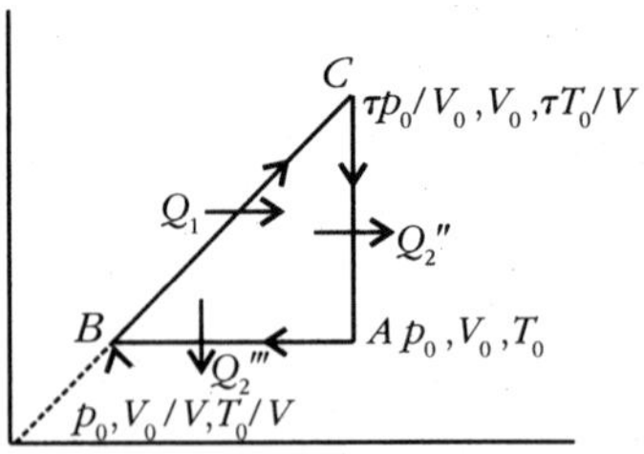

We have $$\frac{\tau}{\nu} = \nu \quad \text{or} \quad \nu = \sqrt{\tau}$$

Here,
$$Q_2'' = C_v T_0(\sqrt{\tau} - 1),$$

$$Q_2''' = C_p T_0\left(1 - \frac{1}{\sqrt{\tau}} = C_p \frac{T_0}{\sqrt{\tau}}(\sqrt{\tau} - 1)\right)$$

Thus,
$$Q_2' = Q_2'' + Q_2''' = \frac{RT_0}{\gamma - 1}\left(\sqrt{\tau} - 1\right)\left(1 + \frac{\gamma}{\sqrt{\tau}}\right)$$

Along BC, the specific heat C is given by

$$CdT = C_v dT + pdV = C_v dT + d\left(\frac{1}{2}\alpha V^2\right) = \left(C_v + \frac{1}{2}R\right)dT$$

Thus,
$$Q_1 = \frac{1}{2}RT_0 \frac{\gamma + 1}{\gamma - 1}\frac{\tau - 1}{\sqrt{\tau}}$$

Finally $$\eta = 1 - \frac{Q_2'}{Q_1} = 1 - 2\frac{\sqrt{\tau}+\gamma}{\sqrt{\tau}+1}\frac{1}{\gamma+1} = \frac{(\gamma-1)(\sqrt{\tau}-1)}{(\gamma+1)(\sqrt{\tau}+1)}$$

2.128 We write Claussius inequality in the form

$$\int \frac{đ_1 Q}{T} - \int \frac{đ_2 Q}{T} \le 0$$

where $đ_1 Q$ is the heat transferred to the system but $đ_2 Q$ is heat rejected by the system, both are positive and this explains the minus sign before $đ_2 Q$.

In this inequality $T_{max} > T > T_{min}$ and we can write

$$\int \frac{đ_1 Q}{T_{max}} - \int \frac{đ_2 Q}{T_{min}} < 0$$

Thus, $$\frac{Q_1}{T_{max}} < \frac{Q_2'}{T_{min}} \quad \text{or} \quad \frac{T_{min}}{T_{max}} < \frac{Q_2'}{Q_1}$$

or $$\eta = 1 - \frac{Q_2'}{Q_1} < 1 - \frac{T_{min}}{T_{max}} = \eta_{cannot}$$

2.129 We consider an infinitesimal carnot cycle with isothermal process at temperatures $T + dT$ and T. Let δA be the work done in the cycle and δQ, be the heat received at the higher temperature. Then by Carnot's theorem

$$\frac{\delta A}{\delta Q_1} = \frac{dT}{T}$$

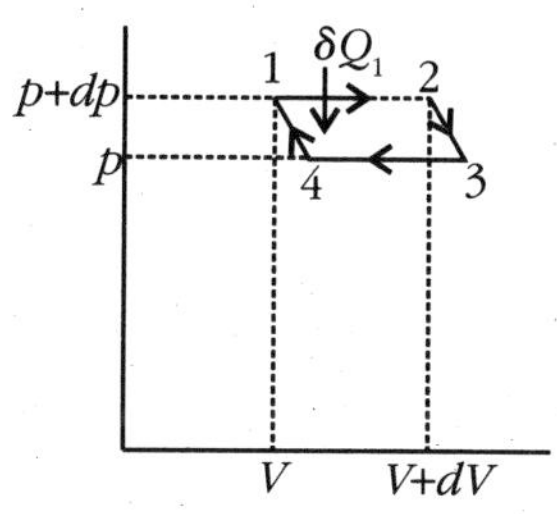

On the other hand,

$$\delta A = dp dV = \left(\frac{\partial p}{\partial T}\right)_V dT dV$$

while, $$\delta Q_1 = dU_{12} + p dV = \left[\left(\frac{\partial U}{\partial T}\right)_T + p\right] dV$$

Hence, $$\left(\frac{\partial U}{\partial V}\right)_T + p = T\left(\frac{\partial p}{\partial T}\right)_V$$

2.130 (a) In an isochoric process, the entropy change will be

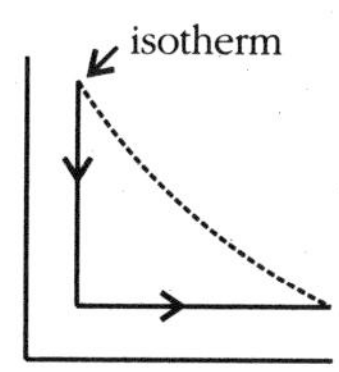

$$\Delta S = \int_{T_i}^{T_f} \frac{C_V dT}{T} = C_V \ln\frac{T_f}{T_i} = C_V \ln n = \frac{R \ln n}{\gamma - 1}$$

For carbon dioxide $\gamma = 1.30$, so

$$\Delta S = 19.2 \text{ J/K mol}$$

(b) For an isobaric process, entropy change will be

$$\Delta S = C_p \ln\frac{T_f}{T_i} = C_p \ln n = \frac{\gamma R \ln n}{\gamma - 1}$$

$$= 25 \text{ J/K mol}$$

2.131 In an isothermal expansion, entropy change is

$$\Delta S = \nu R \ln \frac{V_f}{V_i}$$

So,
$$\frac{V_f}{V_i} = e^{\Delta S/\nu R} = 2.0 \text{ times}$$

2.132 The entropy change depends on the final and initial states only, so we can calculate it directly along the isotherm, it is $\Delta S = 2\ R \ln n = 20$ J/K, assuming that the final volume is n times the initial volume.

2.133 If the initial temperature is T_0 and volume is V_0 then in an adiabatic expansion,

$$TV^{\gamma-1} = T_0 V_0^{\gamma-1}$$

So,
$$T = T_0 n^{1-\gamma} = T_1 \quad (\text{where } n = V_1/V_0)$$

V_1 being the volume at the end of the adiabatic process. There is no entropy change in this process. Next the gas is compressed isobarically and the net entropy change is

$$\Delta S = \left(\frac{m}{M} C_p\right) \ln \frac{T_f}{T_1}$$

But,
$$\frac{V_1}{T_1} = \frac{V_0}{T_f} \quad \text{or} \quad T_f = T_1 \frac{V_0}{V_1} = T_0 n^{-\gamma}$$

So,
$$\Delta S = \left(\frac{m}{M} C_p\right) \ln \frac{1}{n} = -\frac{m}{M} C_p \ln n = -\frac{m}{M} \frac{R\gamma}{\gamma - 1} \ln n = -9.7 \text{ J/K}$$

2.134 The entropy change depends on the initial and final state only so can be calculated for any process. We choose to evaluate the entropy change along the pair of lines shown in the figure. Then

$$\Delta S = \int_{T_0}^{T_0/\beta} \frac{\nu C_V dT}{T} + \int_{T_0/\beta}^{\alpha T_0/\beta} \nu C_p \frac{dT}{T}$$

$$= (-C_V \ln \beta + C_p \ln \alpha)\nu = \frac{\nu R}{\gamma - 1}(\gamma \ln \alpha - \ln \beta)$$

$$\approx -11 \text{ J/K}$$

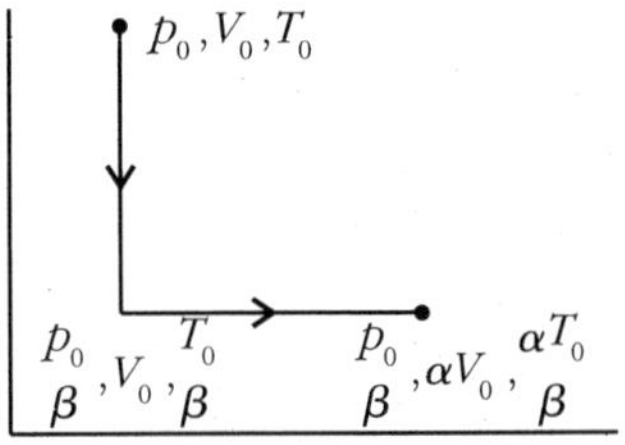

2.135 To calculate the required entropy difference we only have to calculate the entropy difference for a process in which the state of the gas in vessel 1 is changed to that in vessel 2.

$$\Delta S = \nu\left(\int_{T_1}^{T_1/\alpha\beta} C_V \frac{dT}{T} + \int_{T_1/\alpha\beta}^{T_1/\beta} C_p \frac{dT}{T}\right)$$

$$= \nu(C_p \ln \alpha - C_V \ln \alpha\beta)$$

$$= \nu\left(R \ln \alpha - \frac{R}{\gamma - 1} \ln \beta\right) = \nu R\left(\ln \alpha - \frac{\ln \beta}{\gamma - 1}\right)$$

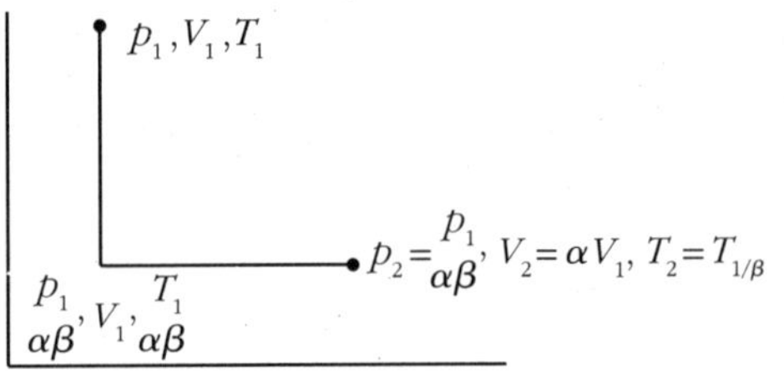

$$\Delta S = 0.85 \text{ J/K}$$

(using $\gamma = 5/3, \alpha = 2, \beta = 1.5$ and $\nu = 1.2$).

2.136 For the polytropic process with index n, pV^n = constant. Along this process (as in solution of Problem 2.122),

$$C = R\left(\frac{1}{\gamma-1} - \frac{1}{n-1}\right) = \frac{n-\gamma}{(\gamma-1)(n-1)} \cdot R$$

So, $$\Delta S = \int_{T_0}^{\tau T_0} C\frac{dT}{T} = \frac{n-\gamma}{(\gamma-1)(n-1)} R\ln\tau$$

2.137 The process in question may be written as

$$\frac{p}{p_0} = \alpha\frac{V}{V_0}$$

(where α is a constant and p_0, V_0 are some reference values).

For this process (from solution of Problem 2.127), the specific heat is

$$C = C_V + \frac{1}{2}R = R\left(\frac{1}{\gamma-1} + \frac{1}{2}\right) = \frac{1}{2}R\frac{\gamma+1}{\gamma-1}$$

Along the line volume increases α times then so does the pressure. The temperature must then increase α^2 times. Thus,

$$\Delta S = \int_{T_0}^{\alpha^2 T_0} \nu C\frac{dT}{T} = \frac{\nu R}{2}\frac{\gamma+1}{\gamma-1}\ln\alpha^2 = \nu R\frac{\gamma+1}{\gamma-1}\ln\alpha$$

$$\Delta S = 46.1 \text{ J/K}.$$

(using $\nu = 2$, $\gamma = \frac{5}{3}$ and $\alpha = 2$)

2.138 Let p_1, V_1 be a reference point on the line and let (p, V) be any other point.

Then, $$p = p_0 - \alpha V$$

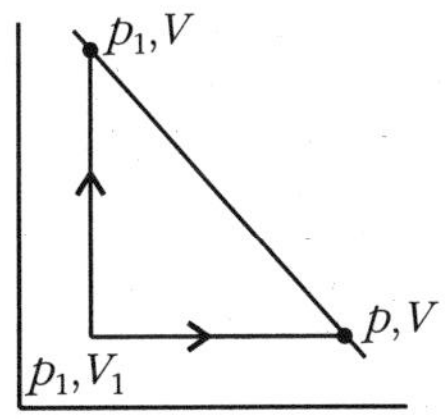

The entropy difference

$$\Delta S = S(p,V) - S(p_1,V_1)$$
$$= C_V \ln\frac{p}{p_1} + C_p \ln\frac{V}{V_1}$$
$$= C_V \ln\frac{p_0-\alpha V}{p_1} + C_p \ln\frac{V}{V_1}$$

For an extremum of ΔS

$$\frac{\partial\Delta S}{\partial V} = \frac{-\alpha C_V}{p_0-\alpha V} + \frac{C_p}{V} = 0$$

or $$C_p(p_0-\alpha V) - \alpha V C_V = 0$$

or $$\gamma(p_0-\alpha V) - \alpha V = 0 \quad\text{or}\quad V = V_m = \frac{\gamma p_0}{\alpha(\gamma+1)}$$

This gives a maximum of ΔS because $\dfrac{\partial^2\Delta S}{\partial V^2} < 0$.

Note that a maximum of ΔS is a maximum of $S\,(p, V)$.

2.139 For the process, $$S = aT + C_V \ln T$$

and specific heat $$C = T\frac{dS}{dT} = aT + C_V$$

On the other hand $dQ = CdT = C_V dT + pdV$ for an ideal gas.

Thus, $$pdV = \frac{RT}{V}dV = aTdT$$

or $$\frac{RdV}{aV} = dT \quad \text{or} \quad \frac{R}{a}\ln V + \text{ constant} = T$$

Using $$T = T_0 \quad \text{when} \quad V = V_0$$

we get $$T = T_0 + \frac{R}{a}\ \ln\ \frac{V}{V_0}$$

2.140 For a Van der Waal gas

$$\left(p + \frac{a}{V^2}\right)(V - b) = RT$$

The entropy change along an isotherm can be calculated from

$$\Delta S = \int_{V_1}^{V_2}\left(\frac{\partial S}{\partial V}\right)_T dV$$

It follows from solution of Problem 2.129 that

$$\left(\frac{\partial S}{\partial V}\right)_T = \left(\frac{\partial p}{\partial T}\right)_V = \frac{R}{V - b}$$

(assuming a, b to be known constants).

Thus, $$\Delta S = R\ln\frac{V_2 - b}{V_1 - b}$$

2.141 We use, $$\Delta S = \int_{V_1,T_1}^{V_2,T_2} dS(V,T) = \int_{T_1}^{T_2}\left(\frac{\partial S}{\partial T}\right)_{V_1} dT + \int_{V_1}^{V_2}\left(\frac{\partial S}{\partial V}\right)_{T=T_2} dV$$

$$= \int_{T_1}^{T_2}\frac{C_V dT}{T} + \int_{V_1}^{V_2}\frac{R}{V - b}dV = C_V\ \ln\frac{T_2}{T_1} + R\ln\frac{V_2 - b}{V_1 - b}$$

(assuming CV, a, b to be known constants).

2.142 We can take $S \to 0$ as $T \to 0$.

Then, $$S = \int_0^T C\frac{dT}{T} = \int_0^T aT^2 dT = \frac{1}{3}aT^3$$

2.143 $$\Delta S = \int_{T_1}^{T_2}\frac{CdT}{T} = \int_{T_1}^{T_2}\frac{m(a + bT)}{T}dT = mb(T_2 - T_1) + ma\ln\frac{T_2}{T_1}$$

$= 2.0$ kJ (on substituting values)

2.144 Here, $$T = aS^n \quad \text{or} \quad S = \left(\frac{T}{a}\right)^{\frac{1}{n}}$$

Then, $$C = T\frac{1}{n}\frac{T^{\frac{1}{n}-1}}{a^{1/n}} = \frac{S}{n}$$

Clearly, $$C < 0 \quad \text{if} \quad n < 0$$

2.145 We know,

$$S - S_0 = \int_{T_0}^{T}\frac{CdT}{T} = C\ln\frac{T}{T_0}$$

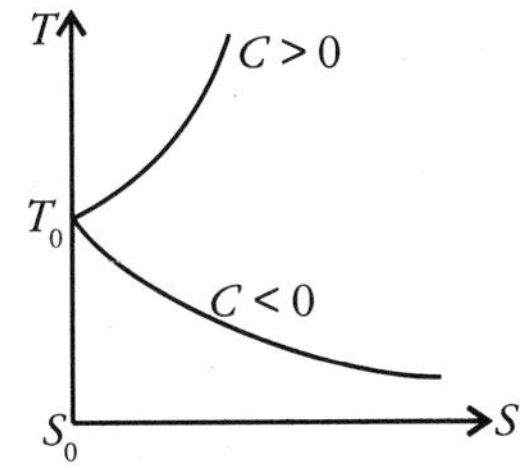

(assuming C to be a known constant).

Then, $$T = T_0 \exp\left(\frac{S - S_0}{C}\right)$$

2.146 (a) $C = T\dfrac{dS}{dT} = -\dfrac{\alpha}{T}$

(b) $Q = \displaystyle\int_{T_1}^{T_2} CdT = \alpha \ln\frac{T_1}{T_2}$

(c) $W = \Delta Q - \Delta U$

$$= \alpha\ln\frac{T_1}{T_2} + C_V(T_1 - T_2)$$

Since for an ideal gas C_V is constant and $\Delta U = C_V(T_2 - T_1)$ (as U does not depend on V).

2.147 (a) We have from the definition

$$Q = \int TdS = \text{area under the curve}$$

$$Q_1 = T_0\,(S_1 - S_0)$$

$$Q_2' = \frac{1}{2}(T_0 + T_1)(S_1 - S_0)$$

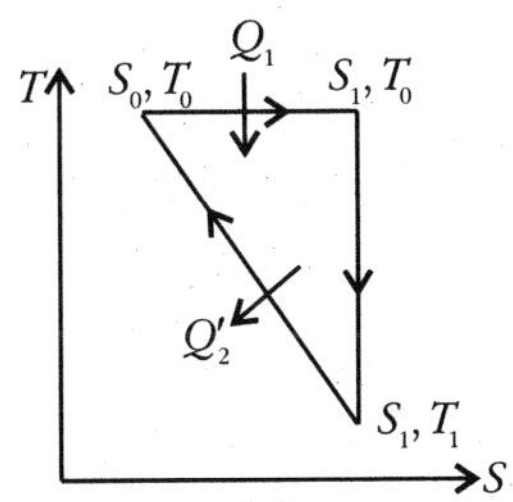

(a)

Thus, using $$T_1 = \frac{T_0}{n}$$

We get $$\eta = 1 - \frac{T_0 + T_1}{2T_0} = 1 - \frac{1 + 1/n}{2} = \frac{n-1}{2n}$$

(b) Here, $$Q_1 = \frac{1}{2}(S_1 - S_0)(T_1 + T_0)$$

$$Q_2' = T_1(S_1 - S_0)$$

$$\eta = 1 - \frac{2T_1}{T_1 + T_0} = \frac{T_0 - T_1}{T_0 + T_1} = \frac{n-1}{n+1}$$

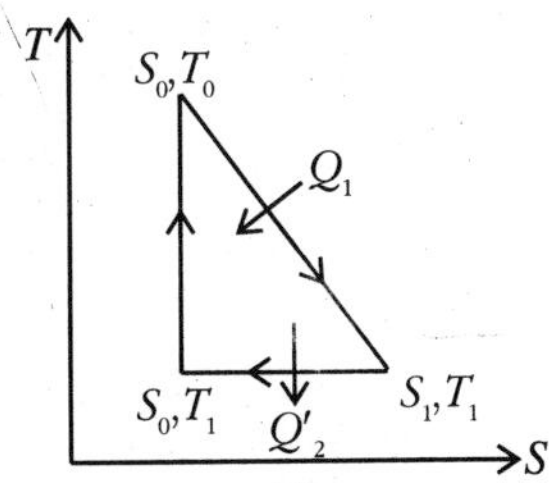

(b)

2.148 In this case, known as free expansion, no work is done and no heat is exchanged. So internal energy must remain unchanged, i.e., $U_f = U_i$. For an ideal gas this implies constant temperature, i.e., $T_f = T_i$. The process is irreversible but the entropy change can be calculated by considering a reversible isothermal process. Then, as before

$$\Delta S = \int \frac{dQ}{T} = \int_{V_1}^{V_2} \frac{pdV}{T} = \nu R \ln n = 20.1 \text{ J/K}$$

2.149 The process consists of two parts. The first part is a free expansion in which $U_f = U_i$. The second part is adiabatic compression in which work done results in change of internal energy. Obviously,

$$0 = U_F - U_f + \int_{V_f}^{V_0} pdV \quad \text{and} \quad V_f = 2V_0$$

Now in the first part

$$p_f = \frac{1}{2}p_0 \quad \text{and} \quad V_f = 2V_0$$

(because there is no change of temperature).

In the second part,

$$pV^\gamma = \frac{1}{2}p_0(2V_0)^\gamma = 2^{\gamma-1}p_0V_0^\gamma$$

$$\int_{2V_0}^{V_0} pdV = \int_{2V_0}^{V_0} \frac{2^{\gamma-1}p_0V_0^\gamma}{V^\gamma}dV = \left[\frac{2^{\gamma-1}p_0V_0^\gamma}{-\gamma+1}V^{1-\gamma}\right]_{2V_0}^{V_0}$$

$$= 2^{\gamma-1}p_0V_0^\gamma V_0^{-\gamma+1}\frac{(2^{-\gamma+1}-1)}{\gamma-1}$$

$$= \frac{RT_0}{\gamma-1}(1-2^{\gamma-1})$$

Thus,
$$\Delta U = U_F - U_i = \frac{RT_0}{\gamma-1}(2^{\gamma-1}-1)$$

The entropy change $\Delta S = \Delta S_1 + \Delta S_{II}$.

$\Delta S_1 = R \ln 2$ and $\Delta S_{II} = 0$ as the process is reversible adiabatic. Thus, $\Delta S = R \ln 2$.

2.150 In all adiabatic processes, by virtue of first law of thermodynamics

$$Q = U_f - U_i + A = 0$$

Thus,
$$U_f = U_i - A$$

For a slow process,

$$A' = \int_{V_0}^{V} pdV \text{ (where for a quasistatic adiabatic process } pV^\gamma = \text{constant)}$$

On the other hand for a fast process, the external work done is $A'' < A'$. In fact $A'' = 0$ for free expansion. Thus U'_f (slow) $< U''_f$ (fast).

Since U depends on temperature only, if $T'_f < T''_f$, consequently $p''_f > p'_f$ from the ideal gas equation $pV = RT$.

The pressure will be higher after the fast expansion.

2.151 Let $V_1 = V_0$, $V_2 = nv_0$.

Since the temperature is the same, the required entropy change can be calculated by considering isothermal expansion of the gas in either parts into the whole vessel.

Thus,
$$\Delta S = \Delta S_I + \Delta S_{II}$$
$$= \nu_1 R \ln \frac{V_1 + V_2}{V_1} + \nu_2 R \ln \frac{V_1 + V_2}{V_2}$$
$$= \nu_1 R \ln(1+n) + \nu_2 R \ln\left(\frac{1+n}{n}\right)$$
$$= 5.1 \text{ J/K}$$

2.152 Let c_1 be the specific heat of copper and c_2 be the specific heat of water.

Then,
$$\Delta S = \int_{7+273}^{T_0} \frac{c_2 m_2 dT}{T} - \int_{T_0}^{97+273} \frac{m_1 c_1 dT}{T}$$
$$= m_2 c_2 \ln \frac{T_0}{280} - m_1 c_1 \ln \frac{370}{T_0}$$

T_0 can be found from,
$$c_2 m_2 (T_0 - 280) = m_1 c_1 (370 - T_0) \text{ or } T_0 = \frac{280 m_2 c_2 + 370 m_1 c_1}{c_2 m_2 + m_1 c_1}$$

Using $c_1 = 0.39$ J/g K, $c_2 = 4.18$ J/g K,
$$T_0 \approx 300 \text{ K} \quad \text{and} \quad \Delta S = 28.4 - 24.5 \cong 3.9 \text{ J/K}$$

2.153 For an ideal gas the internal energy depends on temperature only. We can consider the process in question to be one of simultaneous free expansion, then the total energy $U = U_1 + U_2$.

Since $U_1 = C_V T_1$, $U_2 = C_V T_2$ so, $U = 2C_V (T_1 + T_2)/2$ and $(T_1 + T_2)/2$ is the final temperature.

The entropy change is obtained by considering isochoric process because in effect, the gas remains confined to its vessel.

So,
$$\Delta S = \int_{T_1}^{(T_1+T_2)/2} \frac{C_V dT}{T} - \int_{(T_1+T_2)/2}^{T_2} C_V \frac{dT}{T} = C_V \ln \frac{(T_1 + T_2)^2}{4T_1 T_2}$$

Since,
$$(T_1 + T_2)^2 = (T_1 - T_2)^2 + 4T_1 T_2,$$

We get
$$\Delta S > 0$$

2.154 (a) Each atom has probability 1/2 to be in either compartment. Thus $P = 2^{-N}$.

(b) Typical atomic velocity at room temperature is ~ 10^5 cm/s so it takes an atom 10^{-5} s to cross the vessel. This is the relevant time scale for our problem. Let $T = 10^{-5}$ s, then in time T there will be t/T crossing or arrangements of the atoms. This will be large enough to produce the given arrangement if
$$\frac{t}{\tau} 2^{-N} \approx 1 \quad \text{or} \quad N \approx \frac{\ln t/\tau}{\ln 2} \approx 75$$

2.155 The statistical weight is
$$N_{C_{N/2}} = \frac{N!}{N/2! \frac{N}{2}!} = \frac{10 \times 9 \times 8 \times 7 \times 6}{5 \times 4 \times 3 \times 2} = 252$$

The probability distribution is

$$N_{C_{N/2}} 2^{-N} = 252 \times 2^{-10} = 24.6\%$$

2.156 The probabilities that the half A contains n molecules is

$$N_{C_n} \times 2^{-N} = \frac{N!}{n!(N-n)!} 2^{-N}; \quad \frac{1}{32}, \frac{5}{32}, \frac{10}{32}, \frac{10}{32}, \frac{5}{32}, \frac{1}{32}, \text{ respectively.}$$

2.157 The probability of one molecule being confined to the marked volume is

$$P = \frac{V}{V_0}$$

We can choose this molecule in many (N_{C_1} ways. The probability that n molecules get confined to the marked volume is clearly

$$N_{C_n} P^n (1-P)^{N-n} = \frac{N!}{n!(N-n)!} p^n (1-P)^{N-n}$$

2.158 In a sphere of diameter d, the number of molecules are

$$N = \frac{\pi d^3}{6} n_0$$

where n_0 = Loschmidt's number = number of molecules per unit volume (1 cc) under NTP.

The relative fluctuation in this number is

$$\frac{\partial N}{N} = \frac{\sqrt{N}}{N} = \frac{1}{\sqrt{N}} = \eta$$

or
$$\frac{1}{\eta^2} = \frac{\pi}{6} d^3 n_0 \text{ or } d^3 = \frac{6}{\pi n_0 \eta^2} \Rightarrow d = \left(\frac{6}{\pi \eta^2 n_0}\right)^{1/3} = 0.41 \ \mu\text{m}$$

The average number of molecules in this sphere is

$$\frac{1}{\eta^2} = 10^6$$

2.159 For a monoatomic gas, $C_V = 3/2R$ per mole.

The entropy change in the process is

$$\Delta S = S - S_0 = \int_{T_0}^{T_0+\Delta T} C_V \frac{dT}{T} = \frac{3}{2} R \ln\left(1 + \frac{\Delta T}{T_0}\right)$$

Now from the Boltzmann equation, $S = k \ln \Omega$.

So,
$$\frac{\Omega}{\Omega_0} = e^{(S-S_0)/k} = \left(1 + \frac{\Delta T}{T_0}\right)^{\frac{3N_A}{2}} = \left(1 + \frac{1}{300}\right)^{\frac{3\times 6}{2}\times 10^{23}} = 10^{1.31\times 10^{21}}$$

Thus the statistical weight increases by this factor.

2.5 Liquids. Capillary Effects

2.160 (a) $\Delta p = \alpha\left(\frac{1}{d/2}+\frac{1}{d/2}\right)=\frac{4\alpha}{d}$

$$= \frac{4\times 490\times 10^{-3}}{1.5\times 10^{-6}}\frac{N}{m^2} = 1.307\times 10^{6}\frac{N}{m^2} = 13 \text{ atm}$$

(b) The soap bubble has two surfaces.

So, $$\Delta p = 2\alpha\left(\frac{1}{d/2}+\frac{1}{d/2}\right)=\frac{8\alpha}{d}$$

$$= \frac{8\times 45}{3\times 10^{-3}} = 1.2\times 10^{-3} \text{ atm}$$

2.161 The pressure just inside the hole will be less than the outside pressure by $4\alpha/d$. This can support a height h of Hg where,

$$\rho\, gh = \frac{4\alpha}{d} \quad \text{or} \quad h = \frac{4\alpha}{\rho g d}$$

So, $$h = \frac{4\times 490\times 10^{-3}}{13.6\times 10^{3}\times 9.8\times 70\times 10^{-6}} = \frac{200}{13.6\times 70} \approx 0.21 \text{ m of Hg}$$

2.162 By Boyle's law

$$\left(p_0+\frac{8\alpha}{d}\right)\frac{4\pi}{3}\left(\frac{d}{2}\right)^3 = \left(\frac{p_0}{n}+\frac{8\alpha}{\eta d}\right)\frac{4\pi}{3}\left(\frac{\eta d}{2}\right)^3$$

or $$p_0\left(1-\frac{\eta^3}{n}\right) = \frac{8\alpha}{d}(\eta^2-1)$$

Thus, $$\alpha = \frac{1}{8}p_0 d\left(1-\frac{\eta^3}{n}\right)\Big/(\eta^2-1)$$

2.163 The pressure has terms due to hydrostatic pressure and capillarity and they add as

$$p = p_0 + \rho g h + \frac{4\alpha}{d}$$

$$= \left(1+\frac{5\times 9.8\times 10^{3}}{10^{5}}+\frac{4\times 0.73\times 10^{-3}}{4\times 10^{-6}}\times 10^{-5}\right)\text{atm} = 2.22 \text{ atm}$$

2.164 By Boyle's law

$$\left(p_0 + hg\rho + \frac{4\alpha}{d}\right)\frac{\pi}{6}d^3 = \left(p_0+\frac{4\alpha}{nd}\right)\frac{\pi}{6}n^3 d^3$$

or $$hg\rho - p_0(n^3-1) = \frac{4\alpha}{d}(n^2-1)$$

or $$h = \left[p_0(n^3-1)+\frac{4\alpha}{d}(n^2-1)\right]\Big/g\rho = 4.98 \text{ m of water.}$$

2.165 Clearly

$$\Delta h\rho g = 4\alpha\,|\cos\theta|\left(\frac{1}{d_1}-\frac{1}{d_2}\right)$$

$$\Delta h = \frac{4\alpha\,|\cos\theta|\,(d_2-d_1)}{d_1 d_2 \rho g} = 11 \text{ mm}$$

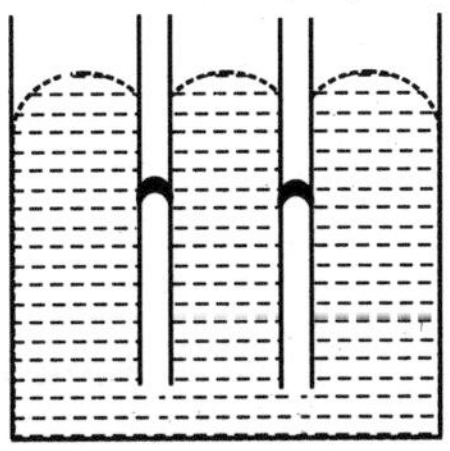

2.166 In a capillary with diameter $d = 0.5$ mm, water will rise to a height

$$h = \frac{2\alpha}{\rho g r} = \frac{4\alpha}{\rho g d} = \frac{4 \times 73 \times 10^{-3}}{10^3 \times 9.8 \times 0.5 \times 10^{-3}} = 59.6 \text{ mm}$$

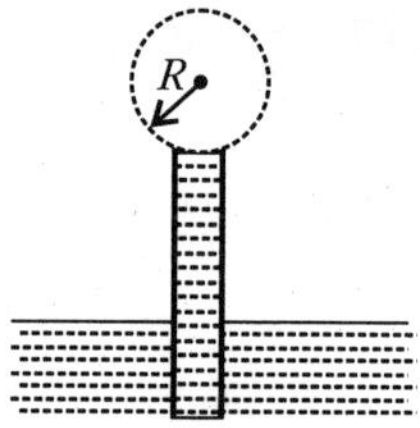

Since this is greater than the height (= 25 mm) of the tube, a meniscus of radius R will be formed at the top of the tube, where

$$R = \frac{2\alpha}{\rho g h} = \frac{2 \times 73 \times 10^{-3}}{10^3 \times 9.8 \times 25 \times 10^{-3}} \approx 0.6 \text{ mm}$$

2.167 Initially the pressure of air in the capillary is p_0 and it's length is l. When submerged under water, the pressure of air in the portion above water must be $p_0 + 4\alpha/d$, since the level of water inside the capillary is the same as the level outside. Thus by Boyle's law

$$\left(p_0 + \frac{4\alpha}{d}\right)(l - x) = p_0 l$$

or
$$\frac{4\alpha}{d}(l - x) = p_0 x$$

or
$$x = \frac{l}{1 + p_0 d/4\alpha} = 1.4 \text{ cm (on substituting values)}$$

2.168 We have by Boyle's law

$$\left(p_0 - \rho g h + \frac{4\alpha\cos\theta}{d}\right)(l - h) = p_0 l$$

or
$$\frac{4\alpha\cos\theta}{d} = \rho g h + \frac{p_0 h}{l - h}$$

Hence,
$$\alpha = \left(\rho g h + \frac{p_0 h}{l - h}\right)\frac{d}{4\cos\theta}$$

2.169 Suppose the liquid rises to a height h. Then the total energy of the liquid in the capillary is

$$E(h) = \frac{\pi}{4}\left(d_2^2 - d_1^2\right)h \times \rho g \times \frac{h}{2} - \pi(d_2 + d_1)\,\alpha h$$

Minimising E, we get
$$h = \frac{4\alpha}{\rho g(d_2 - d_1)} = 6 \text{ cm}$$

2.170 Let h be the height of the water level at a distance x from the edge. Then the total energy of water in the wedge above the level outside is

$$E = \int x\delta\varphi \cdot dx \cdot h \cdot \rho g \frac{h}{2} - 2\int dx \cdot h \cdot \alpha\cos\theta$$

$$= \int dx \frac{1}{2} x\rho g\delta\varphi\left(h^2 - 2\frac{2\alpha\cos\theta}{x\rho g\delta\varphi}h\right)$$

$$= \int dx \frac{1}{2} x\rho g\delta\varphi\left[\left(h - \frac{2\alpha\cos\theta}{x\rho g\delta\varphi}\right)^2 - \frac{4\alpha^2\cos^2\theta}{x^2\rho^2 g^2\delta\varphi^2}\right]$$

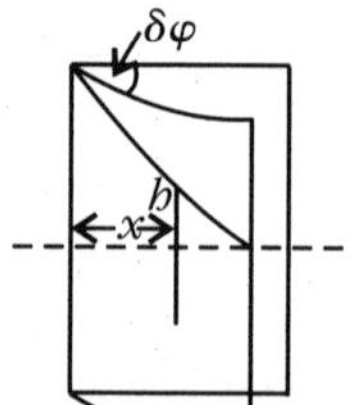

This is minimum when $h = \dfrac{2\alpha\cos\theta}{x\rho g\delta\varphi}$

2.171 From the equation of continuity

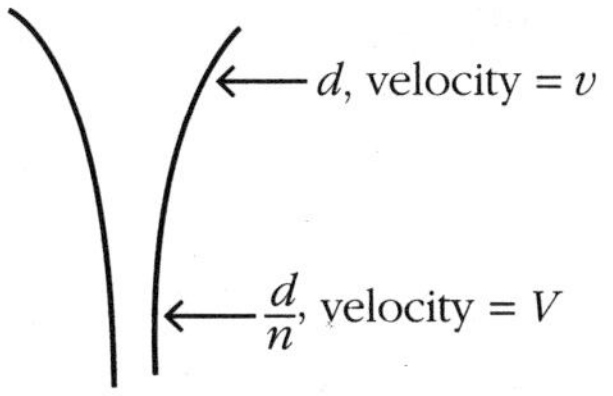

$$\frac{\pi}{4} d^2 \cdot v = \frac{\pi}{4}\left(\frac{d}{n}\right)^2 \cdot V \quad \text{or} \quad V = n^2 v$$

We then apply Bernoulli's theorem

$$\frac{p}{\rho} + \frac{1}{2} v^2 + \Phi = \text{constant}$$

The pressure p differs from the atmospheric pressure by capillary effects. At the upper section, $p = p_0 + 2\alpha/d$ neglecting the curvature in the vertical plane.

Thus,
$$\frac{p_0 + 2\alpha/d}{\rho} + \frac{1}{2} v^2 + gl = \frac{p_0 + 2n\alpha/d}{\rho} + \frac{1}{2} n^4 v^2$$

or
$$v = \sqrt{\frac{2gl - (4\alpha/\rho d)(n-1)}{n^4 - 1}}$$

Finally, the liquid coming out per second is,

$$V = \frac{1}{4} \pi d^2 \sqrt{\frac{2gl - (4\alpha/\rho d)(n-1)}{n^4 - 1}} = 0.9 \text{ cm}^3/\text{s (on substituting values)}$$

2.172 The radius of curvature of the drop is R_1 at the upper end of the drop and R_2 at the lower end. Then the pressure inside the drop is $p_0 + (2\alpha/R_1)$ at the top end and $p_0 + (2\alpha/R_2)$ at the bottom end. Hence,

$$p_0 + \frac{2\alpha}{R_1} = p_0 + \frac{2\alpha}{R_2} + \rho g h \quad \text{or} \quad \frac{2\alpha(R_2 - R_1)}{R_1 R_2} = \rho g h$$

To a first approximation $R_1 \approx R_2 \approx h/2$.

So,
$$R_2 - R_1 \approx 1/8 \rho g h^3/\alpha \approx 0.20 \text{ mm}$$

and if
$$h = 2.3 \text{ mm}, \ \alpha = 73 \text{ mN/m}$$

2.173 We must first calculate the pressure difference inside the film from that outside. This is given by

$$p = \alpha\left(\frac{1}{r_1} + \frac{1}{r_2}\right)$$

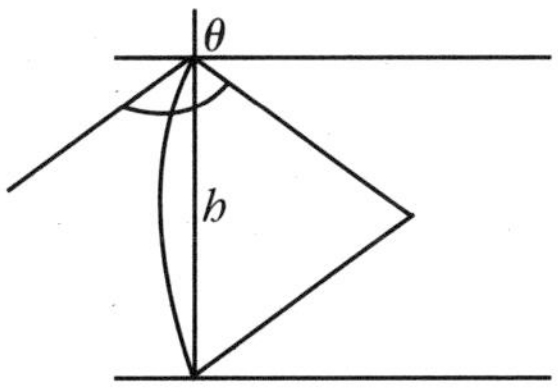

Here, $2r_1|\cos\theta| = h$ and $r_2 \approx R$, where R is the radius of the tablet and can be neglected. Thus the total force exerted by mercury drop on the upper glass plate is typically,

$$\frac{2\pi R^2 \alpha |\cos\theta|}{h}$$

We should put h/n for h because the tablet is compressed n times. Then since Hg is nearly incompressible, $\pi R^2 h$ = constant, so $R \to R\sqrt{n}$. Thus, total force

$$= \frac{2\pi R^2 \alpha |\cos\theta|}{h} n^2$$

Part of the force is needed to keep the Hg in the shape of a tablet rather than in the shape of infinitely thin sheet. This part can be calculated being putting $n = 1$ above. Thus

$$mg + \frac{2\pi R^2 \alpha |\cos\theta|}{h} = \frac{2\pi R^2 \alpha |\cos\theta|}{h} n^2$$

or
$$m = \frac{2\pi R^2 \alpha |\cos\theta|}{hg}(n^2 - 1) = 0.7 \text{ kg}$$

2.174 The pressure inside the film is less than that outside by an amount

$$p = \alpha\left(\frac{1}{r_1} + \frac{1}{r_2}\right)$$

(where r_1 and r_2 are the principal radii of curvature of the meniscus).

One of these is small, being given by $h = 2r_1 \cos\theta$ while the other is large and will be ignored.

Then, $$F \approx \frac{2A\cos\theta}{h}\alpha$$

(where A = area of the water film between the plates)

Now, $$A = \frac{m}{\rho h}$$

So, $$F = \frac{2m\alpha}{\rho h^2} \text{ [when } \theta \text{ (the angle of contact) = 0]}$$

$$= 1.0 \text{ N (on substituting values)}$$

2.175 This is analogous to the previous problem except that, $A = \pi R^2$.

So, $$F = \frac{2\pi R^2\alpha}{h} = 0.6 \text{ kN}$$

2.176 The energy of the liquid between the plates is

$$E = ldh\rho g\frac{h}{2} - 2\,\alpha lh = \frac{1}{2}\rho gldh^2 - 2\alpha lh$$

$$= \frac{1}{2}\rho gld\left(h - \frac{2\alpha}{\rho gd}\right)^2 - \frac{2\alpha^2 l}{\rho gd}$$

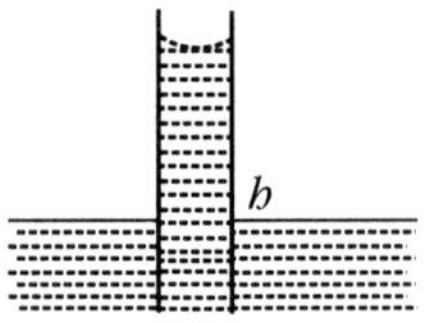

This energy is minimum when, $h = 2\alpha/\rho gd$ and the minimum potential energy is then

$$E_{\min} = -\frac{2\alpha^2 l}{\rho gd}$$

The force of attraction between the plates can be obtained from this as

$$F = \frac{-\partial E_{\min}}{\partial d} = -\frac{2\alpha^2 l}{\rho gd^2} \text{ (minus sign means the force is attractive)}$$

Thus, $$F = \frac{\alpha lh}{d} = 13 \text{ N}$$

2.177 Suppose the radius of bubble is x at some instant. Then the pressure inside is $p_0 + 4\alpha/x$. The flow through the capillary is given by Poiseuille's equation,

$$Q = \frac{\pi r^4}{8\eta l}\frac{4\alpha}{x} = -4\pi x^2\frac{dx}{dt}$$

Integrating we get,

$$\frac{\pi r^4\alpha}{2\eta l}t = \pi(R^4 - x^4)$$

(where we have used the fact that $t = 0$ when $x = R$).

The life time of the bubble corresponding to $x = 0$ is

$$t = \frac{2\eta lR^4}{\alpha r^4}$$

2.178 If the liquid rises to a height h, the energy of the liquid column becomes

$$E = \rho g \pi r^2 h \cdot \frac{h}{2} - 2\pi r h \alpha = \frac{1}{2}\rho g \pi \left(rh - 2\frac{\alpha}{\rho g}\right)^2 - \frac{2\pi\alpha^2}{\rho g}$$

This is minimum when $rh = 2\alpha/\rho g$ and that is relevant height to which water must rise.

At this point, $$E_{\min} = -\frac{2\pi\alpha^2}{\rho g}$$

Since $E = 0$ in the absence of surface tension, heat liberated must be

$$Q = \frac{2\pi\alpha^2}{\rho g}$$

2.179 (a) The free energy per unit area being α,

$$F = \pi\alpha d^2 = 3\ \mu\text{J (on substituting values)}$$

(b) $F = 2\pi\alpha d^2$ because the soap bubble has two surfaces.
Substitution gives, $F = 10\ \mu$J.

2.180 When two mercury drops each of diameter d merge, the resulting drop has diameter d_1

where, $$\frac{\pi}{6}d_1^3 = \frac{\pi}{6}d^3 \times 2 \quad \text{or} \quad d_1 = 2^{1/3}d$$

The increase in free energy is

$$\Delta F = \pi\, 2^{2/3}\, d^2\alpha - 2\pi d^2\alpha = 2\pi d^2\alpha(2^{-1/3} - 1) = -1.43\ \mu\text{J}$$

2.181 Work must be done to stretch the soap film and compress the air inside. The former is simply $2\alpha \times 4\pi R^2 = 8\pi R^2\alpha$, there being two sides of the film. To get the latter we note that the compression is isothermal and work done is

$$-\int_{V_i = V_0}^{V_f = V} p\,dV$$

where, $$V_0 p_0 = \left(p_0 + \frac{4\alpha}{R}\right)\cdot V \quad \text{and} \quad V = \frac{4\pi}{3}R^3$$

or $$V_0 = \frac{pV}{p_0} \quad \text{and} \quad p = p_0 + \frac{4\alpha}{R}$$

and minus sign is needed because we are calculating work done on the system. Thus since pV remains constant, the work done is

$$pV \ln\frac{V_0}{V} = p\,V \ln\frac{p}{p_0}$$

So, $$A' = 8\pi R^2\alpha + pV\ln\frac{p}{p_0}$$

$$A' = 8\pi R^2\alpha + pV\ln\frac{p}{p_0}$$

2.182 When heat is given to a soap bubble, the temperature of the air inside rises and the bubble expands but unless the bubble bursts, the amount of air inside does not change. Further we shall neglect the variation of the surface tension with temperature. Then from the gas equations

$$\left(p_0 + \frac{4\alpha}{r}\right)\frac{4\pi}{3}r^3 = \nu RT \quad (\nu = \text{constant})$$

Differentiating we get,

$$\left(p_0 + \frac{8\alpha}{3r}\right) 4\pi r^2 dr = \nu R dT$$

or
$$dV = 4\pi r^2 dr = \frac{\nu R dT}{p_0 + \frac{8\alpha}{3r}}$$

Now from the first law of thermo dynamics

$$đQ = \nu C dT = \nu C_V dT + \frac{\nu R dT}{p_0 + 8\alpha/3r}\left(p_0 + \frac{4\alpha}{r}\right)$$

or
$$C = C_V + R\frac{p_0 + 4\alpha/r}{p_0 + 8\alpha/3r}$$

using $C_p = C_V + R$, we get

$$C = C_p + \frac{1/2\ R}{1 + 3p_0 r/8\alpha}$$

2.183 Consider an infinitesimal Carnot cycle with isotherms at $T - dT$ and T. Let A be the work done during the cycle. Then

$$A = [\alpha(T - dT) - \alpha(T)]\ \delta\sigma = -\frac{d\alpha}{dT} dT \delta\sigma$$

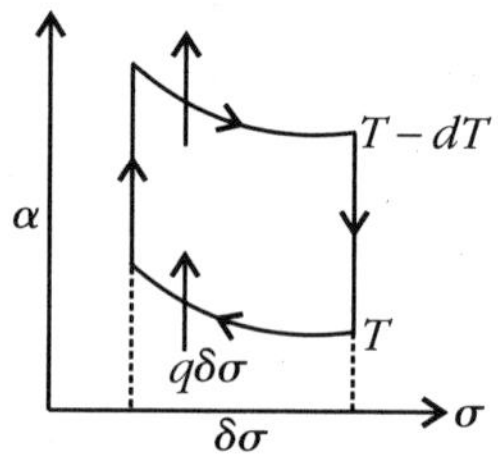

(where $\delta\sigma$ is the change in the area of film; we are considering only one surface here).

Then by Carnot theorem

$$\eta = \frac{A}{Q_1} = \frac{dT}{T}$$

or
$$\frac{-\frac{d\alpha}{dT} dT \delta\sigma}{q\delta\sigma} = \frac{dT}{T} \quad \text{or} \quad q = -T\frac{d\alpha}{dT}$$

2.184 As before we can calculate the heat required. It is taking into account two sides of the soap film. So,

$$\delta q = -T\frac{d\alpha}{dT}\delta\sigma \times 2$$

Thus,
$$\Delta S = \frac{\delta q}{T} = -2\frac{d\alpha}{dT}\delta\sigma$$

Now,
$$\Delta F = 2\alpha\delta\sigma$$

So,
$$\Delta U = \Delta F + T\Delta S = 2\left(\alpha - T\frac{d\alpha}{dT}\right)\delta\sigma$$

2.6 Phase Transformations

2.185 The condensation takes place at constant pressure and temperature and the work done is $A = p\Delta V$, where ΔV is the volume of the condensed vapor phase. It is

$$p\Delta V = \frac{\Delta m}{M} RT = 1.2 \text{ J}$$

(where $M = 18$ g is the molecular weight of water).

2.186 The specific volume of water (the liquid) will be written as V_l'. Since $V'_{\text{vap}} >> V_l'$, most of the weight is due to water. Thus if m_l is mass of the liquid and m_{vap} that of the vapor then

$$m = ml + m_{\text{vap}}$$

$$V = m_l V_l' + m_{\text{vap}} V'_{\text{vap}} \quad \text{or} \quad V - mV_l' = m_{\text{vap}}(V'_{\text{vap}} - V_l')$$

So,
$$m_{\text{vap}} = \frac{V - mV_l'}{V'_{\text{vap}} - V_l'} = 20 \text{ g}$$

Its volume is $V'_{\text{vap}} = 1.0$ l

2.187 The volume of the condensed vapor was originally $V_0 - V$ at temperature $T = 373$ K. Its mass will be given by

$$p_0(V_0 - V) = \frac{m}{M} RT \quad \text{or} \quad m = \frac{Mp_0(V_0 - V)}{RT} = 2 \text{ g}$$

(where p_0 = standard atmospheric pressure).

2.188 We let V_l' = specific volume of liquid. $V'_{\text{vap}} = NV_l'$ = specific volume of vapor.

Let V = original volume of the vapor.

Then,
$$M\frac{pV}{RT} = m_1 + m_{\text{vap}} = \frac{V}{NV_l'} \quad \text{or} \quad \frac{V}{n} = (m_l + Nm_{\text{vap}})V_l'$$

So,
$$(N-1)m_l V_l' = V\left(1 - \frac{1}{n}\right) = \frac{V}{n}(n-1) \quad \text{or} \quad \eta = \frac{m_l V_l'}{V/n} = \frac{n-1}{N-1}$$

In the case when the final volume of the substance corresponds to the midpoint of a horizontal portion of the isothermal line in the pV diagram, the final volume must be $(1 + N)\, V_l'/2$ per unit mass of the substance. Of this, the volume of the liquid is $V_l'/2$ per unit total mass of the substance.

Thus,
$$\eta = \frac{1}{1+N}$$

2.189 From the first law of thermodynamics

$$\Delta U + A = Q = mq$$

(where q is the specific latent heat of vaporization).

So, increment in entropy is given by

$$\Delta S = \frac{Q}{T} = \frac{mq}{T} = 6.0 \text{ kJ/K} \quad \text{(on substituting values)}$$

Now, $$A = p(V'_{\text{vap}} - V'_l)\, m = m\frac{RT}{M}$$

Thus, $$\Delta U = m\left(q - \frac{RT}{M}\right)$$

For water, this gives $\approx 2.08 \times 10^6$ J.

2.190 Some of the heat used in heating water to the boiling temperature $T = 100°\text{C} = 373$ K. The remaining heat $= Q - mc\Delta T$ (here c = specific heat of water, $\Delta T = 100$ K) is used to create vapor. If the piston rises to a height h then the volume of vapor will be $\approx Sh$ (neglecting water). Its mass will be

$$\frac{p_0 Sh}{RT} \times M$$

Heat of vaporization will be

$$\frac{p_0 Sh\, Mq}{RT}$$

To this must be added the work done in creating the saturated vapor $- p_0 Sh$.

Thus, $$Q - mc\Delta T \approx p_0 Sh\left(1 + \frac{qM}{RT}\right) \quad \text{or} \quad h \approx \frac{Q - mc\Delta T}{p_0 S\left(1 + \frac{qM}{RT}\right)} = 20 \text{ cm}$$

2.191 A quantity $mc\,(T - T_0)/q$ of saturated vapor must condense to heat the water to boilling point $T = 373$ K. (Here c = specific heat of water, $T_0 = 295$ K = initial water temperature.)

The work done in lowering the piston will then be

$$\frac{mc\,(T - T_0)}{q} \times \frac{RT}{M} = 25 \text{ J}$$

Since work done per unit mass of the condensed vapor is $pV = RT/M$.

2.192 Given

$$\Delta P = \frac{\rho_{\text{vap}}}{\rho_l}\frac{2\alpha}{r} = \frac{\rho_{\text{vap}}}{\rho_l} \times \frac{4\alpha}{d} = \eta p_{\text{vap}}$$
$$= \eta P_{\text{vap}} = \eta\frac{(m/M)RT}{V_{\text{vap}}} = \frac{\eta RT}{M}\rho_{\text{vap}}$$

or $$d = \frac{4\alpha M}{\rho_l RT\eta}$$

For water, $\alpha = 73$ dynes/cm, $M = 18$ g, $\rho_l = 1$ g/cc and $T = 300$ K. Using there values and with $\eta = 0.01$, we get $\approx 0.2\ \mu$m.

2.193 At equilibrium the number of "liquid" molecules evaporating must equal the number of "vapor" molecules condensing. By kinetic theory, this number is

$$\eta \times \frac{1}{4} n \langle v \rangle = \eta \times \frac{1}{4} n \times \sqrt{\frac{8}{\pi}\frac{kT}{m}}$$

The required mass is

$$\mu = m \times \eta \times n \times \sqrt{\frac{kT}{2\pi m}} = \eta\ nkT\sqrt{\frac{m}{2\pi kT}}$$

$$= \eta p_0\sqrt{\frac{M}{2\pi RT}} = 0.35\ \text{g/cm}^2$$

where p_0 is standard atmospheric pressure, $T = 373$ K and M = molecular weight of water.

2.194 Here we must assume that μ is also the rate at which the tungsten filament loses mass when in an atmosphere of its own vapor at this temperature and that η (of the previous problem) ≈ 1. Then

$$p = \mu\sqrt{\frac{2\pi RT}{M}} = 0.9\ \text{nPa}$$

where p = pressure of the saturated vapor.

2.195 From the Van der Waal's equation

$$p = \frac{RT}{V-b} - \frac{a}{V^2}$$

(where V = volume of one gram mole of the substances).

For water V = 18 cm³ per mol = 1.8×10^{-2} l mol^{-1} and a = 5.47 atm l²/mol².

If molecular attraction vanishes, the equation will be

$$p' = \frac{RT}{V-b} \quad \text{(for the same specific volume)}$$

Thus,

$$\Delta p = \frac{a}{V^2} = \frac{5.47}{1.8 \times 1.8} \times 10^4\ \text{atm} \approx 1.7 \times 10^4\ \text{atm}$$

2.196 The internal pressure being a/V^2, the work done in condensation is

$$\int_{V_l}^{V_g} \frac{a}{V^2}dV = \frac{1}{V_l} - \frac{a}{V_g} \approx \frac{a}{V_l}$$

This by assumption is Mq, M being the molecular weight, q being specific latent heat of vaporization and V_l, V_g being the molar volumes of the liquid and gas, respectively

Thus,

$$p_i = \frac{a}{V_l^2} = \frac{Mq}{V_l} = \rho q$$

(where ρ is the density of the liquid).

For water

$$p_i = 2 \times 10^4\ \text{atm}$$

2.197 The Van der Waal's equation can be written as (for one mole)

$$p(V) = \frac{RT}{V-b} - \frac{a}{V^2}$$

At the critical point $\left(\frac{\partial p}{\partial V}\right)_T$ and $\left(\frac{\partial^2 p}{\partial V^2}\right)_T$ vanish. Thus,

$$0 = \frac{RT}{(V-b^2)} - \frac{2a}{V^3} \quad \text{or} \quad \frac{RT}{(V-b)^2} = \frac{2a}{V^3}$$

$$0 = \frac{2RT}{(V-b)^3} - \frac{6a}{V^4} \quad \text{or} \quad \frac{RT}{(V-b)^3} = \frac{3a}{V^4}$$

Solving these simultaneously, we get on division

$$V - b = \frac{2}{3}V, V = 3\,b \approx V_{MCr}$$

This is the critical molar volume. Putting this back, we get

$$\frac{RT_{Cr}}{4b^2} = \frac{2a}{27b^3} \quad \text{or} \quad T_{Cr} = \frac{8a}{27bR}$$

Finally

$$p_{cr} = \frac{RT_{Cr}}{V_{MCr} - b} - \frac{a}{V_{MCr}^2} = \frac{4a}{27b^2} - \frac{a}{9b^2} = \frac{a}{27b^2}$$

From these we see that

$$\frac{p_{Cr}V_{MCr}}{RT_{Cr}} = \frac{a/9\text{b}}{8a/27b} = \frac{3}{8}$$

2.198 We have,

$$\frac{p_{Cr}}{RT_{Cr}} = \frac{a/27b^2}{8a/27b} = \frac{1}{8b}$$

Thus,

$$b = R\frac{T_{Cr}}{8p_{Cr}} = \frac{0.082 \times 304}{73 \times 8} = 0.043 \text{ l/mol}$$

and

$$\frac{(RT_{Cr})^2}{p_{Cr}} = \frac{64a}{27} \quad \text{or} \quad a = \frac{27}{64}\frac{(RT_{Cr})^2}{p_{Cr}} = 3.59 \text{ atm l}^2/\text{mol}^2$$

2.199 Specific volume is molar volume divided by molecular weight. Thus

$$V'_{Cr} = \frac{V_{MCr}}{M} = \frac{3RT_{Cr}}{8Mp_{Cr}} = \frac{3 \times 0.082 \times 562}{8 \times 78 \times 47} \text{ l/g} = 4.71 \text{ cm}^3/\text{g}$$

2.200 We have,

$$\left(p + \frac{a}{V_M^2}\right)(V_M - b) = RT$$

or

$$\frac{p + a/V_M^2}{p_{Cr}} \times \frac{V_M - b}{V_{Mcr}} = \frac{8}{3}\frac{T}{T_{Cr}}$$

or

$$\left(\pi + \frac{a}{p_{Cr}V_M^2}\right) \times \left(\nu - \frac{b}{V_{Mcr}}\right) = \frac{8}{3}\tau$$

$$\left(\text{where } \pi = \frac{p}{p_{Cr}},\ \nu = \frac{V_M}{V_{MCr}},\ \tau = \frac{T}{T_{Cr}}\right)$$

or
$$\left(\pi + \frac{27b^2}{V_M^2}\right)\left(\nu - \frac{1}{3}\right) = \frac{8}{3}\tau \quad \text{or} \quad \left(\pi + \frac{3}{\nu^2}\right)\left(\nu - \frac{1}{3}\right) = \frac{8}{3}\tau$$

When $\pi = 12$ and $\nu = 1/2$, we get

$$\tau = \frac{3}{8} \times 24 \times \frac{1}{6} = \frac{3}{2}$$

2.201 (a) The critical volume V_{MCr} is the maximum volume in the liquid phase and the minimum volume in the gaseous. Thus,

$$V_{\max} = \frac{1000}{18} \times 3 \times 0.0301 = 5l$$

(b) The critical pressure is the maximum possible pressure in the vapor phase in equilibrium with liquid phase. Thus,

$$p_{\max} = \frac{a}{27b^2} = \frac{5.47}{27 \times 0.03 \times 0.03} = 225 \text{ atm}$$

2.202 We have,
$$T_{Cr} = \frac{8}{27}\frac{a}{bR} = \frac{8}{27} \times \frac{3.62}{0.043 \times 0.082} \approx 304 \text{ K}$$
$$\rho_{Cr} = \frac{M}{3b} = \frac{44}{3 \times 43}\text{g/cm}^3 = 0.34 \text{ g/cm}^3$$

2.203 The vessel is such that either vapor or liquid of mass m occupies it at critical point. Then its volume will be

$$V_{Cr} = \frac{m}{M}V_{MCr} = \frac{3}{8}\frac{RT_{Cr}}{p_{Cr}}\frac{m}{M}$$

The corresponding volume in liquid phase at room temperature is

$$V = \frac{m}{\rho}$$

(where ρ = density of liquid ether at room temperature).

Thus,
$$\eta = \frac{V}{V_{Cr}} = \frac{8Mp_{Cr}}{3RT_{Cr}\rho} \approx 0.254$$

(using the given data and ρ = 720 g/l).

2.204 We apply the following relation, at constant T to the cycle 1234531

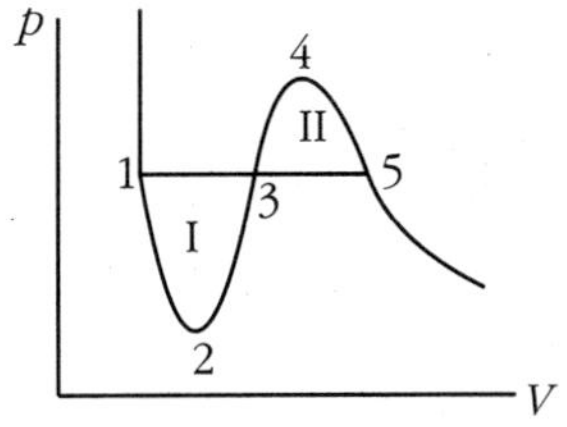

$$T\oint ds = \oint dU + \oint pdV$$

Here,
$$dS = \oint dU = 0$$

So,
$$\oint pdV = 0$$

This implies that the areas I and II are equal. This reasoning is inapplicable to the cycle 1231, for example. This cycle is irreversible because it involves the irreversible transition from a single phase to a two-phase state at the point 3.

2.205 When a portion of super-cooled water turns into ice some heat is liberated, which should heat it upto ice point. Neglecting the variation of specific heat of water, the fraction of water turning into ice is clearly

$$f = \frac{c|t|}{q} = 0.25$$

(where c = specific heat of water and q = latent heat of fusion of ice).

Clearly $f = 1$ at $t = -80°C$.

2.206 From the Clausius–Clapeyron (C–C) equation

$$\frac{dT}{dp} = \frac{T(V_2' - V_1')}{q_{12}}$$

Here, q_{12} is the specific latent heat absorbed in $1 \to 2$ and 1 = solid, 2 = liquid.

$$\Delta T = \frac{T(V_2' - V_1')}{q_{12}} \Delta p = -\frac{273 \times 0.091}{333} \times 1 \text{ atm cm}^3\text{ K/J}$$

$$1 \frac{\text{atm} \times \text{cm}^3}{\text{J}} = \frac{10^5\text{ N/m}^2 \times 10^{-6}\text{ m}^3}{\text{J}} = 10^{-1}\text{ Nm/J}, \ \Delta T = 0.0075\text{ K}$$

2.207 Here 1 = liquid, 2 = steam.

So,

$$\Delta T = \frac{T(V_2' - V_1')}{q_{12}} \Delta p$$

or

$$V_2' \approx \frac{q_{12}}{T}\frac{\Delta T}{\Delta p} = \frac{2250}{373} \times \frac{0.9}{3.2} \times 10^{-3}\text{ m}^3/\text{g} = 1.7\text{ m}^3/\text{kg}$$

2.208 From C–C equation

$$\frac{dp}{dT} = \frac{q_{12}}{T(V_2' - V_1')} = \frac{q_{12}}{TV_2'}$$

Assuming the saturated vapor to be ideal gas, we can write

$$\frac{1}{V_2'} = \frac{MP}{RT}$$

Thus,

$$\Delta p = \frac{Mq}{RT^2} p\Delta T$$

and

$$p = p_0\left(1 + \frac{Mq}{RT^2}\Delta T\right) \approx 1.04\text{ atm}$$

2.209 From C–C equation, neglecting the volume of the liquid

$$\frac{dp}{dT} \approx \frac{q_{12}}{TV_2'} \approx \frac{Mq}{RT^2} p \quad (\text{as } q = q_{12})$$

or

$$\frac{dp}{p} = \frac{Mq}{RT}\frac{dT}{T}$$

Now, $$pV = \frac{m}{M}RT \quad \text{or} \quad m = \frac{MpV}{RT} \quad \text{(for an ideal gas)}$$

So, $$\frac{dm}{m} = \frac{dp}{p} - \frac{dT}{T} \quad (V \text{ is constant} = \text{specific volume})$$

$$= \left(\frac{Mq}{RT} - 1\right)\frac{dT}{T} = \left(\frac{18 \times 2250}{373 - 1} - 1\right) \times \frac{1.5}{375} \approx 4.85\%$$

2.210 From *C-C* equation

$$\frac{dp}{dT} \approx \frac{q}{TV_2'} \approx \frac{Mq}{RT^2}p$$

Integrating $$\ln p = \text{constant} - \frac{Mq}{RT}$$

So, $$p = p_0 \exp\left[\frac{Mq}{R}\left(\frac{1}{T_0} - \frac{1}{T}\right)\right]$$

This is reasonable for $|T - T_0| << T_0$, and far below critical temperature.

2.211 As before from solution of Problem 2.206, the lowering of melting point is given by

$$\Delta T = -\frac{T\Delta V'}{q}p$$

The superheated ice will then melt in part. The fraction that will melt is

$$\eta = \frac{CT\Delta V'}{q^2}p = 0.03$$

where C is the specific heat capacity of ice, $T = 273$ K and q is the specific heat of melting.

2.212 (a) The equations of the transition lines are

$$\log p = 9.05 - \frac{1800}{T} \quad \text{(solid-gas)}$$

$$= 6.78 - \frac{1310}{T} \quad \text{(liquid-gas)}$$

At the triple point they intersect. Thus,

$$2.27 = \frac{490}{T_{tr}} \quad \text{or} \quad T_{tr} = \frac{490}{2.27} = 216 \text{ K}$$

The corresponding p_{tr} is 5.14 atm.

In the formula $\log p = a - b/T$, we compare b with the corresponding term in the equation in Problem 2.210. Then

$$\ln p = a \times 2.303 - \frac{2.303\, b}{T} \Rightarrow 2.303b = \frac{Mq}{R}$$

or $$q_{\text{solid-gas}} = \frac{2.303 \times 1800 \times 8.31}{44} = 783 \text{ J/g}$$

$$q_{\text{liquid-gas}} = \frac{2.303 \times 1310 \times 831}{44} = 570 \text{ J/g}.$$

Finally $$q_{\text{solid-liquid}} = 213 \text{ J/g (on subtraction)}$$

2.213
$$\Delta S = \int_{T_1}^{T_2} mc\frac{dT}{T} + \frac{mq}{T_2} \approx m\left(c\ln\frac{T_2}{T_1} + \frac{q}{T_2}\right)$$
$$= 10^3\left(4.18\ln\frac{373}{283} + \frac{2250}{373}\right) \approx 7.2 \text{ kJ/K}$$

2.214
$$\Delta S \approx \frac{q_m}{T_1} + c\ln\frac{T_2}{T} + \frac{q_v}{T_2}$$
$$= \frac{333}{273} + 4.18\ln\frac{373}{273} + \frac{2250}{373} \approx 8.56 \text{ J/K}$$

2.215 Specific heat of copper, $c = 0.39$ J/gK. Suppose all the ice does not melt, then

heat rejected $= 90 \times 0.39\ (90 - 0) = 3159$ J

and heat gained by ice $= 50 \times 2.09 \times 3 + x \times 333$

This gives, $x = 8.5$ g

Thc hypothesis is correct and final temperature will be $T = 273$ K.

Hence change in entropy of copper piece

$$\Delta S = mc\ln\frac{T}{T_1} = mc\ln\frac{273}{363} = -10 \text{ J/K}$$

2.216 (a) Here $t_2 = 60$°C. Suppose the final temperature is t°C. Then,

heat lost by water $= m_2 c\ (t_2 - t)$

heat gained by ice $= m_1 q_m + m_1 c\ (t - t_1)$ (if all the ice melts)

In this case, $m_1 q_m = m_2 \times 4.18\ (60 - t)$ (for $m_1 = m_2$)

So the final temperature will be 0°C and only some ice will melt.

Then $100 \times 4.18(60) = m_1' \times 333$

or $m_1' = 75.3\,\text{g} = \text{amount of ice that will melt}$

Finally
$$\Delta S = 75.3 \times \frac{333}{273} + 100 \times 4.18\ \ln\frac{273}{333}$$
$$\Delta S = \frac{m_1' q_m}{T_1} + m_2 c\ln\frac{T_1}{T_2}$$
$$= m_2 c\frac{(T_2 - T_1)}{T_1} - m_2 c\ln\frac{T_2}{T_1}$$
$$= m_2 c\left[\frac{T_2}{T_1} - 1 - \ln\frac{T_2}{T_1}\right] = 8.8 \text{ J/K}$$

(b) If $m_2 c t_2 > m_1 q_m$, then all the ice will melt as one can check and the final temperature can be obtained like this

$$m_2 c(T_2 - T) = m_1 q_m + m_1 c(T - T_1)$$
$$(m_2 T_2 + m_1 T_1)c - m_1 q_m = (m_1 + m_2)cT$$

or $$T = \frac{m_2T_2 + m_1T_1 - \frac{m_1q_m}{c}}{m_1 + m_2} \approx 280 \text{ K}$$

and $$\Delta S = \frac{m_1 q}{T_1} + c\left(m_1 \ln\frac{T}{T_1} - m_2 \ln\frac{T_2}{T}\right) = 19 \text{ J/K}$$

2.217 $$\Delta S = \frac{mq_2}{T_2} - mc\ln\frac{T_2}{T_1} + \frac{Mq_{\text{ice}}}{T_1}$$

where, $$mq_{\text{ice}} = m\left(q_2 + c\left(T_2 - T_1\right)\right)$$

So, $$\Delta S = mq_2\left(\frac{1}{T_1} - \frac{1}{T_2}\right) + mc\left(\frac{T_2}{T_1} - 1 - \ln\frac{T_2}{T_1}\right)$$
$$= 0.2245 + 0.2564 = 0.48 \text{ J/K}$$

2.218 When heat dQ is given to the vapor its temperature will change by dT, pressure by dp and volume by dV, it being assumed that the vapor remains saturated.

Then by *C–C* equation,

$$\frac{dp}{dT} = \frac{q}{TV'}(V'_{\text{vap}} >> V'_l) \quad \text{or} \quad dp = \frac{q}{TV'}dT$$

On the other hand, $$pV' = \frac{RT}{M}$$

So, $$pdV' + V'dp = \frac{RdT}{M}$$

Hence, $$pdV' = \left(\frac{R}{M} - \frac{q}{T}\right)dT$$

Finally $$dQ = CdT = dU + pdV'$$
$$= C_v dT + \left(\frac{R}{M} - \frac{q}{T}\right)dT = C_p dT - \frac{q}{T}dT$$

(C_p, C_V refer to unit mass here).

Thus, $$C = C_p - \frac{q}{T}$$

For water, $$C_p = \frac{R\gamma}{\gamma - 1}\cdot\frac{1}{M} \quad (\text{with } \gamma = 1.32 \text{ and } M = 18)$$

So, $$C_p = 1.90 \text{ J/g K}$$

and $$C = -4.13 \text{ J/g K} = -74 \text{ J/mol K}$$

2.219 The required entropy change can be calculated along a process in which the water is heated from T_1 to T_2 and then allowed to evaporate. The entropy change for this is

$$\Delta S = C_p \ln\frac{T_2}{T_1} + \frac{qM}{T_2}$$

where q = specific latent heat of vaporization.

2.7 Transport Phenomena

2.220 (a) The fraction of gas molecules which traverses distances exceeding the mean free path without collision is just the probability to traverse the distance $s = \lambda$ without collision.

Thus,
$$P = e^{-1} = \frac{1}{e} = 0.37$$

Hence the sought fraction $\eta = 0.37$.

(b) This probability is
$$P = e^{-1} - e^{-2} = 0.23$$

Hence the sought fraction $\eta = 0.23$.

2.221 From the formula $\dfrac{1}{\eta} = e^{\Delta l/\lambda}$ or $\lambda = \dfrac{\Delta l}{\ln \eta}$.

2.222 (a) Let $P(t)$ = probability of no collision in the interval $(0, t)$. Then
$$P(t + dt) = P(t)\,(1 - \alpha dt)$$

or
$$\frac{dP}{dt} = -\alpha P(t) \quad \text{or} \quad P(t) = e^{-\alpha t}$$

where we have used $P(0) = 1$.

(b) The mean interval between collision is also the mean interval of no collision. Then,

$$<t> = \frac{\int_0^\infty t e^{-\alpha t}\,dt}{\int_0^\infty e^{-\alpha t}\,dt} = \frac{1}{\alpha}\frac{\Gamma(2)}{\Gamma(1)} = \frac{1}{\alpha}$$

2.223 (a)
$$\lambda = \frac{1}{\sqrt{2}\pi d^2 n} = \frac{kT}{\sqrt{2}\pi d^2 p}$$

$$= \frac{1.38\times 10^{-23}\times 273}{\sqrt{2}\pi\,(0.37\times 10^{-9})^2\times 10^5} = 6.2\times 10^{-8}\,\text{m}$$

$$\tau = \frac{\lambda}{<v>} = \frac{6.2\times 10^{-8}}{454}\,s = 0.136\,\text{ns}$$

(b)
$$\lambda = 6.2\times 10^{6}\ \text{m}$$
$$\tau = 1.36\times 10^{4}\text{s} = 3.8\ \text{h}$$

2.224 The mean distance between molecules is of the order

$$\left(\frac{22.4\times 10^{-3}}{6.0\times 10^{23}}\right)^{1/3} = \left(\frac{224}{6}\right)^{1/3}\times 10^{-9}\ \text{m} \approx 3.34\times 10^{-9}\ \text{m}$$

This is about 18.5 times smaller than the mean free path calculated in Problem 2.223(a) above.

2.225 We know that the Van der Waal's constant b is four times the molecular volume. Thus,

$$b = 4N_A \frac{\pi}{6} d^3 \quad \text{or} \quad d = \left(\frac{3d}{2\pi N_A}\right)^{1/3}$$

Hence, $$\lambda = \left(\frac{kT_0}{\sqrt{2}\pi P_0}\right)\left(\frac{2\pi N_A}{3b}\right)^{2/3} = 84 \text{ nm (on substituting values)}$$

2.226 The velocity of sound in N_2 is

$$\sqrt{\frac{\gamma p}{\rho}} = \sqrt{\frac{\gamma RT}{M}}$$

So, $$\frac{1}{\nu} = \sqrt{\frac{\gamma RT_0}{M}} = \frac{RT_0}{\sqrt{2}\pi d^2 p_0 N_A}$$

$$= 5.5 \text{ GHz (on substituting values)}$$

2.227 (a) $\lambda > l$ if $p < \dfrac{kT}{\sqrt{2}\pi d^2 l}$

≈ 0.7 Pa for O_2

(b) The corresponding molecular concentration n is obtained by dividing by kT and is 1.84×10^{20} per $m^3 = 1.84 \times 10^{14}$ per cm^3 and the corresponding mean distance is given by,

$$\frac{l}{n^{1/3}} = \frac{10^{-2}}{(0.184)^{1/3} \times 10^5} = 1.8 \times 10^{-7}\,\text{m} \approx 0.18\,\mu\text{m}$$

2.228 (a) $$\nu = \frac{1}{\tau} = \frac{1}{\lambda/<v>} = \frac{<v>}{\lambda}$$

$$= \sqrt{2}\pi d^2\, n <v> = 0.74 \times 10^{10}\ \text{s}^{-1}$$

where $$n = \frac{P_0}{kT_0} \quad \text{and} \quad <v> = \sqrt{\frac{8RT}{\pi M}}$$

(b) Total number of collisions is

$$\frac{1}{2} n\nu = 1.0 \times 10^{29}\ \text{s}^{-1}\ \text{cm}^{-3}$$

Note, the factor 1/2, which come because when two molecules collide we must not count it twice.

2.229 (a) $$\lambda = \frac{1}{\sqrt{2}\pi d^2 n}$$

d is a constant and n is a constant for an isochoric process, so λ is constant for an isochoric process.

$$\nu = \frac{<v>}{2} = \sqrt{\frac{8RT}{M\pi}} \Big/ \lambda$$

i.e., $$\nu \propto \sqrt{T}$$

(b)
$$\lambda = \frac{1}{\sqrt{2}\pi d^2}\frac{kT}{P} \propto T \quad \text{(for an isobaric process)}$$

$$\nu = \frac{<v>}{\lambda} \propto \frac{\sqrt{T}}{T} = \frac{1}{\sqrt{T}} \quad \text{(for an isobaric process)}$$

2.230 (a) In an isochoric process λ is constant and

$$\nu \propto \sqrt{T} \propto \sqrt{pV} \propto \sqrt{p} \propto \sqrt{n}$$

so ν increases n times

(b) $\lambda = \dfrac{kT}{\sqrt{2}\pi d^2 p}$ must decrease n times in an isothermal process and ν must increase n times because $<v>$ is constant in an isothermal process.

2.231 (a)
$$\lambda \propto \frac{1}{n} \geq \frac{1}{N/V} = \frac{V}{N}$$

Thus,
$$\lambda \propto V \quad \text{and} \quad \nu \propto \frac{T^{1/2}}{V}$$

But in an adiabatic process $TV^{\gamma-1}$ = constant, so $TV^{2/5}$ = constant (as $\gamma = 7/5$ here).

or
$$T^{1/2} \propto V^{-1/5}$$

Thus,
$$\nu \propto V^{-6/5}$$

(b) Given
$$\lambda \propto \frac{T}{P}$$

But,
$$p\left(\frac{T}{p}\right)^{\gamma} = \text{constant}$$

or
$$\frac{T}{p} \propto p^{-1/\gamma} \quad \text{or} \quad T \propto p^{1-1/\gamma}$$

Thus,
$$\lambda \propto p^{-1/\gamma} = p^{-5/7}$$

when,
$$\nu = \frac{<v>}{\lambda} \propto \frac{p}{\sqrt{T}} \propto p^{1/2+1/2\gamma} = p^{\gamma+1/2\gamma} = p^{6/7}$$

(c) Given
$$\lambda \propto V$$

But,
$$TV^{2/5} = \text{constant} \quad \text{or} \quad V \propto T^{-5/2}$$

Thus,
$$\lambda \propto T^{-5/2}$$

$$\nu \propto \frac{T^{1/2}}{V} \propto T^3$$

2.232 In the polytropic process of index n, pV^n = constant, TV^{n-1} = constant and $p^{1-n}\,T^n$ = constant.

(a) When $\lambda \propto V$,

$$\nu \propto \frac{T^{1/2}}{V} \propto V^{\frac{1-n}{2}} V^{-1} = V^{-\frac{n+1}{2}}$$

(b) When $\lambda \propto \dfrac{T}{p}$,

$$T^n \propto p^{n-1} \quad \text{or} \quad T \propto p^{1-1/n}$$

So,

$$\lambda \propto p^{-1/n}$$

$$\nu = \frac{<v>}{\lambda} \propto \frac{p}{\sqrt{T}} \propto p^{1-1/2+1/2n} = p^{(n+1)/2n}$$

(c) When $\lambda \propto \dfrac{T}{p}$,

$$p \propto T^{n/n-1}$$

$$\lambda \propto T^{1-n/n-1} = T^{-1/n-1} = T^{1/1-n}$$

$$\nu \propto \frac{p}{\sqrt{T}} \propto T^{(n/n-1)-(1/2)} = T^{(n+1)/2(n-1)}$$

2.233 (a) The number of collision between the molecules in a unit volume is

$$\frac{1}{2} n\nu = \frac{1}{\sqrt{2}} \pi d^2 \, n^2 <v> \propto \frac{\sqrt{T}}{V^2}$$

This remains constant is polytropic process, pV^{-3} = constant.

Using solution of Problem 2.122, the molar specific heat for the polytropic process, pV^{α} = constant, is

$$C = R\left(\frac{1}{\gamma-1} - \frac{1}{\alpha-1}\right)$$

Thus,

$$C = R\left(\frac{1}{\gamma-1} + \frac{1}{4}\right) = R\left(\frac{5}{2} + \frac{1}{4}\right) = \frac{11}{4} R$$

It can also be written as $\dfrac{1}{4} R(1+2i)$ (where $i = 5$)

On substituting values $C = 23$ J/K mol.

(b) In this case $\dfrac{\sqrt{T}}{V}$ = constant and so pV^{-1} = constant.

So,

$$C = R\left(\frac{1}{\gamma-1} + \frac{1}{2}\right) = R\left(\frac{5}{2} + \frac{1}{2}\right) = 3R$$

which can also be written as $\dfrac{R}{2}(i+1)$ (where $i = 5$).

On substituting values $C = 29$ J/K mol.

2.234 We can assume that all molecules, incident on the hole, leak out. Then,

$$-dN = -d(nV) = \frac{1}{4} n <v> S dt$$

or

$$dn = -n \frac{dt}{4V/S<v>} = -n \frac{dt}{\tau}$$

Integrating, $$n = n_0\, e^{-t/\tau}$$

Hence, $$<v> = \sqrt{\frac{8RT}{\pi M}}$$

2.235 If the temperature of the compartment 2 is η times more than that of compartment 1, it must contain $1/\eta$ times less number of molecules since pressure must be the same when the big hole is open. If M = mass of the gas in 1 then the mass of the gas in 2 must be M/η. So immediately after the big hole is closed,

$$n_1^0 = \frac{M}{mV} \quad \text{and} \quad n_2^0 = \frac{M}{mV\eta}$$

where m = mass of each molecule and n_1^0, n_2^0 are concentrations in 1 and 2. After the big hole is closed the pressures will differ and concentration will become n_1 and n_2, where

$$n_1 + n_2 = \frac{M}{mV\eta}(1+\eta)$$

On the other hand

$$n_1 <v_1> = n_2 <v_2>, \quad \text{i.e.,} \quad n_1 = \sqrt{\eta}\, n_2$$

Thus, $$n_2(1+\sqrt{\eta}) = \frac{m}{mV\eta}(1+\eta) = n_2^0(1+\eta)$$

So, $$n_2 = n_2^0 \frac{1+\eta}{1+\sqrt{\eta}}$$

2.236 We know

$$\eta = \frac{1}{3} <v>\lambda\rho = \frac{1}{3} <v> \frac{1}{\sqrt{2}\pi d^2} m\alpha \sqrt{T}$$

Thus η changing α times implies T changing α^2 times.

On the other hand

$$D = \frac{1}{3} <v>\lambda = \frac{1}{3}\sqrt{\frac{8kT}{\pi m}} \frac{kT}{\sqrt{2}\pi d^2 p}$$

Thus D changing β times means $\frac{T^{3/2}}{p}$ changing β times.

So p must change $\frac{\alpha^3}{\beta}$ times = 2 times (on substituting values).

2.237 $D \propto \frac{\sqrt{T}}{n} \propto V\sqrt{T}$ and $\eta \propto \sqrt{T}$

(a) D will increase n times, and η will remain constant if T is constant.

(b) $D \propto \frac{T^{3/2}}{p} \propto \frac{(pV)^{3/2}}{p} = p^{1/2}\, V^{3/2}$

Also, $$\eta \propto \sqrt{pV}$$

Thus D will increase $n^{3/2}$ times, η will increase $n^{1/2}$ times, if p is constant.

2.238 $D \propto V\sqrt{T}$ and $\eta \propto \sqrt{T}$

In an adiabatic process

$$TV^{\gamma-1} = \text{constant} \quad \text{or} \quad T \propto V^{1-\gamma}$$

Now V is decreased $1/n$ times.

Thus, $$D \propto V^{\frac{3-\gamma}{2}} = \left(\frac{1}{n}\right)^{\frac{3-\gamma}{2}} = \left(\frac{1}{n}\right)^{4/5}$$

$$\eta \propto V^{\frac{1-\gamma}{2}} = \left(\frac{1}{n}\right)^{-1/5} = n^{1/5}$$

So D decreases $n^{4/5}$ times and η increase $n^{1/5}$ times $\approx$ 6.3 and 1.6 times, respectively.

2.239 (a) $D \propto V\sqrt{T} \propto \sqrt{pV^3}$

Thus D remains constant in the process pV^3 = constant. So polytropic index of the process $n = 3$.

(b) $\eta \propto \sqrt{T} \propto \sqrt{pV}$

So η remains constant in the isothermal process. pV = constant, $n = 1$, here.

(c) Heat conductivity $\kappa = \eta C_V$ and C_V is a constant for the ideal gas.

Thus $n = 1$ here also.

2.240 $$\eta = \frac{1}{3}\sqrt{\frac{8kT}{\pi m}}\frac{m}{\sqrt{2}\pi d^2} = \frac{2}{3}\sqrt{\frac{mkT}{\pi^3}}\frac{1}{d^2}$$

or $$d = \left(\frac{2}{3\eta}\right)^{1/2}\left(\frac{mkT}{\pi^3}\right)^{1/4} = \left(\frac{2}{3\times 18.9}\times 10^6\right)^{1/2}\left(\frac{4\times 8.31\times 273\times 10^{-3}}{\pi^3\times 36\times 10^{46}}\right)^{1/4}$$

$$= 10^{-10}\left(\frac{2}{3\times 18.9}\right)^{1/2}\left(\frac{4\times 83.1\times 273}{\pi^3\times 0.36}\right)^{1/4} \approx 0.178\,\text{nm}$$

2.241
$$\kappa = \frac{1}{3}\langle v\rangle\lambda\rho c_V$$
$$= \frac{1}{3}\sqrt{\frac{8kT}{m\pi}}\frac{1}{\sqrt{2}\pi d^2 n}mn\frac{C_V}{M}$$

c_V is the specific heat capacity which is C_V/M. Now C_V is the same for all monoatomic gases such as He and A.

Thus,
$$\kappa \propto \frac{1}{\sqrt{M}d^2}$$

or
$$\frac{\kappa_{\text{He}}}{\kappa_A} = 8.7 = \frac{\sqrt{M_A}\,d_A^2}{\sqrt{M_{\text{He}}}\,d_{\text{He}}^2} = \sqrt{10}\,\frac{d_A^2}{d_{\text{He}}^2}$$

$$\frac{d_A}{d_{\text{He}}} = \sqrt{\frac{8.7}{\sqrt{10}}} = 1.658 \approx 1.7$$

2.242 In this case

$$N_1\frac{r_2^2 - r_1^2}{r_1^2 r_2^2} = 4\pi\eta\omega$$

or
$$N_1\frac{2R\Delta R}{R^4} \approx 4\pi\eta\omega \quad \text{or} \quad N_1 = \frac{2\pi\eta\omega R^3}{\Delta R}$$

To decrease N_1 n times, η must be decreased n times. Now η does not depend on pressure until the pressure is so low that the mean free path equals, say, 1/2 ΔR. Then the mean free path is fixed and η decreases with pressure. The mean free path equals 1/2 ΔR, when

$$\frac{1}{\sqrt{2}\pi d^2 n_0} = \Delta R \quad (\text{where } n_0 = \text{concentration})$$

Corresponding pressure is

$$p_0 = \frac{\sqrt{2}kT}{\pi d^2 \Delta R}$$

The sought pressure is n times less, given by

$$p = \frac{\sqrt{2}kT}{\pi d^2 n\Delta R} = 70.7 \times \frac{10^{-23}}{10^{-18} \times 10^{-3}} \approx 0.71 \text{ Pa}$$

The answer is qualitative and depends on the choice 1/2 ΔR for the mean free path.

2.243 We neglect the moment of inertia of the gas in a shell. Then the moment of friction forces on a unit length of the cylinder must be a constant as a function of r.

So,
$$2\pi r^3\eta\frac{d\omega}{dr} = N_1 \quad \text{or} \quad \omega(r) = \frac{N_1}{4\pi\eta}\left(\frac{1}{r_1^2} - \frac{1}{r_2^2}\right)$$

and $$\omega = \frac{N_1}{4\pi\eta}\left(\frac{1}{r_1^2} - \frac{1}{r_2^2}\right) \quad \text{or} \quad \eta = \frac{N_1}{4\pi\omega}\left(\frac{1}{r_1^2} - \frac{1}{r_2^2}\right)$$

2.244 We consider two adjoining layers. The angular velocity gradient is ω/h. So the moment of the frictional force is

$$N = \int_0^a r 2\pi r dr \eta r \frac{\omega}{h} = \frac{\pi\eta a^4\omega}{2h}$$

2.245 In the ultra-rarefied gas we must determine η by taking $\lambda = (1/2)h$.

Then, $$\eta = \frac{1}{3}\sqrt{\frac{8kT}{m\pi}} \times \frac{1}{2}h \times \frac{mp}{kT} = \frac{1}{3}\sqrt{\frac{2M}{\pi RT}}hp$$

So, $$N = \frac{1}{3}\omega a^4 p\sqrt{\frac{\pi M}{2RT}}$$

2.246 Take an infinitesimal section of length dx and apply Poiseuille's equation to this.

Then, $$\frac{dV}{dt} = \frac{-\pi a^4}{8\eta}\frac{\partial p}{\partial x}$$

From the formula, $$pV = RT \cdot \frac{m}{M}$$

$$pdV = \frac{RT}{M}dm$$

or $$\frac{dm}{dt} = \mu = -\frac{\pi a^4 M}{8\eta RT}\frac{pdp}{dx}$$

This equation implies that if the flow is isothermal, then

$$p\frac{dp}{dx} = \frac{|p_2^2 - p_1^2|}{2l} = \text{constant}$$

Thus, $$\mu = \frac{\pi a^4 M}{16\eta RT}\frac{|p_2^2 - p_1^2|}{l}$$

2.247 Let T be the temperature of the interface. Then heat flowing from left is equal to the heat flowing into right in equilibrium.

Thus, $$\kappa_1 = \frac{T_1 - T}{l_1} = \kappa_2 = \frac{T - T_2}{l_2} \quad \text{or} \quad T = \frac{(\kappa_1 T_1)/l_1 + (\kappa_2 T_2)/l_2}{\kappa_1/l_1 + \kappa_2/l_2}$$

2.248 We have

$$\kappa_1 = \frac{T_1 - T}{l_1} = \kappa_2\frac{T - T_2}{l_2} = \kappa\frac{T_1 - T_2}{l_1 - l_2}$$

or using the previous result

$$\frac{\kappa_1}{l_1}\left(T_1 - \frac{\dfrac{\kappa_1 T_1}{l_1} + \dfrac{\kappa_2 T_2}{l_2}}{\dfrac{\kappa_1}{l_1} + \dfrac{\kappa_2}{l_2}}\right) = \kappa \frac{T_1 - T_2}{l_1 + l_2}$$

or

$$\frac{\kappa_1}{l_1} \frac{\dfrac{\kappa_2}{l_2}(T_1 - \mathrm{T}_2)}{\dfrac{\kappa_1}{l} + \dfrac{\kappa_2}{l_2}} = \kappa \frac{T_1 - T_2}{l_1 + l_2} \Rightarrow \kappa = \frac{l_1 + l_2}{\dfrac{l_1}{\kappa_1} + \dfrac{l_2}{\kappa_2}}$$

2.249 By definition the heat flux per unit area is

$$\dot{Q} = -K\frac{dT}{dx} = -\alpha\frac{d}{dx}\ln T = \text{constant} = +\frac{\alpha}{l}\ln T_1/T_2$$

Integrating $$\ln T = \frac{x}{l}\ln\frac{T_2}{T_1} + \ln T_1$$

(where T_1 = temperature at the end $x = 0$)

So, $$T = T_1\left(\frac{T_2}{T_1}\right)^{x/l} \quad \text{and} \quad \dot{Q} = \frac{\alpha}{l}\ln T_1/T_2$$

2.250 Suppose the chunks have temperatures T_1, T_2 at time t and $T_1 - dT_1$, $T_2 + dT_2$ at time $dt + t$.

Then, $$C_1 dT_1 = C_2 dT_2 = \frac{kS}{l}(T_1 - \mathrm{T}_2)dt$$

Thus, $$d\Delta T = -\frac{\kappa S}{l}\left(\frac{1}{C_1} + \frac{1}{C_2}\right)\Delta T\, dt \,(\text{where } \Delta T = T_1 - T_2)$$

Hence, $$\Delta T = (\Delta T)_0 e^{-t/\gamma}\left(\text{where } \frac{1}{\tau} = \frac{\kappa \cdot S}{l}\left(\frac{1}{C_1} + \frac{1}{C_2}\right)\right)$$

2.251 $$\dot{Q} = \kappa\frac{\partial T}{\partial x} = -A\sqrt{T}\frac{\partial T}{\partial x}(\text{where } A = \text{constant})$$

$$= -\frac{2}{3}A\frac{\partial T^{3/2}}{\partial x}$$

$$= \frac{2}{3}A\frac{(T_1^{3/2} - T_2^{3/2})}{l}$$

Thus, $$T^{3/2} = \text{constant} - \frac{x}{l}(T_1^{3/2} - T_2^{3/2})$$

Using $T = T_1$, at $x = 0$, we get

$$T^{3/2} = T_1^{3/2} + \frac{x}{l}\left(T_2^{3/2} - T_1^{3/2}\right) \quad \text{or} \quad \left(\frac{T}{T_1}\right)^{3/2} = 1 + \frac{x}{l}\left(\left(\frac{T_2}{T}\right)^{3/2} - 1\right)$$

$$T = T_1\left[1 + \frac{x}{l}\left\{\left(\frac{T_2}{T_1}\right)^{3/2} - 1\right\}\right]^{2/3}$$

where x is the distance from the plate maintained at the temperature T_1.

2.252 $\kappa = \frac{1}{3}\sqrt{\frac{8RT}{\pi M}} - \frac{1}{\sqrt{2}\pi d^2 n} mn \frac{R(i/2)}{M} = \frac{R^{3/2} i T^{3/2}}{3\pi^{3/2} d^2 \sqrt{M} N_A}$

Then from the previous problem

$$q = \frac{2iR^{3/2}\left(T_2^{3/2} - T_1^{3/2}\right)}{9\pi^{3/2} d^2 \sqrt{M} N_A l},$$

$i = 3$ here and d is the effective diameter of helium atom.

2.253 At this pressure and average temperature (= 27°C = 300 K)

$$T = \frac{(T_1 + T_2)}{2}$$

$$\lambda = \frac{1}{\sqrt{2}\pi d^2} \frac{\kappa T}{p} = 2330 \times 10^{-5} \text{ m} = 23.3 \text{ mm} \gg 5.0 \text{ mm} = l$$

The gas is ultra-thin and we write $\lambda = (1/2)\, l$ here.

Then, $$q = \kappa \frac{dT}{dx} = \kappa \frac{T_2 - T_1}{l}$$

where, $$\kappa = \frac{1}{3} <v> \times \frac{1}{2} l \times \frac{MP}{RT} \times \frac{R}{\gamma - 1} \times \frac{1}{M} = \frac{p<v>}{6T(\gamma - 1)} l$$

and $$q = \frac{p<v>}{6T(\gamma - 1)}(T_2 - T_1) = 22 \text{ W/m}^2 \text{ (on subituting values)}$$

where, $$<v> = \sqrt{\frac{8RT}{M\pi}}.$$

We have used $T_2 - T_1 \ll \frac{T_2 + T_1}{2}$ here.

2.254 At equilibrium, $2\pi\gamma\kappa \frac{dT}{dr} = -A = \text{constant.}$

So, $$T = B - \frac{A}{2\pi\kappa} \ln r$$

But $T = T_1$, when $r = R_1$ and $T = T_2$, when $r = R_2$.

From this we find $T = T_1 + \frac{T_2 - T_1}{\ln R_2/R_1} \ln \frac{r}{R_1}$

2.255 At equilibrium $4\pi r^2 \kappa \frac{dT}{dr} = -A = \text{constant}$

$$T = B + \frac{A}{4\pi\kappa} \frac{1}{r}$$

Using $T = T_1$ when $r = R_1$ and $T = T_2$ when $r = R_2$,

$$T = T_1 + \frac{T_2 - T_1}{(1/R_2) - (1/R_1)} \left(\frac{1}{r} - \frac{1}{R_1} \right)$$

2.256 The heat flux vector is $-\kappa \nabla T$ and its divergence equals w. Thus,

$$\nabla^2 T = -\frac{w}{\kappa}$$

or $$\frac{1}{r}\left(\frac{\partial}{\partial r}\right)\left(r\frac{\partial T}{\partial r}\right) = -\frac{w}{\kappa} \text{ in cylindrical coordinates.}$$

or $$T = B + A \ln r - \frac{w}{4\kappa} r^2$$

Since T is finite at $r = 0$, $A = 0$. Also $T = T_0$ at $r = R$

So, $$B = T_0 + \frac{w}{4\kappa} R^2$$

Thus, $$T = T_0 + \frac{w}{4\kappa}(R^2 - r^2)$$

where r is the distance from the axis of the wire also called axial radius.

2.257 Here again

$$\nabla^2 T = -\frac{w}{\kappa}$$

So in spherical polar coordinates,

$$\frac{1}{r^2}\frac{\partial}{\partial r}\left(r^2 \frac{\partial T}{\partial r}\right) = -\frac{w}{\kappa} \quad \text{or} \quad r^2 \frac{\partial T}{\partial r} = -\frac{w}{3\kappa} r^3 + A$$

or $$T = B - \frac{A}{r} - \frac{w}{6\kappa} r^2$$

Again $$A = 0 \quad \text{and} \quad B = T_0 + \frac{w}{6\kappa} R^2$$

So finally, $$T = T_0 + \frac{w}{6\kappa} \times (R^2 - r^2)$$

ELECTRODYNAMICS 3

3.1 Constant Electric Field in Vacuum

3.1 F_{ele} (for electrons) $= \dfrac{q^2}{4\pi\varepsilon_0 r^2}$ and $F_{gr} = \dfrac{\gamma m^2}{r^2}$

Thus, $$\frac{F_{ele}}{F_{gr}} \text{ (for electrons)} = \frac{q^2}{4\pi\varepsilon_0 \gamma m^2}$$

$$= \frac{(1.602\times 10^{-19}\text{ C})^2}{\left(\dfrac{1}{9\times 10^9}\right)\times 6.67\times 10^{-11}\text{ m}^3/(\text{kg}\cdot\text{s}^2)\times(9.11\times 10^{-31}\text{ kg})^2} = 4\times 10^{42}$$

Similarly $$\frac{F_{ele}}{F_{gr}} \text{ (for proton)} = \frac{q^2}{4\pi\varepsilon_0 \gamma m^2}$$

$$= \frac{(1.602\times 10^{-19}\text{ C})^2}{\left(\dfrac{1}{9\times 10^9}\right)\times 6.67\times 10^{-11}\text{ m}^3/(\text{kg}\cdot\text{s}^2)\times(1.672\times 10^{-27}\text{ kg})^2} = 1\times 10^{36}$$

For $F_{ele} = F_{gr}$,

$$\frac{q^2}{4\pi\varepsilon_0 r^2} = \frac{\gamma m^2}{r^2} \quad \text{or} \quad \frac{q}{m} = \sqrt{4\pi\varepsilon_0\gamma}$$

$$= \sqrt{\frac{6.67\times 10^{-11}\text{ m}^3(\text{kg}\cdot\text{s}^2)}{9\times 10^9}} = 0.86\times 10^{-10}\text{ C/kg}$$

3.2 Total number of atoms in the sphere of mass 1 gram $= \dfrac{1}{63.54}\times 6.023\times 10^{23}$.

So the total nuclear charge $\lambda = \dfrac{6.023\times 10^{23}}{63.54}\times 1.6\times 10^{-19}\times 29$.

Now the charge on the sphere = total nuclear charge – total electronic charge

$$= \frac{6.023\times 10^{23}}{63.54}\times 1.6\times 10^{-19}\times 29\left(1-\frac{1}{100}\right) = 4.298\times 10^2\text{ C}$$

Hence, force of interaction between these two spheres,

$$F = \frac{1}{4\pi\varepsilon_0}\cdot\frac{[4.398\times 10^2]^2}{1^2}\text{ N} = 9\times 10^9\times 10^4\times 19.348\text{ N} = 1.74\times 10^{15}\text{ N}$$

3.3 Let the balls be deviated by an angle θ from the vertical when separation between them equals x.

Applying Newton's second law of motion for any one of the spheres, we get

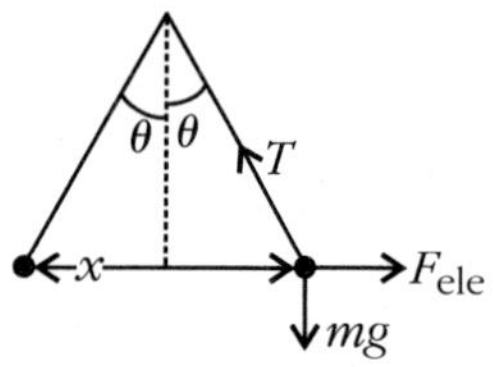

$$T\cos\theta = mg \quad (1)$$

and

$$T\sin\theta = F_{ele} \quad (2)$$

From the Eqs. (1) and (2)

$$\tan\theta = \frac{F_{ele}}{mg} \quad (3)$$

But from the figure

$$\tan\theta = \frac{x}{2\sqrt{l^2 - \left(\frac{x}{2}\right)^2}} \cong \frac{x}{2l} \text{ as } x \ll l \quad (4)$$

From Eqs. (3) and (4)

$$F_{ele} = \frac{mg\,x}{2l} \quad \text{or} \quad \frac{q^2}{4\pi\varepsilon_0 x^2} = \frac{mg\,x}{2l}$$

Thus

$$q^2 = \frac{2\pi\varepsilon_0 mgx^3}{l} \quad (5)$$

Differentiating Eq. (5) with respect to time

$$2q\frac{dq}{dt} = \frac{2\pi\varepsilon_0 mg}{l}3x^2\frac{dx}{dt}$$

According to the problem $\frac{dx}{dt} = v = \frac{a}{\sqrt{x}}$ $\left(\text{approach velocity is } \frac{dx}{dt}\right)$

so,

$$\left(\frac{2\pi\varepsilon_0 mg}{l}x^3\right)^{1/2}\frac{dq}{dt} = \frac{3\pi\varepsilon_0 mg}{l}x^2\frac{a}{\sqrt{x}}$$

Hence,

$$\frac{dq}{dt} = \frac{3}{2}a\sqrt{\frac{2\pi\varepsilon_0 mg}{l}}$$

Note: *If the plane of figure is x-y plane and point of suspension is O, Eq. (3) can be obtained more easily using net torque equal to zero about OZ-axis, i.e.,* $\tau_{0Z} = 0$.

3.4 Let us choose coordinate axes as shown in the figure and fix three charges, q_1, q_2 and q_3 having position vectors $\mathbf{r}_1$, $\mathbf{r}_2$ and $\mathbf{r}_3$, respectively.

Now, for the equilibrium of q_3

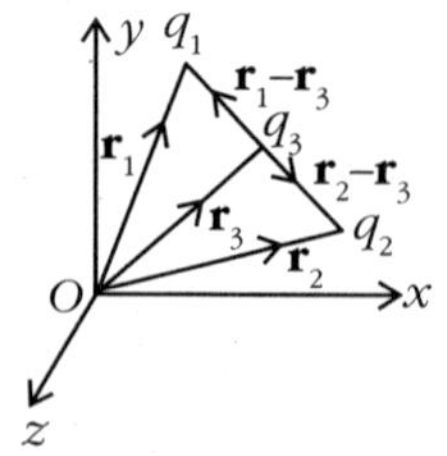

$$\frac{q_2q_3(\mathbf{r}_2-\mathbf{r}_3)}{|\mathbf{r}_2-\mathbf{r}_3|^3} + \frac{q_1q_3(\mathbf{r}_1-\mathbf{r}_3)}{|\mathbf{r}_1-\mathbf{r}_3|^3} = 0$$

or

$$\frac{q_2}{|\mathbf{r}_2-\mathbf{r}_3|^2} = \frac{q_1}{|\mathbf{r}_1-\mathbf{r}_3|^2}$$

because

$$\frac{\mathbf{r}_2-\mathbf{r}_3}{|\mathbf{r}_2-\mathbf{r}_3|} = -\frac{\mathbf{r}_1-\mathbf{r}_3}{|\mathbf{r}_1-\mathbf{r}_3|}$$

So,

$$\sqrt{q_2}(\mathbf{r}_1-\mathbf{r}_3) = \sqrt{q_1}(\mathbf{r}_3-\mathbf{r}_2)$$

or

$$\mathbf{r}_3 = \frac{\sqrt{q_2}\mathbf{r}_1 + \sqrt{q_1}\mathbf{r}_2}{\sqrt{q_1}+\sqrt{q_2}}$$

Also for the equilibrium of q_1

$$\frac{q_3(\mathbf{r}_3-\mathbf{r}_1)}{|\mathbf{r}_3-\mathbf{r}_1|^3}+\frac{q_2(\mathbf{r}_2-\mathbf{r}_1)}{|\mathbf{r}_2-\mathbf{r}_1|^3}=0$$

or

$$q_3=\frac{-q_2}{|\mathbf{r}_2-\mathbf{r}_1|^2}|\mathbf{r}_1-\mathbf{r}_3|^2$$

Substituting the value of $\mathbf{r}_3$, we get

$$q_3=\frac{-q_1q_2}{\left(\sqrt{q_1}+\sqrt{q_2}\right)^2}$$

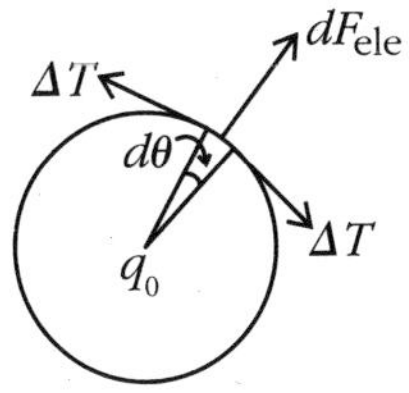

3.5 When the charge q_0 is placed at the center of the ring, the wire gets stretched. The component of extra tension ΔT towards the center will balance the electric force dF_{ele} due to the charge q_0 on a differential part of the ring which subtends angle $d\theta$ at the center, having the charge

$$dq=\left(\frac{q}{2\pi}\right)d\theta$$

$$2\Delta T\sin\frac{d\theta}{2}=dF_{ele}$$

As $d\theta$ is very small

$$2\Delta T\sin\frac{d\theta}{2}\approx Td\theta$$

Using

$$dF_{ele}=\frac{q_0dq}{4\pi\varepsilon_0r^2}=\frac{1}{4\pi\varepsilon_0}\frac{q_0}{r^2}\left(\frac{q}{2\pi}\right)d\theta$$

$$\Delta\,Td\,\theta=\frac{1}{4\pi\varepsilon_0}\frac{q_0}{r^2}\left(\frac{q}{2\pi}\right)d\theta$$

Hence,

$$\Delta\,T=\frac{qq_0}{8\pi^2\varepsilon_0r^2}$$

3.6 Sought field strength $\mathbf{E}=\frac{1}{4\pi\varepsilon_0}\frac{q}{|(\mathbf{r}-\mathbf{r}_0)|^3}\cdot(\mathbf{r}-\mathbf{r}_0)=2.7\mathbf{i}-3.6\mathbf{j}$ kV/m

So,

$$|\mathbf{E}|=\frac{1}{4\pi\varepsilon_0}\frac{q}{|\mathbf{r}-\mathbf{r}_0|^2}$$

$= 4.5$ kV/m (on substituting values)

3.7 Electric field strength due to a point charge (q) at a field point (P) in vacuum is given by

$$E=\frac{1}{4\pi\varepsilon_0}\frac{q\mathbf{r}}{r^3}$$

$$=\frac{1}{4\pi\varepsilon_0}\frac{q}{|(\mathbf{r}-\mathbf{r}_0)|^3}$$

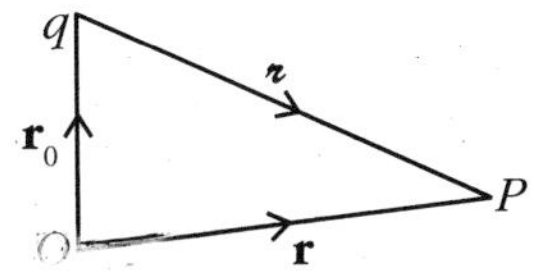

Taking into account the expression for E, let us fix the coordinate system by taking the point of intersection of the diagonals as the origin and let $\mathbf{k}$ be the unit vector directed normally, emerging from the plane of figure.

Hence the sought field strength:

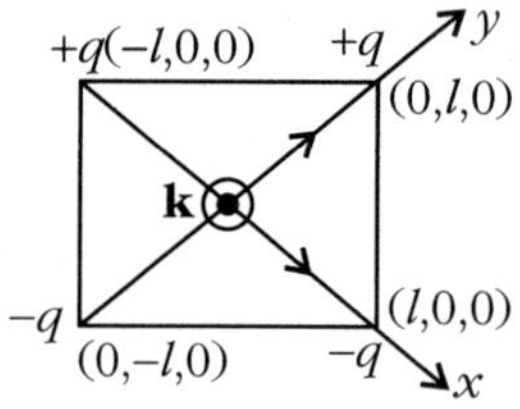

$$\mathbf{E} = \frac{q}{4\pi\varepsilon_0}\frac{l\mathbf{i} + x\mathbf{k}}{(l^2 + x^2)^{3/2}} + \frac{-q}{4\pi\varepsilon_0}\frac{l(-\mathbf{i}) + x\mathbf{k}}{(l^2 + x^2)^{3/2}} + \frac{-q}{4\pi\varepsilon_0}\frac{l\mathbf{j} + x\mathbf{k}}{(l^2 + x^2)^{3/2}} + \frac{q}{4\pi\varepsilon_0}\frac{l(-\mathbf{j}) + x\mathbf{k}}{(l^2 + x^2)^{3/2}}$$

$$= \frac{q}{4\pi\varepsilon_0(l^2 + x^2)^{3/2}}[2\,l\mathbf{i} - 2l\mathbf{j}]$$

Thus,

$$E = \frac{ql}{\sqrt{2}\pi\varepsilon_0\,(l^2 + x^2)^{3/2}}$$

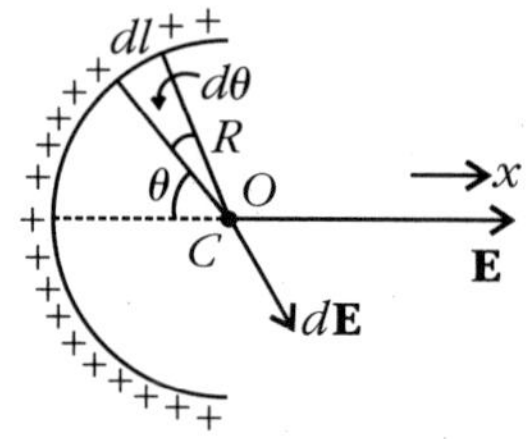

3.8 From the symmetry of the problem the sought field strength

$$E = \int dE_x$$

where the projection of field strength along x-axis duc to an elemental charge is

$$dE_x = \frac{dq\cos\theta}{4\pi\varepsilon_0 R^2} = \frac{q\,R\cos\theta\,d\theta}{4\pi^2\varepsilon_0 R^3}$$

Hence,

$$E = \frac{q}{4\pi^2\varepsilon_0 R^2}\int_{-\pi/2}^{\pi/2}\cos\theta\,d\theta = \frac{q}{2\pi^2\varepsilon_0 R^2}$$

$$= 0.10\text{ kV/m} \quad \text{(on substituting values)}$$

3.9 It can be easily shown that vector $\mathbf{E}$ in this case must be directed along the axis of the ring (see figure). Let us isolate differential linear charge element dq on the ring in the vicinity of point A. We write the expression for the component dE_n of the field created by this element at a field point P as

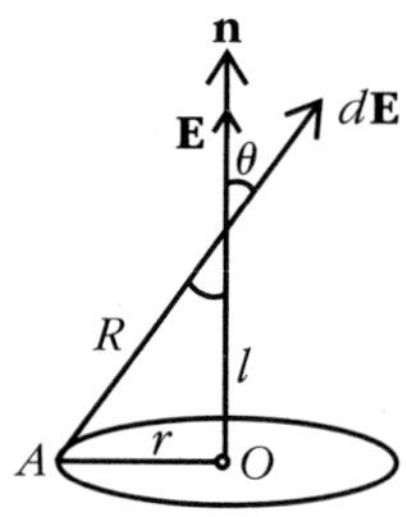

$$dE_n = \frac{1}{4\pi\varepsilon_0}\frac{dq}{R^2}\cos\theta = \frac{dq}{4\pi\varepsilon_0}\frac{l}{(r^2 + l^2)^{3/2}}$$

(where $\mathbf{n}$ is the unit vector perpendicular to the plane of ring or along the axis of the ring).

Hence,

$$E = \int dE_n = \frac{1}{4\pi\varepsilon_0}\frac{l}{(r^2 + l^2)^{3/2}}\int dq = \frac{q}{4\pi\varepsilon_0}\frac{l}{(r^2 + l^2)^{3/2}}$$

It can be seen that for $l >> r$, the field $E \approx q/4\pi\varepsilon_0 l^2$, i.e., at large distances the system behaves as a point charge.

From the expression of $E = E(l)$, the electric field is zero at the center of the ring ($l = 0$) and also zero when l is very large. Hence, there must be a value of l for which the electric field is maximum. This value of l can be obtained by setting the first derivative of E to zero as follows

$$\frac{dE}{dl} = 0 \quad \text{or} \quad \frac{q(r^2 - 2l^2)}{4\pi\varepsilon_0(r^2 + l^2)^{5/2}} = 0$$

Which yields,

$$l = \frac{r}{\sqrt{2}}$$

$$E_{\max} = \frac{q}{6\sqrt{3}\pi\varepsilon_0 r^2}$$

3.10 The electric potential at a distance x from the given ring is given by

$$\varphi(x) = \frac{q}{4\pi\varepsilon_0 x} - \frac{q}{4\pi\varepsilon_0 (R^2 + x^2)^{1/2}}$$

Hence, the field strength along x-axis (which is the net field strength in our case) is

$$E_x = -\frac{d\varphi}{dx} = \frac{1}{4\pi\varepsilon_0}\left[\frac{q}{x^2} - \frac{qx}{(R^2+x^2)^{3/2}}\right]$$

$$= \frac{\dfrac{q}{4\pi\epsilon_0} x^3 \left[\left(1+\dfrac{R^2}{x^2}\right)^{3/2} - 1\right]}{x^2(R^2+x^2)^{3/2}}$$

$$= \frac{\dfrac{q}{4\pi\varepsilon_0} x^3 \left[1 + \dfrac{3}{2}\dfrac{R^2}{x^2} + \dfrac{3}{8}\dfrac{R^4}{x^4} + \cdots\right]}{x^2(R^2+x^2)^{3/2}}$$

Neglecting the higher power of R/x, as $x >> R$

$$E = \frac{3qR^2}{8\pi\varepsilon_0 x^4}$$

Note: *Instead of $\varphi(x)$, we may write $E(x)$ directly using solution of problem 3.9.*

3.11 From the solution of Problem 3.9, the electric field strength due to ring at a point on its axis (say x-axis) at distance x from the center of the ring is given by

$$E(x) = \frac{qx}{4\pi\varepsilon_0 (R^2+x^2)^{3/2}}$$

and from symmetry, **E** at every point on the axis is directed along the x-axis (see figure).

Let us consider an element dx on thread which carries the charge λdx. The electric force experienced by the element in the field of ring

$$dF = (\lambda\, dx)\, E(x) = \frac{\lambda\, qx dx}{4\pi\varepsilon_0 \,(R^2+x^2)^{3/2}}$$

Thus the sought interaction

$$F = \int_0^\infty \frac{\lambda\, qx dx}{4\pi\varepsilon_0 (R^2+x^2)^{3/2}}$$

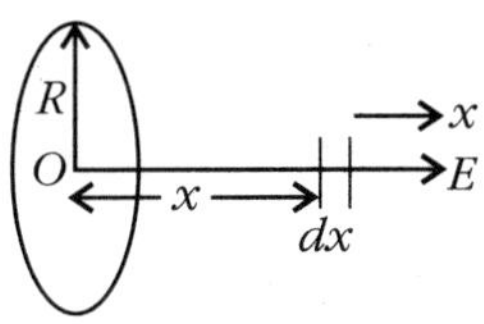

On integrating we get

$$F = \frac{\lambda q}{4\pi\varepsilon_0 R}$$

3.12 (a) The given charge distribution is shown in the figure. The symmetry of this distribution implies that vector **E** at the point O is directed to the right, and its magnitude is equal to the sum of the projection onto the direction of **E** of vectors **dE** from elementary charges dq. The projection of vector **dE** onto vector **E** is

$$dE \cos\varphi = \frac{1}{4\pi\varepsilon_0}\frac{dq}{R^2}\cos\varphi$$

where $\quad dq = \lambda R d\varphi = \lambda_0 R \cos\varphi d\varphi. \quad (1)$

Integrating Eq. (1) over φ between 0 and 2π we find the magnitude of the vector **E** as

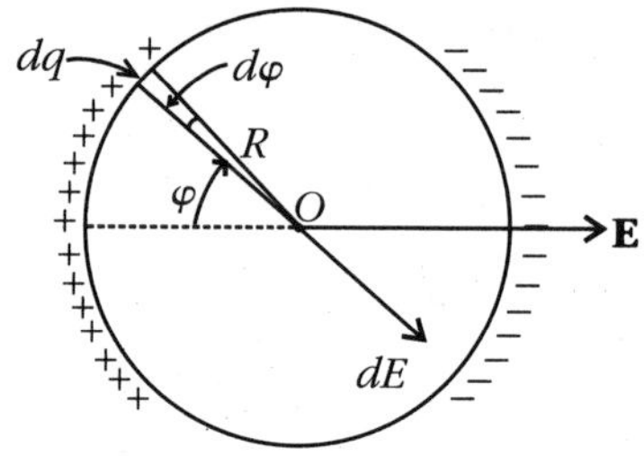

$$E = \frac{\lambda_0}{4\pi\varepsilon_0 R}\int_0^{2\pi} \cos^2 \varphi d\varphi = \frac{\lambda_0}{4\varepsilon_0 R}$$

It should be noted that this integral is evaluated in the most simple way if we take into account that $<\cos^2 \varphi> = 1/2$. Then

$$\int_0^{2\pi} \cos^2\varphi d\varphi = <\cos^2 \varphi> 2\pi = \pi$$

(b) Let us take differential length element of the ring at an azimuthal angle φ from the x-axis, the element subtends an angle $d\varphi$ at the center, and carries charge $dq = \lambda R d\varphi = (\lambda_0 \cos \varphi) R d\varphi$.

Taking the plane of ring as x-y plane and center of the ring as origin O, locations of field point **r**, of charge element **R** and of field point relative to charge element $\boldsymbol{\imath}$ are shown in the figure.

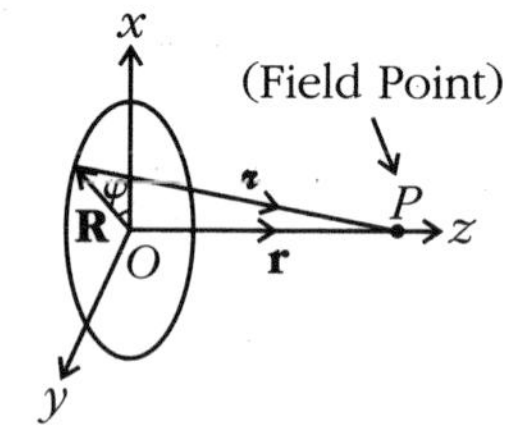

Electric field strength at the field point due to considered charge element

$$d\mathbf{E} = \frac{1}{4\pi\varepsilon_0}\frac{dq}{\imath^3}\boldsymbol{\imath}$$

Using $\boldsymbol{\imath} = \mathbf{r} - \mathbf{R} = x\mathbf{k} - (R\cos \varphi\mathbf{i} + R\sin \varphi\mathbf{j})$ and $\imath^3 = (R^2 + x^2)^{3/2}$

$$d\mathbf{E} = \frac{1}{4\pi\varepsilon_0}\frac{(\lambda_0 \cos\varphi)Rd\varphi}{(R^2+x^2)^{3/2}}\{x\mathbf{k} - (R\cos\varphi\,\mathbf{i} + R\sin\varphi\,\mathbf{j})\}$$

So, sought net electric field strength

$$\mathbf{E} = \int d\mathbf{E}$$

$$= \frac{\lambda_0 R}{4\pi\varepsilon_0(R^2+x^2)^{3/2}}\left[x\mathbf{k}\int_0^{2\pi} \cos\varphi d\varphi - R\mathbf{i}\int_0^{2\pi} \cos\varphi d\varphi - R\mathbf{j}\int_0^{2\pi} \sin 2\varphi d\varphi\right]$$

Taking into account

$$\int_0^{2\pi} \cos\varphi d\varphi = 0, \int_0^{2\pi} \sin\varphi d\varphi = 0 \quad \text{and} \quad \int_0^{2\pi} \cos^2 \varphi d\varphi = \pi$$

We get $$\mathbf{E} = \frac{1}{4\pi\varepsilon_0}\frac{\pi\lambda_0 R^2(-\mathbf{i})}{(R^2+x^2)^{3/2}}$$

Hence, $$E = \frac{\lambda_0 R^2}{4\varepsilon_0(R^2+x^2)^{3/2}}$$

For $x >> R$, $$E_x = \frac{p}{4\pi\varepsilon_0 x^3} \quad \text{where } p = \lambda_0 \pi R^2$$

Alternate:

Take an element S at an azimuthal angle φ from the x-axis, the element subtending an angle $d\varphi$ at the center. The elementary field at P due to the element is

$$\frac{\lambda_0 \cos\varphi d\varphi R}{4\pi\varepsilon_0 (x^2+R^2)} \quad \text{(along } SP \text{ with components)}$$

$$\frac{\lambda_0 \cos\varphi d\varphi \; R}{4\pi\varepsilon_0 (x^2+R^2)} \times \{\cos\theta \text{ along } OP, \sin\theta \text{ along } OS\}$$

where $$\cos\theta = \frac{x}{(x^2+R^2)^{1/2}}$$

and $$\sin\theta = \frac{R}{(x^2+R^2)^{1/2}}$$

The component along OP vanishes on integration as $\int_0^{2\pi} \cos\varphi d\varphi = 0$.

The component along OS can be broken into the parts along Ox and Oy and given by

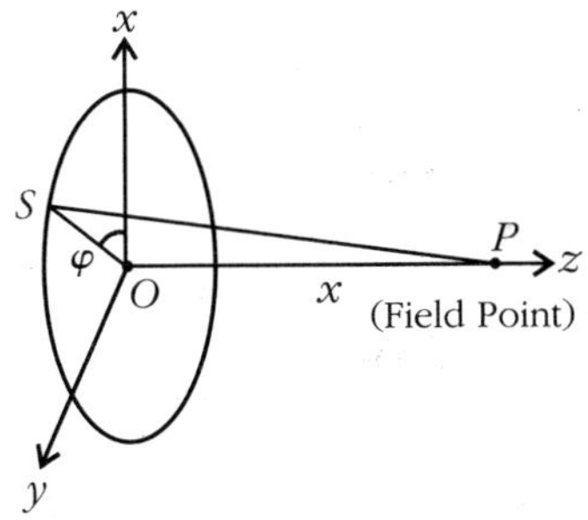

$$\frac{\lambda_0 R^2 \cos\varphi d\varphi}{4\pi\varepsilon_0 (x^2+R^2)^{3/2}} \times \{\cos\varphi \text{ along } Ox, \sin\varphi \text{ along } Oy\}$$

On integration, the part along Oy vanishes.

So, $$E_x = \frac{\lambda_0 R^2}{4\pi\varepsilon_0 (x^2+R^2)^{3/2}} \int_0^{2R} \cos^2\varphi d\varphi$$

as $$\int_0^{2R} \cos^2\varphi \sin\varphi d\varphi = 0$$

Finally $$E = E_x = \frac{\lambda_0 R^2}{4\pi\varepsilon_0 (x^2+R^2)^{3/2}}$$

3.13 (a) It is clear from symmetry considerations that vector **E** must be directed as shown in the figure. This shows the way of solving this problem: we must find the component dE_r of the field created by the element dl of the rod, having the charge dq and then integrate the result over all the elements of the rod. In this case

$$dE_r = dE \cos\alpha = \frac{1}{4\pi\varepsilon_0} \frac{\lambda dl}{r_0^2} \cos\alpha$$

where $\lambda = q/2a$ is the linear charge density. Let us reduce this equation to the form convenient for integration. Figure shows that $dl \cos\alpha = r_0 d\alpha$ and $r_0 = r/\cos\alpha$.

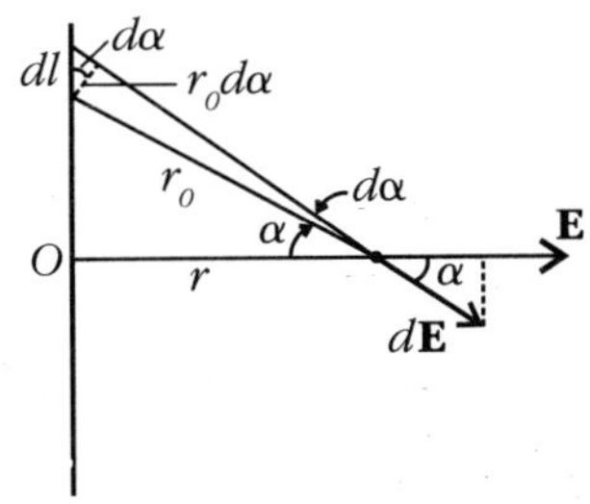

Consequently,

$$dE_r = \frac{1}{4\pi\varepsilon_0} \frac{\lambda r_0 d\alpha}{r_0^2} = \frac{\lambda}{4\pi\varepsilon_0 r} \cos\alpha \, d\alpha$$

This expression can be easily integrated to yield

$$E = \frac{\lambda}{4\pi\varepsilon_0 r} 2\left(\int_0^{\alpha_0} \cos\alpha d\alpha\right) = \frac{\lambda}{4\pi\varepsilon_0 r^2}(2\sin\alpha_0)$$

where α_0 is the maximum value of the angle α.

$$\sin\alpha_0 = \frac{a}{\sqrt{a^2+r^2}}$$

Thus,

$$E = \frac{q/2a}{4\pi\varepsilon_0 r}\left(2\frac{a}{\sqrt{a^2+r^2}}\right)$$

$$= \frac{q}{4\pi\varepsilon_0 r\sqrt{a^2+r^2}}$$

Note that in this case also $E \approx \frac{q}{4\pi\varepsilon_0 r}$ for $r >> a$, as for the field of a point charge.

(b) Let, us consider the element of length dl at a distance l from the center of the rod, as shown in the figure.

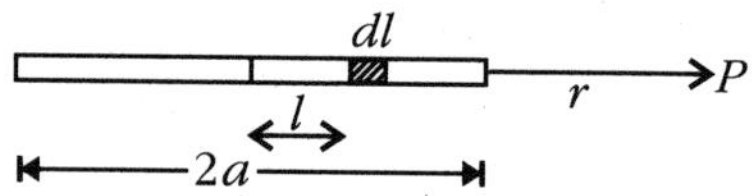

Then field at P, due to this element is given by

$$dE = \frac{\lambda dl}{4\pi\varepsilon_0 (r-l)^2}$$

if the element lies on the side as shown in the figure,

and

$$dE = \frac{\lambda dl}{4\pi\varepsilon_0 (r+l)^2}$$

if the element lies on the other side.

Hence,

$$E = \int dE = \int_0^a \frac{\lambda\, dl}{4\pi\varepsilon_0 (r-l)^2} + \int_0^a \frac{\lambda\, dl}{4\pi\varepsilon_0 (r+l)^2}$$

On integrating and putting $\lambda = \frac{q}{2a}$, we get

$$E = \frac{q}{4\pi\varepsilon_0}\frac{1}{(r^2-a^2)}$$

For $r >> a$,

$$E \approx \frac{q}{4\pi\varepsilon_0 r^2}$$

3.14 The problem is reduced to finding E_x and E_y, which are the projections of **E**, under the assumption $\lambda > 0$ (see figure).

Let us start with E_x. The contribution to E_x from the charge element of the segment dx is

$$dE_x = \frac{1}{4\pi\varepsilon_0}\frac{\lambda\, dx}{r^2}\sin\alpha \qquad (1)$$

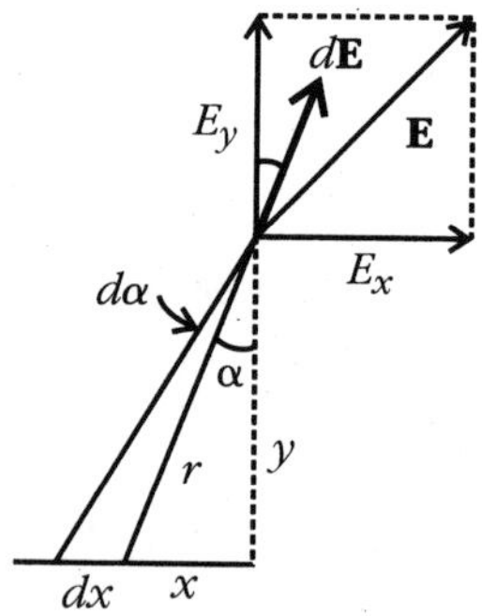

Let us reduce this expression to the form convenient for integration. Thus,

$$dx = \frac{r\, d\alpha}{\cos\alpha},\ r = \frac{y}{\cos\alpha}$$

Then,

$$dE_x = \frac{\lambda}{4\pi\varepsilon_0 y}\sin\alpha\, d\alpha$$

Integrating this expression over α between 0 and $\pi/2$, we find

$$E_x = \frac{\lambda}{4\pi\varepsilon_0 y}$$

In order to find the projection E_y it is sufficient to recall that dE_y differs from dE_x in that $\sin\alpha$ in Eq. (1) is simply replaced by $\cos\alpha$.

This gives

$$dE_y = (\lambda \cos\alpha \, d\alpha)/4\pi\varepsilon_0 y \quad \text{and} \quad E_y = \lambda/4\pi\varepsilon_0 y$$

We have obtained an interesting result that $E_x = E_y$ independently of y, i.e., **E** is oriented at the angle of 45° to the rod. The modulus of **E** is

$$\mathbf{E} = \sqrt{E_x^2 + E_y^2} = \frac{\lambda\sqrt{2}}{4\pi\varepsilon_0 y}$$

3.15 For Fig. (a): Using the solution of Problem 3.14, the net electric field strength at the point O due to straight parts of the thread equals zero. For the curved part (arc), let us derive a general expression i.e. let us calculate the field strength at the center of arc of radius R and linear charge density λ and which subtends angle θ_0 at the center. From the symmetry the sought field strength will be directed along the bisector of the angle θ_0 and is given by

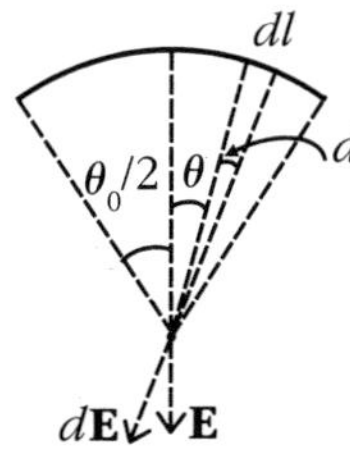

$$E = \int_{-\theta_0/2}^{+\theta_0/2} \frac{\lambda\,(R d\theta)}{4\pi\varepsilon_0 R^2}\cos\theta = \frac{\lambda}{2\pi\varepsilon_0 R}\sin\frac{\theta_0}{2}$$

In our problem $\theta_0 = \pi/2$, thus the field strength due to the turned part at the point

$$E_0 = \frac{\sqrt{2}\lambda}{4\pi\varepsilon_0 R}$$

which is also the sought result. For Fig. (b): Using the solution of Problem 3.14 (a), net field strength at O due to straight parts equals

$$\sqrt{2}\left(\frac{\sqrt{2}\lambda}{4\pi\varepsilon_0 R}\right) = \frac{\lambda}{2\pi\varepsilon_0 R} \quad \text{(directed vertically downward)}$$

Now using the solution of Problem 3.15 (a) field strength due to the given curved part (semi-circle) at the point O becomes

$$\frac{\lambda}{2\pi\varepsilon_0 R} \quad \text{(directed vertically upward)}$$

Hence the sought net field strength becomes zero.

3.16 Given charge distribution on the surface $\sigma = \mathbf{a} \cdot \mathbf{r}$ is shown in the figure. Symmetry of this distribution implies that the sought **E** at the center O of the sphere is opposite to **a**. For the ring element

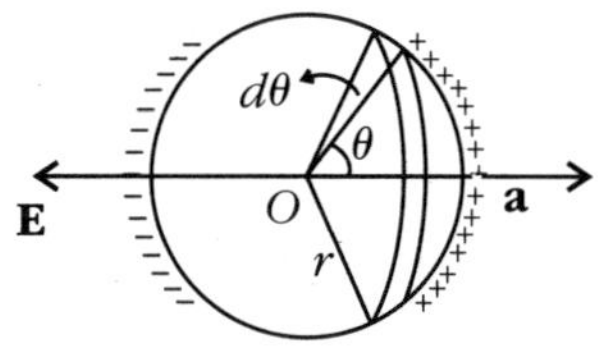

$$dq = \sigma(2\pi r \sin\theta)\, r d\theta$$
$$= (\mathbf{a}\cdot\mathbf{r})\, 2\pi r^2 \sin\theta\, d\theta = 2\pi a r^3 \sin\theta\cos\theta\, d\theta$$

Again from symmetry, field strength due to any ring element $d\mathbf{E}$ is also opposite to **a**, i.e., $d\mathbf{E} \uparrow\downarrow \mathbf{a}$. Hence,

$$d\mathbf{E} = \frac{dq\, r\cos\theta}{4\pi\varepsilon_0 (r^2\sin^2\theta + r^2\cos^2\theta)^{3/2}}\frac{-\mathbf{a}}{a} \quad \text{(Using the result of Problem 3.9)}$$

$$= \frac{(2\pi a r^3 \sin\theta\cos\theta\, d\theta)\, r\cos\theta}{4\pi\varepsilon_0 r^3}\frac{(-\mathbf{a})}{a}$$

$$= \frac{-\mathbf{a}r}{2\varepsilon_0}\sin\theta\cos^2\theta\, d\theta$$

Thus, $$\mathbf{E} = \int d\mathbf{E} = \frac{(-\mathbf{a})r}{2\varepsilon_0}\int_0^\pi \sin\theta\cos^2\theta\, d\theta$$

Integrating, we get

$$\mathbf{E} = \frac{\mathbf{a}r}{2\varepsilon_0}\frac{2}{3} = -\frac{\mathbf{a}r}{3\varepsilon_0}$$

3.17 We start from two charged spherical balls each of radius R with equal and opposite charge densities $+\rho$ and $-\rho$. The center of the balls are at $+\mathbf{a}/2$ and $-\mathbf{a}/2$, respectively, so the equation of their surfaces are

$$\left|\mathbf{r} - \frac{\mathbf{a}}{2}\right| = R \quad \text{or} \quad r - \frac{a}{2}\cos\theta \cong R$$

and $$r + \frac{a}{2}\cos\theta \cong R$$

considering a to be small. The distance between the two surfaces in the radial direction at angle is θ $|a\cos\theta|$ and does not depend on the azimuthal angle. It is seen from the diagram that the surface of the sphere has in effect a surface density $\sigma_0 = \sigma_0 \cos\theta$ when $\sigma_0 = \rho a$.

Inside any uniformly charged spherical ball, the field is radial and has the magnitude given by Gauss's theorem

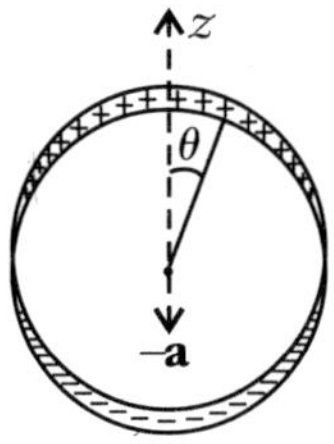

$$4\pi r^2 E = \frac{4\pi}{3} r^3 \rho/\varepsilon_0$$

or $$E = \frac{\rho r}{3\varepsilon_0}$$

In vector notation, using the fact that V must be measured from the center of the ball, we get, for the present case

$$\mathbf{E} = \frac{\rho}{3\varepsilon_0}\left(\mathbf{r} - \frac{\mathbf{a}}{2}\right) - \frac{\rho}{3\varepsilon_0}\left(\mathbf{r} + \frac{\mathbf{a}}{2}\right)$$
$$= -\rho\,\mathbf{a}/3\varepsilon_0 = -\frac{\sigma_0}{3\varepsilon_0}\mathbf{k}$$

where $\mathbf{k}$ is the unit vector along the polar axis from which θ is measured.

3.18 Let us consider an elemental spherical shell of thickness dr. Thus surface charge density of the shell $\sigma = \rho dr = (\mathbf{a}\cdot\mathbf{r})\, dr$.

Thus using the solution of Problem 3.16, field strength due to this spherical shell

$$d\mathbf{E} = -\frac{\mathbf{a}r}{3\varepsilon_0} dr$$

Hence, the sought field strength

$$\mathbf{E} = -\frac{\mathbf{a}}{3\varepsilon_0}\int_0^R r\,dr = -\frac{\mathbf{a}R^2}{6\varepsilon_0}$$

3.19 Due to the line charge

$$\mathbf{E} = \frac{\lambda}{4\pi\varepsilon_0\rho}(-\mathbf{k}) + \frac{\lambda}{4\pi\varepsilon_0\rho}(\mathbf{e}_\rho) \text{ (see solution of Problem 3.14)}$$

Distribution of field **E** allows to take a finite ring element of radius $0 < \rho < R$ and thickness $d\rho$. Hence area vector of the taken ring element $d\mathbf{S} = (2\pi\rho d\rho)(-\mathbf{k})$.

Now flux across the considered ring element

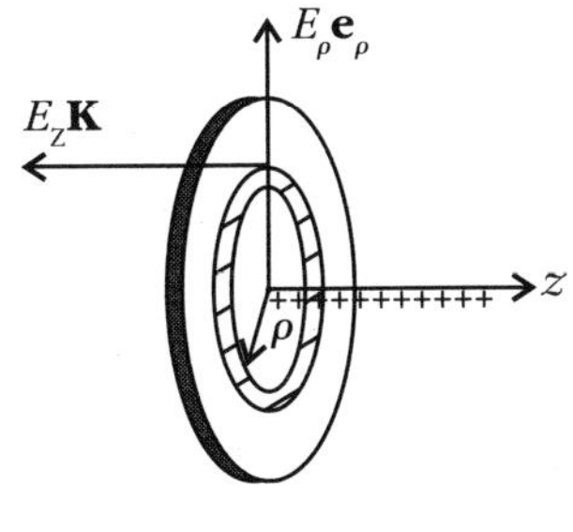

$$d\Phi = \mathbf{E}\cdot d\mathbf{S} = \frac{\lambda}{4\pi\varepsilon_0}\left[\frac{-\mathbf{k}}{\rho} + \frac{\mathbf{e}_\rho}{\rho}\right]\cdot 2\pi\rho d\rho\,(-\mathbf{k})$$

$$d\Phi = +\frac{2\pi\lambda\, d\rho}{4\pi\varepsilon_0} \text{ (because } \mathbf{e}_\rho \perp \mathbf{k})$$

So,

$$\Phi = \int d\Phi = \frac{\lambda R}{2\varepsilon_0}$$

3.20 Take the line joining the point charges as z-axis, In this problem at a normal distance ρ (from z-axis) **E** is the same and is directed towards positive z-axis. So, symmetry of the field **E** allows us to take a finite ring element of radius ρ and thickness $d\rho$.

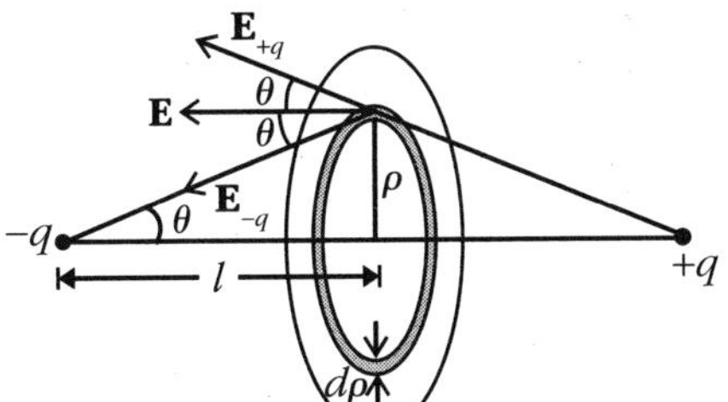

E at all the locations of taken ring element is

$$\mathbf{E} = \frac{2q}{4\pi\varepsilon_0(l^2+\rho^2)}\cos\theta\,\mathbf{k} = \frac{2ql}{4\pi\varepsilon_0(l^2+\rho^2)^{3/2}}\mathbf{k}$$

So,
$$d\Phi = \mathbf{E}\cdot d\mathbf{S} = \frac{2ql}{4\pi\varepsilon_0(l^2+\rho^2)^{3/2}}\mathbf{k}\cdot 2\pi\rho d\rho\,\mathbf{k} = \frac{ql\rho d\rho}{\varepsilon_0(l^2+\rho^2)^{3/2}}$$

Hence,
$$\Phi = \int d\Phi = \frac{ql}{\varepsilon_0}\int_0^R \frac{\rho d\rho}{(l^2+\rho^2)^{3/2}} = \frac{q}{\varepsilon_0}\left[1 - \frac{l}{\sqrt{l^2+R^2}}\right]$$

It can also be solved by considering a ring element or by using solid angle.

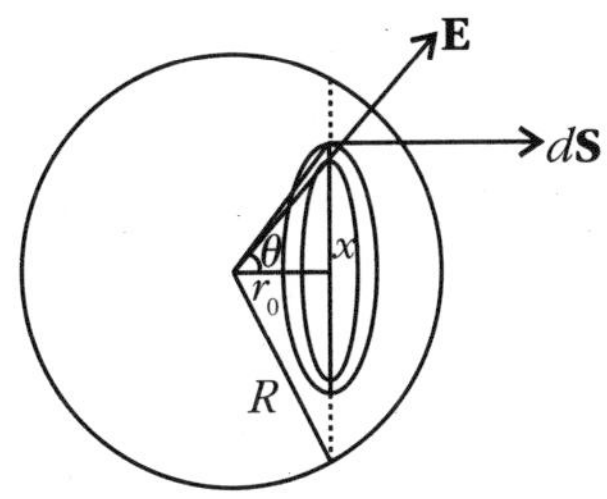

3.21 From Gauss' theorem, electric field strength at an inside point at a distance r from the center of uniformly charged ball of volume density ρ is $\mathbf{E} = \dfrac{\rho\mathbf{r}}{3\varepsilon_0}$.

So **E** is not uniform over the given section (a disk of radius $\sqrt{R^2 - r_0^2}$) of the ball. Thus the section is considered to be made up of ring elements whose radii varies from zero to $\sqrt{R^2 - r_0^2}$. Let us consider a ring element of radius x and thickness dx, as shown in the figure, so that magnitude of **E** takes same value over the considered ring element.

Flux over the consider d ring element

$$d\Phi = \mathbf{E}\cdot d\mathbf{S} = E_r dS\cos\theta$$

Using $E_r = \rho r/3\varepsilon_0$, $ds = 2\pi x dx$, and $\cos\theta = r_0/r$, we get

$$d\Phi = \frac{\rho r}{3\varepsilon_0}2\pi x\,dx\frac{r_0}{r} = \frac{\rho r_0}{3\varepsilon_0}2\pi x dx$$

Hence, sought flux

$$\Phi = \frac{2\pi\rho r_0}{3\varepsilon_0}\int_0^{\sqrt{R^2-r_0^2}} x\,dx = \frac{2\pi\rho r_0}{3\varepsilon_0}\frac{(R^2-r_0^2)}{2} = \frac{\pi\rho r_0}{3\varepsilon_0}(R^2-r_0^2)$$

3.22 The field at P due to each of the threads at A and B are the same having the magnitude

$$E_0 = \frac{\lambda}{2\pi\varepsilon_0(x^2+l^2/4)^{1/2}} \quad \text{(directed along } AP \text{ and } BP\text{)}$$

The resultant is along OP, given by

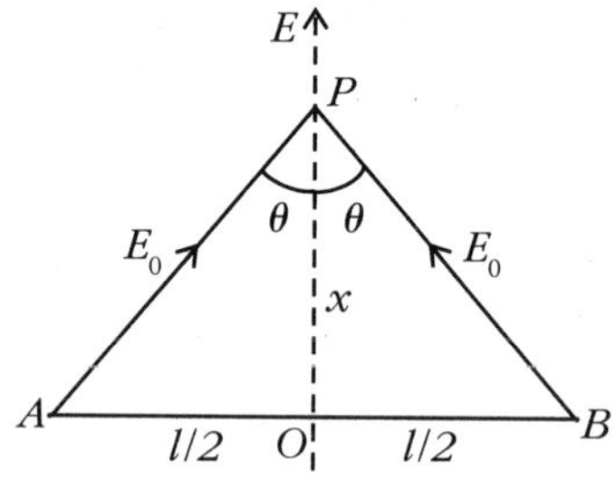

$$E = 2E_0\cos\theta$$

$$= \frac{\lambda x}{\pi\varepsilon_0(x^2+l^2/4)}$$

$$= \frac{\lambda}{\pi\varepsilon_0\left[x+\frac{l^2}{4x}-2\cdot\frac{l}{2\sqrt{x}}\cdot\sqrt{x}+l\right]}$$

$$= \frac{\lambda}{\pi\varepsilon_0\left[\left(\sqrt{x}-\frac{1}{2\sqrt{x}}\right)^2+l\right]}$$

This is the maximum when $x = \frac{l}{2}$ and $E = E_{max} = \frac{\lambda}{\pi\varepsilon_0 l}$.

3.23 Take a section of the cylinder perpendicular to its axis through the point where the electric field is to be calculated. (All points on the axis are equivalent.) Consider an element S with azimuthal angle φ. The length of the element is $Rd\varphi$, R being the radius of cross section of the cylinder. The element itself is a section of an infinite strip. The electric field at O due to this strip is

$$\frac{\sigma_0\cos\varphi\,(Rd\varphi)}{2\pi\varepsilon_0 R}\text{along } SO$$

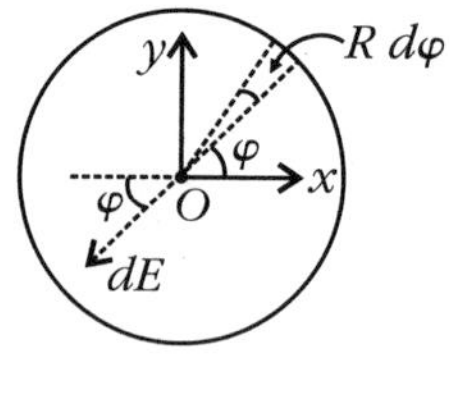

This can be resolved into

$$\frac{\sigma_0\cos\varphi\,d\varphi}{2\pi\varepsilon_0}\begin{Bmatrix}\cos\varphi \text{ along } Ox\\ \sin\varphi \text{ along } yO\end{Bmatrix}$$

On integration the component along yO vanishes. What remains is

$$\int_0^{2x}\frac{\sigma_0\cos^2\varphi\,d\varphi}{2\pi\varepsilon_0} = \frac{\sigma_0}{2\varepsilon_0}$$

along xO, i.e., along the direction $\varphi = \pi$.

3.24 Since the field is axis-symmetric (as the field of a uniformly charged filament), we conclude that the flux through the sphere of radius R is equal to the flux through the lateral surface of a cylinder having the same radius and the height $2R$, as shown in the figure.

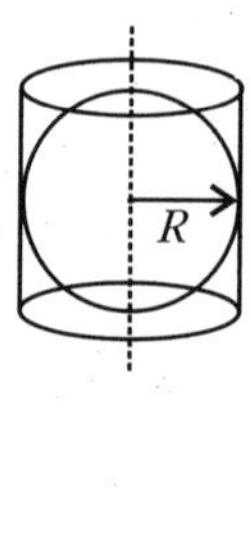

Now, $$\Phi = \int \mathbf{E}\cdot d\mathbf{S} = E_\rho S$$

But, $$E_\rho = \frac{a}{R}$$

Thus, $$\Phi = \frac{a}{R}S = \frac{a}{R}2\pi\, R \cdot 2R = 4\pi aR$$

Alternate:

The distribution of electric field is given by

$$\mathbf{E} = \frac{a(x\mathbf{i} + y\mathbf{j})}{x^2 + y^2} = \frac{a\boldsymbol{\rho}}{\rho^2}$$

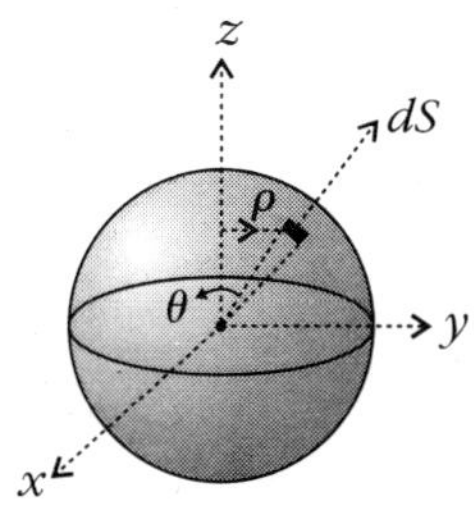

where $\mathbf{p} = x\mathbf{i} + y\mathbf{j}$, which is vector component of position vector $\mathbf{r}$ normal to the z-axis. Let us take a differential area element dS on the surface of the given sphere of radius R, as shown in the figure.

Then, $$d\Phi = \mathbf{E} \cdot d\mathbf{S} = \frac{a\boldsymbol{\rho}}{\rho^2} \cdot d\mathbf{S} = \frac{a}{\rho}\, dS \sin\theta = \frac{a\, dS}{\rho}\left(\frac{\rho}{R}\right) = \frac{a}{R} dS$$

So, $$\Phi = \frac{a}{R}\int dS = \frac{a}{R}(4\pi R^2) = 4\pi aR$$

3.25 (a) Let us consider a sphere of radius $r < R$, then charge enclosed by the considered sphere

$$q_{\text{enclosed}} = \int_0^r 4\pi r^2 dr \rho = \int_0^r 4\pi r^2 \rho_0 \left(1 - \frac{r}{R}\right) dr \tag{1}$$

Now, applying Gauss' theorem

$$E_r 4\pi r^2 = \frac{q_{\text{enclosed}}}{\varepsilon_0}$$

(where E_r is the projection of electric field along the radial line)

$$= \frac{\rho_0}{\varepsilon_0}\int_0^r 4\pi r^2 \left(1 - \frac{r}{R}\right) dr$$

or $$E_r = \frac{\rho_0 r}{3\varepsilon_0}\left[1 - \frac{3r}{4R}\right]$$

For a point outside the sphere $r > R$, so

$$q_{\text{enclosed}} = \int_0^R 4\pi r^2 dr \rho_0 \left(1 - \frac{r}{R}\right) \quad \text{(as there is no charge outside the ball)}$$

Again from Gauss' theorem

$$E_r 4\pi r^2 = \int_0^R \frac{4\pi r^2 dr \rho_0 \left(1 - \frac{r}{R}\right)}{\varepsilon_0}$$

or $$E_r = \frac{\rho_0}{r^2 \varepsilon_0}\left[\frac{R^3}{3} - \frac{R^4}{4R}\right] = \frac{\rho_0 R^3}{12\, r^2 \varepsilon_0}$$

(b) As magnitude of electric field decreases with increasing r for $r > R$, field will be maximum for $r < R$. Now, for E_r to be maximum

$$\frac{d}{dr}\left(r - \frac{3r^2}{4R}\right) = 0 \quad \text{or} \quad 1 - \frac{3r}{2R} = 0 \quad \text{or} \quad r = r_m = \frac{2R}{3}$$

Hence,
$$E_{\max} = \frac{\rho_0 R}{9\varepsilon_0}$$

3.26 Let the charge carried by the sphere be q, then using Gauss' theorem for a spherical surface having radius $r > R$, we can write

$$E4\pi r^2 = \frac{q_{\text{enclosed}}}{\varepsilon_0} = \frac{q}{\varepsilon_0} + \frac{1}{\varepsilon_0}\int_R^r \frac{\alpha}{r} 4\pi r^2 dr$$

On integrating, we get

$$E4\pi r^2 = \frac{(q - 2\pi\alpha R^2)}{\varepsilon_0} + \frac{4\pi\alpha r^2}{2\varepsilon_0}$$

The intensity E does not depend on r when the expression in the parentheses is equal to zero. Hence,

$$q = 2\pi\alpha R^2 \quad \text{and} \quad E = \frac{\alpha}{2\varepsilon_0}$$

3.27 Let us consider a spherical layer of radius r and thickness dr, having its center coinciding with the center of the system. Then using Gauss' theorem for this surface,

$$E_r 4\pi r^2 = \frac{q_{\text{enclosed}}}{\varepsilon_0} = \int_0^r \frac{\rho dV}{\varepsilon_0}$$
$$= \frac{1}{\varepsilon_0}\int_0^r \rho_0 e^{-\alpha r^3} 4\pi r^2 dr$$

After integration

$$E_r 4\pi r^2 = \frac{\rho 4\pi}{3\varepsilon_0\alpha}[1 - e^{-\alpha r^3}]$$

or
$$E_r = \frac{\rho_0}{3\varepsilon_0\alpha r^2}[1 - e^{-\alpha r^3}]$$

now when $\alpha r^3 \gg 1$,
$$E_r \approx \frac{\rho_0}{3\varepsilon_0\alpha r^2}$$

and when $\alpha r^3 \ll 1$,
$$E_r \approx \frac{\rho_0 r}{3\varepsilon_0}$$

3.28 Using Gauss' theorem we can easily show that the electric field strength within a uniformly charged sphere is

$$\mathbf{E} = \left(\frac{\rho}{3\varepsilon_0}\right)\mathbf{r}$$

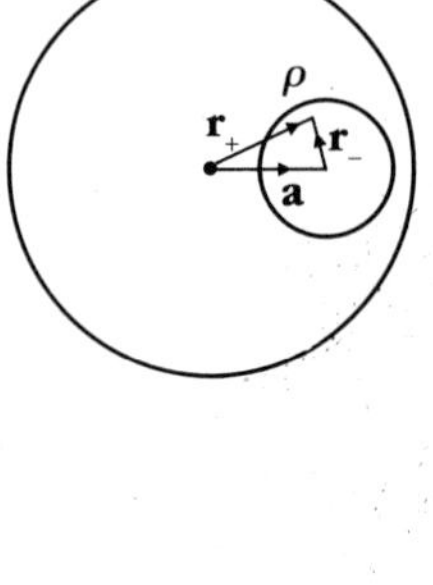

The cavity, in our problem, may be considered as the superposition of two balls, one with the charge density ρ and the other with $-\rho$. Let P be a point inside the cavity such that its position vector with respect to the center of cavity be $\mathbf{r}_-$ and with respect to the center of the ball be $\mathbf{r}_+$. Then from the principle of superposition, field inside the cavity, at an arbitrary point P is given by

$$\mathbf{E} = \mathbf{E}_+ + \mathbf{E}_-$$

$$= \frac{\rho}{3\varepsilon_0}(\mathbf{r}_+ - \mathbf{r}_-) = \frac{\rho}{3\varepsilon_0}\mathbf{a}$$

Note: The *expression* obtained for ***E*** *shows that it is valid regardless of the ratio* between the *radii of the sphere and the distance between their* centers.

3.29 Let us consider a cylindrical Gaussian surface of radius r and height h inside an infinitely long charged cylinder with charge density ρ. Now from Gauss' theorem

$$\mathbf{E}_r 2\pi rh = \frac{q_{\text{enclosed}}}{\varepsilon_0}$$

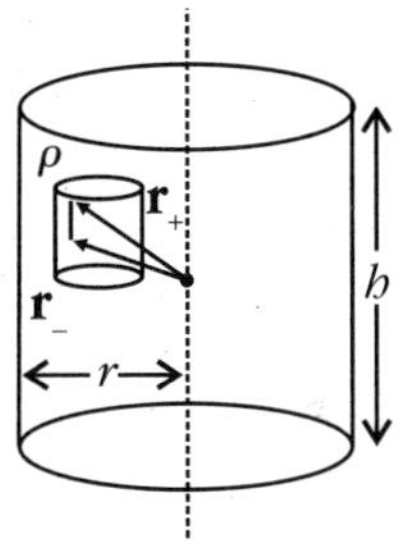

(where E_r is the field inside the cylinder at a distance r from its axis)

or $$E_r\, 2\pi rh = \frac{\rho\pi r^2 h}{\varepsilon_0} \Rightarrow E_r = \frac{\rho r}{2\varepsilon_0}$$

Now, using the method of Problem 3.28, field at a point P inside the cavity is

$$\mathbf{E} = \mathbf{E}_+ + \mathbf{E}_- = \frac{\rho}{2\varepsilon_0}(\mathbf{r}_+ - \mathbf{r}_-) = \frac{\rho}{2\varepsilon_0}\mathbf{a}$$

3.30 The arrangement of the rings is as shown in the figure. Now, potential at the point 1 is equal to sum of potential at 1 due to the ring 1 and potential at 1 due to the ring 2. So,

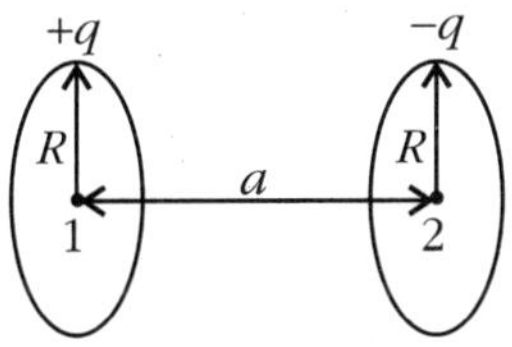

$$\varphi_1 = \frac{q}{4\pi\varepsilon_0 R} + \frac{-q}{4\pi\varepsilon_0(R^2 + a^2)^{1/2}}$$

Similarly, the potential at point 2

$$\varphi_2 = \frac{-q}{4\pi\varepsilon_0 R} + \frac{q}{4\pi\varepsilon_0(R^2 + a^2)^{1/2}}$$

Hence, the sought potential difference

$$\varphi_1 - \varphi_2 = \Delta\varphi = 2\left(\frac{q}{4\pi\varepsilon_0 R}\right) + \frac{-q}{4\pi\varepsilon_0(R^2 + a^2)^{1/2}}$$

$$= \frac{q}{2\pi\varepsilon_0 R}\left(1 - \frac{1}{\sqrt{1 + (a/R)^2}}\right)$$

3.31 We know from Gauss' theorem that the electric field due to an infinitely long straight wire, at a perpendicular distance **r** from it is

$$E_r = \frac{\lambda}{2\pi\varepsilon_0 r}.$$

So, the work done is

$$\int_1^2 E_r dr = \int_x^{\eta x} \frac{\lambda}{2\pi\varepsilon_0 r} dr$$

(where x is perpendicular distance from the thread by which point 1 is removed from it).

Hence,
$$\Delta\varphi_{12} = \frac{\lambda}{2\pi\varepsilon_0}\ln\eta$$
$$= 5 \text{ kV (on substituting values)}$$

3.32 Let us consider a ring element as shown in the figure. Then the charge carried by the element, $dq = (2\pi R \sin\theta)\, Rd\theta\sigma$.

Hence, potential due to the considered element at center of the hemisphere is

$$d\varphi = \frac{1}{4\pi\varepsilon_0}\frac{dq}{R} = \frac{2\pi\sigma R\sin\theta\, d\theta}{4\pi\varepsilon_0} = \frac{\sigma R}{2\varepsilon_0}\sin\theta\, d\theta$$

So potential due to the whole hemisphere is

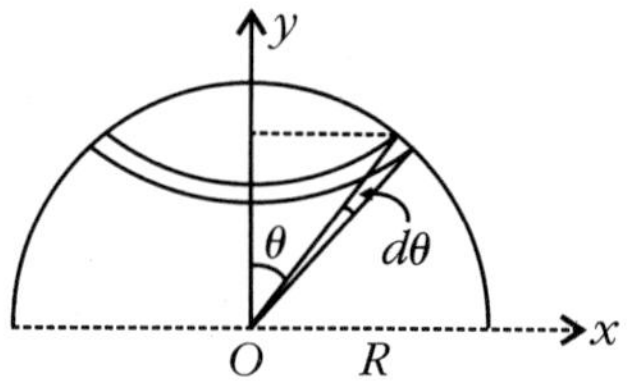

$$\varphi = \frac{R\sigma}{2\varepsilon_0}\int_0^{\pi/2}\sin\theta\, d\theta = \frac{\sigma R}{2\varepsilon_0}$$

Now from the symmetry of the problem, net electric field of the hemisphere is directed towards the negative y-axis. So, we have

$$dE_y = \frac{1}{4\pi\varepsilon_0}\frac{dq\cos\theta}{R^2} = \frac{\sigma}{2\varepsilon_0}\sin\theta\cos\theta\, d\theta$$

Thus,
$$E = E_y' = \frac{\sigma}{2\varepsilon_0}\int_0^{\pi/2}\sin\theta\cos\theta\, d\theta = \frac{\sigma}{4\varepsilon_0}\int_0^{\pi/2}\sin 2\theta\, d\theta = \frac{\sigma}{4\varepsilon_0}\text{(along } yO)$$

3.33 Let us consider an elementary ring of thickness dy and radius y as shown in the figure. Then potential at a point P, at distance l from the center of the disk is

$$d\varphi = \frac{\sigma\, 2\pi y dy}{4\pi\varepsilon_0(y^2 + l^2)^{1/2}}$$

Hence potential due to the whole disk is given by

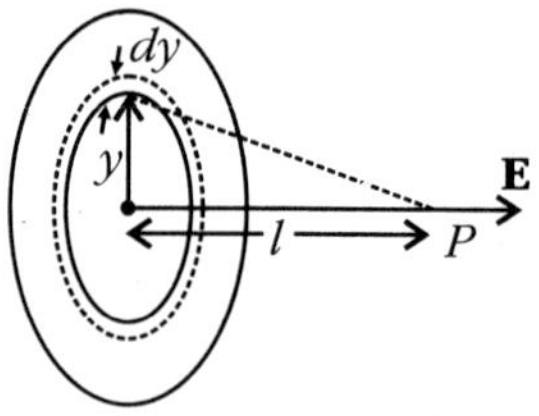

$$\varphi = \int_0^R \frac{\sigma\, 2\pi y dy}{4\pi\varepsilon_0(y^2 + l^2)^{1/2}} = \frac{\sigma l}{2\varepsilon_0}\left(\sqrt{1 + (R/l)^2} - 1\right)$$

From symmetry

$$E = E_l = -\frac{d\varphi}{dl}$$
$$= -\frac{\sigma}{2\varepsilon_0}\left[\frac{2l}{2\sqrt{R^2 + l^2}} - 1\right] = \frac{\sigma}{2\varepsilon_0}\left[1 - \frac{1}{\sqrt{1 + (R/l)^2}}\right]$$

when $l \to 0$, $\varphi = \dfrac{\sigma R}{2\varepsilon_0}$ and $E = \dfrac{\sigma}{2\varepsilon_0}$

when $l >> R$,
$$\varphi \approx \frac{\sigma R^2}{4\varepsilon_0 l} \quad \text{and} \quad E \approx \frac{\sigma R^2}{4\varepsilon_0 l^2}$$

3.34 By definition the potential in the case of a surface charge distribution is defined by

$$\varphi = \frac{1}{4\pi\varepsilon_0}\int\frac{\sigma dS}{r} \qquad (1)$$

In order to simplify integration, we shall choose the area element dS in the form of a part of the ring of radius r and width dr (see figure). Then $dS = 2\theta r\, dr$, $r = 2R\cos\theta$ and $dr = -2R\sin\theta\, d\theta$. After substituting these expresion into integral in Eq. (1), we obtain the expresion for φ at the point O as

$$\varphi = -\frac{\sigma R}{\pi\varepsilon_0}\int_{\pi/2}^{0}\theta\sin\theta\, d\theta$$

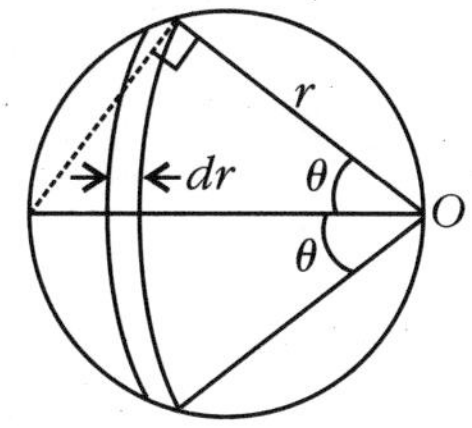

We integrate by parts, denoting $\theta = u$ and $\sin\theta\, d\theta = dv$ to get

$$\int\theta\sin\theta\, d\theta = -\theta\cos\theta + \int\cos\theta\, d\theta$$

$$= -\theta\cos\theta + \sin\theta$$

which gives -1 after substituting the limits of integration. As a result, we obtain

$$\varphi = \frac{\sigma R}{\pi\varepsilon_0}$$

3.35 In accordance with the problem $\varphi = \mathbf{a}\cdot\mathbf{r}$. Thus from the equation $\mathbf{E} = -\nabla\varphi$, we get

$$\mathbf{E} = -\left[\frac{\partial}{\partial x}(a_x x)\mathbf{i} + \frac{\partial}{\partial y}(a_y y)\mathbf{j} + \frac{\partial}{\partial z}(a_z z)\mathbf{k}\right] = -[a_x\mathbf{i} + a_y\mathbf{j} + a_z\mathbf{k}] = -a$$

3.36 (a) Given, $\varphi = a(x^2 - y^2)$

So, $\mathbf{E} = -\nabla\varphi = -2a\,(x\mathbf{i} - y\mathbf{j})$

The sought shape of field lines is as shown in the figure (a) of answer sheet, assuming $a > 0$.

(b) Since, $\varphi = axy$

So, $\mathbf{E} = -\nabla\varphi = -ay\mathbf{i} - ax\mathbf{j}$

The plot of field lines is as shown in the figure (b) of answer sheet.

3.37 Given, $\varphi = a\,(x^2 + y^2) + bz^2$

So, $\mathbf{E} = -\nabla\varphi = -[2ax\mathbf{i} + 2ay\mathbf{j} + 2bz\mathbf{k}]$

Hence, $|\mathbf{E}| = 2\sqrt{a^2(x^2 + y^2) + b^2z^2}$

Shape of the equipotential surface:

Put $\rho = x\mathbf{i} + y\mathbf{j}$ or $\rho^2 = x^2 + y^2$

Then the equipotential surface has the equation

$$a\rho^2 + bz^2 = \text{constant} = \varphi$$

If $a > 0$, $b > 0$, then $\varphi > 0$ and the equation of the equipotential surface is

$$\frac{\rho^2}{\varphi/a} + \frac{z^2}{\varphi/b} = 1$$

which is an ellipse in ρ, z coordinates. In three dimensions, the surface is an ellipsoid of revolution with semi-axes $\sqrt{\varphi/a}$ and $\sqrt{\varphi/b}$.

If $a > 0$, $b < 0$, then φ can be ≥ 0. If $\varphi > 0$ then the equation is

$$\frac{\rho^2}{\varphi/a} - \frac{z^2}{\varphi/|b|} = 1$$

This is a single cavity hyperboloid of revolution about z axis.

If $\varphi = 0$, then the equation can be written as

$$a\rho^2 - |b|z^2 = 0$$

or

$$z = \pm\sqrt{\frac{a}{|b|}}\rho$$

This is the equation of a right circular cone.

If $\varphi < 0$, then the equation can be written as

$$|b|z^2 - a\rho^2 = |\varphi|$$

or

$$\frac{z^2}{|\varphi|/|b|} - \frac{\rho^2}{|\varphi|/a} = 1$$

This is a two cavity hyperboloid of revolution about z-axis.

3.38 From Gauss' theorem, intensity at a point inside the sphere, at a distance r from the center is given by, $E_r = \dfrac{\rho r}{3\varepsilon_0}$ and outside it, is given by $E_r = \dfrac{1}{4\pi\varepsilon_0}\dfrac{q}{r^2}$.

(a) Potential at the center of the sphere

$$\varphi_0 = \int_0^\infty \mathbf{E}\cdot d\mathbf{r} = \int_0^R \frac{\rho r}{3\varepsilon_0}dr + \int_R^\infty \frac{q}{4\pi\varepsilon_0 r^2}dr = \frac{\rho}{3\varepsilon}\frac{R^2}{2} + \frac{q}{4\pi\varepsilon_0 R}$$

$$= \frac{q}{8\pi\varepsilon_0 R} + \frac{q}{4\pi\varepsilon_0 R} = \frac{3q}{8\pi\varepsilon_0 R} \quad \left(\text{as } \rho = \frac{3q}{4\pi R^3}\right)$$

(b) Now, potential at any point inside the sphere, at a distance r from it's center

$$\varphi(r) = \int_r^R \frac{\rho}{3\varepsilon_0} r\,dr + \int_r^\infty \frac{q}{4\pi\varepsilon_0}\frac{dr}{r^2}$$

On integration,

$$\varphi(r) = \frac{3q}{8\varepsilon_0 R}\left[1 - \frac{r^2}{3R^2}\right] = \varphi_0\left[1 - \frac{r^2}{3R^2}\right]$$

3.39 Let two charges $+q$ and $-q$ be separated by a distance l. Then electric potential at a point at distance $r >> l$ from this dipole

$$\varphi(r) = \frac{+q}{4\pi\varepsilon_0 r_+} + \frac{-q}{4\pi\varepsilon_0 r_-} = \frac{q}{4\pi\varepsilon_0}\left(\frac{r_- - r_+}{r_+ r_-}\right) \qquad (1)$$

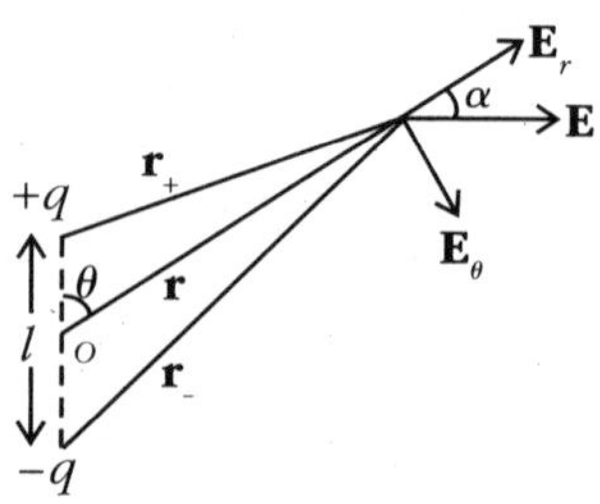

But $\quad r_- - r_+ \approx l\cos\theta \quad$ and $\quad r_+ r_- \approx r^2 \qquad (2)$

From Eqs. (1) and (2),

$$\varphi(r) = \frac{ql\cos\theta}{4\pi\varepsilon_0 r^2} = \frac{p\cos\theta}{4\pi\varepsilon_0 r^2}\ \varphi = \frac{\mathbf{p}\cdot\mathbf{r}}{4\pi\varepsilon_0 r^3}$$

where $\mathbf{p}$ is the electric moment vector.

Now,

$$E_r = \frac{-\partial\varphi}{\partial r} = \frac{2p\cos\theta}{4\pi\varepsilon_0 r^3}$$

and $$E_\theta = -\frac{\partial\varphi}{r\partial\theta} = \frac{p\sin\theta}{4\pi\varepsilon_0 r^3}$$

So, $$E = \sqrt{E_r^2 + E_\theta^2} = \frac{p}{4\pi\varepsilon_0 r^3}\sqrt{4\cos^2\theta + \sin^2\theta} = \frac{p}{4\pi\varepsilon_0 r^3}\sqrt{1+3\cos^2\theta}$$

3.40 From the results obtained in the previous problem

$$E_r = \frac{2p\cos\theta}{4\pi\varepsilon_0 r^3} \quad \text{and} \quad E_\theta = \frac{p\sin\theta}{4\pi\varepsilon_0 r^3}$$

From the given figure in the problem book, it is clear that,

$$E_z = E_r\cos\theta - E_\theta\sin\theta = \frac{p}{4\pi\varepsilon_0 r^3}(3\cos^2\theta - 1)$$

and $$E_\perp = E_r\sin\theta + E_\theta\cos\theta = \frac{3p\sin\theta\cos\theta}{4\pi\varepsilon_0 r^3}$$

When, $$\vec{\mathbf{E}}_\perp\vec{\mathbf{p}},\ |\mathbf{E}| = \mathbf{E}_\perp \quad \text{and} \quad E_z = 0$$

So, $$3\cos^2\theta = 1 \quad \text{and} \quad \cos\theta = \frac{1}{\sqrt{3}}$$

Thus $\vec{\mathbf{E}}_\perp\vec{\mathbf{p}}$ at the points located on the lateral surface of the cone, with its axis coinciding with the direction of z-axis and semi vertex angle $\theta = \cos^{-1} 1/\sqrt{3}$.

3.41 Let us assume that the dipole is at the center of one equipotential surface which is spherical (see figure). On an equipotential surface, the net electric field strength along the tangent of it becomes zero. Thus,

$$-E_0\sin\theta + E_\theta = 0 \quad \text{or} \quad -E_0\sin\theta + \frac{p\sin\theta}{4\pi\varepsilon_0 r^3} = 0$$

where E is electric field generated by the dipole.

Hence, $$r = \left(\frac{p}{4\pi\varepsilon_0 E_0}\right)^{1/3}$$

Alternate:

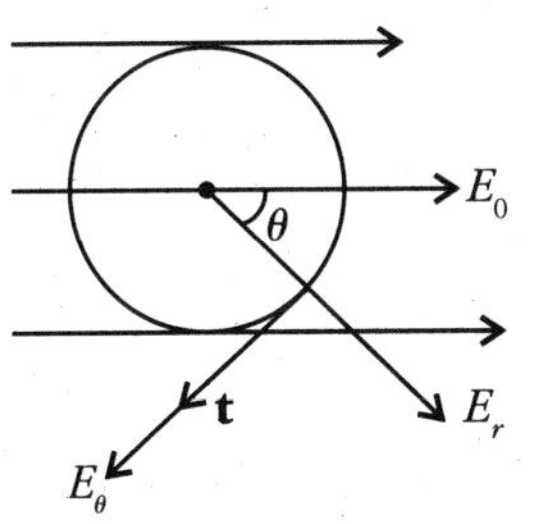

Potential at the point, near the dipole is given by

$$\varphi = \frac{\mathbf{p}\cdot\mathbf{r}}{4\pi\varepsilon_0 r^3} - \mathbf{E}_0\cdot\mathbf{r} + \text{constant},$$

$$= \left(\frac{p}{4\pi\varepsilon_0 r^3} - E_0\right)\cos\theta + \text{constant}$$

For φ to be constant,

$$\frac{p}{4\pi\varepsilon_0 r^3} - E_0 = 0 \quad \text{or} \quad \frac{p}{4\pi\varepsilon_0 r^3} = E_0$$

Thus, $$r = \left(\frac{p}{4\pi\varepsilon_0 E_0}\right)^{1/3}$$

3.42 Let P be a point, at distance $r >> l$ and at an angle θ to the vector $\mathbf{l}$ (see figure). Thus $\mathbf{E}$ at P

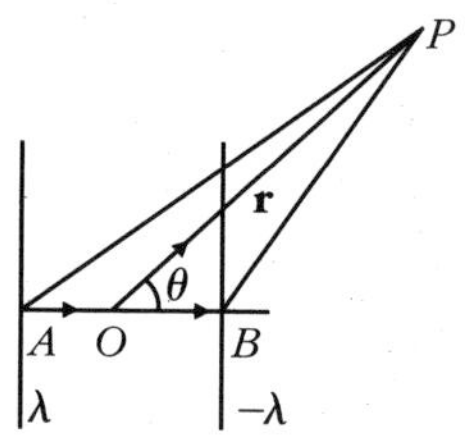

$$= \frac{\lambda}{2\pi\varepsilon_0}\frac{\mathbf{r}+\frac{\mathbf{l}}{2}}{\left|\mathbf{r}+\frac{\mathbf{l}}{2}\right|^2} - \frac{\lambda}{2\pi\varepsilon_0}\frac{\mathbf{r}-\frac{\mathbf{l}}{2}}{\left|\mathbf{r}-\frac{\mathbf{l}}{2}\right|^2}$$

$$= \frac{\lambda}{2\pi\varepsilon_0}\left[\frac{\mathbf{r}+\mathbf{l}/2}{r^2+\frac{l^2}{4}+rl\cos\theta} - \frac{\mathbf{r}-\mathbf{l}/2}{r^2+\frac{l^2}{4}-rl\cos\theta}\right]$$

$$= \frac{\lambda}{2\pi\varepsilon_0}\left(\frac{\mathbf{l}}{r^2} - \frac{2l\mathbf{r}}{r^3}\cos\theta\right)$$

Hence,
$$E = |\mathbf{E}| = \frac{\lambda l}{2\pi\varepsilon_0 r^2} \quad (\text{for } r >> l)$$

Also,
$$\varphi = \frac{\lambda}{2\pi\varepsilon_0}\ln|\mathbf{r}+\mathbf{l}/2| - \frac{\lambda}{2\pi\varepsilon_0}\ln|\mathbf{r}-\mathbf{l}/2|$$

$$= \frac{\lambda}{4\pi\varepsilon_0}\ln\frac{r^2+rl\cos\theta+l^2/4}{r^2-rl\cos\theta+l^2/4} = \frac{\lambda l\cos\theta}{2\pi\varepsilon_0 r} \quad (\text{for } r >> l)$$

3.43 The potential can be calculated by superposition. Choose the plane of the upper ring as $x = l/2$ and that of the lower ring as $x = -l/2$.

Then,
$$\varphi = \frac{q}{4\pi\varepsilon_0[R^2+(x-l/2)^2]^{1/2}} - \frac{q}{4\pi\varepsilon_0[R^2+(x+l/2)^2]^{1/2}}$$

$$\approx \frac{q}{4\pi\varepsilon_0[R^2+x^2-lx]^{1/2}} - \frac{q}{4\pi\varepsilon_0[R^2+x^2+lx]^{1/2}}$$

$$\approx \frac{q}{4\pi\varepsilon_0(R^2+x^2)^{1/2}}\left(1+\frac{lx}{2(R^2+x^2)}\right) - \frac{q}{4\pi\varepsilon_0(R^2+x^2)^{1/2}}\left(1-\frac{lx}{2(R^2+x^2)}\right)$$

$$\approx \frac{qlx}{4\pi\varepsilon_0(R^2+x^2)^{3/2}}$$

For $|x| >> R$, $\varphi \approx \dfrac{ql}{4\pi\varepsilon_0 x^2}$.

The electric field is

$$E = -\frac{\partial\varphi}{\partial x}$$

$$= -\frac{ql}{4\pi\varepsilon_0(R^2+x^2)^{3/2}} + \frac{3}{2}\frac{ql}{(R^2+x^2)^{5/2}4\pi\varepsilon_0}\times 2x = \frac{ql(2x^2-R^2)}{4\pi\varepsilon_0(R^2+x^2)^{5/2}}$$

For $|x| >> R$, $E \approx \dfrac{ql}{2\pi\varepsilon_0 x^3}$.

The plot is as shown in the answer sheet.

3.44 The field of a pair of oppositely charged sheets with holes can by superposition be reduced to that of a pair of uniform opposite charged sheets and disks with opposite charges. Now the charged sheets do not contribute any field outside them. Thus, using the result of the previous problem

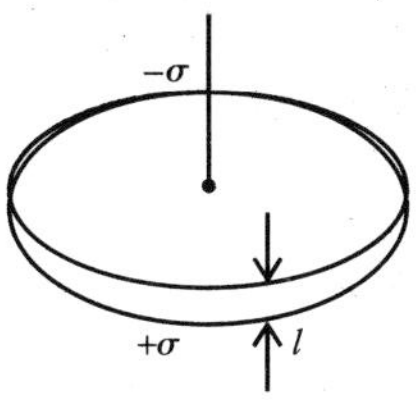

$$\varphi = \int_0^R \frac{(-\sigma)\, l 2\pi r dr\, x}{4\pi\varepsilon_0 (r^2 + x^2)^{3/2}} = \frac{\sigma x l}{4\varepsilon_0} \int_{x^2}^{R^2+x^2} \frac{dy}{y^{3/2}} \text{ (putting } y = x^2 + r^2)$$

$$= \frac{\sigma x l}{2\varepsilon_0 \sqrt{R^2 + x^2}}$$

$$E_x = -\frac{\partial \varphi}{\partial x} = -\frac{\sigma l}{2\varepsilon_0}\left[\frac{1}{\sqrt{R^2+x^2}} - \frac{x^2}{(R^2+x^2)^{3/2}}\right] = -\frac{\sigma l R^2}{2\varepsilon_0 (R^2+x^2)^{3/2}}$$

The plot is as shown in the answer sheet.

3.45 For $x > 0$, we can use the result as given above and write

$$\varphi \approx \pm \frac{\sigma l}{2\varepsilon_0}\left(1 - \frac{|x|}{(R^2 + x^2)^{1/2}}\right)$$

for the solution that vanishes at α. There is a discontinuity in potential for $|x| = 0$. The solution for negative x is obtained by $\sigma \to -\sigma$. Thus,

$$\varphi = -\frac{\sigma l x}{2\varepsilon_0 (R + x^2)^{1/2}} + \text{constant}$$

Hence, ignoring the jump

$$E = -\frac{\partial \varphi}{\partial x} = \frac{\sigma l R^2}{2\varepsilon_0 (R^2 + x^2)^{3/2}}$$

For large $|x|$,

$$\varphi \approx \pm \frac{p}{4\pi\varepsilon_0 x^2} \quad \text{and} \quad E \approx \frac{p}{2\pi\varepsilon_0 |x|^3} \quad (\text{where } p = \pi R^2\, \sigma l)$$

3.46 In cylindrical coordinates r, θ, z,

$$E_r = \frac{\lambda}{2\pi\varepsilon_0 r},\ E_\theta = 0,\ E_\varphi = 0$$

and

$$\mathbf{F} = p\frac{\partial \mathbf{E}}{\partial l}$$

or

$$\vec{\mathbf{F}} = (\mathbf{p}\cdot\nabla)\mathbf{E}$$

(a) For **p** along the thread, $\mathbf{F} = 0$

as **E** does not change as the point of observation is moved along the thread.

(b) For **p** along **r**

$$\mathbf{F} = p_r \frac{\partial(E_r \mathbf{e}_r)}{\partial r} = p\left(\frac{\partial E_r}{\partial r}\right)\mathbf{e}_r \left(\text{because } \frac{\partial \mathbf{e}_r}{\partial r} = 0\right)$$

$$= \frac{\lambda p}{2\pi\varepsilon_0 r^2}\mathbf{e}_r = -\frac{\lambda \mathbf{p}}{2\pi\varepsilon_0 r^2}$$

(c) For $\mathbf{p}$ along $\mathbf{e}_\theta$ $\quad \mathbf{F} = p_\theta \dfrac{\partial(E_r \mathbf{e}_r)}{r\partial\theta} = p\dfrac{\partial}{r\partial\theta}\left(\dfrac{\lambda}{2\pi\varepsilon_0 r}\mathbf{e}_r\right)$

$$= \frac{p\lambda}{2\,\pi\varepsilon_0 r^2}\frac{\partial \mathbf{e}_r}{\partial\theta} = \frac{p\lambda}{2\pi\varepsilon_0 r^2}\mathbf{e}_\theta = \frac{\mathbf{p}\lambda}{2\pi\varepsilon_0 r^2}\left(\text{because}\,\frac{\partial \mathbf{e}_r}{\partial\theta} = \mathbf{e}_\theta\right)$$

3.47 Force on a dipole of moment p is given by

$$F = \left|p\frac{\partial \mathbf{E}}{\partial l}\right|$$

In our problem, field due to a dipole at a distance l is given by

$$|\mathbf{E}| = \frac{p}{2\pi\varepsilon_0 l^3}$$

Hence, the force of interaction

$$F = \frac{3p^2}{2\pi\varepsilon_0 l^4} = 2.1\times10^{-16}\,\text{N}$$

3.48 Given, $\quad -d\varphi = \mathbf{E}\cdot d\mathbf{r} = a(y\,dx + xdy) = ad\,(xy)$

On integrating, $\quad \varphi = -a\,xy + \text{constant}$

3.49 Given, $\quad -d\varphi = \mathbf{E}\cdot d\mathbf{r} = [2\,axy\mathbf{i} + 2(x^2 - y^2)\mathbf{j}]\cdot[dx\mathbf{i} + dy\mathbf{j}]$

or $\quad d\varphi = 2a\,x\,y\,dx + a(x^2 - y^2)dy = ad(x^2y) - ay^2\,dy$

On integrating, $\quad \varphi = ay\left(\dfrac{y^2}{3} - x^2\right) + \text{constant}$

3.50 Given,

$$-d\varphi = \mathbf{E}\cdot d\mathbf{r} = (ay\mathbf{i} + (ax + bz)\mathbf{j} + by\mathbf{k})\cdot(dx\mathbf{i} + dy\mathbf{j} + dz\mathbf{k})$$
$$= a\,(y\,dx + ax\,dy) + b\,(zdy + ydz) = ad\,(xy) + bd\,(yz)$$

On integrating, $\quad \varphi = -(a\,xy + b\,yz) + \text{constant}$

3.51 Field intensity along x-axis

$$E_x = -\frac{\partial\varphi}{\partial x} = 3\,ax^2$$

Then using Gauss' theorem in differential form

$$\frac{\varphi E_x}{\varphi x} = \frac{\rho(x)}{\varepsilon_0} \quad \text{so,} \quad \rho(x) = 6a\,\varepsilon_0 x$$

3.52 In the space between the plates, we have by the Poisson equation

$$\frac{\partial^2\varphi}{\partial x^2} = -\frac{\rho_0}{\varepsilon_0}$$

or $\quad \varphi = -\dfrac{\rho_0}{2\varepsilon_0}x^2 + Ax + B$

where ρ_0 is the constant space charge density between the plates.

We can choose $\varphi(0) = 0$ so, $B = 0$

Then $$\varphi(d) = \Delta\varphi = Ad - \frac{\rho_0 d^2}{2\varepsilon_0} \quad \text{or} \quad A = \frac{\Delta\varphi}{d} + \frac{\rho_0 d}{2\varepsilon_0}$$

Now, $$E = \frac{\partial\varphi}{\partial x} = \frac{\rho_0}{\varepsilon_0}x - A = 0 \quad (\text{for } x = 0)$$

if $$A = \frac{\Delta\varphi}{d} + \frac{\rho_0 d}{2\varepsilon_0} = 0$$

then, $$\rho_0 = -\frac{2\varepsilon_0\Delta\varphi}{d^2}$$

Also, $$E(d) = \frac{\rho_0 d}{\varepsilon_0}$$

3.53 As $\varphi = \varphi(r)$, so field intensity is along the radial line and is given by

$$E_r = -\frac{\partial\varphi}{\partial r} = -2ar \tag{1}$$

Due to spherical symmetry of field of **E**, electric flux over a spherical Gaussian surface of radius r is $4\pi r^2 E_r$.

From Gauss' theorem,

$$4\pi r^2 E_r = \frac{q}{\varepsilon_0}$$

On differenting,

$$4\pi\, d(r^2 E_r) = \frac{dq}{\varepsilon_0}$$

where dq is the charge contained between the sphere of radii r and $r + dr$.

Therefore, $$dq = \rho \cdot 4\pi r^2 dr$$

So, $$4\pi(r^2 dE_r + 2rE_r\, dr) = \frac{\rho \cdot 4\pi r^2 dr}{\varepsilon_0}$$

$$\frac{\partial E_r}{\partial r} + \frac{2}{r}E_r = \frac{\rho}{\varepsilon_0}$$

$$-2a + \frac{2}{r}(-2ar) = \frac{\rho}{\varepsilon_0}$$

Hence, $$\rho = -6\varepsilon_0 a$$

Alternate:

Here $\mathbf{E} = -\nabla\varphi = -2ar\mathbf{e}_r = E_r\mathbf{e}_r$, where $\mathbf{e}_r$ is unit vector towards radius vector and $E_r = -2ar$.

From Gauss' law in the differential form

$$\nabla \cdot \mathbf{E} = \frac{\rho}{\varepsilon_0}$$

$$\frac{1}{r^2}\frac{\partial}{\partial r}(r^2 E_r) = \frac{\rho}{\varepsilon_0}$$

$$-\frac{1}{r^2}\frac{\partial}{\partial r}(r^2\, 2ar) = \frac{\rho}{\varepsilon_0} \quad (\text{using } E_r = -2ar)$$

Hence, $$\rho = -6\varepsilon_0 a$$

3.2 Conductors and Dielectrics in an Electric Field

3.54 When the ball is charged, for the equilibrium of ball, electric force on it must counter balance the excess spring force exerted on the ball due to the extension in the spring.

Thus,
$$F_{\text{ele}} = F_{\text{spring}}$$

or
$$\frac{q^2}{4\pi\varepsilon_0 (2l)^2} = \kappa x$$

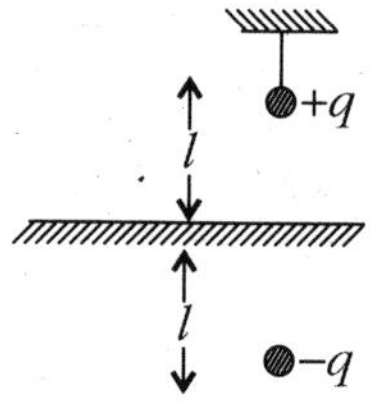

(the force on the charge q might be considered to arise from attraction by the electrical image),

or
$$q = 4/\sqrt{\pi\varepsilon_0 \kappa x}$$

which is the sought charge on the sphere.

3.55 By definition, the work of this force done upon an elementary displacement dx (see figure) is given by

$$dA = F_x dx = -\frac{q^2}{4\pi\varepsilon_0 (2x)^2} dx,$$

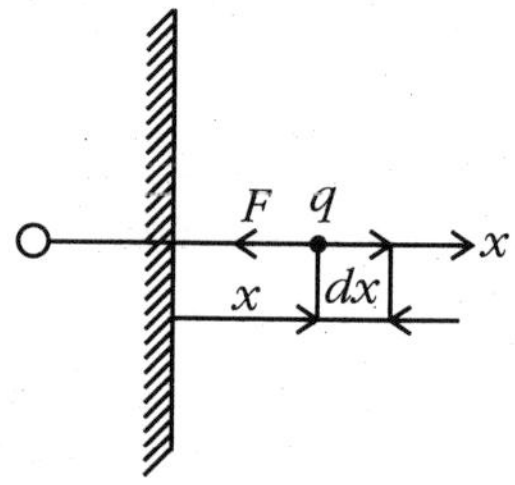

where the expression for the force is obtained with the help of the image method, Integrating this equation over x between l and ∞, we find

$$A = -\frac{q^2}{16\pi\varepsilon_0} \int_l^\infty \frac{dx}{x^2} = -\frac{q^2}{16\pi\varepsilon_0 l}$$

3.56 (a) Using the concept of electrical image, it is clear that the magnitude of the force acting on each charge,

$$|\mathbf{F}| = \sqrt{2}\frac{q^2}{4\pi\varepsilon_0 l^2} - \frac{q^2}{4\pi\varepsilon_0 \left(\sqrt{2}\,l\right)^2}$$

$$= \frac{q^2}{8\pi\varepsilon_0 l^2}\left(2\sqrt{2} - 1\right)$$

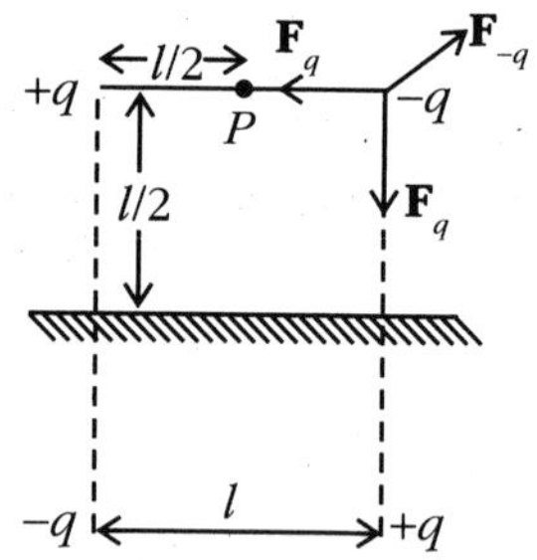

(b) Also, from the figure, magnitude of electrical field strength at P

$$E = 2\left(1 - \frac{1}{5\sqrt{5}}\right)\frac{q}{\pi\varepsilon_0 l^2}$$

3.57 Using the concept of electrical image, it is easily seen that the force on the charge q is

$$F = \frac{\sqrt{2}\,q^2}{4\pi\varepsilon_0 (2l)^2} + \frac{(-q)^2}{4\pi\varepsilon_0 \left(2\sqrt{2}\,l\right)^2}$$

$$= \frac{(2\sqrt{2} - 1)\,q^2}{32\pi\varepsilon_0 l^2} \quad \text{(It is an attractive force.)}$$

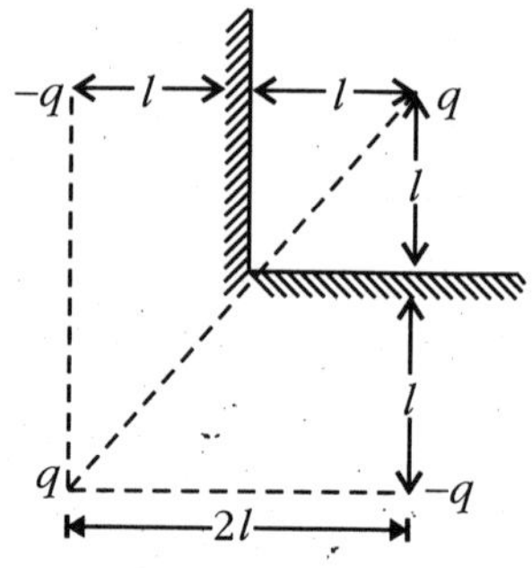

3.58 Using the concept of electrical image, force on the dipole **p**,

$$\mathbf{F} = p\frac{\partial \mathbf{E}}{\partial l}$$

(where **E** is field at the location of **p** due to (−**p**))

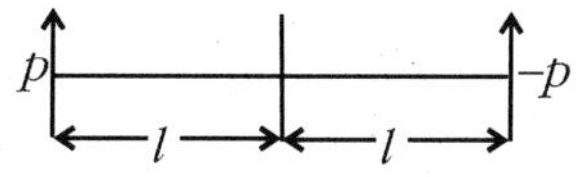

or $$|\mathbf{F}| = p\left|\frac{\partial \mathbf{E}}{\partial l}\right| = \frac{3p^2}{32\pi\varepsilon_0 l^4}$$

as $$|\mathbf{E}| = \frac{p}{4\pi\varepsilon_0(2l)^3}$$

3.59 To find the surface charge density, we must know the electric field at the point P (see figure) which is at a distance r from the point O.

Using the image mirror method, the field at P,

$$E = 2E\cos\alpha = 2\frac{q}{4\pi\varepsilon_0 x^2}\frac{l}{x} = \frac{ql}{2\pi\varepsilon_0(l^2+r^2)^{3/2}}$$

Now from Gauss' theorem, the surface charge density on conductor is connected with the electric field near its surface (in vaccum) through the relation $\sigma = \varepsilon_0 E_n$, where E_n is the projection of **E** onto the outward normal **n** (with respect to the conductor).

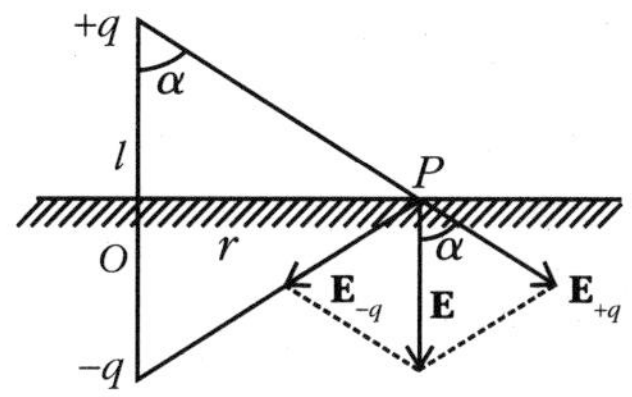

As our field strength **E** ↑↓ **n**, so

$$\sigma = -\varepsilon_0 E = -\frac{ql}{2\pi(l^2+r^2)^{3/2}}$$

3.60 (a) The force F_1 on unit length of the thread is given by

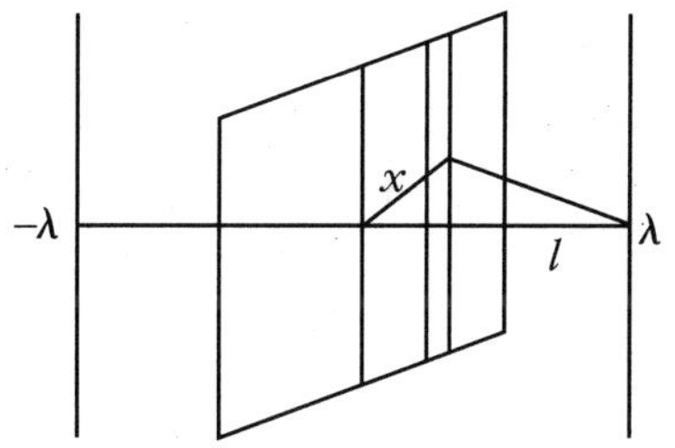

$$F_1 = \lambda E_1$$

where E_1 is the field at the thread due to image charge given by

$$E_1 = \frac{-\lambda}{2\pi\varepsilon_0(2l)}$$

Thus, $$F_1 = -\frac{\lambda^2}{4\pi\varepsilon_0 l}$$

Here, minus sign means that the force is one of attraction.

(b) There is an image thread with charge density −λ behind the conducting plane. We calculate the electric field on the conductor considering the thread and its image. It is

$$E(x) = E_n(x) = \frac{\lambda l}{\pi\varepsilon_0(x^2+l^2)}$$

Thus, $$\sigma(x) = \varepsilon_0 E_n = \frac{\lambda l}{\pi(x^2+l^2)}$$

3.61 (a) At O

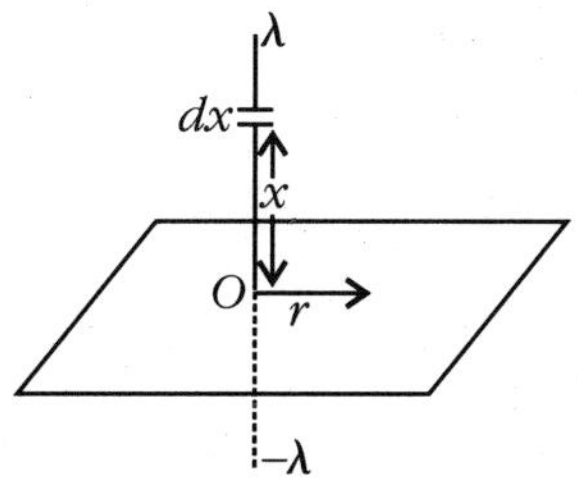

$$E_n(O) = 2\int_l^\infty \frac{\lambda\,dx}{4\pi\varepsilon_0}x_2 = \frac{\lambda}{2\pi\varepsilon_0 l}$$

So, $$\sigma(O) = \varepsilon_0 E_n = \frac{\lambda}{2\pi l}$$

(b)
$$E_n(r) = 2\int_l^\infty \frac{\lambda dx}{4\pi\varepsilon_0(x^2+r^2)} \frac{x}{(x^2+r^2)^{1/2}} = \frac{\lambda}{2\pi\varepsilon_0}\int_l^\infty \frac{x\,dx}{(x^2+r^2)^{3/2}}$$
$$= \frac{\lambda}{4\pi\varepsilon_0}\int_{l^2+r^2}^\infty \frac{dy}{y^{3/2}} \quad \text{(putting } y = x^2+r^2\text{)}$$
$$= \frac{\lambda}{2\pi\varepsilon_0\sqrt{l^2+r^2}}$$

Hence,
$$\sigma(r) = \varepsilon_0 E_n = \frac{\lambda}{2\pi\sqrt{l^2+r^2}}$$

3.62 (a) It can be easily seen that in accordance with the image method, a charge $-q$ must be located on a similar ring but on the other side of the conducting plane (see figure) at the same perpendicular distance. From the solution of Problem 3.9, electric field strength on axis of a ring of radius R at a distance l from it center is
$$\frac{ql}{4\pi\varepsilon_0(R^2+l^2)^{3/2}}$$

In our case this expression must be doubled. Hence, net field strength at point O
$$\mathbf{E} = 2\frac{ql(-\mathbf{n})}{4\pi\varepsilon_0\left(R^2+l^2\right)^{3/2}}$$

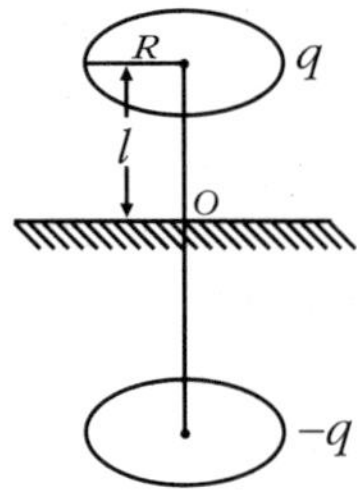

where $\mathbf{n}$ is unit vector outward normal to the conducting plane.

Now,
$$E_n = \frac{\sigma}{\varepsilon_0}$$

Hence,
$$\sigma = \frac{-ql}{2\pi(R^2+l^2)^{3/2}}$$

where minus sign indicates that the induced charge is opposite in sign to that of charge $q > 0$.

(b) The net field strength at the center of the ring is vector sum of field strengths setup by the charges q and $-q$ and is given by
$$\mathbf{E} = 0 + \frac{q(2l)}{4\pi\varepsilon_0[R^2+(2l)^2]^{3/2}}(-\mathbf{n}) \quad \text{(using result of Problem 3.9)}$$

The potential at the center of the ring is equal to the algebraic sum of the potentials at this point created by the charges q and $-q$, given by
$$\varphi = \frac{q}{4\pi\varepsilon_0 R} - \frac{q}{4\pi\varepsilon_0\sqrt{R^2+(2l)^2}}$$
$$= \frac{1}{4\pi\varepsilon_0}\left(\frac{q}{R} - \frac{q}{\sqrt{R^2+4l^2}}\right)$$

3.63 Potential φ is the same for all the points of sphere. Thus we can calculate its value at the center O of the sphere, because only for this point, it can be calculated in the most simple way.

So,
$$\varphi = \frac{1}{4\pi\varepsilon_0}\frac{q}{l} + \varphi' \qquad (1)$$

where the first term is the potential of the charge q, while the second is the potential due to the charges induced on the surface of the sphere. But since all induced charges are at the same distance equal to the radius of the circle from the point C and the total induced charge is equal to zero, so $\varphi' = 0$. Thus Eq. (1) is reduced to the form,

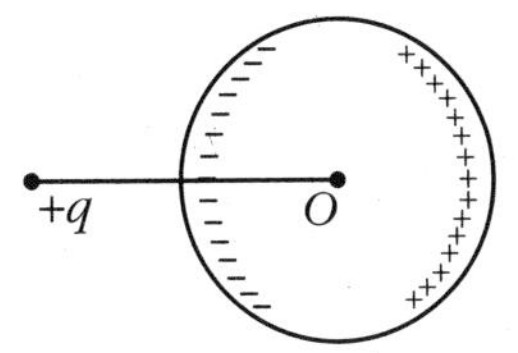

$$\varphi = \frac{1}{4\pi\varepsilon_0}\frac{q}{l}$$

3.64 Point charge q is not placed at the center of the sphere, so the induced negative charge $-q$ on the inner surface is not distributed uniformly. The left over charge $+q$ on the outer surface of the conductor distributes itself uniformly so that the field strength due to the left over charge at point inside of the conductor becomes zero. Hence, the potential at O is given by

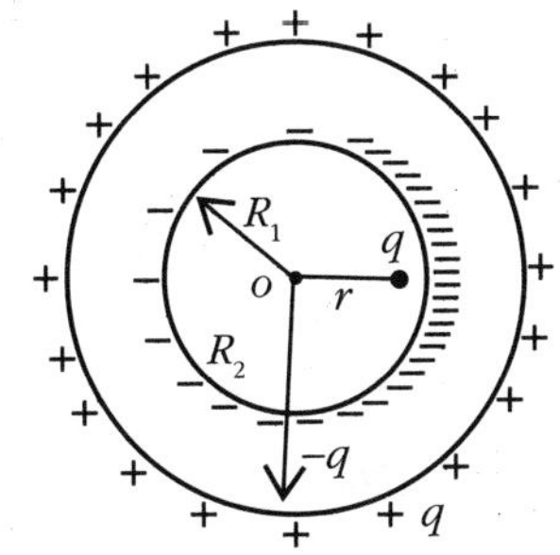

$$\varphi_0 = \frac{q}{4\pi\varepsilon_0 r} + \frac{\int(-\sigma)ds}{4\pi\varepsilon_0 R_1} + \frac{q}{4\pi\varepsilon_0 R_2}$$

$$\varphi_0 = \frac{q}{4\pi\varepsilon_0 r} + \frac{(-q)}{4\pi\varepsilon_0 R_1} + \frac{q}{4\varepsilon_0 \pi R_2}$$

It should be noticed that the potential can be found in such a simple way only at O, since all the induced charges on the inner surface are at the same distance from this point, and their distribution (which is unknown to us) does not play any role.

3.65 Potential at the inside sphere is given by

$$\varphi_a = \frac{q_1}{4\pi\varepsilon_0 a} + \frac{q_2}{4\pi\varepsilon_0 b}$$

Obviously, $\varphi_a = 0 \quad \text{for} \quad q_2 = -\frac{b}{a}q_1$ (1)

When $r \geq b$,

$$\varphi_r = \frac{q_1}{4\pi\varepsilon_0 r} + \frac{q_2}{4\pi\varepsilon_0} = \frac{q_1}{4\pi\varepsilon_0}\left(1 - \frac{b}{a}\right)\Big/ r \quad \text{(using Eq. 1)}$$

And when $r \leq b$,

$$\varphi_r = \frac{q_1}{4\pi\varepsilon_0 r} + \frac{q_2}{4\pi\varepsilon_0 b} = \frac{q_1}{4\pi\varepsilon_0}\left(\frac{1}{r} - \frac{1}{a}\right)$$

3.66 (a) As the metallic plates 1 and 4 are isolated and connected by means of a conductor, $\varphi_1 = \varphi_4$. Plates 2 and 3 have the same amount of positive and negative charges and due to induction, plates 1 and 4 are respectively negatively and positively charged and in addition to it all the four plates are located at a small but equal distance d relative to each other. Hence, electric field strength between 1–2 and 3–4 is the same (say $\mathbf{E}$). Let $\mathbf{E}'$ be the field strength between the plates 2 and 3, which is directed from 2 to 3. Hence $\mathbf{E}' \uparrow\downarrow \mathbf{E}$ (see figure).

According to the problem

$$E'd = \Delta\varphi = \varphi_2 - \varphi_3 \tag{1}$$

In addition,

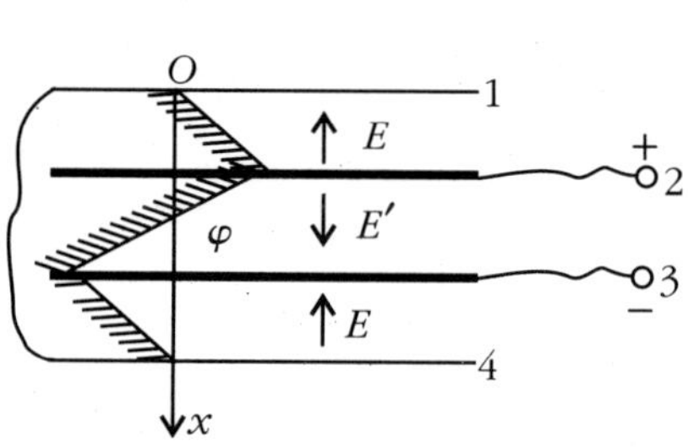

$$\varphi_1 - \varphi_4 = (\varphi_1 - \varphi_2) + (\varphi_2 - \varphi_3) + (\varphi_3 - \varphi_4)$$

or $$0 = -Ed + \Delta\varphi - Ed$$

or $$\Delta\varphi = 2Ed \Rightarrow E = \frac{\Delta\varphi}{2d} = \frac{E'}{2} \qquad (2)$$

Distribution of potential φ is also shown in the figure.

(b) Since $E \propto \sigma$, we can state according to Eq. (2) for part (a), that the charge on the plate 2 is divided into two parts; such that 1/3rd of it lies on the upper side and 2/3rd on its lower face.

Thus charge density of upper face of plate 2 or of plate 1 or plate 4 and lower face of 3 is

$$\sigma = \varepsilon_0 E = \frac{\varepsilon_0 \Delta\varphi}{d}$$

and charge density of lower face of 2 or upper face of 3 is

$$\sigma' = \varepsilon_0 E' = \varepsilon_0 \frac{\Delta\varphi}{d}$$

Hence, the net charge density of plate 2 or 3 becomes

$$\sigma + \sigma' = \frac{3\varepsilon_0 \Delta\varphi}{2d}$$

which is obvious from the argument.

3.67 The problem of point charge between two conducting planes is more easily tackled (if we want only the total charge induced on the planes) if we replace the point charge by a uniformly charged plane sheet. Let σ be the charge density on this sheet and E_1, E_2 be the outward electric field on the two sides of this sheet.

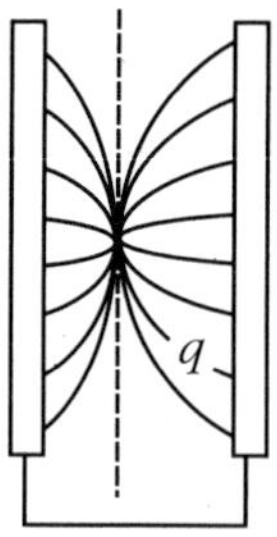

Then, $$E_1 + E_2 = \frac{\sigma}{\varepsilon_0}$$

The conducting planes will be assumed to be grounded.

Then, $$E_1 x = E_2 (l - x)$$

Hence, $$E_1 = \frac{\sigma}{l\varepsilon_0}(l - x) \quad \text{and} \quad E_2 = \frac{\sigma}{l\varepsilon_0} x$$

This means that the induced charge densities on the plane conductors are

$$\sigma_1 = -\frac{\sigma}{l}(l - x) \quad \text{and} \quad \sigma_2 = -\frac{\sigma}{l} x$$

Hence, $$q_1 = -\frac{q}{l}(l - x) \quad \text{and} \quad q_2 = -\frac{q}{l} x$$

3.68 Near the conductor

$$E = E_n = \frac{\sigma}{\varepsilon_0}$$

This field can be written as the sum of two parts E_1 and E_2. E_1 is the electric field due to an infinitesimal area dS. Very near it

$$E_1 = \pm \frac{\sigma}{2\varepsilon_0}$$

The remaining part contributes

$$E_2 = \frac{\sigma}{2\varepsilon_0} \quad \text{(on both sides)}$$

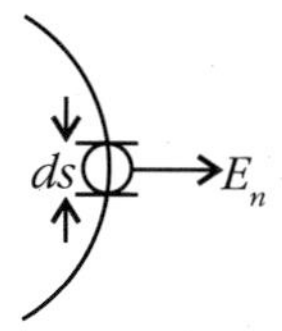

In calculating the force on the element dS, we drop E_1 (because it is a self-force). Thus,

$$\frac{dF}{ds} = \sigma\frac{\sigma}{2\varepsilon_0} = \frac{\sigma^2}{2\varepsilon_0}$$

3.69 The total force on the hemisphere is

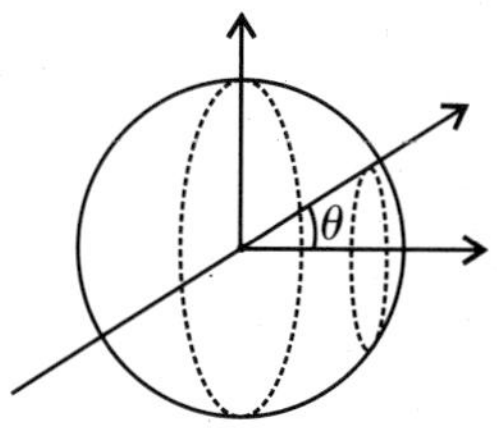

$$F = \int_0^{\pi/2} \frac{\sigma^2}{2\varepsilon_0}\cos\theta \cdot 2\pi R\sin\theta R\,d\theta$$

$$= \frac{2\pi R^2\sigma^2}{2\varepsilon_0}\int_0^{\pi/2}\cos\theta\sin\theta\,d\theta$$

$$= \frac{2\pi R^2}{2\varepsilon_0}\times\frac{1}{2}\times\left(\frac{q}{4\pi R^2}\right)^2 = \frac{q^2}{32\pi\varepsilon_0 R^2}$$

$= 0.5$ kN (on substituting values)

3.70 We know that the force acting on the area element dS of a conductor is,

$$d\mathbf{F} = \frac{\sigma^2}{2\varepsilon_0}d\mathbf{S} \quad (1)$$

It follows from symmetry considerations that the resultant force **F** is directed along the z-axis, and hence, it can be represented as the sum (integral) of the projection of elementary forces given in Eq. (1) onto the z-axis, so

$$dF_z = dF\cos\theta \quad (2)$$

For simplicity, let us consider an element area $dS = 2\pi R\sin\theta\,d\theta$ (see figure).

Now, Eq. (2) takes the form

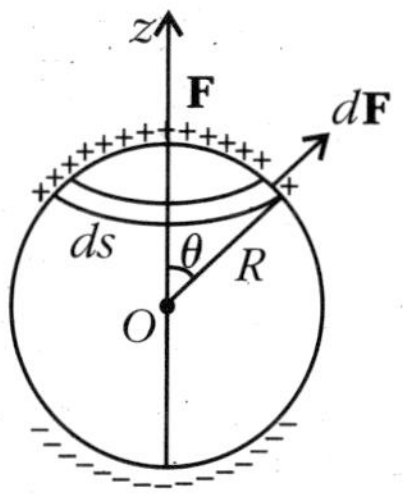

$$dF_z = \frac{\pi\sigma^2 R^2}{\varepsilon_0}\sin\theta\cos\theta\,d\theta$$

$$= -\left(\frac{\pi\sigma_0^2 R^2}{\varepsilon_0}\right)\cos^3\theta\,d\cos\theta$$

Integrating this expression over the half sphere, i.e., with respect to cos θ between 1 and 0, we obtain

$$F = F_z = \frac{\pi\sigma_0^2 R^2}{4\varepsilon_0}$$

3.71 The total polarization is $P = (\varepsilon - 1)\varepsilon_0 E$. This must be equal to $n_0 p/N$, where n_0 is the concentration of water molecules. Thus,

$$N = \frac{n_0 P}{(\varepsilon - 1)\varepsilon_0 E} = 2.93\times 10^3 \text{ (on substituting values)}$$

3.72 From the general formula in vector form

$$\mathbf{E} = \frac{1}{4\pi\varepsilon_0}\frac{(3\mathbf{p}\cdot\mathbf{e}_r)\mathbf{e}_r - \mathbf{p}}{r^3}$$

$$\mathbf{E} = \frac{1}{4\pi\varepsilon_0}\frac{2\mathbf{p}}{l^3} \quad (\text{where } r = l \text{ and } \mathbf{r} \uparrow\uparrow \mathbf{p})$$

This will cause the induction of a dipole moment

$$\mathbf{p}_{\text{ind}} = \beta\frac{1}{4\pi\varepsilon_0}\frac{2\mathbf{p}}{l^3} \times \varepsilon_0$$

Thus, the force

$$|\mathbf{F}| = \frac{\beta}{4\pi}\frac{2p}{l^3}\frac{\delta}{\delta l}\frac{1}{4\pi\varepsilon_0}\frac{2p}{l^3} = \frac{3\beta p^2}{4\pi^2\varepsilon_0 l^7}$$

3.73 The electric field E at distance x from the center of the ring is

$$E(x) = \frac{qx}{4\pi\varepsilon_0(R^2 + x^2)^{3/2}}$$

The induced dipole moment is

$$p = \beta\varepsilon_0 E = \frac{q\beta x}{4\pi(R^2 + x^2)^{3/2}}$$

The force on this molecule is

$$F = p\frac{\partial}{\partial x}E = \frac{q\beta x}{4\pi(R^2 + x^2)^{3/2}}\frac{q}{4\pi\varepsilon_0}\frac{\partial}{\partial x}\frac{x}{(R^2 + x^2)^{3/2}} = \frac{q^2\beta}{16\pi^2\varepsilon_0}\frac{x(R^2 - 2x^2)}{(R^2 + x^2)^4}$$

This vanishes for $x = \dfrac{\pm R}{\sqrt{2}}$ (apart from $x = 0$, $x = \infty$).

It is maximum when

$$\frac{\partial}{\partial x}\frac{x(R^2 - x^2 \times 2)}{(R^2 + x^2)^4} = 0$$

or
$$(R^2 - 2x^2)(R^2 + x^2) - 4x^2(R^2 + x^2) - 8x^2(R^2 - 2x^2) = 0$$

or
$$R^4 - 13x^2R^2 + 10x^4 = 0 \quad \text{or} \quad x^2 = \frac{R^2}{20}(13 \pm \sqrt{129})$$

or
$$x = \frac{R}{\sqrt{20}}\sqrt{13 \pm \sqrt{129}} \text{ (on either side).}$$

Plot of $F_x(x)$ is as shown in the answer sheet.

3.74 Inside the ball

$$\mathbf{D}(\mathbf{r}) = \frac{q}{4\pi}\frac{\mathbf{r}}{r^3} = \varepsilon\varepsilon_0\mathbf{E}$$

Also
$$\varepsilon_0\mathbf{E} + \mathbf{P} = \mathbf{D} \quad \text{or} \quad \mathbf{P} = \frac{\varepsilon - 1}{\varepsilon}\mathbf{D} = \frac{\varepsilon - 1}{\varepsilon}\frac{q}{4\pi}\frac{\mathbf{r}}{r^3}$$

Also,
$$q' = -\oint \mathbf{P}\cdot d\mathbf{S} = -\frac{\varepsilon - 1}{\varepsilon}\frac{q}{4\pi}\int d\Omega = -\frac{\varepsilon - 1}{\varepsilon}q$$

3.75 $D_{\text{diel}} = \varepsilon\varepsilon_0 E_{\text{diel}} = D_{\text{conductor}} = \sigma$ or $E_{\text{diel}} = \dfrac{\sigma}{\varepsilon\varepsilon_0}$

$$P_n = (\varepsilon - 1)\varepsilon_0 E_{\text{diel}} = \frac{\varepsilon - 1}{\varepsilon}\sigma$$

$$\sigma' = -P_n = -\frac{\varepsilon - 1}{\varepsilon}\sigma$$

This is the surface density of bound charges.

3.76 From the solution of the Problem 3.74, charge on the interior surface of the conductor is given by,

$$q'_{\text{in}} = \frac{\varepsilon - 1}{\varepsilon}\int \sigma dS = -\frac{\varepsilon - 1}{\varepsilon}q$$

Since the dielectric as a whole is neutral, there must be a total charge equal to

$$q'_{\text{outer}} = +\frac{\varepsilon - 1}{\varepsilon}q$$

on the outer surface of the dielectric.

3.77 (a) Positive extraneous charge is distributed uniformly over the internal surface layer. Let σ_0 be the surface density of the charge.

Clearly, for $r < a$ $\qquad E = 0$

By Gauss' theorem:

For $a < r b$ $\qquad \varepsilon_0 E \times 4\pi r^2 = 4\pi a^2 \sigma_0$

or $\qquad E = \dfrac{\sigma_0}{\varepsilon_0 \varepsilon}, \left(\dfrac{a}{r}\right)^2 \quad (a < r < b)$

For $r > b$, similarly

$$E = \frac{\sigma_0}{\varepsilon_0}\left(\frac{a}{r}\right)^2$$

Now, $\qquad E = -\dfrac{\partial\varphi}{\partial r}$

So by integration from infinity, where $\varphi(\infty) = 0$, we get

$$\varphi = \frac{\sigma_0 a^2}{\varepsilon_0 r} \quad (r > b)$$

For $a < r < b$, $\qquad \varphi = \dfrac{\sigma_0 a^2}{\varepsilon_0 \varepsilon r} + B$ (where B is a constant)

By continuity, $\qquad \varphi = \dfrac{\sigma_0 a^2}{\varepsilon_0 \varepsilon}\left(\dfrac{1}{r} - \dfrac{1}{b}\right) + \dfrac{\sigma_0 a^2}{\varepsilon_0 b}$

For $r < a$, $\qquad \varphi = A = \text{constant}$

By continuity, $\qquad \varphi = \dfrac{\sigma_0 a^2}{\varepsilon_0 \varepsilon}\left(\dfrac{1}{a} - \dfrac{1}{b}\right) + \dfrac{\sigma_0 a^2}{\varepsilon_0 b}$

Plot of $\varphi - r$ is given in the answer sheet of problem book.

(b) Positive extraneous charge is distributed uniformly over the internal volume of the dielectric. Let ρ_0 = volume density of the charge in the dielectric, for $a < r < b$. Then,

$$E = 0 \quad (\text{for } r < a)$$

and
$$\varepsilon_0 \varepsilon 4\pi r^2 E = \frac{4\pi}{3}(r^3 - a^3)\rho_0 \quad (\text{for } a < r < b)$$

or
$$E = \frac{\rho_0}{3\varepsilon_0 \varepsilon}\left(r - \frac{a^3}{r^2}\right)$$

$$E = \frac{4\pi}{3}\frac{(b^3 - a^3)\rho_0}{\varepsilon_0 4\pi r^2} \quad (\text{for } r > b)$$

or
$$E = \frac{(b^3 - a^3)\rho_0}{3\varepsilon_0 r^2} \quad (\text{for } r > b)$$

By integration,

$$\varphi = \frac{(b^3 - a^3)\rho_0}{3\varepsilon_0 r} \quad (\text{for } r > b)$$

or
$$\varphi = B - \frac{\rho_0}{3\varepsilon_0 \varepsilon}\left(\frac{r^2}{2} + \frac{a^3}{r}\right) \quad (\text{for } a < r < b)$$

By continuity,

$$\frac{b^3 - a^3}{3\varepsilon_0 b}\rho_0 = B - \frac{\rho_0}{3\varepsilon_0 \varepsilon}\left(\frac{b^2}{2} + \frac{a^3}{b}\right)$$

or
$$B = \frac{\rho_0}{3\varepsilon_0 \varepsilon}\left\{\frac{\varepsilon(b^3 - a^3)}{b} + \left(\frac{b^2}{2} + \frac{a^3}{b}\right)\right\}$$

Finally
$$\varphi = B - \frac{\rho_0}{3\varepsilon_0 \varepsilon}\left\{\left(\frac{a^2}{2} + a^2\right)\right\} = B - \frac{\rho_0 a^2}{2\varepsilon_0 \varepsilon} \quad (\text{for } r < a)$$

On the basis of obtained expressions, $E(r)$ and $(\varphi)(r)$ can be plotted as shown in the answer sheet.

3.78 Let us take points 1 and 2 just left and right of the interface. Point 2 can be treated like A given in the problem book. Continuity of tangential component of **E** at the interface gives

$$E_{1t} = E_{2t}$$
$$E_{1t} = E_0 \sin \alpha_0$$

As there is no free charge at the interface, so normal component of **D** has continuity at the interface.

Hence,
$$D_{1n} = D_{2n}$$
$$\varepsilon_0 \varepsilon E_{1n} = \varepsilon_0 E_0 \cos \alpha_0$$
$$E_{1n} = \frac{E_0 \cos \alpha_0}{\varepsilon}$$

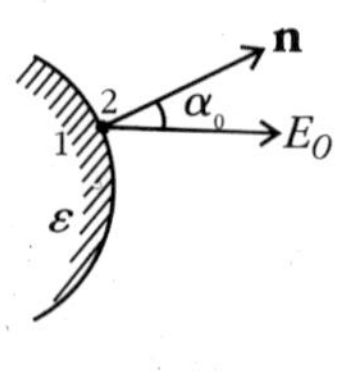

So,
$$E_1 = \sqrt{E_{1t}^2 + E_{1n}^2}$$
$$\tan \alpha = \frac{E_{1t}}{E_{1n}} = \varepsilon \tan \alpha_0$$

From the boundary condition of polarization vector **P**, we get

$$P_{2n} - P_{1n} = -\sigma'$$
$$0 - P_{1n} = -\sigma'$$
$$P_{1n} = \sigma'$$
$$\varepsilon_0(\varepsilon - 1)E_{1n} = \sigma'$$
$$\varepsilon_0(\varepsilon - 1)\frac{E_0 \cos\alpha_0}{\varepsilon} = \sigma'$$
$$\sigma' = \frac{\varepsilon_0(\varepsilon - 1)}{\varepsilon} E_0 \cos\alpha_0$$

3.79 From the previous problem, we know

$$\sigma' = \varepsilon_0 \frac{\varepsilon - 1}{\varepsilon} E_0 \cos\theta$$

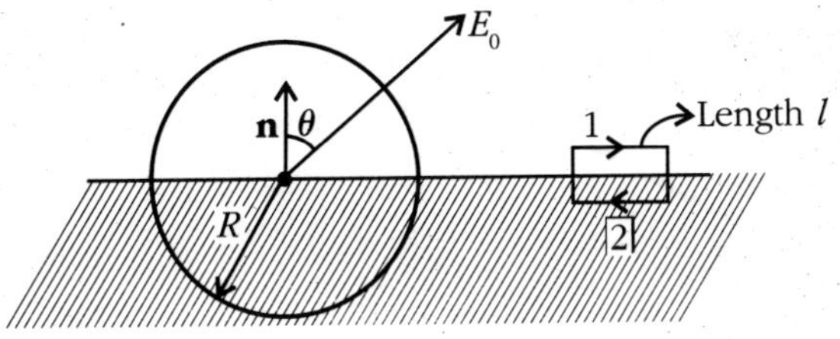

(a) Then, $$\oint \mathbf{E} \cdot d\mathbf{S} = \frac{1}{\varepsilon_0} Q = \pi R^2 E_0 \cos\theta \frac{\varepsilon - 1}{\varepsilon}$$

(b) $$\oint \mathbf{D} \cdot d\mathbf{r} = (D_{1t} - D_{2t})l = (\varepsilon_0 E_0 \sin\theta - \varepsilon\varepsilon_0 E_0 \sin\theta)$$
$$= -(\varepsilon - 1)\varepsilon_0 E_0 l \sin\theta$$

3.80 (a) $$\operatorname{div} D = \frac{\partial D_x}{\partial x} = \rho \quad \text{and} \quad D = \rho l$$

$$E_x = \frac{\rho l}{\varepsilon\varepsilon_0} \quad (\text{for } l < d)$$

and $$E_x = \frac{\rho d}{\varepsilon_0} = \text{constant } (\text{for } l > d)$$

Also, $$\varphi(x) = -\frac{\rho l^2}{2\varepsilon\rho\varepsilon_0} \quad (\text{for } l < d) \quad \text{and} \quad \varphi(x) = A - \frac{\rho l d}{\varepsilon_0} \quad (\text{for } l > d)$$

By continuity, $$\varphi(x) = \frac{\rho d}{\varepsilon_0}\left(d - \frac{d}{2\varepsilon} - l\right)$$

On the basis of obtained expressions, $E_x(x)$ and $\varphi(x)$ can be plotted as shown in the figure of answer sheet.

(b) $$\rho' = -\operatorname{div} \mathbf{P} = -\operatorname{div}(\varepsilon - 1)\varepsilon_0 \mathbf{E} = -\rho\frac{(\varepsilon - 1)}{\varepsilon}$$

$$\sigma' = P_{1n} - P_{2n} \text{ (where } n \text{ is the normal from 1 to 2)}$$
$$= P_{1n} (\mathbf{P}_2 = 0, \text{ as 2 is vacuum})$$
$$= (\rho d - \rho d/\varepsilon) = \rho d \frac{\varepsilon - 1}{\varepsilon}$$

3.81 (a)

$$\text{div}\,\mathbf{D} = \frac{l}{r^2}\frac{\partial}{\partial r}(r^2 D_r) = \rho$$

$$r^2 D_r = \rho\frac{r^3}{3} + A,\ D_r = \frac{1}{3}\rho r + \frac{A}{r^2},\ r < R$$

$$A = 0 \text{ as } D_r \neq \infty \text{ at } r = 0,$$

Thus,

$$E_r = \frac{\rho r}{3\varepsilon\varepsilon_0}$$

For $r > R$,

$$D_r = \frac{B}{r^2}$$

By continuity of D_r at $r = R$,

$$B = \frac{\rho r^3}{3}$$

So,

$$E_r = \frac{\rho R^3}{3\varepsilon_0 r^2} \quad (\text{for } r > R)$$

$$\varphi = \frac{\rho R^3}{3\varepsilon_0 r} \quad (\text{for } r > R) \quad \text{and} \quad \varphi = -\frac{\rho r^2}{6\varepsilon\varepsilon_0} + C \quad (\text{for } r < R)$$

By continuity of φ $C = +\dfrac{\rho R^2}{3\varepsilon_0} + \dfrac{\rho R^2}{6\varepsilon\varepsilon_0}$

See answer sheet for graphs of $E(r)$ and $\varphi(r)$.

(b) $$\rho' = -\text{div}\,\mathbf{P} = -\frac{1}{r^2}\frac{\partial}{\partial r}\left\{\frac{r^3}{3}\rho\left(1 - \frac{1}{\varepsilon}\right)\right\} = -\frac{\rho(\varepsilon - 1)}{\varepsilon}$$

$$\sigma' = P_{1r} - P_{2r} = P_{1r} = \frac{1}{3}\rho R\left(1 - \frac{1}{\varepsilon}\right)$$

3.82 Because there is a discontinuity in polarization at the boundary of the dielectric disk, a bound surface charge appears, which is the source of the electric field inside and outside the disk. We have for the electric field at the origin,

$$\mathbf{E} = -\int \frac{\sigma' dS}{4\pi\varepsilon_0 r^3}\mathbf{r}$$

where $\mathbf{r}$ = radius vector from the origin to the element dS. $\sigma' = P_n = P\cos\theta$ on the curved surface (as $P_n = 0$ on the flat surface). Here θ = angle between $\mathbf{r}$ and $\mathbf{P}$. By symmetry, $\mathbf{E}$ will be parallel to $\mathbf{P}$.

Thus,

$$E \cong -\int_0^{2\pi} \frac{P\cos\theta\, Rd\theta\cdot\cos\theta}{4\pi\varepsilon_0 R^2}\cdot d$$

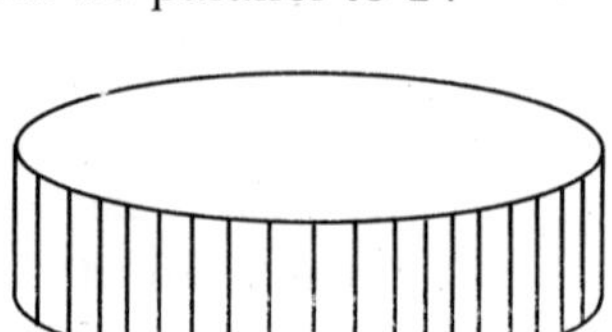

where, $r = R$, if $d \ll R$.

So,

$$E \cong -\frac{Pd}{4\varepsilon_0 R}$$

and

$$\mathbf{E} = -\frac{\mathbf{P}d}{4\varepsilon_0 R}$$

3.83 Since there are no free extraneous charges anywhere

$$\text{div } \mathbf{D} = \frac{\partial D_x}{\partial x} = 0 \quad \text{or} \quad D_x = \text{constant}$$

But $\qquad D_x = 0 \text{ at } \infty \quad \text{so,} \quad D_x = 0, \text{ everywhere}$

Thus, $\qquad \mathbf{E} = -\frac{\mathbf{P}_0}{\varepsilon_0}\left(1 - \frac{x^2}{d^2}\right) \quad \text{or} \quad E_x = -\frac{P_0}{\varepsilon_0}\left(1 - \frac{x^2}{d^2}\right)$

So, $\qquad \varphi = \frac{P_0 x}{\varepsilon_0} - \frac{P_0 x^3}{3\varepsilon_0 d^2} + \text{constant}$

Hence, $\qquad \varphi(+d) - \varphi(-d) = \frac{2P_0 d}{\varepsilon_0} - \frac{2P_0 d^3}{3d^2 \varepsilon_0} = \frac{4P_0 d}{3\varepsilon_0}$

3.84 (a) We have $\qquad D_1 = D_2 \quad \text{or} \quad \varepsilon E_2 = E_1$

Also, $\qquad E_1 \frac{d}{2} + E_2 \frac{d}{2} = E_0 d \Rightarrow E_1 + E_2 = 2E_0$

Hence, $\qquad E_2 = \frac{2E_0}{\varepsilon + 1} \quad \text{and} \quad E_1 = \frac{2\varepsilon E_0}{\varepsilon + 1}$

and $\qquad D_1 = D_2 = \frac{2\varepsilon\varepsilon_0 E_0}{\varepsilon + 1}$

(b) We have $\qquad D_1 = D_2 \quad \text{or} \quad \varepsilon E_2 = E_1 = \frac{\sigma}{\varepsilon_0} = E_0$

Thus, $\qquad E_1 = E_0 \quad \text{and} \quad E_2 = \frac{E_0}{\varepsilon}$

and $\qquad D_1 = D_2 = \varepsilon_0 E_0$

3.85 (a) Constant voltage across the plates implies

$$E_1 = E_2 = E_0,\ D_1 = \varepsilon_0 E_0,\ D_2 = \varepsilon_0 \varepsilon E_0$$

(b) Constant charge across the plates implies

$$E_1 = E_2 D_1 = \varepsilon_0 E_1,\ D_2 = \varepsilon \varepsilon_0 E_2 = \varepsilon D_1$$

So, $\qquad E_1(1 + \varepsilon) = 2E_0 \quad \text{or} \quad E_1 = E_2 = \frac{2E_0}{\varepsilon + 1}$

3.86 At the interface of the dielectric and vacuum,

$$E_{1l} = E_{2l}$$

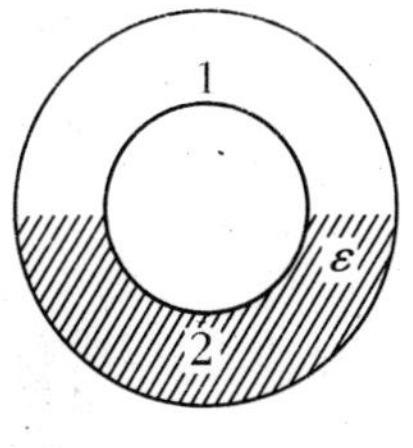

The electric field must be radial and

$$E_1 = E_2 = \frac{A}{\varepsilon_0 \varepsilon r^2} \quad (\text{for } a < r < b)$$

Now, $\qquad q = \frac{A}{R^2}(2\pi R^2) + \frac{A}{\varepsilon R^2}(2\pi R^2)$

$$= A\left(1 + \frac{1}{\varepsilon}\right)2\pi$$

or $\qquad E_1 = E_2 = \frac{q}{2\pi\varepsilon_0 r^2(1 + \varepsilon)}$

3.87 In air the forces are as shown. In *K*–oil,

$$F \to F' = F/\varepsilon \quad \text{and} \quad mg \to mg\left(1-\frac{\rho_0}{\rho}\right)$$

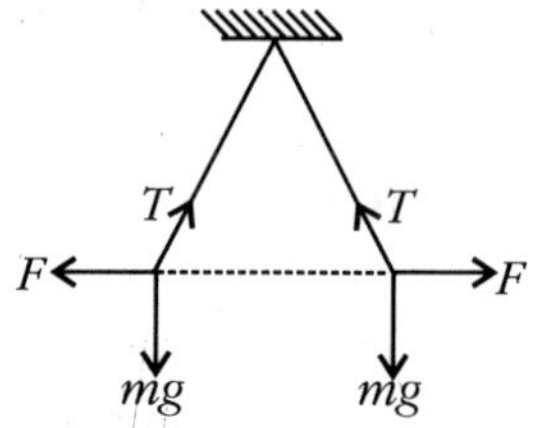

where ρ_0 is the density of *K*–oil and ρ that of the material of which the balls are made.

Since the inclinations do not change

$$\frac{1}{\varepsilon} = 1 - \frac{\rho_0}{\rho}$$

or
$$\frac{\rho_0}{\rho} = 1 - \frac{1}{\varepsilon} = \frac{\varepsilon - 1}{\varepsilon}$$

or
$$\rho = \rho_0 \frac{\varepsilon}{\varepsilon - 1}$$
$$= 1.6 \text{ g/cm}^3 \text{ (on substituting values)}$$

3.88 Within the ball, the electric field can be resolved into normal and tangential components,

$$E_n = E\cos\theta \quad \text{and} \quad E_t = E\sin\theta$$

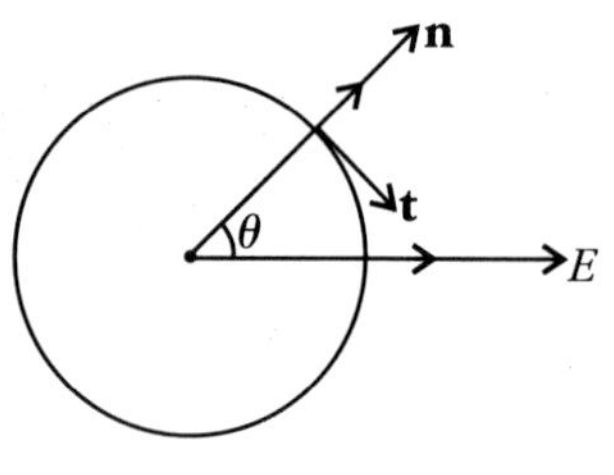

Then,
$$D_n = \varepsilon\varepsilon_0 E\cos\theta$$

and
$$P_n = (\varepsilon - 1)\varepsilon_0 E\cos\theta$$

or
$$\sigma' = (\varepsilon - 1)\varepsilon_0 E\cos\theta$$

So,
$$\sigma_{\max} = (\varepsilon - 1)\varepsilon_0 E,$$
$$= 3.5 \text{ nC/m}^2 \text{ (on substituting values)}$$

and total charge of one sign,

$$q' = \int_0^1 (\varepsilon - 1)\varepsilon_0 E\cos\theta \, 2\pi R^2 d(\cos\theta)$$
$$= \pi R^2 \varepsilon_0(\varepsilon - 1)E = 10 \text{ pC (on substituting values)}$$

(Since we are interested in the total charge of one sign, we must integrate $\cos\theta$ from 0 to 1 only).

3.89 The charge is at *A* in the medium 1 and has an image point at *A'* in the medium 2. The electric field in the medium 1 is due to the actual charge *q* at *A* and the image charge *q'* at *A'*. The electric field in 2 is due to a corrected charge *q''* at *A*. Thus, on the boundary between 1 and 2

$$E_{1n} = \frac{q'}{4\pi\varepsilon_0 r^2}\cos\theta - \frac{q}{4\pi\varepsilon_0 r^2}\cos\theta$$

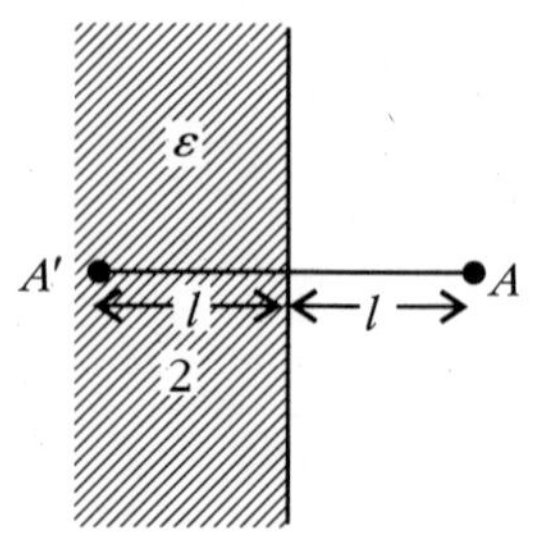

$$E_{2n} = \frac{-q''}{4\pi\varepsilon_0 r^2}\cos\theta$$

$$E_{1t} = \frac{q'}{4\pi\varepsilon_0 r^2}\sin\theta + \frac{q}{4\pi\varepsilon_0 r^2}\sin\theta$$

$$E_{2t} = \frac{q''}{4\pi\varepsilon_0 r^2}\sin\theta$$

The boundary conditions are

$$D_{1n} = D_{2n} \quad \text{and} \quad E_{1t} = E_{2t}$$

$$\varepsilon q'' = q - q'$$

$$q'' = q + q'$$

So,
$$q'' = \frac{2q}{\varepsilon+1} \quad \text{and} \quad q' = -\frac{\varepsilon-1}{\varepsilon+1}q$$

(a) The surface density of the bound charge on the surface of the dielectric is

$$\sigma' = P_{2n} = D_{2n} - \varepsilon_0 E_{2n} = (\varepsilon-1)\varepsilon_0 E_{2n}$$

$$= -\frac{\varepsilon-1}{\varepsilon+1}\frac{q}{2\pi r^2}\cos\theta = -\frac{\varepsilon-1}{\varepsilon+1}\frac{ql}{2\pi r^3}$$

(b) Total bound charge is

$$-\frac{\varepsilon-1}{\varepsilon+1}q\int_0^\infty \frac{l}{2\pi(l^2+x^2)^{3/2}}2\pi x dx = -\frac{\varepsilon-1}{\varepsilon+1}q$$

3.90 The force on the point charge q is due to the bound charges. This can be calculated from the field at this change after extracting out the self field. This image field is

$$E_{\text{image}} = \frac{\varepsilon-1}{\varepsilon+1}\frac{q}{4\pi\varepsilon_0(2l)^2}$$

Thus,
$$F = \frac{\varepsilon-1}{\varepsilon+1}\frac{q^2}{16\pi\varepsilon_0 l^2}$$

3.91 Given
$$\mathbf{E}_P = \frac{q\mathbf{r}_1}{4\pi\varepsilon_0 r_1^3} + \frac{q'\mathbf{r}_2}{4\pi r_2^3\varepsilon_0} \quad \text{(for } P \text{ in 1)}$$

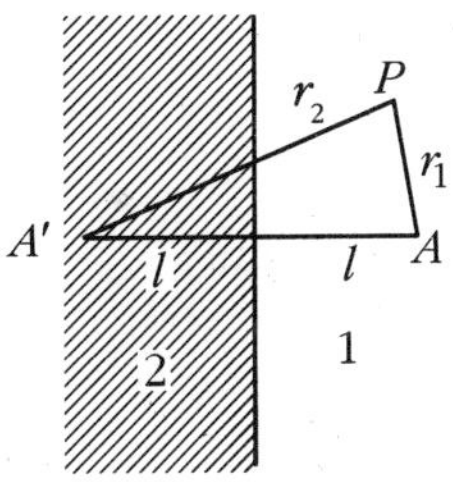

and
$$\mathbf{E}_P = \frac{q''\mathbf{r}_1}{4\pi\varepsilon_0 r_1^3} \quad \text{(for } P \text{ in 2)}$$

where
$$q'' = \frac{2q}{\varepsilon+1} \quad \text{and} \quad q' = q'' - q$$

In the limit $\mathbf{1} \to 0$

$$\mathbf{E}_p = \frac{(q+q')\mathbf{r}}{4\pi\varepsilon_0 r^3} = \frac{q\mathbf{r}}{2\pi\varepsilon_0(1+\varepsilon)r^3} \quad \text{(in either part)}$$

Thus,
$$E_P = \frac{q}{2\pi\varepsilon_0(1+\varepsilon)r^2}$$

$$\varphi = \frac{q}{2\pi\varepsilon_0(1+\varepsilon)r}$$

$$D = \frac{q}{2\pi\varepsilon_0(1+\varepsilon)r^2} \times \begin{cases} 1 \text{ in vacuum} \\ \varepsilon \text{ in dielectric} \end{cases}$$

3.92
$$\mathbf{E}_P = \frac{q\mathbf{r}_2}{4\pi\varepsilon_0\varepsilon r_2^3} + \frac{q'\mathbf{r}_1}{4\pi\varepsilon_0 r_1^3} \text{(for } P \text{ in 2)}$$

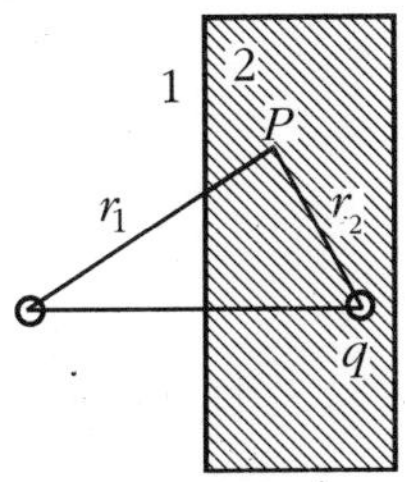

$$\mathbf{E}_P = \frac{q''\mathbf{r}_2}{4\pi\varepsilon_0 r_2^3} \quad \text{(for } P \text{ in 1)}$$

Using the boundary conditions,

$$E_{1n} = \varepsilon E_{2n},\ E_{1t} = E_{2t}$$

This implies

$$q - \varepsilon q' = q'' \quad \text{and} \quad q + \varepsilon q' = \varepsilon q''$$

So,

$$q'' = \frac{2q}{\varepsilon + 1} \quad \text{and} \quad q' = \frac{\varepsilon - 1}{\varepsilon + 1}\frac{q}{\varepsilon}$$

Then, as earlier,

$$\sigma' = \frac{ql}{2\pi r^3} \cdot \left(\frac{\varepsilon - 1}{\varepsilon + 1}\right) \cdot \frac{1}{\varepsilon}$$

3.93 To calculate the electric field, first we note that an image charge will be needed to ensure that the electric field on the metal boundary is normal to the surface.

The image charge must have magnitude $-q/\varepsilon$, so that the tangential component of the electric field may vanish. Now,

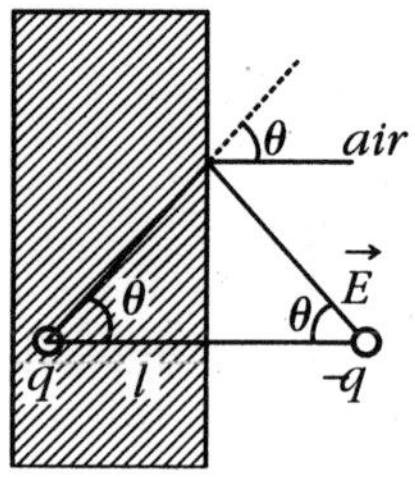

$$E_n = \frac{1}{4\pi\varepsilon_0}\left(\frac{q}{\varepsilon r^2}\right) 2\cos\theta = \frac{ql}{2\pi\varepsilon_0 \varepsilon r^3}$$

Then,

$$P_n = D_n - \varepsilon_0 E_n = \frac{(\varepsilon - 1)ql}{2\pi\varepsilon r^3} = \sigma'$$

This is the density of bound charge on the surface.

3.94 Since the condenser plates are connected,

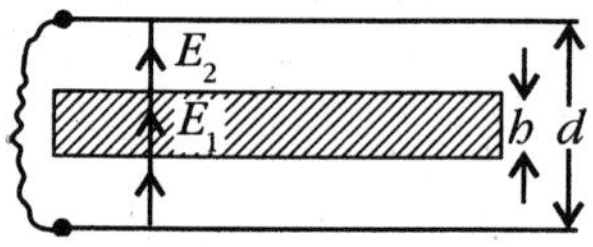

$$E_1 b + E_2(d - b) = 0$$

and

$$P + \varepsilon_0 E_1 = \varepsilon_0 E_2$$

or

$$E_1 + \frac{P}{\varepsilon_0} = E_2$$

Thus,

$$E_2 d - \frac{Pb}{\varepsilon_0} = 0 \quad \text{or} \quad E_2 = \frac{Pb}{\varepsilon_0 d}$$

$$E_1 = -\frac{P}{\varepsilon_0}\left(1 - \frac{b}{d}\right)$$

3.95 Given $\mathbf{P} = \alpha\mathbf{r}$, where $\mathbf{r}$ = distance from the axis. The space density of charges is given by, $\rho' = -\text{div}\,\mathbf{P} = -2\alpha$. On using these, we get

$$\text{div}\,\mathbf{r} = \frac{1}{r}\frac{\partial(\mathbf{r}\cdot\mathbf{r})}{\partial r} = 2$$

3.96 In a uniformly charged sphere

$$\mathbf{E} = \frac{\rho_0}{3\varepsilon_0}\mathbf{r}$$

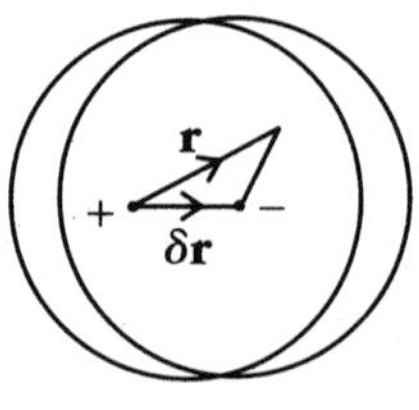

The total electric field is

$$\mathbf{E} = \frac{1}{3\varepsilon_0}\rho_0\mathbf{r} - \frac{1}{3\varepsilon_0}(\mathbf{r} - \delta\mathbf{r})\rho_0$$

$$= \frac{1}{3\varepsilon_0}\rho_0\,\delta\mathbf{r} = -\frac{\mathbf{P}}{3\varepsilon_0} \quad (\text{where } \rho\,\delta\mathbf{r} = -\mathbf{P})$$

The potential outside is

$$\varphi = \frac{1}{4\pi\varepsilon_0}\left(\frac{Q}{|\mathbf{r}|} - \frac{Q}{|\mathbf{r} - \delta\mathbf{r}|}\right) = \frac{\mathbf{P}_0 \cdot \mathbf{r}}{4\pi\varepsilon_0 r^3} \quad (\text{for } r > R)$$

where $\mathbf{P}_0 = -\dfrac{4\pi}{3}R^3\rho_0\partial\mathbf{r}$, is the total dipole moment.

3.97 The electric field $\mathbf{E}_0$ in a spherical cavity in a uniform dielectric of permittivity ε is related to the far away field $\mathbf{E}$, in the following manner. Imagine the cavity to be filled up with the dielectric. Then there will be a uniform field $\mathbf{E}$ everywhere and a polarization $\mathbf{P}$, given by $\mathbf{P} = (\varepsilon - 1)\varepsilon_0\mathbf{E}$.

Now take out the sphere making the cavity, the electric field inside the sphere will be

$$-\frac{\mathbf{P}}{3\varepsilon_0}$$

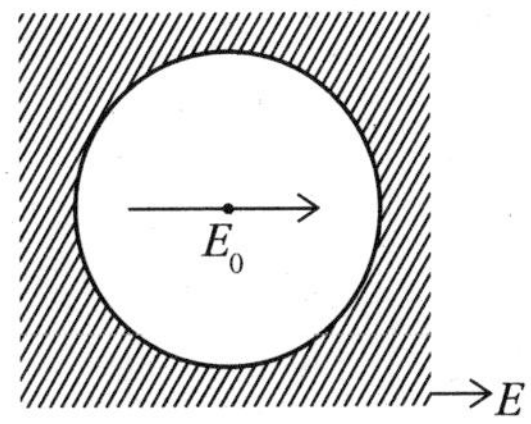

By superposition,

$$\mathbf{E}_0 - \frac{\mathbf{P}}{3\varepsilon_0} = \mathbf{E}$$

or

$$\mathbf{E}_0 = \mathbf{E} + \frac{1}{3}(\varepsilon - 1)\mathbf{E} = \frac{1}{3}(\varepsilon + 2)\mathbf{E}$$

3.98 By superposition the field $\mathbf{E}$ inside the ball is given by

$$\mathbf{E} = \mathbf{E}_0 - \frac{\mathbf{P}}{3\varepsilon_0}$$

On the other hand, if the sphere is not too small, then the macroscopic equation $\mathbf{P} = (\varepsilon - 1)\varepsilon_0\mathbf{E}$ must hold. Thus,

$$\mathbf{E}\left(1 + \frac{1}{3}(\varepsilon - 1)\right) = \mathbf{E}_0 \quad \text{or} \quad \mathbf{E} = \frac{3\mathbf{E}_0}{\varepsilon + 2}$$

Also,

$$\mathbf{P} = 3\varepsilon_0\frac{\varepsilon - 1}{\varepsilon + 2}\mathbf{E}_0$$

3.99 This is to be handled by the same trick as in Problem 3.96. We have effectively a two dimensional situation. For a uniform cylinder full of charge, with charge density ρ_0, the electric field E at an inside point is along the (cylindrical) radius vector $\mathbf{r}$ and is given by

$$\mathbf{E} = \frac{1}{2\varepsilon_0}\rho\mathbf{r}$$

Since

$$\text{div}\,\mathbf{E} = \frac{l}{r}\frac{\partial}{\partial r}(rE_r) = \frac{\rho}{\varepsilon_0} \quad \text{hence,} \quad E_r = \frac{\rho}{2\varepsilon_0}r$$

Therefore the polarized cylinder can be thought of as two equal and opposite charge distributions displaced with respect to each other. So,

$$\mathbf{E} = \frac{1}{2\varepsilon_0}\rho\mathbf{r} - \frac{1}{2\varepsilon_0}\rho(\mathbf{r} - \delta\mathbf{r}) = \frac{1}{2\varepsilon_0}\rho\,\delta\mathbf{r} = -\frac{\mathbf{P}}{2\varepsilon_0}$$

using $\mathbf{P} = -\rho\delta\mathbf{r}$ (direction of electric dipole moment vector being from the negative charge to positive charge).

3.100 As in solution of Problem 3.98, we can write

$$\mathbf{E} = \mathbf{E}_0 - \frac{\mathbf{P}}{2\varepsilon_0}$$

Also,

$$\mathbf{P} = (\varepsilon - 1)\varepsilon_0\mathbf{E}$$

So,

$$\mathbf{E}\left(\frac{\varepsilon + 1}{2}\right) = \mathbf{E}_0$$

or

$$\mathbf{E} = \frac{2\mathbf{E}_0}{\varepsilon + 1} \quad \text{and} \quad \mathbf{P} = 2\varepsilon_0\frac{\varepsilon - 1}{\varepsilon + 1}\mathbf{E}_0$$

3.3 Electric Capacitance. Energy of an Electric Field

3.101 Let us impart an imaginary charge q to the conductor, then

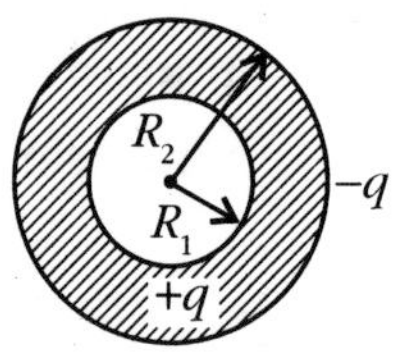

$$\varphi_+ - \varphi_- = \int_{R_1}^{R_2} \frac{q}{4\pi\varepsilon_0 \varepsilon r^2} dr + \int_{R_2}^{\infty} \frac{q}{4\pi\varepsilon_0 r^2} dr$$

$$= \frac{q}{4\pi\varepsilon_0\varepsilon}\left[\frac{1}{R_1} - \frac{1}{R_2}\right] + \frac{q}{4\pi\varepsilon_0}\frac{1}{R_2}$$

$$= \frac{q}{4\pi\varepsilon_0\varepsilon}\left[\frac{(\varepsilon-1)}{R_2} + \frac{1}{R_1}\right]$$

Hence the sought capacitance,

$$C = \frac{q}{\varphi_+ - \varphi_-} = \frac{q 4\pi\varepsilon_0\varepsilon}{q\left[\frac{(\varepsilon-1)}{R_2} + \frac{1}{R_1}\right]} = \frac{4\pi\varepsilon_0\varepsilon R_1}{(\varepsilon-1)\frac{R_1}{R_2} + 1}$$

3.102 From the symmetry of the problem, the voltage across each capacitor $\Delta\varphi = \xi/2$ and charge on each capacitor $q = C\xi/2$ in the absence of dielectric.

Now when the dielectric is filled up in one of the capacitors, the equivalent capacitance of the system is given by

$$C_0' = \frac{C\varepsilon}{1+\varepsilon}$$

and the potential difference across the capacitor, which is filled with dielectric is

$$\Delta\varphi' = \frac{q'}{\varepsilon C} = \frac{C\varepsilon}{(1+\varepsilon)}\frac{\xi}{C\varepsilon} = \frac{\xi}{(1+\varepsilon)}$$

But $\varphi \propto E$.

So, as φ decreases $1/2\,(1+\varepsilon)$ times, the field strength also decreases by the same factor and flow of charge

$$\Delta q = q' - q$$

$$= \frac{C\varepsilon}{(1+\varepsilon)}\xi - \frac{C}{2}\xi = \frac{1}{2}C\xi\frac{(\varepsilon-1)}{(\varepsilon+1)}$$

3.103 (a) Since it is a series combination of two capacitors,

$$\frac{1}{C} = \frac{d_1}{\varepsilon_0\varepsilon_1 S} + \frac{d_2}{\varepsilon_0\varepsilon_2 S} \quad \text{or} \quad C = \frac{\varepsilon_0 S}{(d_1/\varepsilon_1) + (d_2/\varepsilon_2)}$$

(b) Let σ be the initial surface charge density, then density of bound charge on the boundary plane is given by

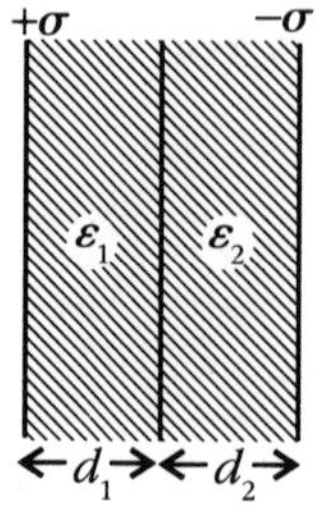

$$\sigma' = \sigma\left(1 - \frac{1}{\varepsilon_1}\right) - \sigma\left(1 - \frac{1}{\varepsilon_2}\right) = \sigma\left(\frac{1}{\varepsilon_2} - \frac{1}{\varepsilon_1}\right)$$

But, $$\sigma = \frac{q}{S} = \frac{CV}{S} = \frac{\varepsilon_0 S\varepsilon_1\varepsilon_2}{\varepsilon_2 d_1 + \varepsilon_1 d_2}\frac{V}{S}$$

So, $$\sigma' = \frac{\varepsilon_0 V(\varepsilon_1 - \varepsilon_2)}{\varepsilon_2 d_1 + \varepsilon_1 d_2}$$

3.104 (a) We point the x-axis towards the right and place the origin on the left hand side plate. The left hand side plate is assumed to be positively charged. Since ε varies linearly, we can write,

$$\varepsilon(x) = a + bx$$

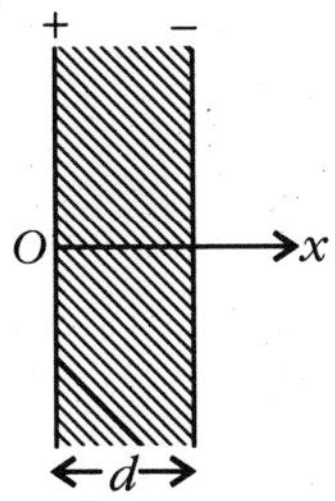

where a and b can be determined from the boundary condition. We have,

$$\varepsilon = \varepsilon_1 \text{ at } x = 0 \quad \text{and} \quad \varepsilon = \varepsilon_2 \text{ at } x = d$$

Thus, $$\varepsilon(x) = \varepsilon_1 + \left(\frac{\varepsilon_2 - \varepsilon_1}{d}\right)x$$

Now potential difference between the plates

$$\varphi_+ - \varphi_- = \int_0^d \mathbf{E}\cdot d\mathbf{r} = \int_0^d \frac{\sigma}{\varepsilon_0 \varepsilon(x)} dx$$

$$= \int_0^d \frac{\sigma}{\varepsilon_0\left(\varepsilon_1 + \frac{\varepsilon_2 - \varepsilon_1}{d}x\right)} dx = \frac{\sigma d}{(\varepsilon_2 - \varepsilon_1)\varepsilon_0} \ln \frac{\varepsilon_2}{\varepsilon_1}$$

Hence, the sought capacitance,

$$C = \frac{\sigma S}{\varphi_+ - \varphi_-} = \frac{(\varepsilon_2 - \varepsilon_1)\varepsilon_0 S}{(\ln \varepsilon_2/\varepsilon_1) d}$$

(b) Since, $$D = \frac{q}{S} \quad \text{and} \quad P = \frac{q}{S} - \frac{q}{S\varepsilon(x)}$$

and the space density of bound charge is

$$\rho' = -div\, P = -\frac{q(\varepsilon_2 - \varepsilon_1)}{Sd\varepsilon^2(x)}$$

3.105 Let us impart an imaginary charge q to the conductor. Now, the potential difference between the plates will be

$$\varphi_+ - \varphi_- = \int_{R_1}^{R_2} \mathbf{E}\cdot d\mathbf{r}$$

$$= \int_{R_1}^{R_2} \frac{q}{4\pi\varepsilon_0 a/r} \frac{1}{r^2} dr = \frac{q}{4\pi\varepsilon_0 a} \ln R_2/R_1$$

Hence, the sought capacitance

$$C = \frac{q}{\varphi_+ - \varphi_-} = \frac{q 4\pi\varepsilon_0 a}{q \ln R_2/R_1} = \frac{4\pi\varepsilon_0 a}{\ln R_2/R_1}$$

3.106 Let λ be the linear charge density.

Then, $$E_{1m} = \frac{\lambda}{2\pi\varepsilon_0 R_1 \varepsilon_1} \tag{1}$$

and $$E_{2m} = \frac{\lambda}{2\pi\varepsilon_0 R_2 \varepsilon_2} \tag{2}$$

The breakdown in either case will occur at the smaller value of r for a simultaneous breakdown of both dielectrics. From Eqs. (1) and (2)

$$E_{1m} R_1 \varepsilon_1 = E_{2m} R_2 \varepsilon_2$$

which is the sought relationship.

3.107 Let λ be the linear charge density, then, the sought potential difference

$$\varphi_+ - \varphi_- = \int_{R_1}^{R_2} \frac{\lambda}{2\pi\varepsilon_0\varepsilon_1 r} dr + \int_{R_2}^{R_3} \frac{\lambda}{2\pi\varepsilon_0\varepsilon_2 r} dr$$

$$= \frac{\lambda}{2\pi\varepsilon_0}\left[\frac{1}{\varepsilon_1}\ln R_2/R_1 + \frac{1}{\varepsilon_2}\ln R_3/R_2\right]$$

As, $E_1R_1\varepsilon_1 < E_2R_2\varepsilon_2$, so

$$\frac{\lambda}{2\pi\varepsilon_0} = E_1R_1\varepsilon_1$$

is the maximum acceptable value, and for values greater than $E_1R_1\varepsilon_1$, dielectric breakdown will take place. Hence, the maximum potential difference between the plates

$$\varphi_+ - \varphi_- = E_1R_1\varepsilon_1\left[\frac{1}{\varepsilon_1}\ln R_2/R_1 + \frac{1}{\varepsilon_2}\ln R_3/R_2\right]$$

$$= E_1R_1\left[\ln R_2/R_1 + \frac{\varepsilon_1}{\varepsilon_2}\ln R_3/R_2\right]$$

3.108 Let us suppose that linear charge density of the wires be λ, then the potential difference will be $\varphi_+ - \varphi_- = \varphi - (-\varphi) = 2\varphi$. The intensity of the electric field created by one of the wires at a distance x from its axis can be easily found with the help of the Gauss' theorem as

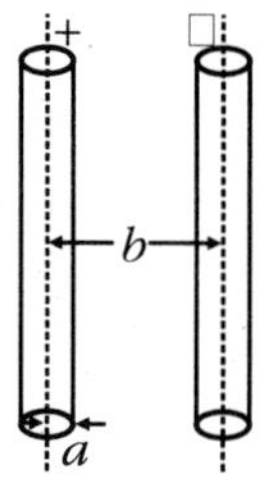

$$E = \frac{\lambda}{2\pi\varepsilon_0 x}$$

Then,
$$\varphi = \int_a^{b-a} E\,dx = \frac{\lambda}{2\pi\varepsilon_0}\ln\frac{b-a}{a}$$

Hence, capacitance per unit length is given by

$$\frac{\lambda}{\varphi_+ - \varphi_-} = \frac{\lambda}{2\varphi} = \frac{\pi\varepsilon_0}{\ln(b-a)/a} \cong \frac{\pi\varepsilon_0}{\ln b/a} \text{ (as } b \gg a)$$

3.109 The field in the regions between the conducting plane and the wire can be obtained by using an oppositely charged wire as an image on the other side.

Then the potential difference between the wire and the plane is

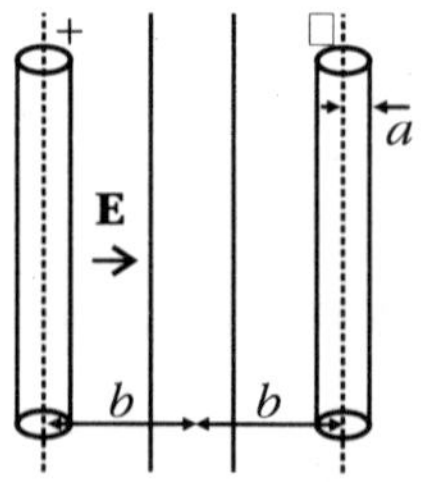

$$\varphi_+ - \varphi_- = \int \mathbf{E}\cdot d\mathbf{r}$$

$$= \int_a^b \left[\frac{\lambda}{2\pi\varepsilon_0 r} + \frac{\lambda}{2\pi\varepsilon_0(2b-r)}\right]dr$$

$$= \frac{\lambda}{2\pi\varepsilon_0}\ln\frac{b}{a} - \frac{\lambda}{2\pi\varepsilon_0}\ln\frac{b}{2b-a}$$

$$= \frac{\lambda}{2\pi\varepsilon_0}\ln\frac{2b-a}{a}$$

$$= \frac{\lambda}{2\pi\varepsilon_0}\ln\frac{2b}{a} \quad \text{(as } b \gg a)$$

Hence, the sought mutual capacitance of the system per unit length of the wire

$$\frac{\lambda}{\Delta\varphi} = \frac{2\pi\varepsilon_0}{\ln 2b/a}$$

3.110 When $b >> a$, the charge distribution on each spherical conductor is practically unaffected by the presence of the other conductor. Then, the potential $\varphi_+(\varphi_-)$ on the positive (respectively negative) charged conductor is

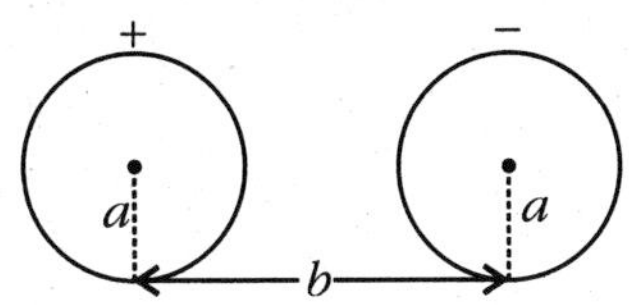

$$+\frac{q}{4\pi\varepsilon_0\varepsilon a}\left(-\frac{q}{4\pi\varepsilon_0\varepsilon a}\right)$$

Thus, $$\varphi_+ - \varphi_- = \frac{q}{2\pi\varepsilon\varepsilon_0 a} \quad \text{and} \quad C = \frac{q}{\varphi_+ - \varphi_-} = 2\pi\varepsilon_0\varepsilon a$$

Note: *If we require terms which depend on a/b, we have to take account of distribution of charge on the conductors.*

3.111 Let us apply the method of image. Then the potential difference between the positively charged sphere and the conducting plane is one half the nominal potential difference between the sphere and its image and is defined as

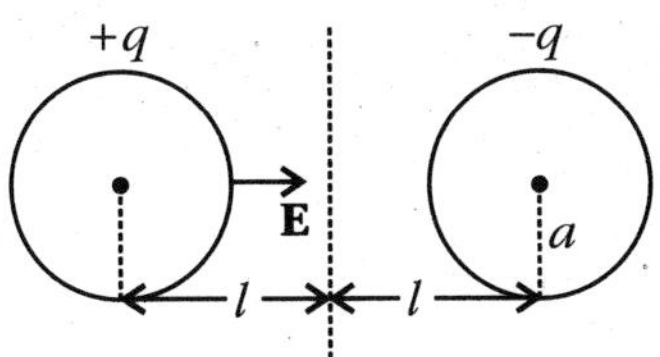

$$\Delta\varphi = \frac{1}{2}(\varphi_+ - \varphi_-) \cong \frac{q}{4\pi\varepsilon_0 a}$$

Thus, $$C = \frac{q}{\Delta\varphi} \approx 4\pi\varepsilon_0 a \quad (\text{for } l \gg a)$$

3.112 (a) Since $\varphi_1 = \varphi_B$ and $\varphi_2 = \varphi_A$. The arrangement of capacitors shown in the problem is equivalent to the arrangement shown in the figure.

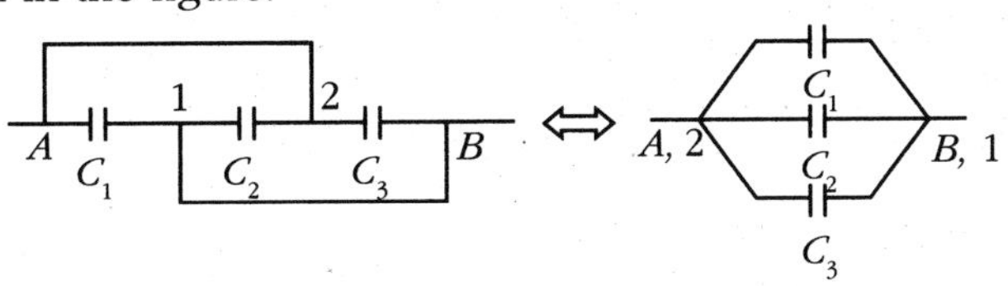

Hence, the capacitance between A and B is

$$C_0 = C_1 + C_2 + C_3$$

(b) From the symmetry of the problem, there is no potential difference between D and E. So, the combination reduces to a simple arrangement shown in the figure.

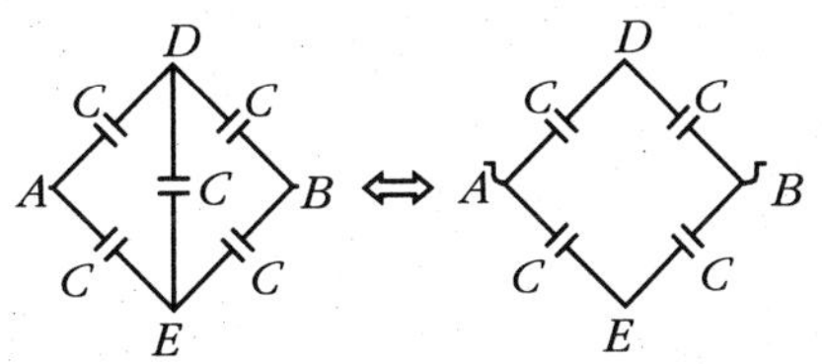

Hence, the net capacitance between A and B is

$$C_0 = \frac{C}{2} + \frac{C}{2} = C$$

3.113 (a) In the given arrangement, we have three capacitors of equal capacitance $C = \varepsilon_0 S/d$ and the first and third plates are at the same potential.

Hence, we can resolve the network into a simple form using series and parallel grouping of capacitors, as shown in the figure.

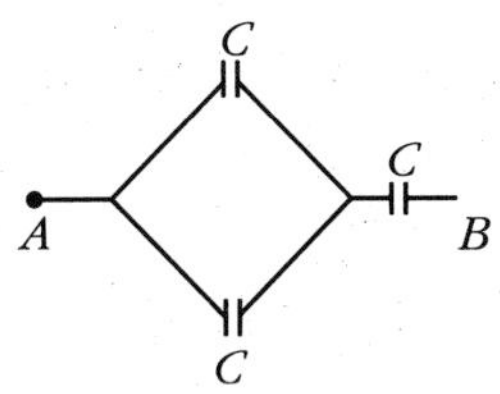

Thus, the equivalent capacitance is

$$C_0 = \frac{(C+C)C}{(C+C)+C} = \frac{2}{3}C = \frac{2}{3}\frac{\varepsilon_0 S}{d}$$

(b) Let us imagine that plates 1 and 2 have the charges $+q$ and $-q$ and these are distributed to the other plates using charge conservation and electric induction (see figure).

Since the potential difference between the plates 1 and 2 is zero,

$$-\frac{q_1}{C}+\frac{q_2}{C}-\frac{q_1}{C}=0 \quad \left(\text{where } C=\frac{\varepsilon_0 S}{d}\right)$$

or
$$q_2 = 2q_1$$

The potential difference between A and B

$$\varphi = \varphi_A - \varphi_B = \frac{q_2}{C}$$

Hence, the sought capacitance

$$C_0 = \frac{q}{\varphi} = \frac{q_1+q_2}{q_2/C} = \frac{3q_1}{2q_1/C} = \frac{3}{2}C = \frac{3\varepsilon_0 S}{2d}$$

3.114 The amount of charge that the capacitor of capacitance C_1 can withstand would be $q_1 = C_1V_1$ and similarly the charge that the capacitor of capacitance C_2 can withstand will be $q_2 = C_2V_2$. But in a series combination, charge on both the capacitors will be same, so, q_{max} that the combination can withstand $= C_1V_1$.

Also, $C_1V_1 < C_2V_2$, from the numerical data given.

Now, net capacitance of the system

$$C_0 = \frac{C_1C_2}{C_1+C_2}$$

Hence,
$$V_{max} = \frac{q_{max}}{C_0} = \frac{C_1V_1}{C_1C_2/C_1+C_2} = V_1\left(1+\frac{C_1}{C_2}\right) = 9 \text{ kV}$$

3.115 Let us distribute the charges, as shown in the figure. Now, we know that in a closed circuit, $-\Delta\varphi = 0$.

So, in the loop $DCFED$,

$$\frac{q_1}{C_1}-\frac{q_2}{C_1}-\frac{q_2}{C_2}=\xi \text{ or } q_1 = C_1\left[\xi + q_2\left(\frac{1}{C_1}+\frac{1}{C_2}\right)\right] \quad (1)$$

Again in the loop $DGHED$,

$$\frac{q_1}{C_1}+\frac{q_1+q_2}{C_2}=\xi \quad (2)$$

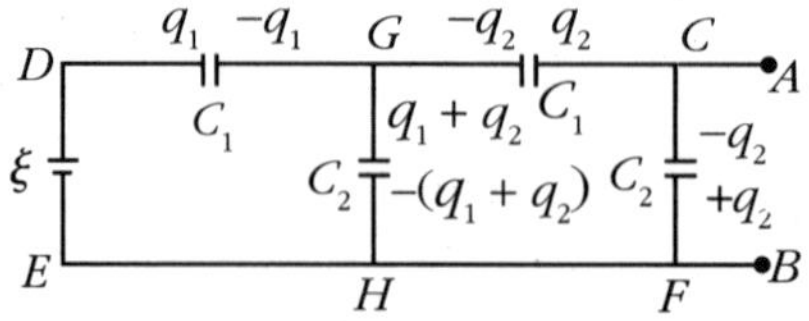

Using Eqs. (1) and (2), we get

$$q_2\left[\frac{1}{C_1}+\frac{3}{C_2}+\frac{C_1}{C_2^2}\right] = -\frac{\xi C_1}{C_2}$$

Now,
$$\varphi_A - \varphi_B = \frac{-q_2}{C_2} = \frac{\xi}{C_2^2/C_1}\frac{1}{\left[\frac{1}{C_1}+\frac{3}{C_2}+\frac{C_1}{C_2^2}\right]}$$

or
$$\varphi_A - \varphi_B = \frac{\xi}{\left[\frac{C_2^2}{C_1^2}+\frac{3C_2}{C_1}+1\right]} = \frac{\xi}{\eta^2+3\eta+1} = 10 \text{ V}$$

3.116 The infinite circuit may be reduced to the circuit shown in the figure where C_0 is the net capacitance of the combination.

So,
$$\frac{1}{C+C_0}+\frac{1}{C}=\frac{1}{C_0}.$$

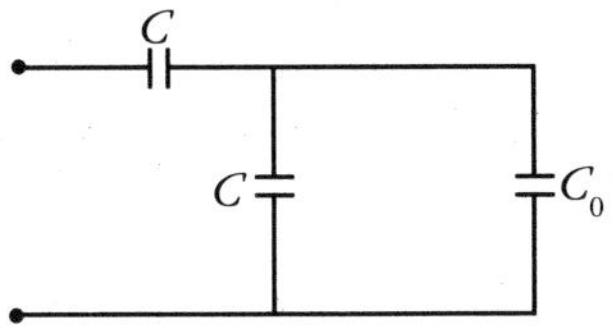

Solving the quadratic equation,
$$CC_0+C_0^2-C^2=0$$

we get,
$$C_0=\frac{(\sqrt{5}-1)}{2}C \quad \text{(taking only } +ve \text{ value as } C_0 \text{ cannot be negative)}$$
$$=0.62\ \text{C}$$

3.117 Let us make the charge distribution as shown in the figure.

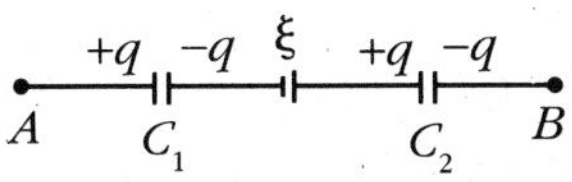

Now,
$$\varphi_A-\varphi_B=\frac{q}{C_1}-\xi+\frac{q}{C_2}$$

or
$$q=\frac{(\varphi_A-\varphi_B)+\xi}{C_1+C_2}C_1C_2$$

Hence, voltage across the capacitor C_1
$$=\frac{q}{C_1}=\frac{(\varphi_A-\varphi_B)+\xi}{C_1+C_2}C_2=10\ \text{V}$$

and voltage across the capacitor, C_2
$$=\frac{q}{C_2}=\frac{(\varphi_A-\varphi_B)+\xi}{C_1+C_2}C_1=5\ \text{V}$$

3.118 Let $\xi_2>\xi_1$, then using $-\Delta\varphi=0$ in the closed circuit (see figure),

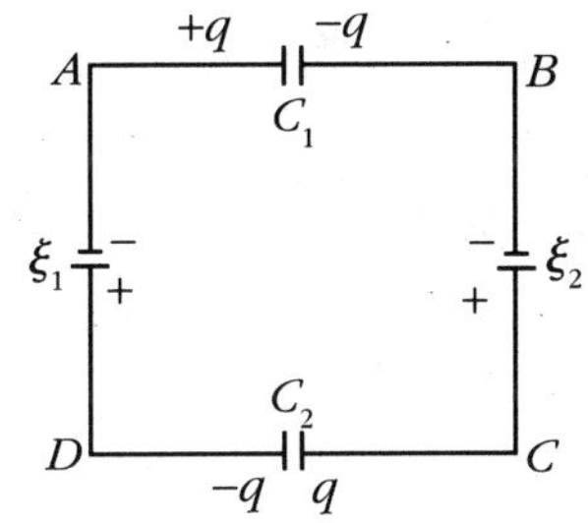

$$\frac{-q}{C_1}+\xi_2-\frac{q}{C_2}-\xi_1=0$$

or
$$q=\frac{(\xi_2-\xi_1)C_1C_2}{(C_1+C_2)}$$

Hence the potential difference (P.D.) across the left and right plates of capacitors,
$$\varphi_1=\frac{q}{C_1}=\frac{(\xi_2-\xi_1)C_2}{C_1+C_2}$$

and similarly,
$$\varphi_2=\frac{-q}{C_2}=\frac{(\xi_1-\xi_2)C_1}{C_1+C_2}$$

3.119 Using the solution of the foregoing problem, the amount of charge on each capacitor is given by
$$|q|=\frac{|\xi_2-\xi_1|C_1C_2}{C_1+C_2}$$

3.120 Make the charge distribution as shown in the figure. In the circuit 12561, $-\Delta\varphi=0$

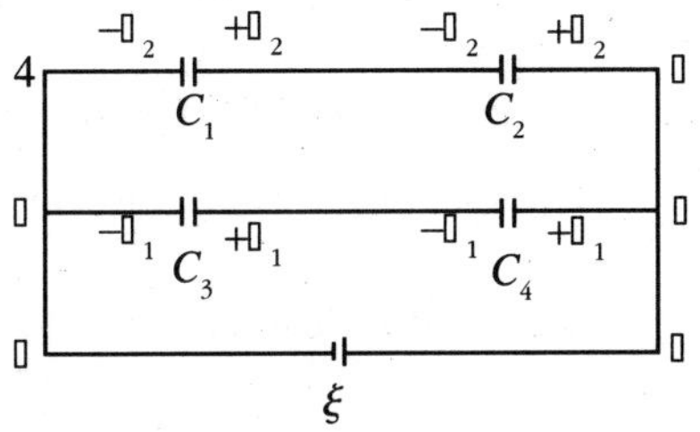

$$\frac{q_1}{C_4}-\frac{q_1}{C_3}-\xi=0 \quad \text{or} \quad q_1=\frac{\xi C_3C_4}{C_3+C_4}$$

and in the circuit 13461,

$$\frac{q_1}{C_2}+\frac{q_2}{C_1}-\xi=0 \quad \text{or} \quad q_2=\frac{\xi C_1C_2}{C_1+C_2}$$

Now,

$$\varphi_A-\varphi_B=\frac{q_2}{C_1}-\frac{q_1}{C_3}$$

$$=\xi\left[\frac{C_2}{C_1+C_2}-\frac{C_4}{C_3+C_4}\right]=\xi\left[\frac{C_2C_3-C_1C_4}{(C_1+C_2)(C_3+C_4)}\right]$$

It becomes zero, when

$$(C_2C_3-C_1C_4)=0 \quad \text{or} \quad \frac{C_1}{C_2}=\frac{C_3}{C_4}$$

3.121 Let us assume the charge q flows through the connecting wires, then at the state of equilibrium, charge distribution will be as shown in the figure. In the closed circuit, using $-\Delta\varphi=0$, we get

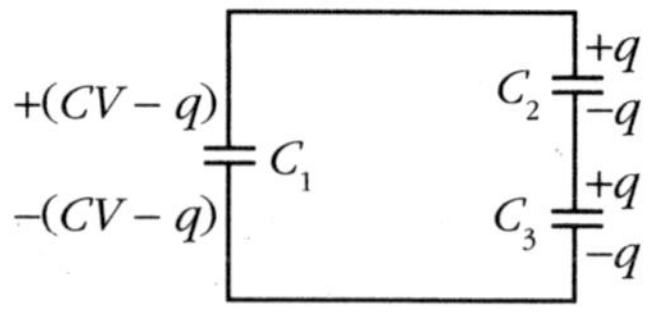

$$-\frac{(C_1V-q)}{C_1}+\frac{q}{C_2}+\frac{q}{C_3}=0$$

or

$$q=\frac{V}{(1/C_1+1/C_2+1/C_3)}=0.06 \text{ mC}$$

3.122 Initially, charge on the capacitor C_1 or C_2, is given by

$$q=\frac{\xi C_1C_2}{C_1+C_2}$$

since they are in a series combination Fig. (a).

When the switch is closed, in the circuit $CDEFC$, from $-\Delta\varphi=0$ Fig. (b).

$$\xi-\frac{q_2}{C_2}=0 \quad \text{or} \quad q_2=C_2\xi \tag{1}$$

And in the closed loop $BCFAB$, from $-\Delta\varphi=0$

$$\frac{-q_1}{C_1}+\frac{q_2}{C_2}-\xi=0 \tag{2}$$

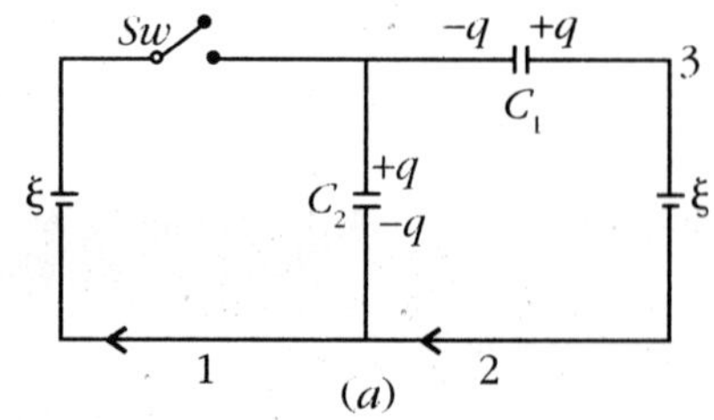

(a)

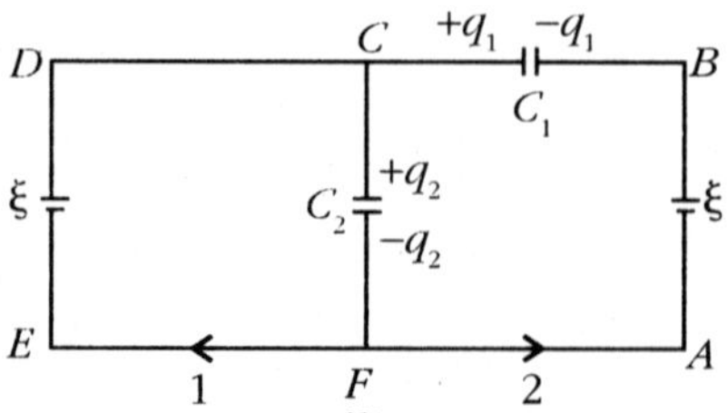

(b)

From Eqs. (1) and (2), $q_1=0$.

Now, charge flowing through section 1 $=(q_1+q_2)-0=C_2\xi$

and charge flowing through section $2=-q_1-q=\dfrac{\xi C_1C_2}{C_1+C_2}$

3.123 When the switch is open Fig. (a),

$$q_0 = \frac{2\xi C_1 C_2}{C_1 + C_2}$$

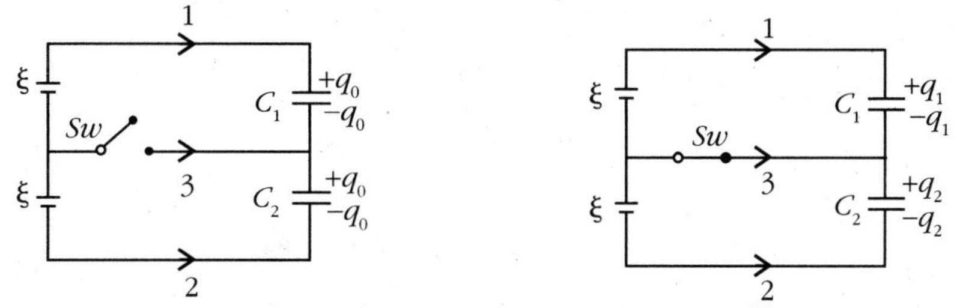

(a) (b)

and when the switch is closed Fig. (b),

$$q_1 = \xi C_1 \quad \text{and} \quad q_2 = \xi C_2$$

Hence, the flow of charge, due to the shorting of switch

$$\text{through section } 1 = q_1 - q_0 = \xi C_1 \left[\frac{C_1 - C_2}{C_1 + C_2}\right] = -24\ \mu\text{C}$$

$$\text{through section } 2 = -q_2 - (q_0) = \xi C_2 \left[\frac{C_1 - C_2}{C_1 + C_2}\right] = -36\ \mu\text{C}$$

and through the section $3 = q_2 - (q_2 - q_1) - 0 = \xi\,(C_2 - C_1) = 60\ \mu\text{C}$

3.124 First of all, make the charge distribution as shown in the figure.

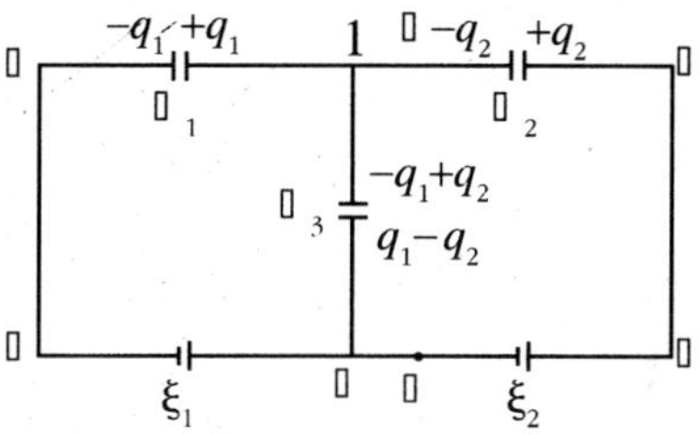

In the loop 12341, using $-\Delta\varphi = 0$

$$\frac{q_1}{C_1} - \xi_1 + \frac{q_1 - q_2}{C_3} = 0 \qquad (1)$$

Similarly in the loop 61456, using $-\Delta\varphi = 0$

$$\frac{q_2}{C_2} + \frac{q_2 - q_1}{C_3} - \xi_2 = 0 \qquad (2)$$

From Eqs. (1) and (2) we have

$$q_2 - q_1 = \frac{\xi_2 C_2 - \xi_1 C_1}{\frac{C_2}{C_3} + \frac{C_1}{C_3} + 1}$$

Hence, $$\varphi_A - \varphi_B = \frac{q_2 - q_1}{C_3} = \frac{\xi_2 C_2 - \xi_1 C_1}{C_1 + C_2 + C_3}$$

3.125 In the loop $ABDEA$, using $-\Delta\varphi = 0$

$$-\xi_3 + \frac{q_1}{C_3} + \frac{q_1 + q_2}{C_1} + \xi_1 = 0 \qquad (1)$$

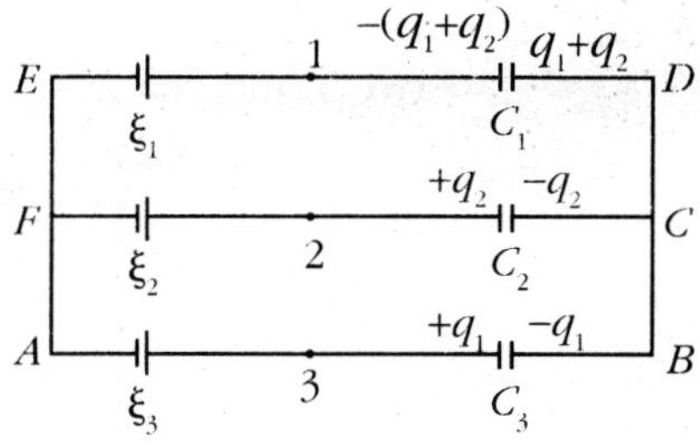

Similarly in the loop $CDEFC$

$$\frac{q_1 + q_2}{C_1} + \xi_1 - \xi_2 + \frac{q_2}{C_2} = 0 \qquad (2)$$

Solving Eqs. (1) and (2), we get

$$q_1 + q_2 = \frac{\xi_2 C_2 - \xi_1 C_2 - \xi_1 C_3 + \xi_3 C_3}{\frac{C_3}{C_1} + \frac{C_2}{C_1} + 1}$$

Now, $$\varphi_1 - \varphi_0 = \varphi_1 = -\frac{q_1 + q_2}{C_1} \text{ (as } \varphi_0 = 0)$$
$$= \frac{\xi_1(C_2 + C_3) - \xi_2 C_2 - \xi_3 C_3}{C_1 + C_2 + C_3}$$

Using symmetry, $$\varphi_2 = \frac{\xi_2(C_1 + C_3) - \xi_1 C_1 - \xi_3 C_3}{C_1 + C_2 + C_3}$$

and $$\varphi_3 = \frac{\xi_3(C_1 + C_2) - \xi_1 C_1 - \xi_2 C_2}{C_1 + C_2 + C_3}$$

The answers have wrong sign in the answer sheet.

3.126 Taking advantage of the symmetry of the problem, we can make a diagram for charge distribution as shown in the figure.

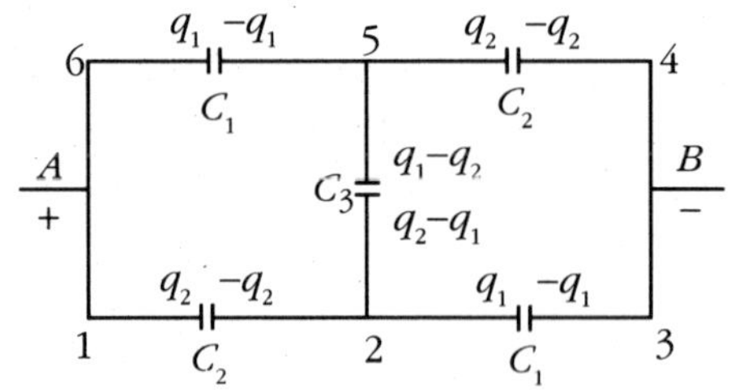

In the loop, 12561, using $-\Delta\varphi = 0$

$$\frac{q_2}{C_2} + \frac{q_2 - q_1}{C_3} - \frac{q_1}{C_1} = 0$$

or $$\frac{q_1}{q_2} = \frac{C_1(C_3 + C_2)}{C_2(C_1 + C_3)} \tag{1}$$

Now, capacitance of the network

$$C_0 = \frac{q_1 + q_2}{\varphi_A - \varphi_B} = \frac{q_1 + q_2}{q_2/C_2 + q_1/C_1}$$
$$= \frac{(1 + q_1/q_2)}{\left(\frac{1}{C_2} + \frac{q_1}{q_2 C_1}\right)} \tag{2}$$

From Eqs. (1) and (2)

$$C_0 = \frac{2C_1C_2 + C_3(C_1 + C_2)}{C_1 + C_2 + 2C_3}$$

3.127 (a) Interaction energy of any two point charges q_1 and q_2 is given by $\frac{q_1 q_2}{4\pi\varepsilon_0 r}$, where r is the separation between the charges.

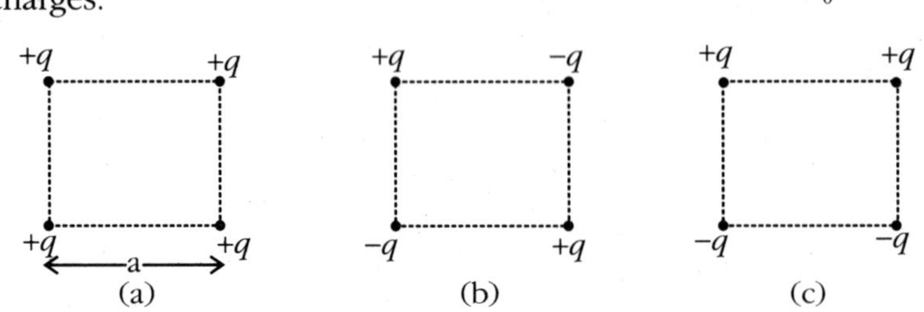

Hence, interaction energy of the system is given by

$$U_a = 4\frac{q^2}{4\pi\varepsilon_0 a} + 2\frac{q^2}{4\pi\varepsilon_0(\sqrt{2}\,a)}$$

$$U_b = 4\frac{-q^2}{4\pi\varepsilon_0 a} + 2\frac{q^2}{4\pi\varepsilon_0(\sqrt{2}\,a)}$$

and

$$U_c = 2\frac{q^2}{4\pi\varepsilon_0 a} - \frac{2q^2}{4\pi\varepsilon_0 a} - \frac{2q^2}{4\pi\varepsilon_0(\sqrt{2}\,a)} = -\frac{\sqrt{2}\,q^2}{4\pi\varepsilon_0 a}$$

3.128 Since the chain is of infinite length, any two charges of the same sign will occur symmetrically to any other charges of the opposite sign. So, interaction energy of each charge with all the others is given by

$$U = -2\frac{q^2}{4\pi\varepsilon_0 a}\left[1 - \frac{1}{2} + \frac{1}{3} - \frac{1}{4}\cdots \text{ up to } \infty\right] \tag{1}$$

But,

$$\ln(1+x) = x - \frac{1}{2}x^2 + \frac{1}{3}x^3 \cdots \text{ up to } \infty$$

Putting $x = 1$, we get

$$\ln 2 = 1 - \frac{1}{2} + \frac{1}{3} + \cdots \text{ up to } \infty \tag{2}$$

From Eqs. (1) and (2),

$$U = \frac{-2q^2 \ln 2}{4\pi\varepsilon_0 a}$$

3.129 Using electrical image method, interaction energy of the charge q with those induced on the plane is given by

$$U = \frac{-q^2}{4\pi\varepsilon_0(2l)} = -\frac{q^2}{8\pi\varepsilon_0 l}$$

3.130 Consider the interaction energy of one of the balls (say 1) and thin spherical shell of the other. This interaction energy can be written as

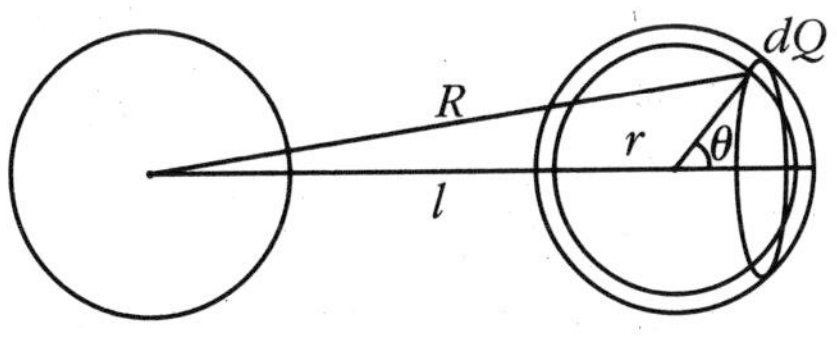

$$\int \varphi\, dq = \int \frac{q_1}{4\pi\varepsilon_0 R}\rho_2(r) 2\pi r^2 \sin\theta\, d\theta\, dr$$

$$= \int_0^{\pi} \frac{\rho_2(r) q_1 r^2 \sin\theta\, d\theta\, dr}{2\varepsilon_0(l^2 + r^2 + 2lr\cos\theta)^{1/2}} = \frac{q_1 r}{2\varepsilon_0 l} dr \int_{l-r}^{l+r} dx\, \rho_2(r)$$

$$= \frac{q_1 r}{2\varepsilon_0 l} dr \xi 2r\rho_2(r)\cdot 2 = \frac{q_1}{4\pi\varepsilon_0 l} 4\pi r^2\, dr\, \rho_2(r)$$

Then, finally integrating

$$U_{\text{int}} = \frac{q_1 q_2}{4\pi\varepsilon_0 l} \quad \text{where } q_2 = \int_0^{\infty} 4\pi r^2 \rho_2(r)\, dr$$

3.131 Charge contained in the capacitor of capacitance C_1 is $q = C_1\varphi$ and the energy, stored in it is given by

$$U_i = \frac{q^2}{2C_1} = \frac{1}{2}C_1\varphi^2$$

Now, when the capacitors are connected in parallel, the equivalent capacitance of the system $C = C_1 + C_2$ and hence, energy stored in the system is

$$U_f = \frac{C_1^2\varphi^2}{2(C_1 + C_2)} \quad \text{(since charge is conserved during the process)}$$

So, increment in the energy

$$\Delta U = \frac{C_1^2\varphi^2}{2}\left(\frac{1}{C_1 + C_2} - \frac{1}{C_1}\right) = \frac{-C_2 C_1 \varphi^2}{2(C_1 + C_2)} = -0.03 \text{ mJ}$$

3.132 The charge on the condensers in position 1 is as shown in Fig. (a).

Here, $$\frac{q}{C} = \frac{q_0}{C_0} = \frac{q + q_0}{C + C_0}$$

and $$(q + q_0)\left(\frac{1}{C + C_0} + \frac{1}{C}\right) = \xi \quad \text{or} \quad q + q_0 = \frac{C(C + C_0)}{C_0 + 2C}\xi$$

Hence, $$q = \frac{C^2\xi}{C_0 + 2C} \quad \text{and} \quad q_0 = \frac{CC_0\xi}{C_0 + 2C}$$

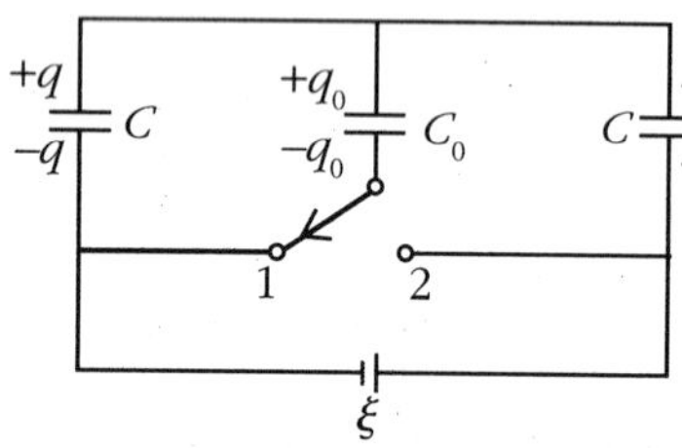

(a)

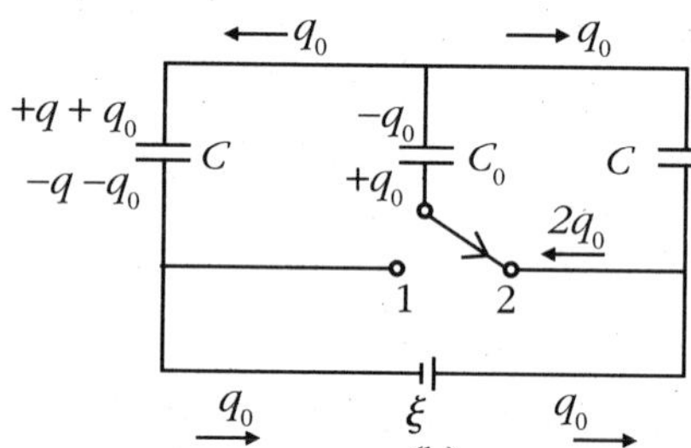

(b)

After the switch is thrown to position 2, the charges change as shown in Fig. (b).

A charge q_0 has flown in the right loop through the two condensers and a charge q_0 through the cell, and because of the symmetry of the problem there is no change in the energy stored in the condensers. Thus,

$$H \text{ (Heat produced)} = \text{Energy delivered by the cell}$$

$$= \Delta q\xi = q_0\xi = \frac{CC_0\xi^2}{C_0 + 2C}$$

3.133 Initially, the charge on the right plate of the capacitor, $q = C(\xi_1 - \xi_2)$ and finally, when switched to the position 2, charge on the same plate of capacitor is $q' = C\xi_1$.

So, $$\Delta q = q' = C\xi_2$$

Now, from energy conservation,

$$\Delta U + \text{Heat liberated} = A_{\text{cell}}$$

(where ΔU is the electrical energy),

or $$\frac{1}{2}C\xi_1^2 - \frac{1}{2}C(\xi_1 - \xi_2)^2 + \text{Heat liberated} = \Delta q\xi_1$$

as only the cell with e.m.f. ξ_1 is responsible for redistribution of the charge.

So, $$C\xi_1\xi_2 - \frac{1}{2}C\xi_2^2 + \text{Heat liberated} = C\xi_2\xi_1$$

Hence, heat liberated $= \frac{1}{2}C\xi_2^2$.

3.134 Self energy of each shell is given by $\frac{q\varphi}{2}$, where φ is the potential of the shell, created only by the charge q on it.

Hence, self energies of the shells 1 and 2 are:

$$W_1 = \frac{q_1^2}{8\pi\varepsilon_0 R_1} \quad \text{and} \quad W_2 = \frac{q_2^2}{8\pi\varepsilon_0 R_2}$$

The interaction energy between the charged shells equals charge q of one shell, multiplied by the potential φ created by other shell, at the point of location of charge q.

So, $$W_{12} = q_1\frac{q_2}{4\pi\varepsilon_0 R_2} = \frac{q_1 q_2}{4\pi\varepsilon_0 R_2}$$

Hence, total energy of the system

$$U = W_1 + W_2 + W_{12}$$

$$= \frac{1}{4\pi\varepsilon_0}\left[\frac{q_1^2}{2R_1} + \frac{q_2^2}{2R_2} + \frac{q_1 q_2}{R_2}\right]$$

3.135 Electric fields inside and outside the sphere is obtained with the help of Gauss' theorem.

$$E_1 = \frac{q(r)}{4\pi\varepsilon_0 r^2} \quad (\text{for } r \le R) \quad \text{and} \quad E_2 = \frac{q}{4\pi\varepsilon_0}\frac{1}{r^2} \quad (\text{for } r > R)$$

Sought self energy of the ball

$$U = W_1 + W_2$$

$$= \int_0^R \frac{\varepsilon_0 E_1^2}{2}4\pi r^2 dr + \int_R^\infty \frac{\varepsilon_0 E_2^2}{2}4\pi r^2 dr = \frac{q^2}{8\pi\varepsilon_0 R}\left(\frac{1}{5}+1\right)$$

Hence, $$U = \frac{3q^2}{4\pi\varepsilon_0 5R} \quad \text{and} \quad \frac{W_1}{W_2} = \frac{1}{5}$$

$$= \frac{3q^2}{20\pi\varepsilon_0 R}$$

3.136 For a spherical layer,

$$\int \frac{1}{2}\varepsilon_0\varepsilon E^2 dV = \int \frac{1}{2}\varepsilon\varepsilon_0 E^2 4\pi r^2 dr$$

To find the electrostatic energy inside the dielectric layer, we have to integrate the upper expression in the limit $[a, b]$. We get,

$$U = \frac{1}{2}\varepsilon_0\varepsilon\int_a^b\left(\frac{q}{4\pi\varepsilon_0\varepsilon r^2}\right)^2 4\pi r^2\, dr = \frac{q^2}{8\pi\varepsilon_0\varepsilon}\left[\frac{1}{a} - \frac{1}{b}\right] = 27 \text{ mJ (on substituting values)}$$

3.137 Since the field is conservative, the total work done by the field force

$$A = U_i - U_f = \frac{1}{2}q(\varphi_1 - \varphi_2)$$

$$= \frac{1}{2}\frac{q^2}{4\pi\varepsilon_0}\left[\frac{1}{R_1} - \frac{1}{R_2}\right] = \frac{q^2}{8\pi\varepsilon_0}\left[\frac{1}{R_1} - \frac{1}{R_2}\right]$$

3.138 Initially, energy of the system, $U_i = W_1 + W_{12}$ where, W_1 is the self energy and W_{12} is the mutual energy.

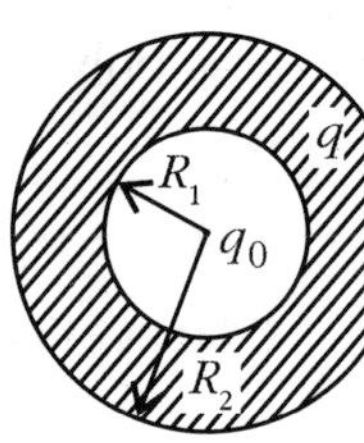

So,
$$U_i = \frac{1}{2}\frac{q^2}{4\pi\varepsilon_0 R_1} + \frac{qq_0}{4\pi\varepsilon_0 R_1}$$

and on expansion, energy of the system is given by

$$U_f = W_1' + W_{12}'$$
$$= \frac{1}{2}\frac{q^2}{4\pi\varepsilon_0 R_1} + \frac{qq_0}{4\pi\varepsilon_0 R_1}$$

Now, work done by the field force, A equals the decrease in electrical energy,

i.e.,
$$A = (U_i - U_f) = \frac{q(q_0 + q/2)}{4\pi\varepsilon_0}\left(\frac{1}{R_1} - \frac{1}{R_2}\right)$$

Alternate:

The work of electric forces is equal to the decrease in electric energy of the system,

$$A = U_i - U_f$$

In order to find the difference $U_i - U_f$, we note that upon expansion of the shell, the electric field and hence the energy localized in it, changed only in the hatched spherical layer consequently (see figure).

$$U_i - U_f = \int_{R_1}^{R_2} \frac{\varepsilon_0}{2}\left(E_1^2 - E_2^2\right)\cdot 4\pi r^2 dr$$

where E_1 and E_2 are the field intensities (in the hatched region at a distance r from the centre of the system) before and after the expansion of the shell. By using Gauss' theorem, we find

$$E_1 = \frac{1}{4\pi\varepsilon_0}\frac{q + q_0}{r^2} \quad \text{and} \quad E_2 = \frac{1}{4\pi\varepsilon_0}\frac{q_0}{r^2}$$

As a result of integration, we obtain

$$A = \frac{q(q_0 + q/2)}{4\pi\varepsilon_0}\left(\frac{1}{R_1} - \frac{1}{R_2}\right)$$

3.139 Energy of the charged sphere of radius r can be obtained from the equation

$$U = \frac{1}{2}q\varphi = \frac{1}{2}q\frac{q}{4\pi\varepsilon_0 r} = \frac{q^2}{8\pi\varepsilon_0 r}$$

If the radius of the shell changes by dr, then work done is

$$4\pi r^2 F_u dr = -dU = \frac{q^2}{8\pi\varepsilon_0 r^2}dr$$

Thus sought force per unit area,

$$F_u = \frac{q^2}{4\pi r^2(8\pi\varepsilon_0 r^2)} = \frac{(4\pi r^2\sigma)^2}{4\pi r^2 \times 8\pi\varepsilon_0 r^2} = \frac{\sigma^2}{2\varepsilon_0}$$

3.140 Initially, there will be induced charges of magnitude $-q$ and $+q$ on the inner and outer surfaces of the spherical layer, respectively. Hence, the total electrical energy of the system is the sum of self energies of spherical shells having radii a and b, and their mutual energies including the point charge q.

$$U_i = \frac{1}{2}\frac{q^2}{4\pi\varepsilon_0 b} + \frac{1}{2}\frac{(-q)^2}{4\pi\varepsilon_0 a} + \frac{-qq}{4\pi\varepsilon_0 a} + \frac{qq}{4\pi\varepsilon_0 b} + \frac{-qq}{4\pi\varepsilon_0 b}$$

or
$$U_i = \frac{q^2}{8\pi\varepsilon_0}\left[\frac{1}{b} - \frac{1}{a}\right]$$

Finally, charge q is at infinity, hence $U_f = 0$.

Now, work done by the agent = increment in the energy

$$= U_f - U_i = \frac{q^2}{8\pi\varepsilon_0}\left[\frac{1}{a} - \frac{1}{b}\right]$$

3.141 (a) Sought work is equivalent to the work performed against the electric field created by one plate held at rest, and moving the other plate away. Therefore the required work,

$$A_{\text{agent}} = qE(x_2 - x_1)$$

where $E = \sigma/2\varepsilon_0$ is the intensity of the field created by one plate at the location of other.

So,
$$A_{\text{agent}} = q\frac{\sigma}{2\varepsilon_0}(x_2 - x_1) = \frac{q^2}{2\varepsilon_0 S}(x_2 - x_1)$$

Alternate:

$$A_{\text{ext}} = \Delta U \text{ (as field is potential)}$$
$$= \frac{q^2}{2\varepsilon_0 S}x_2 - \frac{q^2}{2\varepsilon_0 S}x = \frac{q^2}{2\varepsilon_0 S}(x_2 - x_1)$$

(b) When voltage is kept constant, the force acting on each plate of capacitor will depend on the distance between the plates.

So, elementary work done by agent, in its displacement over a distance dx, relative to the other is given by

$$dA = -F_x dx$$

But,
$$F_x = -\left(\frac{\sigma(x)}{2\varepsilon_0}\right)S\sigma(x) \quad \text{and} \quad \sigma(x) = \frac{\varepsilon_0 V}{x}$$

Hence,
$$A = \int dA = \int_{x_1}^{x_2} \frac{1}{2}\varepsilon_0 \frac{SV^2}{x^2}dx = \frac{\varepsilon_0 SV^2}{2}\left[\frac{1}{x_1} - \frac{1}{x_2}\right]$$

Alternate:

From energy conservation,

$$U_f - U_i = A_{\text{cell}} + A_{\text{agent}}$$

or
$$\frac{1}{2}\frac{\varepsilon_0 S}{x_2}V^2 - \frac{1}{2}\frac{\varepsilon_0 S}{x_1}V^2 = \left[\frac{\varepsilon_0 S}{x_2} - \frac{\varepsilon_0 S}{x_1}\right]V^2 + A_{\text{agent}}$$

(as $A_{\text{cell}} = (q_f - q_i)V = (C_f - C_i)V^2$).

So,
$$A_{\text{agent}} = \left[\frac{1}{x_1} - \frac{1}{x_2}\right]$$

3.142 (a) When a metal plate of thickness ηd is inserted inside the capacitor, the capacitance of the system becomes

$$C_0 = \frac{\varepsilon_0 S}{d(1-\eta)}$$

Initially, charge on the capacitor,

$$q_0 = C_0 V = \frac{\varepsilon_0 SV}{d(1-\eta)}$$

Finally, capacitance of the capacitor,

$$C = \frac{\varepsilon_0 S}{d}$$

As the source is disconnected, charge on the plates will remain same during the process. From energy conservation,

$$U_f - U_i = A_{\text{agent}} \text{ (as the cell does no work)}$$

or

$$\frac{1}{2}\frac{q_0^2}{C} - \frac{1}{2}\frac{q_0^2}{C_0} = A_{\text{agent}}$$

Hence,

$$A_{\text{agent}} = \frac{1}{2}\left[\frac{\varepsilon_0 SV}{d(1-\eta)}\right]^2 \left[\frac{1}{C} - \frac{(1-\eta)}{C}\right] = \frac{1}{2}\frac{CV^2\eta}{(1-\eta)^2} = 1.5 \text{ mJ}$$

(b) Initially, capacitance of the system is given by

$$C_0 = \frac{C\varepsilon}{\eta(1-\varepsilon)+\varepsilon} \quad \text{(since the two capacitors are in series)}$$

So, charge on the plate $q_0 = C_0 V$.

Capacitance of the capacitor, after the glass plate has been removed equals C.

From energy conservation,

$$A_{\text{agent}} = U_f - U_i$$

$$= \frac{1}{2} q_0^2 \left[\frac{1}{C} - \frac{1}{C_0}\right] = \frac{1}{2}\frac{CV^2 \varepsilon\eta(\varepsilon-1)}{[\varepsilon - \eta(\varepsilon-1)]^2} = 0.8 \text{ mJ}$$

3.143 When the capacitor which is immersed in water is connected to a constant voltage source, it gets charged. Suppose σ_0 is the free charge density on the condenser plates. Because water is a dielectric, bound charges also appear in it. Let σ' be the surface density of bound charges. (Because of homogeneity of the medium and uniformity of the field, when we ignore edge effects no volume density of bound charges exists.) The electric field due to free charges is only σ_0/ε_0; that due to bound charges is σ'/ε_0 and the total electric field is $\sigma_0/\varepsilon\varepsilon_0$. Recalling that the sign of bound charges is opposite of the free charges, we have

$$\frac{\sigma_0}{\varepsilon\varepsilon_0} = \frac{\sigma_0}{\varepsilon_0} - \frac{\sigma'}{\varepsilon_0} \quad \text{or} \quad \sigma' = \left(\frac{\varepsilon-1}{\varepsilon}\right)\sigma_0$$

Because of the field that exists due to the free charges (not the total field; the field due to the bound charges must be excluded for this purpose as they only give rise to self energy effects), there is a force attracting the bound charges to the nearby plates. This force is

$$\frac{1}{2}\sigma'\frac{\sigma_0}{\varepsilon_0} = \frac{(\varepsilon-1)}{2\varepsilon\varepsilon_0}\sigma_0^2 \quad \text{(per unit area)}$$

The factor 1/2 needs an explanation. Normally the force on a test charge is qE in an electric field E. But if the charge itself is produced by the electric field then the force must be constructed bit by bit and is

$$F = \int_0^E q(E')dE'$$

if $q(E') \propto E'$, then we get

$$F = \frac{1}{2}q(E)E$$

This factor of 1/2 is well known. For example, the energy of a dipole of moment $\mathbf{p}$ in an electric field $\mathbf{E}_0$ is $-\mathbf{p}\cdot\mathbf{E}_0$, while the energy per unit volume of a linear dielectric in an electric field is $-1/2\,\mathbf{P}\cdot\mathbf{E}_0$

where **P** is the polarization vector (i.e., dipole moment per unit volume). Now the force per unit area manifests itself as excess pressure of the liquid.

Noting that
$$\frac{V}{d} = \frac{\sigma_0}{\varepsilon\varepsilon_0}$$

We get,
$$\Delta p = \frac{\varepsilon_0\varepsilon(\varepsilon-1)V^2}{2d^2}$$

Using $\varepsilon = 81$ for water, gives $\Delta p = 7.17$ kPa = 0.07 atm.

3.144 One way of solving this problem will be exactly as in the previous case, so let us try an alternative method based on energy. Suppose the liquid rises by the distance h. Then let us calculate the extra energy of the liquid as a sum of the polarization energy and the ordinary gravitational energy. The latter is

$$\frac{1}{2}h \cdot \rho g \cdot sh = \frac{1}{2}\rho g S h^2$$

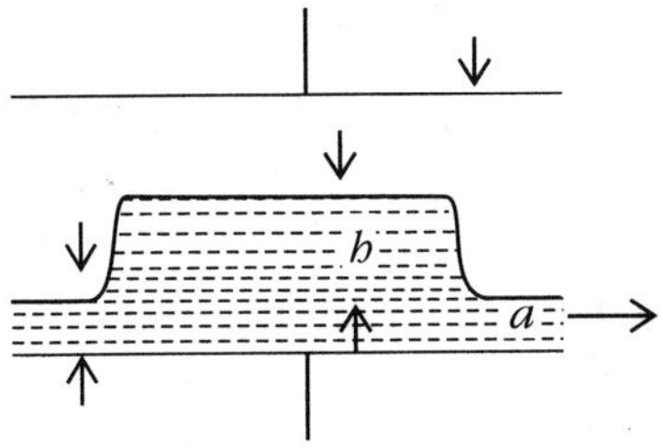

If σ is the free charge surface density on the plate, the bound charge density is, from the previous problem,

$$\sigma' = \frac{\varepsilon-1}{\varepsilon}\sigma$$

This is also the volume density of induced dipole moment, i.e., polarization. Then the energy is, as before

$$-\frac{1}{2}\mathbf{p}\cdot\mathbf{E}_0 = -\frac{1}{2}\cdot\sigma' E_0 = \frac{-1}{2}\cdot\sigma'\frac{\sigma}{\varepsilon_0} = \frac{-(\varepsilon-1)\sigma^2}{2\varepsilon_0\varepsilon}$$

where **p** is polarization vector and the total polarization energy is

$$-S(a+h)\frac{(\varepsilon-1)\sigma^2}{2\varepsilon_0\varepsilon}$$

Then, total energy is

$$U(h) = -S(a+h)\frac{(\varepsilon-1)\sigma^2}{2\varepsilon_0\varepsilon} + \frac{1}{2}\rho g S h^2$$

The actual height to which the liquid rises is determined from the formula

$$\frac{dU}{dh} = U'(h) = 0$$

This gives
$$h = \frac{(\varepsilon-1)\sigma^2}{2\varepsilon_0\varepsilon\rho g}$$

Alternate:

To find the stable height of the liquid, the process of its rise should be assumed to be very slow (in equilibrium). From energy conservation in differential form,

$$dU_{ele} + dU_{gr} = 0$$

So,
$$\frac{dU_{ele}(h)}{dh} + \frac{dU_{gr}(h)}{dh} = 0$$

$$\frac{d}{dh}\left(\frac{q^2}{2C}\right) + \frac{d}{dh}\left(mg\frac{h}{2}\right) = 0$$

Using $m = \rho Sh$ and $q = \sigma S$ = constant, where S is area of each of the plates,

$$\frac{q^2}{2}\frac{d}{dh}\left(\frac{1}{C}\right) + \frac{d}{dh}\left(\rho Shg\frac{h}{2}\right) = 0$$

But, $$\frac{1}{C}=\frac{a}{\varepsilon\varepsilon_0 S}+\frac{h}{\varepsilon\varepsilon_0 S}+\frac{d-h}{\varepsilon_0 S}$$ (where d is separation between the plates)

So, $$\frac{d}{dh}\left(\frac{1}{C}\right)=\frac{d}{dh}\left[\frac{h}{\varepsilon\varepsilon_0 S}+\frac{d-h}{\varepsilon_0 S}\right]$$ (because initial liquid height a is constant)

$$\frac{d}{dh}\left(\frac{1}{C}\right)=\frac{d}{dh}\left[\frac{\varepsilon d-(\varepsilon-1)h}{\varepsilon_0 S\varepsilon}\right]=-\frac{(\varepsilon-1)}{\varepsilon\varepsilon_0 S}$$

Hence, $$-\frac{q^2}{2}\frac{(\varepsilon-1)}{\varepsilon\varepsilon_0 S}+(\rho g S)h=0$$

$$\frac{\sigma^2}{2}\frac{(\varepsilon-1)}{\varepsilon_0}=\rho g h$$

$$h=\frac{\sigma^2(\varepsilon-1)}{2\rho g\varepsilon\varepsilon_0}$$

Note: *Assuming that the lower plate is just immersed, the answer remains the same.*

3.145 We know that the energy of the capacitor,

$$U=\frac{q^2}{2C}$$

Hence, from $$F_x=-\left.\frac{\partial U}{\partial x}\right|_{q=\text{constant}}$$

we have,

$$F_x=\frac{q^2}{2}\frac{\partial C}{\partial x}\Big/C^2 \qquad (1)$$

Now, since $d << R$, the capacitance of the given capacitor can be calculated by the formula of a parallel plate capacitor. Therefore, if the dielectric is introduced upto a depth x and the length of the capacitor is l, we have,

$$C=\frac{2\pi\varepsilon_0\varepsilon Rx}{d}+\frac{2\pi R\varepsilon_0(l-x)}{d} \qquad (2)$$

From Eqs. (1) and (2), we get,

$$F_x=\varepsilon_0(\varepsilon-1)\frac{\pi RV^2}{d}$$

3.146 When the capacitor is kept at a constant potential difference V, the work performed by the moment of electrostatic forces between the plates, when the inner movable plate is rotated by an angle $d\varphi$, equals the increase in the potential energy of the system. This comes about because when changes are made, charges flow from the battery to keep the potential constant and the amount of work done by these charges is twice in magnitude and opposite in sign to the change in energy of the capacitor. Thus,

$$N_z=\frac{\partial U}{\partial\varphi}=\frac{1}{2}V^2\frac{\partial C}{\partial\varphi}$$

Now the capacitor can be thought of as made up of two parts (with and without the dielectric) in parallel.

Thus, $$C=\frac{\varepsilon_0 R^2\varphi}{2d}+\frac{\varepsilon_0\varepsilon(\pi-\varphi)R^2}{2d}$$

as the area of a sector of angle φ is $1/2\ R^2\ \varphi$. Differentiation then gives

$$N_z=-\frac{(\varepsilon-1)\varepsilon_0 R^2V^2}{4d}$$

The negative sign of N_z indicates that the moment of the force is acting clockwise (i.e., trying to suck in the dielectric).

3.4 Electric Current

3.147 The convection current is

$$I = \frac{dq}{dt} \tag{1}$$

here, $dq = \lambda dx$, where λ is the linear charge density.

But, from the Gauss' theorem, electric field at the surface of the cylinder

$$E = \frac{\lambda}{2\pi\varepsilon_0 a}$$

Hence, substituting the value of λ and subsequently of dq in Eq. (1), we get

$$I = \frac{2E\pi\varepsilon_0 a dx}{dt}$$
$$= 2\pi\varepsilon_0 Eav \left(\text{as} \frac{dx}{dt} = v\right)$$
$$= 0.5\,\mu\text{A (on substituting values)}$$

3.148 Since $d \ll r$, the capacitance of the given capacitor can be calculated using the formula for a parallel plate capacitor. Therefore, if water (permittivity ε) is introduced up to a height x and the capacitor is of length l, we have

$$C = \frac{\varepsilon\varepsilon_0 \cdot 2\pi rx}{d} + \frac{\varepsilon_0(l-x)2\pi r}{d} = \frac{\varepsilon_0 2\pi r}{d}(\varepsilon x + l - x)$$

Hence charge on the plate at that instant, $q = CV$.

Again we know that the electric current intensity,

$$I = \frac{dq}{dt} = \frac{d(CV)}{dt}$$
$$= \frac{V\varepsilon_0 2\pi r}{d}\frac{d(\varepsilon x + l - x)}{dt} = \frac{V2\pi r\varepsilon_0}{d}(\varepsilon - 1)\frac{dx}{dt}$$

But, $$\frac{dx}{dt} = v$$

So, $$I = \frac{2\pi r\varepsilon_0(\varepsilon - 1)V}{d}v \approx 0.11\,\mu\text{A}$$

3.149 We have,

$$R_t = R_0\,(1 + \alpha t) \tag{1}$$

where R_t and R_0 are resistances at t°C and 0°C, respectively and α is the mean temperature coefficient of resistance.

So, $$R_1 = R_0(1 + \alpha_1 t) \quad \text{and} \quad R_2 = \eta R_0(1 + \alpha_2 t)$$

(a) In case of series combination, $R = \sum R_i$

So, $$R = R_1 + R_2 = R_0[(1+\eta) + (\alpha_1 + \eta\alpha_2)t]$$
$$= R_0(1+\eta)\left[1 + \frac{\alpha_1 + \eta\alpha_2}{1+\eta}t\right] \tag{2}$$

Comparing Eqs. (1) and (2), we conclude that temperature coefficient of resistance of the circuit

$$\alpha = \frac{\alpha_1 + \eta\alpha_2}{1+\eta}$$

(b) In parallel combination

$$R = \frac{R_0(1+\alpha_1 t)R_0\eta(1+\alpha_2 t)}{R_0(1+\alpha_1 t) + \eta R_0(1+\alpha_2 t)} = R'(1+\alpha' t), \text{ where } R' = \frac{\eta R_0}{1+\eta}$$

Now, neglecting the terms proportional to the product of temperature coefficients, as they have very small values, we get

$$\alpha' = \frac{\eta\alpha_1 + \alpha_2}{1+\eta}$$

3.150 Using the property of symmetry, the currents are flowing as shown in the figures.

(a) In the figures, Ohm's law is applied between 1 and 7 via 1487 (say). So,

$$IR_{eq} = \frac{I}{3}R + \frac{I}{6}R + \frac{I}{3}R = \frac{5}{6}RI$$

Thus, $$R_{eq} = \frac{5R}{6}$$

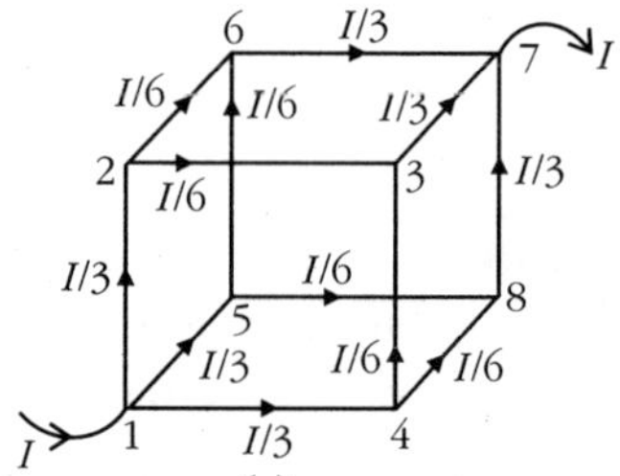

(a)

(b) Between 1 and 2 from the loop 14321,

$$I_1R = 2I_2R + I_3R \quad \text{or} \quad I_1 = I_3 + 2I_2$$

From the loop 48734,

$$(I_2 - I_3)R + 2(I_2 - I_3)R + (I_2 - I_3)R = I_3R$$

or $$4(I_2 - I_3) = I_3 \Rightarrow I_3 = \frac{4}{5}I_2$$

So, $$I_1 = \frac{14}{5}I_2$$

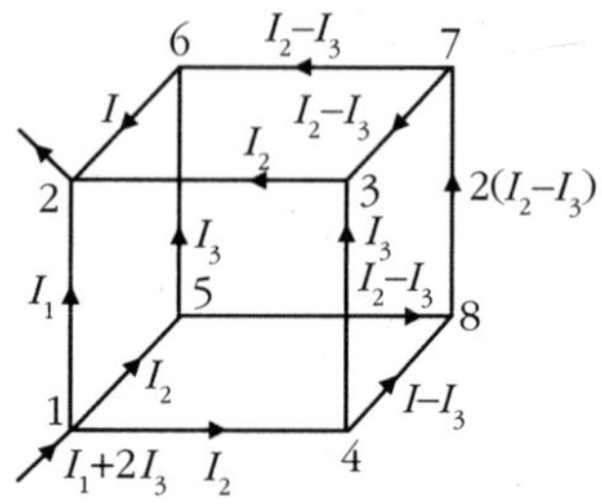

(b)

Then, $$(I_1 + 2I_2)R_{eq} = \frac{24}{5}I_2R_{eq}$$

$$= I_1R = \frac{14}{5}I_2R$$

or $$R_{eq} = \frac{7}{12}R$$

(c) Between 1 and 3 from the loop 15621

$$I_2R = I_1R + \frac{I_1}{2}R \quad \text{or} \quad I_2 = 3\frac{I_1}{2}$$

Then, $$(I_1 + 2I_2)R_{eq} = 4I_1R_{eq}$$

$$= I_2R + I_2R = 3I_1R$$

Hence, $$R_{eq} = \frac{3}{4}R$$

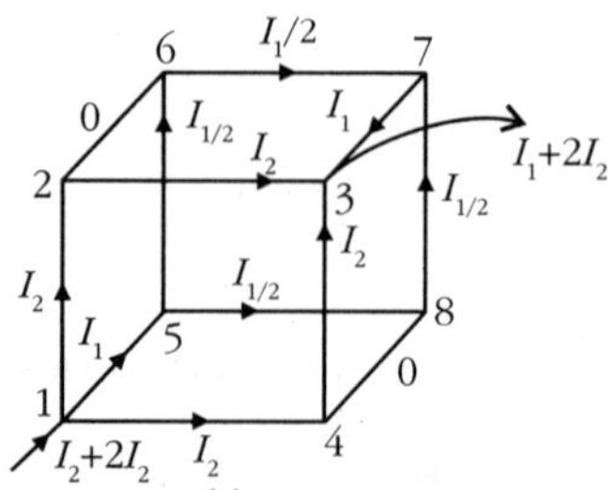

(c)

3.151 Total resistance of the circuit will be independent of the number of cells,

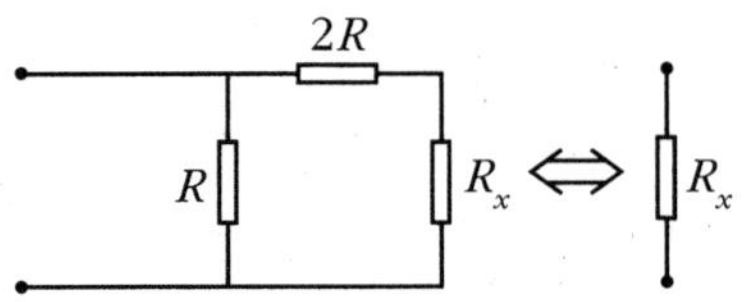

if $$R_x = \frac{(R_x + 2R)R}{R_x + 2R + R}$$

or $$R_x^2 + 2RR_x - 2R^2 = 0$$

On solving and rejecting the negative root of the quadratic equation, we have

$$R_x = R(\sqrt{3} - 1)$$

3.152 Let R_0 be the resistance of the network.

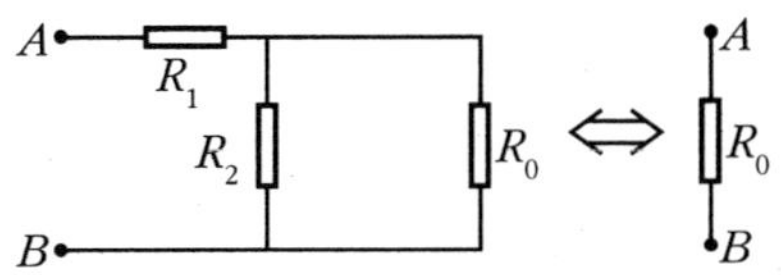

Then, $$R_0 = \frac{R_0 R_2}{R_0 + R_2} + R_1$$

or $$R_0^2 - R_0 R_1 - R_1 R_2 = 0$$

On solving we get,

$$R_0 = \frac{R_1}{2}\left(1 + \sqrt{1 + 4\frac{R_2}{R_1}}\right) + 6\ \Omega$$

3.153 Suppose that the voltage V is applied between the points A and B, then

$$V = IR = I_0 R_0$$

where R is the resistance of the whole grid, I is the current through the grid and I_0 is the current through the segment AB. Now from symmetry, $I/4$ is the part of the current flowing through all the four wire segments, meeting at the point A and similarly the amount of current flowing through the wires meeting at B is also $I/4$. Thus a current $I/2$ flows through the conductor AB, i.e.,

$$I_0 = \frac{I}{2}$$

Hence, $$R = \frac{R_0}{2}$$

3.154 Let us imagine a thin cylindrical layer of inner and outer radii r and $r + dr$, respectively. As lines of current at all the points of this layer are perpendicular to it, such a layer can be treated as a cylindrical conductor of thickness dr and cross-sectional area $2\pi rl$. So, we have,

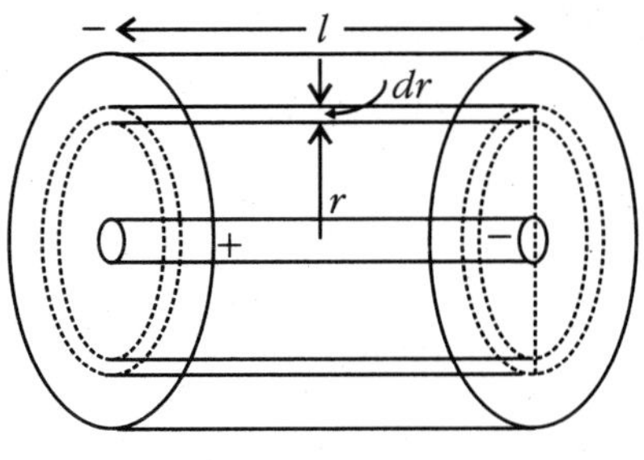

$$dR = \rho\frac{dr}{S(r)} = \rho\frac{dr}{2\pi rl}$$

and integrating between the limit $[a, b]$, we get

$$R = \frac{\rho}{2\pi l}\ \ln\ \frac{b}{a}$$

3.155 Let us imagine a thin spherical layer of inner and outer radii r and $r + dr$. Lines of current at all the points of this layer are perpendicular to it and therefore such a layer can be treated as a spherical conductor of thickness dr and cross sectional area $4\pi r^2$. So,

$$dR = \rho\frac{dr}{4\pi r^2}$$

and integrating between the limits $[a, b]$, we get,

$$R = \frac{\rho}{4\pi}\left[\frac{1}{a} - \frac{1}{b}\right]$$

Now, for $b \to \infty$, we have

$$R = \frac{\rho}{4\pi a}$$

3.156 In this case, resistance of the medium is given by

$$R = \frac{\rho}{4\pi}\left[\frac{1}{a} - \frac{1}{b}\right] \text{ (where } \rho \text{ is the resistivity of the medium)}$$

The current is given by

$$I = \frac{\varphi}{R} = \frac{\varphi}{\frac{\rho}{4\pi}\left[\frac{1}{a} - \frac{1}{b}\right]} \tag{1}$$

Also, $$I = \frac{-dq}{dt} = -\frac{d(C\varphi)}{dt} = -C\frac{d\varphi}{dt} \text{ (as capacitance is constant)} \tag{2}$$

Equating Eqs. (1) and (2), we get

$$\frac{\varphi}{\frac{\rho}{4\pi}\left[\frac{1}{a} - \frac{1}{b}\right]} = -C\frac{d\varphi}{dt}$$

or $$-\int\frac{d\varphi}{\varphi} = \frac{\Delta t}{\frac{C\rho}{4\pi}\left[\frac{1}{a} - \frac{1}{b}\right]}$$

or $$\text{In } \eta = \frac{\Delta t\, 4\pi ab}{C\rho(b-a)}$$

Hence, resistivity of the medium

$$\rho = \frac{4\pi\Delta tab}{C(b-a)\ln\eta}$$

3.157 Let us imagine that the balls have charge $+q$ and $-q$, respectively. The electric field strength at the surface of a ball will be determined only by its own charge and the charge can be considered to be uniformly distributed over the surface, because the other ball is at an infinite distance. Magnitude of the field strength is given by

$$E = \frac{q}{4\pi\varepsilon_0 a^2}$$

So, current density

$$j = \frac{1}{\rho}\frac{q}{4\pi\varepsilon_0 a^2}$$

and electric current

$$I = \int \mathbf{j}\cdot d\mathbf{S} = jS = \frac{q}{\rho 4\pi\varepsilon_0 a^2}\cdot 4\pi a^2 = \frac{q}{\rho\varepsilon_0}$$

But, potential difference between the balls

$$\varphi_+ - \varphi_- = 2\frac{q}{4\pi\varepsilon_0 a}$$

Hence, the sought resistance

$$R = \frac{\varphi_+ - \varphi_-}{I} = \frac{2q/4\pi\varepsilon_0 a}{q/\rho\varepsilon_0} = \frac{\rho}{2\pi a}$$

3.158 (a) The potential in the unshaded region beyond the conductor has the potential of the given charge and its image has the form

$$\varphi = A\left(\frac{1}{r_1} - \frac{1}{r_2}\right)$$

where r_1, r_2 are the distances of the point from the charge and its image, respectively. The potential has been taken to be zero on the conducting plane and on the ball.

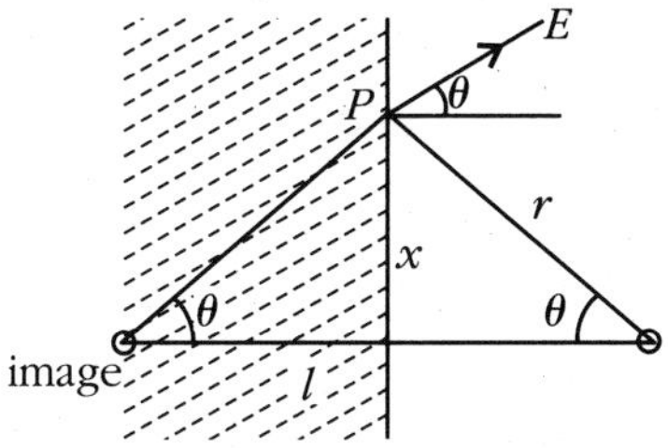

$$\varphi \approx A\left(\frac{1}{a} - \frac{1}{2l}\right) = V$$

So, $A \approx Va$. In this calculation, the condition $a \ll l$, is used to ignore the variation of φ over the ball.

The electric field at P can be calculated similarly. The charge on the ball is

$$Q = 4\pi\varepsilon_0 Va$$

and $$E_p = \frac{Va}{r^2} 2\cos\theta = \frac{2alV}{r^3} \text{ then } j = \frac{1}{\rho}E = \frac{2alV}{\rho r^3}$$

(b) The total current flowing into the conducting plane is

$$I = \int_0^\infty 2\pi x dx j = \int_0^\infty 2\pi x dx \frac{2alV}{\rho(x^2 + l^2)^{3/2}}$$

Putting $y = x^2 + l^2$, we get

$$I = \frac{2\pi alV}{\rho}\int_{l^2}^\infty \frac{dy}{y^{3/2}} = \frac{4\pi aV}{\rho}$$

Hence, $$R = \frac{V}{I} = \frac{\rho}{4\pi a}$$

3.159 (a) The wires can be assumed to be perfect conductors, so the resistance is entirely due to the medium. If the wire is of length L, the resistance R of the medium is $\propto 1/L$ because different sections of the wire are connected in parallel (by the medium) rather than in series. Thus if R_1 is the resistance per unit length of the wire, then $R = R_1/L$. Unit of R_1 is ohm-meter.

The potential at a point P is by symmetry and superposition for ($l >> a$), is given by

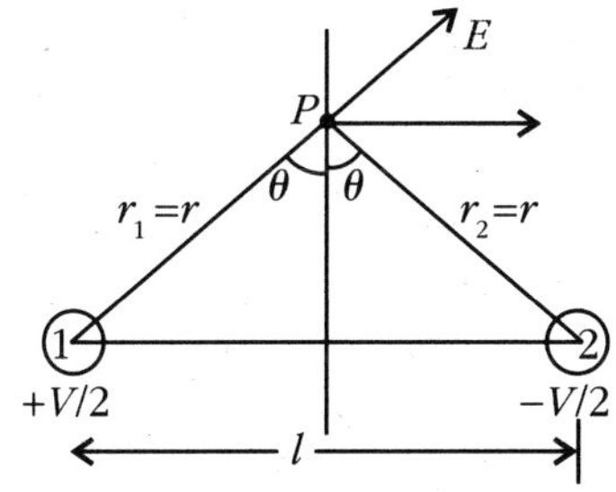

$$\varphi \cong \frac{A}{2}\ln\frac{r_1}{a} - \frac{A}{2}\ln\frac{r_2}{a}$$

$$= \frac{A}{2}\ln\frac{r_1}{r_2}$$

Then, $$\varphi_1 = \frac{V}{2} = \frac{A}{2}\ln\frac{a}{l} \text{ (for the potential of 1)}$$

or $$A = \frac{-V}{\ln l/a}$$

and $$\varphi = -\frac{V}{2\ln l/a}\ln r_1/r_2$$

We then calculate the field at a point P which is equidistant from 1 and 2 and at a distance r from both:

$$E = \frac{V}{2\ln l/a}\left(\frac{1}{r}\right) \times 2\sin\theta$$

$$= \frac{Vl}{2\ln l/a}\frac{1}{r^2}$$

and $$j = \sigma E = \frac{1}{\rho}\frac{Vl}{2r^2 \ln l/a}$$

(b) Near either wire $$E = \frac{V}{2\ln l/a}\frac{1}{a}$$

and $$j = \sigma E = \frac{1}{\rho}\frac{V}{2\ln l/a}$$

Then $$I = \frac{V}{R} = L\frac{V}{R_1} = j2\pi aL$$

which gives $$R_1 = \frac{\rho}{\pi}\ln l/a$$

3.160 Let us imagine that the plates of the capacitor have the charges $+q$ and $-q$.

The capacitance of the network

$$C = \frac{q}{\varphi} = \frac{\varepsilon\varepsilon_0 \int E_n ds}{\varphi} \tag{1}$$

Now, electric current

$$I = \int \mathbf{j}\cdot d\mathbf{S} = \int \sigma E_n dS, \text{ as } \mathbf{j} \uparrow\uparrow \mathbf{E} \tag{2}$$

Hence, using Eq. (1) in (2), we get

$$I = \frac{C\varphi}{\varepsilon\varepsilon_0}\sigma = \frac{C\varphi}{\rho\varepsilon\varepsilon_0} = 1.5\,\mu A$$

3.161 Let us imagine that the conductors have the charges $+q$ and $-q$. As the medium is a poor conductor, the surfaces of the conductors are equipotential and the field configuration is same as in the absence of the medium.

Let us surround, for example, the positively charged conductor, by a closed surface S, just containing the conductor.

Then, $$R = \frac{\varphi}{I} = \frac{\varphi}{\int \mathbf{j}\cdot d\mathbf{S}} = \frac{\varphi}{\int \sigma E_n dS} \quad (\text{as } \mathbf{j} \uparrow\uparrow \mathbf{E})$$

and $$C = \frac{q}{\varphi} = \frac{\varepsilon\varepsilon_0 \int E_n ds}{\varphi}$$

So, $$RC = \frac{\varepsilon\varepsilon_0}{\sigma} = \rho\varepsilon\varepsilon_0$$

3.162 The dielectric ends in a conductor. It is given that, on one side (the dielectric side) the electric displacement D is as shown in figure. Within the conductor, at any point A, there can be no normal component of electric field. For if there were such a field, a current will flow towards depositing charge there which in turn will set up countering electric field causing the normal component to vanish. Then by Gauss' theorem, we can easily derive $\sigma = D_n = D \cos \alpha$, where σ is the surface charge density it A.

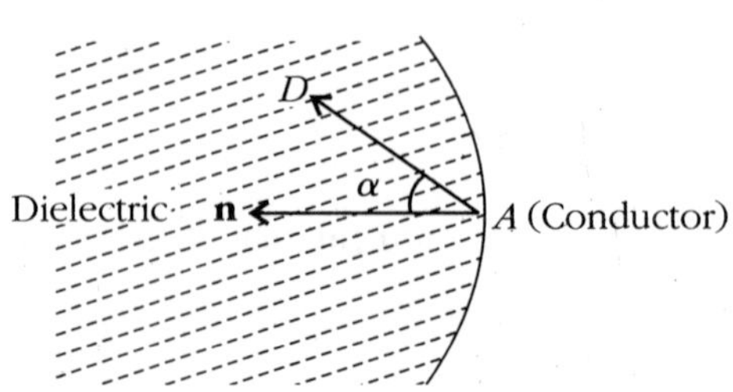

The tangential component is determined from the circulation theorem as

$$\oint \mathbf{E}\cdot d\mathbf{r} = 0$$

It must be continuous across the surface of the conductor. Thus, inside the conductor there is a tangential electric field of magnitude

$$\frac{D\sin\alpha}{\varepsilon_0\varepsilon} \text{ at } A$$

This implies a current, by Ohm's law, of density

$$j = \frac{E}{\rho} = \frac{D\sin\alpha}{\varepsilon_0\varepsilon\rho}$$

3.163 The resistance of a layer of the medium, of thickness dx and at a distance x from the first plate of the capacitor is given by

$$dR = \frac{1}{\sigma(x)}\frac{dx}{S} \tag{1}$$

Now, since σ varies linearly with the distance from the plate, it may be represented as, $\sigma = \sigma_1 + \left(\frac{\sigma_2 - \sigma_1}{d}\right)x$, at a distance x from any one of the plates.

From Eq. (1) $$dR = \frac{1}{\sigma_1 + \left(\frac{\sigma_2-\sigma_1}{d}\right)x}\frac{dx}{S}$$

or $$R = \frac{1}{S}\int_0^d \frac{dx}{\sigma_1 + \left(\frac{\sigma_2-\sigma_1}{d}\right)x} = \frac{d}{S(\sigma_2-\sigma_1)}\ln\frac{\sigma_2}{\sigma_1}$$

Hence, $$I = \frac{V}{R} = \frac{SV(\sigma_2-\sigma_1)}{d\ln\frac{\sigma_2}{\sigma_1}} = 5\,\text{nA}$$

3.164 By charge conservation current density j, leaving the medium 1 must enter the medium 2.

Thus, $$j_1\cos\alpha_1 = j_2\cos\alpha_2$$

Another relation follows from

$$E_{1t} = E_{2t}$$

which is a consequence of

$$\oint \mathbf{E}\cdot d\mathbf{r} = 0.$$

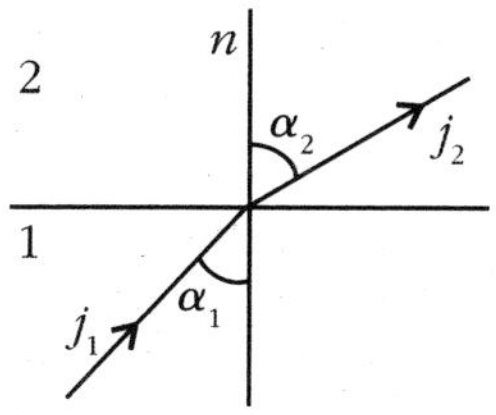

Thus, $$\frac{1}{\sigma_1}j_1\sin\alpha_1 = \frac{1}{\sigma_2}j_2\sin\alpha_2$$

or $$\frac{\tan\alpha_1}{\sigma_1} = \frac{\tan\alpha_2}{\sigma_2}$$

or $$\frac{\tan\alpha_1}{\tan\alpha_2} = \frac{\sigma_1}{\sigma_2}$$

3.165 The electric field in conductor 1 is

$$E_1 = \frac{\rho_1 I}{\pi R^2}$$

and that in 2 is $$E_2 = \frac{\rho_2 I}{\pi R^2}$$

Applying Gauss' theorem to a small cylindrical pill–box at the boundary,

$$-\frac{\rho_1 I}{\pi R^2}ds + \frac{\rho_2 I}{\pi R^2}dS = \frac{\sigma dS}{\varepsilon_0}$$

Thus, $$\sigma = \varepsilon_0(\rho_2 - \rho_1)\frac{I}{\pi R^2}$$

and charge at the boundary $= \varepsilon_0(\rho_2 - \rho_1)I$

3.166 We have, $E_1 d_1 + E_2 d_2 = V$ and by current conservation

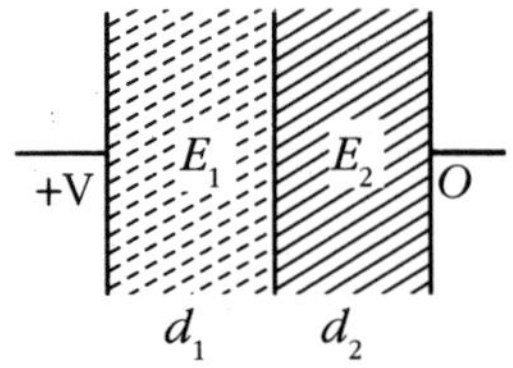

$$\frac{1}{\rho_1}E_1 = \frac{1}{\rho_2}E_2$$

Thus, $$E_1 = \frac{\rho_1 V}{\rho_1 d_1 + \rho_2 d_2}$$

$$E_2 = \frac{\rho_2 V}{\rho_1 d_1 + \rho_2 d_2}$$

At the boundary between the two dielectrics,

$$\sigma = D_2 - D_1 = \varepsilon_0\varepsilon_2 E_2 - \varepsilon_0\varepsilon_1 E_1$$

$$= \frac{\varepsilon_0 V}{\rho_1 d_1 + \rho_2 d_2}(\varepsilon_2\rho_2 - \varepsilon_1\rho_1)$$

Obviously $\sigma = 0$ for $\varepsilon_1\rho_1 = \varepsilon_2\rho_2$.

3.167 By current conservation

$$\frac{E(x)}{\rho(x)} = \frac{E(x) + dE(x)}{\rho(x) + d\rho(x)} = \frac{dE(x)}{d\rho(x)}$$

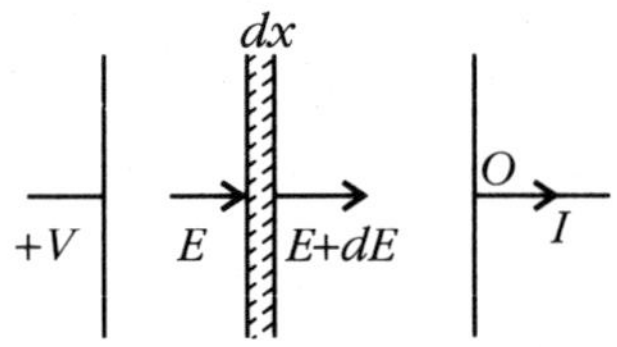

Solving we get,

$$E(x) = C\rho(x) = \frac{I\rho(x)}{A}$$

Hence, charge induced in the slice per unit area is given by

$$d\sigma = \varepsilon_0\frac{I}{A}[\{\varepsilon(x) + d\varepsilon(x)\}\{\rho(x) + d\rho(x)\} - \varepsilon(x)\rho(x)] = \varepsilon_0\frac{I}{A}d[\varepsilon(x)\rho(x)]$$

Thus, $$dQ = \varepsilon_0 I d\,[\varepsilon(x)\rho(x)]$$

Hence total charge induced obtained by integration is

$$Q = \varepsilon_0 I(\varepsilon_2\rho_2 - \varepsilon_1\rho_1)$$

3.168 As in the previous problem,

$$E(x) = C\rho(x) = C(\rho_0 + \rho_1 x)$$

where $$\rho_0 + \rho_1 d = \eta\rho_0 \quad \text{or} \quad \rho_1 = \frac{(\eta - 1)\rho_0}{d}$$

By integration, $$V = \int_0^d C\rho(x)dx = C\rho_0 d\left(1 + \frac{\eta - 1}{2}\right) = \frac{1}{2}C\rho_0 d(\eta + 1)$$

Thus, $$C = \frac{2V}{\rho_0 d(\eta + 1)}$$

Thus volume density of charge present in the medium

$$= \frac{dQ}{Sdx} = \varepsilon_0\, dE(x)/dx$$

$$= \frac{2\varepsilon_0 V}{\rho_0 d(\eta+1)} \times \frac{(\eta-1)\rho_0}{d} = \frac{2\varepsilon_0 V(\eta-1)}{(\eta+1)d^2}$$

3.169 (a) Consider a cylinder of unit length and divide it into shells of radius r and thickness dr. Different sections are parallel. For a typical section

$$d\left(\frac{1}{R_1}\right) = \frac{2\pi r\, dr}{(\alpha/r^2)} = \frac{2\pi r^3\, dr}{\alpha}$$

Integrating,

$$\frac{1}{R_1} = \frac{\pi R^4}{2\alpha} = \frac{S^2}{2\pi\alpha}$$

or

$$R_1 = \frac{2\pi\alpha}{S^2} \quad (\text{where } S = \pi R^2)$$

(b) Suppose the electric field inside is $E_z = E_0$ (z-axis is along the axis of the conductor). This electric field cannot depend on r in steady conditions when other components of E are absent, otherwise it would violate the circulation theorem

$$\oint \mathbf{E}\cdot d\mathbf{r} = 0$$

The current through a section between radii (r and $r + dr$) is

$$2\pi r dr \frac{1}{\alpha/r^2} E = 2\pi r^3 dr \frac{E}{\alpha}$$

Thus,

$$I = \int_0^R 2\pi r^3 dr \frac{E}{\alpha} = \frac{\pi R^4 E}{2\alpha}$$

Hence,

$$E = \frac{2\alpha\pi I}{S^2} \quad (\text{where } S = \pi R^2)$$

3.170 The formula is,

$$q = CV_0(1 - e^{-t/RC})$$

or

$$V = \frac{q}{C} = V_0(1 - e^{-t/RC}) \quad \text{or} \quad \frac{V}{V_0} = 1 - e^{-t/RC}$$

or

$$e^{-t/RC} = 1 - \frac{V}{V_0} = \frac{V_0 - V}{V_0}$$

Hence,

$$t = RC \ln \frac{V_0}{V_0 - V} = RC \ln 10, \text{ if } V = 0.9V_0$$

Thus,

$$t = 0.6\ \mu\text{s}$$

3.171 The charge decays according to the formula

$$q = q_0 e^{-t/RC}$$

Here, RC = mean life = half-life/ln 2

So, half life $T = RC \ln 2$

But,

$$C = \frac{\varepsilon\varepsilon_0 A}{d} \quad \text{and} \quad R = \frac{\rho d}{A}$$

Hence,

$$\rho = \frac{T}{\varepsilon\varepsilon_0 \ln 2} = 1.4\times 10^{13}\ \Omega\text{m}$$

3.172 Suppose q is the charge at time t. Initially $q = C\xi$, at $t = 0$.

Then at time t, from Loop rule

$$IR = \frac{\eta q}{c} - \xi \tag{1}$$

Differentiating Eq. (1) with respect to time and considering that $I = -dq/dt$ (−sign because charge decreases)

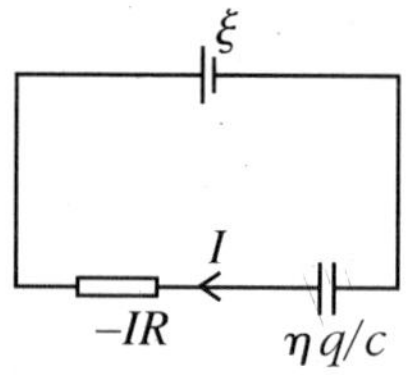

$$R\frac{dI}{dt} = -\frac{\eta}{C} I$$

i.e.,

$$\frac{dI}{I} = \frac{-\eta}{RC} \cdot dt$$

Integration of this equation gives

$$\ln\frac{I}{I_0} = \frac{-\eta t}{RC}$$

So,

$$I = I_0 e^{-\eta t/RC}$$

where I_0 is determined by condition Eq. (1).

Indeed we can write

$$I_0 R = \frac{\eta q_0}{C} - \xi$$

where $q_0 = C\xi$ is the charge of the capacitor before its capacitance has charged.

Therefore,

$$I_0 = (\eta - 1)\frac{\xi}{R}$$

Finally,

$$I = -\frac{dq}{dt} = \frac{\xi(\eta - 1)}{R} e^{-\eta t/RC}$$

3.173 Let r = internal resistance of the battery. We shall take the resistance of the ammeter to be = 0 and that of voltmeter to be G.

Initially,

$$V = \xi - Ir, I = \frac{\xi}{r + G}$$

So,

$$V = \xi\frac{G}{r + G} \tag{1}$$

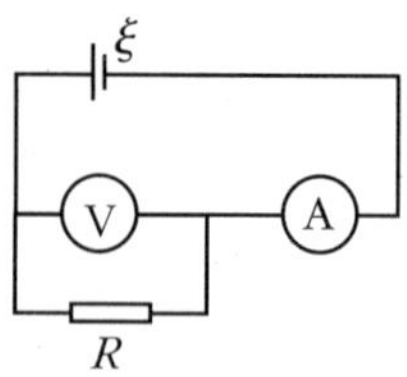

After the voltmeter is shunted

$$\frac{V}{\eta} = \xi - \frac{\xi r}{r + \frac{RG}{R + G}} \text{ (Voltmeter)} \tag{2}$$

and

$$\frac{\xi}{r + \frac{RG}{R + G}} = \eta\frac{\xi}{r + G} \text{ (Ammeter)} \tag{3}$$

From Eqs. (2) and (3) we have

$$\frac{V}{\eta} = \xi - \frac{\eta r \xi}{r + G} \tag{4}$$

From Eqs. (1) and (4)

$$\frac{G}{\eta} = r + G - \eta r \quad \text{or} \quad G = \eta r$$

Then Eq. (1) gives the required reading

$$\frac{V}{\eta} = \frac{\xi}{\eta + 1}$$

$$= 2.0 \text{ V (on substituting values)}$$

3.174 Assume the current flow to be as shown in figure. Then potentials are as shown. Thus,

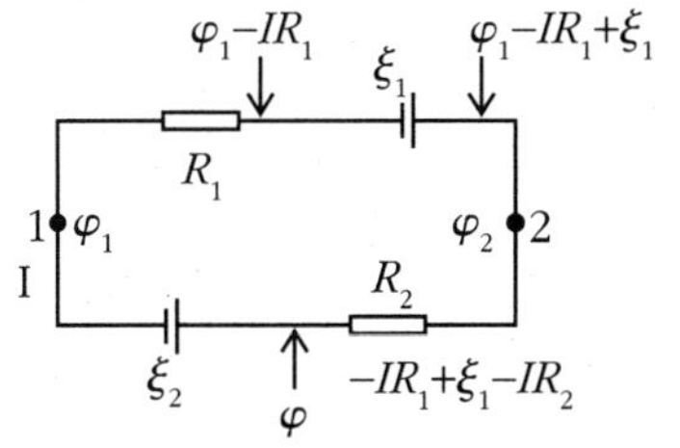

$$\varphi_1 = \varphi_1 - IR_1 + \xi_1 - IR_2 - \xi_2$$

or $$I = \frac{\xi_1 - \xi_2}{R_1 + R_2}$$

and $$\varphi_2 = \varphi_1 - IR_1 + \xi_1$$

So, $$\varphi_1 - \varphi_2 = -\xi_1 + \frac{\xi_1 - \xi_2}{R_1 + R_2} R_1$$

$$= -\frac{(\xi_1 R_2 + \xi_2 R_1)}{(R_1 + R_2)}$$

$$= -4V$$

3.175 Let us consider the current I, flowing through the circuit, as shown in the figure.

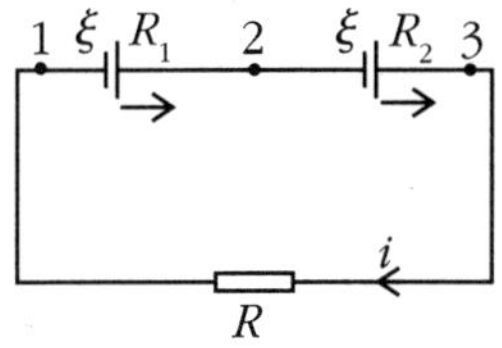

Applying loop rule for the circuit, $-\Delta\varphi = 0$.

So, $$-2\xi + IR_1 + IR_2 + IR = 0$$

or $$2\xi = I(R_1 + R_2 + R)$$

or $$I = \frac{2\xi}{R + R_1 + R_2}$$

Now, if $$\varphi_1 - \varphi_2 = 0$$

$$-\xi + IR_1 = 0$$

or $$\frac{2\xi R_1}{R + R_1 + R_2} = \xi \quad \text{and} \quad 2R_1 = R_2 + R + R_1$$

or $$R = R_1 - R_2 \text{ (which is not possible as } R_2 > R_1)$$

Thus, $$\varphi_2 - \varphi_3 = -\xi + IR_2 = 0$$

or $$\frac{2\xi R_2}{R + R_1 + R_2} = \xi$$

So, $$R = R_2 - R_1$$

which is the required resistance.

3.176 (a) Current, $$I = \frac{N\xi}{NR} = \frac{N\alpha R}{NR} = \alpha \text{ (as } \xi = \alpha R)$$

(b) $$\varphi_A - \varphi_B = n\xi - nIR = n\alpha R - n\alpha R = 0$$

3.177 Since the capacitor is fully charged, no current flows through it. So, current is given by

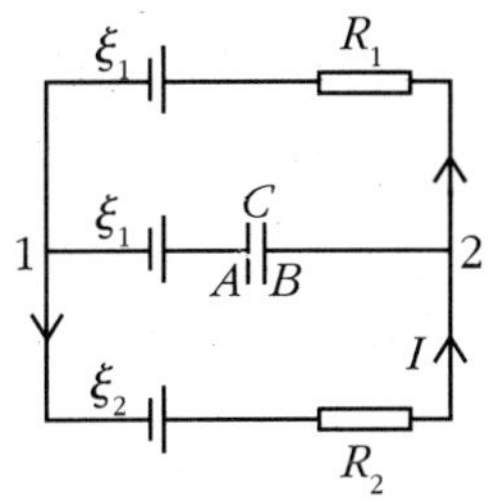

$$I = \frac{\xi_2 - \xi_1}{R_1 + R_2} \text{ (as } \xi_2 > \xi_1)$$

Hence,

$$\varphi_A - \varphi_B = \xi_1 - \xi_2 + IR_2$$
$$= \xi_1 - \xi_2 + \frac{\xi_2 - \xi_1}{R_1 + R_2} R_2$$
$$= \frac{(\xi_2 - \xi_1)R_1}{R_1 + R_2} = -0.5\,\text{V}$$

3.178 Let us make the current distribution, as shown in the figure.

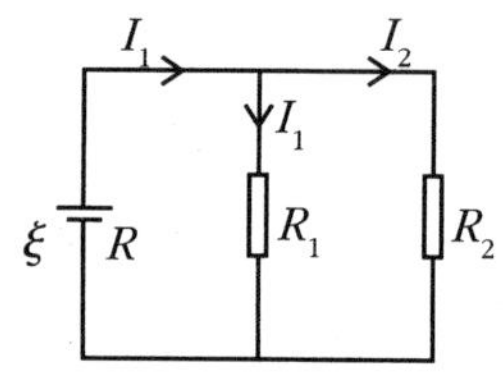

Current $$I = \frac{\xi}{R + \dfrac{R_1 R_2}{R_1 + R_2}} \text{(using loop rule)}$$

So, current through the resistor R_1,

$$I_1 = \frac{\xi}{R + \dfrac{R_1 R_2}{R_1 + R_2}} \frac{R_2}{R_1 + R_2}$$
$$= \frac{\xi R_2}{RR_1 + RR_2 + R_1 R_2} = 1.2\,\text{A}$$

and similarly, current through the resistor R_2,

$$I_2 = \frac{\xi}{R + \dfrac{R_1 R_2}{R_1 + R_2}} \frac{R_1}{R_1 + R_2} = \frac{\xi R_1}{RR_1 + R_1 R_2 + RR_2} = 0.8\,\text{A}$$

3.179 Total resistance

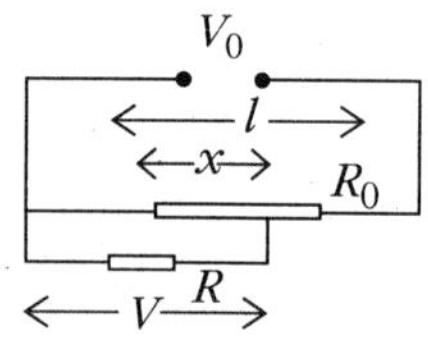

$$= \frac{l - x}{l} R_0 + \frac{R \cdot \dfrac{xR_0}{l}}{R + \dfrac{xR_0}{l}}$$
$$= \frac{l - x}{l} R_0 + \frac{xRR_0}{lR + xR_0}$$
$$= R_0 \left[\frac{l - x}{l} + \frac{xR}{lR + xR_0} \right]$$

Then $$V = V_0 \frac{xR}{lR + xR_0} \Big/ \left(1 - \frac{x}{l} + \frac{xR}{xR_0 + lR} \right)$$
$$= V_0 Rx \Big/ \left\{ lR + R_0 x \left(1 - \frac{x}{l} \right) \right\}$$

For $R \gg R_0$, $$V \cong V_0 \frac{x}{l}$$

3.180 Let us connect a load of resistance R between the points A and B (see figure).

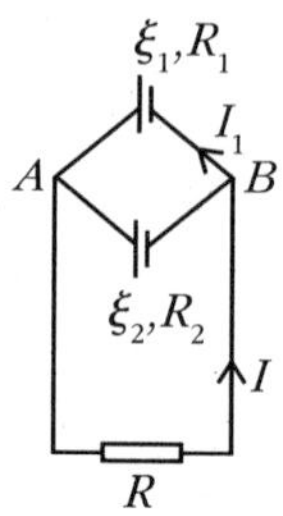

From the loop rule, $\Delta\varphi = 0$, so

$$IR = \xi_1 - I_1 R_1 \qquad (1)$$

and $$IR = \xi_2 - (I - I_1) R_2 \qquad (2)$$

Solving Eqs. (1) and (2), we get

$$I = \frac{\xi_1 R_2 + \xi_2 R_1}{R_1 + R_2} \Big/ R + \frac{R_1 R_2}{R_1 + R_2} = \frac{\xi_0}{R + R_0}$$

where $$\xi_0 = \frac{\xi_1 R_2 + \xi_2 R_1}{R_1 + R_2} \quad \text{and} \quad R_0 = \frac{R_1 R_2}{R_1 + R_2}$$

Thus one can replace the given arrangement of the cells by a single cell having the e.m.f. ξ_0 and internal resistance R_0.

3.181 Make the current distribution, as shown in the figure.

Now, in the loop 12341, applying $-\Delta\varphi = 0$, we get

$$IR + I_1 R_1 + \xi_1 = 0 \qquad (1)$$

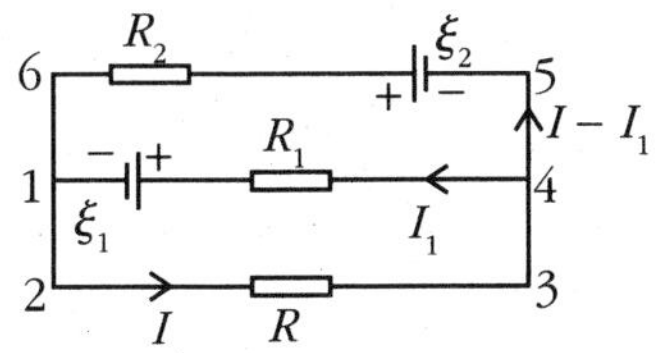

and in the loop 23562,

$$IR - \xi_2 + (I - I_1)R_2 = 0 \qquad (2)$$

Solving Eqs. (1) and (2), we obtain current through the resistance R,

$$I = \frac{(\xi_2 R_1 - \xi_1 R_2)}{RR_1 + RR_2 + R_1 R_2} = 0.02\,\text{A}$$

and it is directed from left to the right.

3.182 At first indicate the currents in the branches using charge conservation (which also includes the point rule).

In the loops 1**BA**61 and **B**34**AB**, from the loop rule using of $-\Delta\varphi = 0$, we get, respectively

$$-\xi_2 + (I - I_1)R_2 + \xi_1 - I_1 R_1 = 0 \qquad (1)$$

$$IR_3 + \xi_3 - (I - I_1)R_2 + \xi_2 = 0 \qquad (2)$$

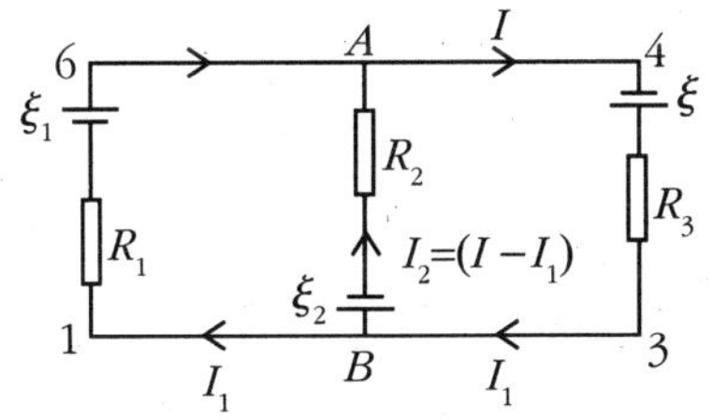

On solving Eqs. (1) and (2), we obtain

$$I_1 = \frac{(\xi_1 - \xi_2)R_3 + R_2(\xi_1 + \xi_3)}{R_1 R_2 + R_2 R_3 + R_3 R_1} \approx 0.06\,\text{A}$$

Thus, $$\varphi_A - \varphi_B = \xi_2 - I_2 R_2 = 0.9\text{V}$$

3.183 Indicate the currents in all the branches using charge conservation as shown in the figure. Applying loop rule and using $-\Delta\varphi = 0$ in the loops 1A781, 1B681 and B456B, respectively, we get

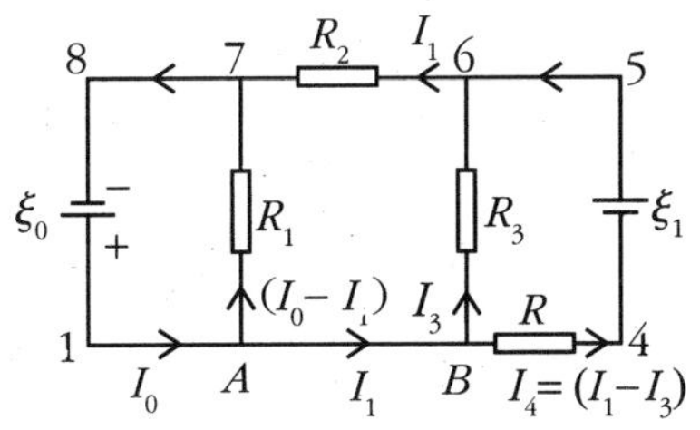

$$\xi_0 = (I_0 - I_1)R_1 \qquad (1)$$

$$I_3 R_3 + I_1 R_2 - \xi_0 = 0 \qquad (2)$$

$$(I_1 - I_3)R - \xi - I_3 R_3 = 0 \qquad (3)$$

Solving Eqs. (1), (2) and (3), we get the sought current

$$(I_1 - I_3) = \frac{\xi(R_2 + R_3) + \xi_0 R_3}{R(R_2 + R_3) + R_2 R_3}$$

3.184 As it is not the case of charging or discharging of capacitor, there is no current through R_4. Indicate the currents in all the branches using charge conservation as shown in the figure.

Applying the loop rule and using $-\Delta\varphi = 0$ in the loops 12341 and 15781, we get

$$-\xi_1 + I_3R_1 - (I_1 - I_3)R_2 = 0 \quad (1)$$

$$-\xi_2 + (I_1 - I_3)R_2 + I_1R_3 = 0 \quad (2)$$

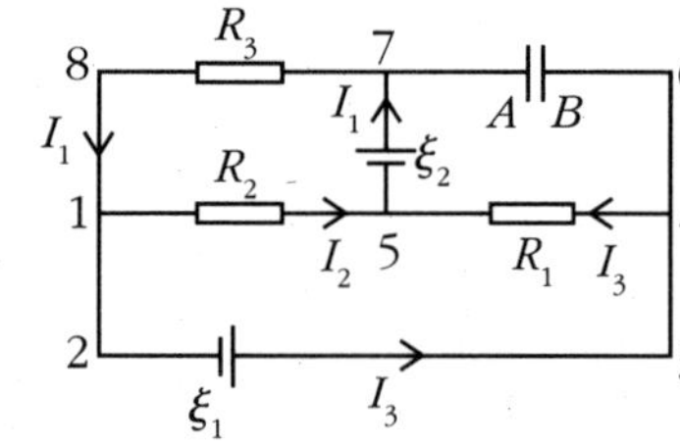

Solving Eqs. (1) and (2), we get

$$I_3 = \frac{\xi_1(R_2 + R_3) + \xi_2R_2}{R_1R_2 + R_2R_3 + R_3R_1}$$

Hence, the sought potential difference

$$\varphi_A - \varphi_B = \xi_2 - I_3R_1$$

$$= \frac{\xi_2R_3(R_1 + R_2) - \xi_1R_1(R_2 + R_3)}{R_1R_2 + R_2R_3 + R_3R_1} = -1 \text{ V (on substituting values)}$$

Alternate:

This problem can also be solved using Kirchhoff's current law.

As potential difference is defined in terms of difference, any point in a network can be assigned as zero level (reference), if no point specifically is earthed or given zero potential already in the problem.

From the diagram given above, for junction 5, having potential V, assuming point 1 to be at 0 potential,

$$\frac{V}{R_2} + \frac{V + \xi_2}{R_3} + \frac{V - \xi_1}{R_1} = 0$$

$$\Rightarrow \frac{V}{20} + \frac{V + 1}{30} + \frac{V - 4}{10} = 0$$

or

$$\frac{V}{2} + \frac{V + 1}{3} + \frac{V - 4}{1} = 0$$

On solving we get $V = 2$ Volt.

Thus the sought potential defference

$$V_A - V_B = \xi_2 + \left(\frac{V - \xi_1}{R_1}\right)R_1 = \xi_2 + V - \xi_1$$

$$= 1 + 2 - 4 = -1 \text{ V.}$$

Note: *One may solve this problem by the application of principle of superposition, in accordance with which the effect of cell in independent in a network.*

3.185 Let us distribute the currents in the paths as shown in figure.

Now, $$\varphi_1 - \varphi_2 = IR_1 + I_1R_2 \quad (1)$$

and $$\varphi_1 - \varphi_3 = IR_1 + (I - I_1)R_3 \quad (2)$$

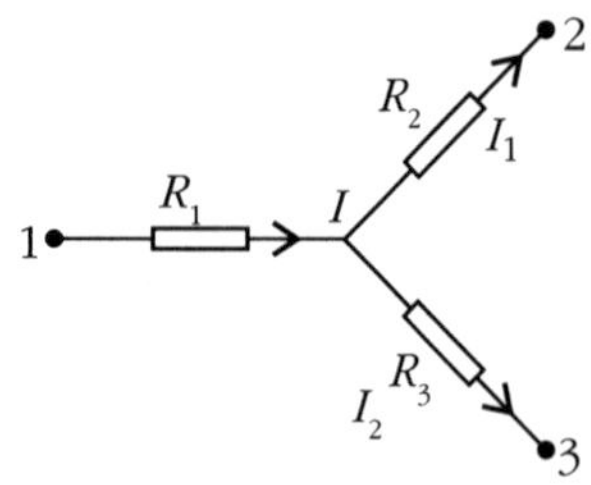

Simplifying Eqs. (1) and (2) we get

$$I = \frac{R_3(\varphi_1 - \varphi_2) + R_2(\varphi_1 - \varphi_3)}{R_1R_2 + R_2R_3 + R_3R_1} = 0.2 \text{ A}$$

3.186 Distribution of current is as shown in figure. From Kirchhoff's second law

$$I_1R_1 = I_2R_3$$

$$I_1R_1 + (I_1 - I_3)R_2 = \text{V}$$

$$I_2R_3 + (I_3 + I_2)R_4 = \text{V}$$

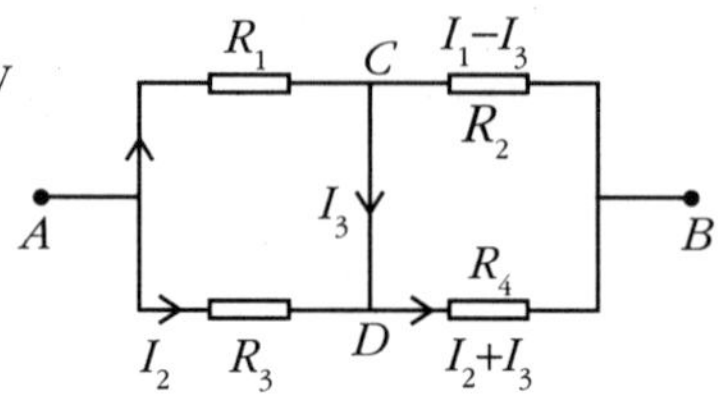

Eliminating I_2, we get

$$I_1(R_1 + R_2) - I_3 I_2 = V$$

$$I_1 \frac{R_1}{R_3}(R_3 + R_4) - I_3 R_4 = V$$

Hence, $$I_3\left[R_4(R_1 + R_2) + \frac{R_1 R_2}{R_3}(R_3 + R_4)\right] = V\left[(R_1 + R_2) - \frac{R_1}{R_3}(R_3 + R_4)\right]$$

$$I_3 = \frac{[R_3(R_1 + R_2) - R_1(R_3 + R_4)]V}{R_3 R_4(R_1 + R_2) + R_1 R_2(R_3 + R_4)}$$

On substitution, we get $I_3 = 1.0$ A from C to D.

3.187 From the symmetry of the problem, current flow is indicated, as shown in the figure.

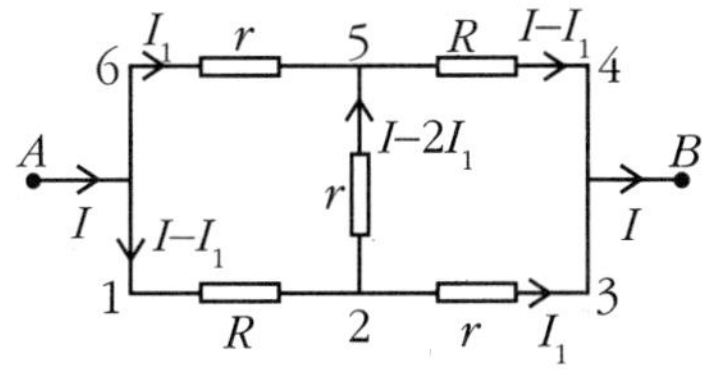

Now, $\varphi_A - \varphi_B = I_1 r + (I - I_1)R$ (1)

In the loop 12561, from $-\Delta\varphi = 0$

$$(I - I_1)R + (I - 2I_1)r - I_1 r = 0$$

or $$I_1 = \frac{(R + r)I}{3r + R} \tag{2}$$

Equivalent resistance between the terminals A and B using Eqs. (1) and (2), is given by

$$R_0 = \frac{\varphi_A - \varphi_B}{I} = \frac{\left[\frac{R + r}{3r + R}(r - R) + R\right]I}{I} = \frac{r(3R + r)}{3r + R}$$

3.188 Let, at any moment of time, charge on the plates be $+q$ and $-q$ respectively, then voltage across the capacitor

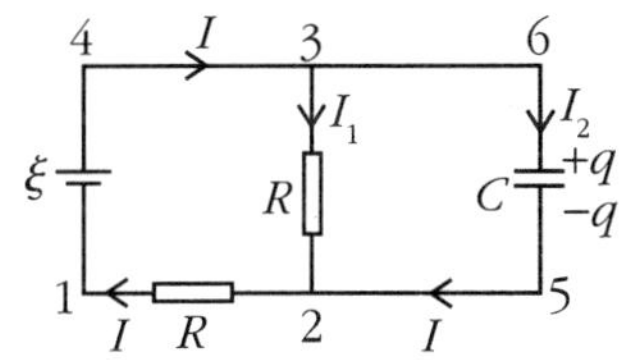

$$\varphi = q/C \tag{1}$$

Now, from charge conservation,

$$I = I_1 + I_2, \text{ where } I_2 = \frac{dq}{dt} \tag{2}$$

In the loop 65146, using $-\Delta\varphi = 0$

$$\frac{q}{C} + \left(I_1 + \frac{dq}{dt}\right)R - \xi = 0 \tag{3}$$

using Eq. (1) and Eq. (2)

In the loop 25632, using $-\Delta\varphi = 0$

$$-\frac{q}{C} + I_1 R = 0 \quad \text{or} \quad I_1 R = \frac{q}{C} \tag{4}$$

Using Eq. (4) in Eq. (3)

$$\frac{dq}{dt}R = \xi_1 - \frac{2q}{C} \quad \text{or} \quad \frac{dq}{\xi - \frac{2q}{C}} = \frac{dt}{R} \tag{5}$$

On integrating Eq. (5) between suitable limits,

$$\frac{\int_0^q dq}{\xi - \frac{2q}{C}} = \frac{1}{R}\int_0^t dt \quad \text{or} \quad -\frac{C}{2}\ln\frac{\xi - \frac{2q}{C}}{\xi} = \frac{t}{R}$$

Thus, $$\frac{q}{C} = V = \frac{1}{2}\xi(1 - e^{-2t/RC})$$

3.189 (a) Since current I is a linear function of time, and at $t = 0$ and Δt, it equals I_0 and zero, respectively, it may be represented as

$$I = I_0\left(1 - \frac{t}{\Delta t}\right)$$

Thus, $$q = \int_0^{\Delta t} I\,dt = \int_0^{\Delta t} I_0\left(1 - \frac{t}{\Delta t}\right)dt = \frac{I_0 \Delta t}{2}$$

So, $$I_0 = \frac{2q}{\Delta t}$$

Hence, $$I = \frac{2q}{\Delta t}\left(1 - \frac{t}{\Delta t}\right)$$

The heat generated is given by

$$H = \int_0^{\Delta t} I^2 R dt = \int_0^{\Delta t}\left[\frac{2q}{\Delta t}\left(1 - \frac{t}{\Delta t}\right)\right]^2 R\,dt = \frac{4q^2 R}{3\Delta t}$$

(b) Obviously the current through the coil is given by

$$I = I_0\left(\frac{1}{2}\right)^{t/\Delta t}$$

Then, $$q = \int_0^{\infty} I dt = \int_0^{\infty} I_0 2^{-t/\Delta t}\,dt = \frac{I_0 \Delta t}{\ln 2}$$

So, $$I_0 = \frac{q \ln 2}{\Delta t}$$

Hence, heat generated in the circuit in the time interval $t[0, \infty]$

$$H = \int_0^{\infty} I^2 R dt = \int_0^{\infty}\left[\frac{q \ln 2}{\Delta t} 2^{-t/\Delta t}\right]^2 R dt = -\frac{q^2 \ln 2}{2\Delta t} R$$

3.190 The equivalent circuit may be drawn as shown in the figure.

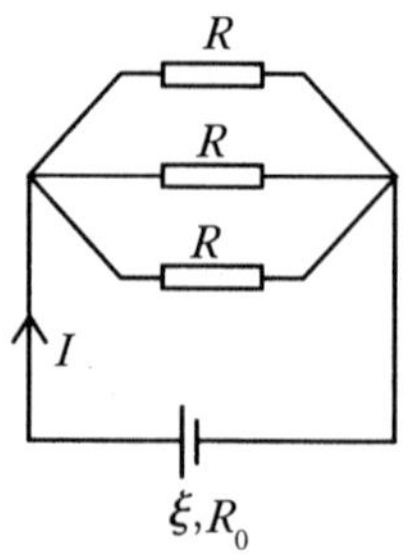

Resistance of the network $= R_0 + (R/3)$.

Let us assume that e.m.f. of the cell is ξ, then current

$$I = \frac{\xi}{R_0 + (R/3)}$$

Now the thermal power generated in the circuit

$$P = I^2 R/3 = \frac{\xi}{(R_0 + (R/3))^2}(R/3)$$

For P to be maximum,

$$\frac{dP}{dR} = 0, \text{ which yields}$$

$$R = 3R_0$$

3.191 We assume current conservation but not Kirchhoff's second law. Then thermal power dissipated is

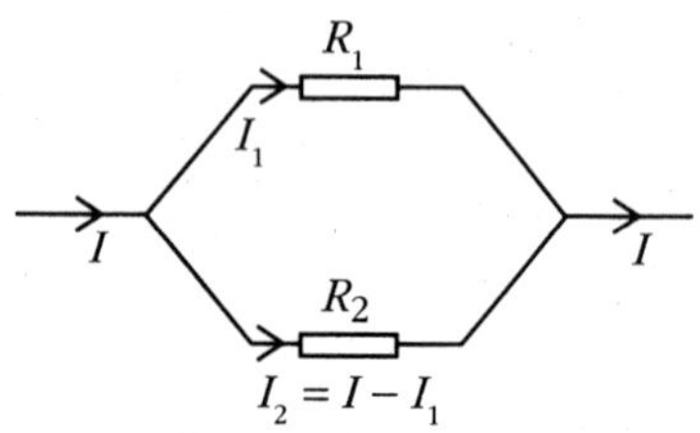

$$P(I_1) = I_1^2 R_1 + (I - I_1)^2 R_2$$
$$= I_1^2(R_1 + R_2) - 2II_1R_2 + I^2R_2$$
$$= [R_1 + R_2]\left[I_1 - \frac{R_2}{R_1 + R_2}I\right]^2 + I^2\frac{R_1R_2}{R_1 + R_2}$$

The resistances being positive, we see that the power dissipated is minimum when

$$I_1 = I\frac{R_2}{R_1 + R_2}$$

This corresponds to the usual distribution of currents over resistances joined in parallel.

3.192 Let internal resistance of the cell be r, then

$$V = \xi - Ir \tag{1}$$

where I is the current in the circuit. We know that thermal power generated in the battery is,

$$Q = I^2r \tag{2}$$

Putting r from Eq. (1) in Eq. (2), we obtain

$$Q = (\xi - V)I = 0.6 \text{ W}$$

In a battery, work is done by electric forces whose origin lies in the chemical processes going on inside the cell. The work so done is stored and used in the electric circuit outside. Its magnitude just equals the power used in the electric circuit. We can say net power developed by the electric force is

$$P = -IV = -2.0 \text{ W}$$

Minus sign means that power is generated and not consumed.

3.193 As far as the motor is concerned, the power delivered is displaced and can be represented by a load R_0. Thus,

$$I = \frac{V}{R + R_0}$$

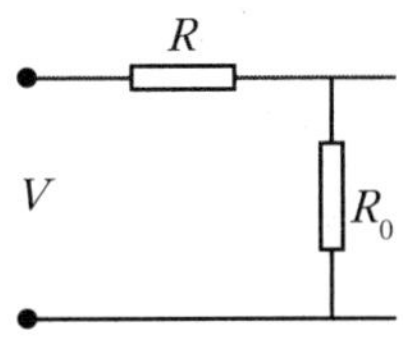

and

$$P = I^2R_0 = \frac{V^2R_0}{(R_0 + R)^2}$$

This is maximum when $R_0 = R$ and the current I is then,

$$I = \frac{V}{2R}$$

The maximum power delivered is

$$\frac{V^2}{4R} = P_{max}$$

The power input is $\frac{V^2}{R + R_0}$ and its value when P is maximum is $\frac{V^2}{2R}$.

The efficiency then is $\frac{1}{2} = 50\%$.

3.194 If the wire diameter decreases by δ, then by the information given, power input

$$P = \frac{V^2}{R} = H \text{ (heat lost through the surface)}$$

Now, $H \propto (1 - \delta)$ like the surface area and $R \propto (1 - \delta)^{-2}$.

So, $$\frac{V^2}{R_0}(1-\delta)^2 = A(1-\delta) \text{ or } V^2(1-\delta) = \text{constant}$$

But, $$V \propto 1 + \eta$$

So, $$(1 + \eta)^2(1 - \delta) = \text{constant} = 1$$

Thus, $$\delta = 2\eta = 2\%$$

3.195 The equation of heat balance is

$$\frac{V^2}{R} - k(T - T_0) = C\frac{dT}{dt}$$

Put $$T - T_0 = \xi$$

So, $$C\dot{\xi}+k\xi=\frac{V^2}{R} \quad \text{or} \quad \dot{\xi}+\frac{k}{C}\xi=\frac{V^2}{CR}$$

or $$\frac{d}{dt}\xi\, e^{kt/C}=\frac{V^2}{CR}e^{kt/C}$$

or $$\xi\, e^{kt/C}=\frac{V^2}{kR}e^{kt/C}+A$$

where A is a constant. Clearly,

$$\xi=0 \text{ at } t=0, \text{ so } A=-\frac{V^2}{kR}$$

Hence, $$T=T_0+\frac{V^2}{kR}(1-e^{-kt/C})$$

3.196 Let $\varphi_A-\varphi_B=\varphi$.

Now, thermal power generated in the resistance R_x

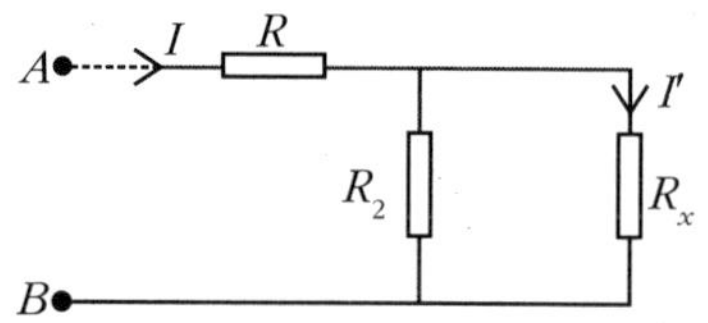

$$P=I'^2R_x=\left[\frac{\varphi}{R_1+\dfrac{R_2R_x}{R_2+R_x}}\frac{R_2}{R_2+R_x}\right]^2R_x$$

For P to be independent of R_x

$$\frac{dP}{dR_x}=0, \text{ which yields}$$

$$R_x=\frac{R_1R_2}{R_1+R_2}=12\,\Omega$$

3.197 Indicate the current flow in circuit as shown in the figure.

Applying loop rule in the closed loop 12561 and using $-\Delta\varphi=0$, we get

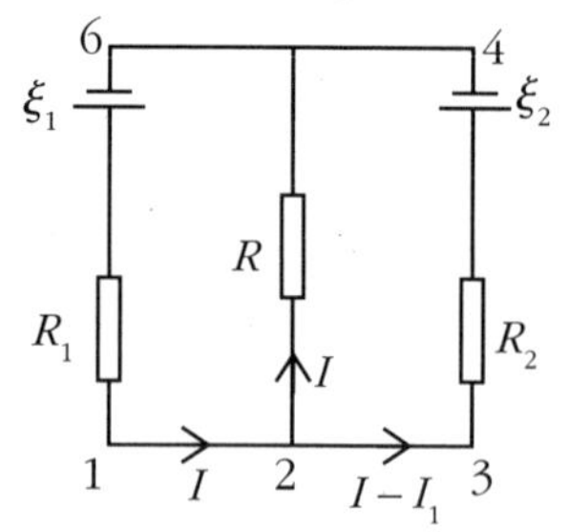

$$I_1R+IR_1=\xi_1 \tag{1}$$

and in the loop 23452,

$$(I-I_1)R_2-I_1R=-\xi_2 \tag{2}$$

Solving Eqs. (1) and (2), we get

$$I_1=\frac{\xi_1R_2+\xi_2R_1}{RR_1+R_1R_2+RR_2}$$

So, thermal power generated in the resistance R is given by

$$P=I_1^2R=\left[\frac{\xi_1R_2+\xi_2R_1}{RR_1+R_1R_2+RR_2}\right]^2R$$

For P to be maximum,

$$\frac{dP}{dR}=0, \text{ which yields}$$

$$R=\frac{R_1R_2}{R_1+R_2}$$

Hence, $$P_{max}=\frac{(\xi_1R_2+\xi_2R_1)^2}{4R_1R_2(R_1+R_2)}$$

3.198 Suppose that there are x number of cells connected in series, in each of the n parallel groups, then

$$nx=N \quad \text{or} \quad x=N/n \tag{1}$$

Now, for any one of the loop, consisting of x cells and the resistor R, we have from the loop rule,

$$IR + \frac{I}{n}xr - x\xi = 0$$

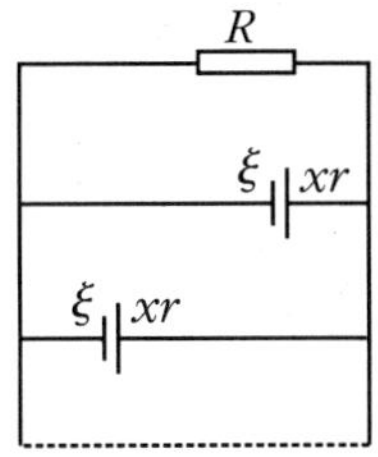

So, $$I = \frac{x\xi}{R + xr/n} = \frac{N/n\xi}{R + Nr/n^2} \quad \text{(using Eq. 1)}$$

Heat generated in the resistor R, is given by

$$Q = I^2R = \left(\frac{Nn\xi}{n^2R + NR}\right)^2 R \quad (2)$$

and for Q to be maximum,

$$\frac{dQ}{dn} = 0, \text{ which yields}$$

$$n = \sqrt{\frac{Nr}{R}} = 3$$

3.199 When switch 1 is closed, maximum charge accumulated on the capacitor, is given by

$$q_{max} = C\xi, \quad (1)$$

and when switch 2 is closed at any arbitrary instant of time,

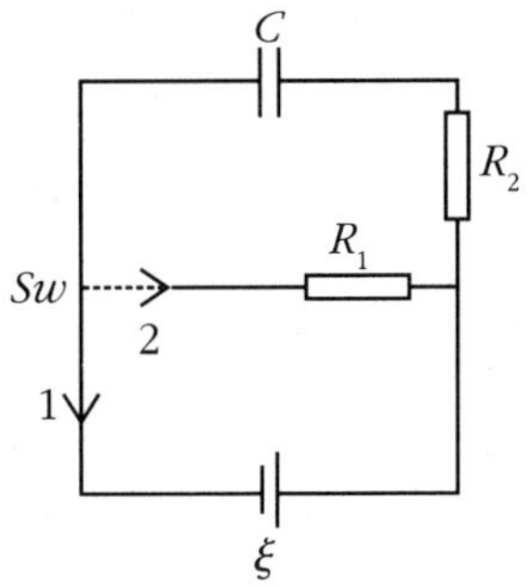

$$(R_1 + R_2)\left(-\frac{dq}{dt}\right) = \frac{q}{c}$$

(because the capacitor is discharging),

or $$\int_{q_{max}}^{q} \frac{1}{q}dq = -\frac{1}{(R_1 + R_2)C}\int_0^t dt$$

Integrating, we get

$$\ln\frac{q}{q_{max}} = \frac{-t}{(R_1 + R_2)C} \text{ or } q = q_{max}e^{\frac{-t}{(R_1+R_2)C}} \quad (2)$$

Differentiating with respect to time, we get

$$I(t) = \frac{dq}{dt} = q_{max}e^{\frac{-t}{(R_1+R_2)C}}\left(-\frac{1}{(R_1 + R_2)C}\right)$$

or $$I(t) = \frac{C\xi}{(R_1 + R_2)C}e^{\frac{-t}{(R_1+R_2)C}}$$

Negative sign is ignored, as we are not interested in the direction of current.

Thus, $$I(t) = \frac{\xi}{(R_1 + R_2)}e^{\frac{-t}{(R_1+R_2)C}} \quad (3)$$

When the switch (Sw) is at position 1, the charge (maximum) accumulated on the capacitor is, $q = C\xi$.

When the switch (Sw) is thrown to position 2, the capacitor starts discharging and as a result, the electric energy stored in the capacitor totally turns into heat energy through the resistors R_1 and R_2 (during a very long interval of time). Thus from the energy conservation, the total heat liberated through the resistors is given by

$$H = U_i = \frac{q^2}{2C} = \frac{1}{2}C\xi^2$$

During the process of discharging of the capacitor, the current through the resistors R_1 and R_2 is the same at all moments of time, thus

$$H_1 \propto R_1 \quad \text{and} \quad H_2 \propto R_2$$

So, $$H_1 = \frac{HR_1}{(R_1 + R_2)} \quad (\text{as } H = H_1 + H_2)$$

Hence, $$H_1 = \frac{1}{2}\frac{CR_1}{R_1 + R_2}\xi^2$$
$$= 60\,\text{mJ (on substituting values)}$$

3.200 When the plate is absent, the capacity of the condenser is

$$C = \frac{\varepsilon_0 S}{d}$$

When the plate is present, the capacity is

$$C' = \frac{\varepsilon_0 S}{d(1-\eta)} = \frac{C}{1-\eta}$$

(a) The energy increment is clearly

$$\Delta U = \frac{1}{2}C'V^2 - \frac{1}{2}CV^2 = -\frac{C\eta}{2(1-\eta)}V^2$$

(b) The charge on the plate is

$$q_i = \frac{CV}{1-\eta} \text{ (initially) and } q_f = CV \text{ (finally)}$$

A charge $CV\,\eta/1 - \eta$ has flown through the battery, charging it and withdrawing $CV^2/1 - \eta$ units of energy from the system into the battery. The energy of thecapacitor has decreased by just half of this. The remaining half, i.e., (1/2) $CV^2\,\eta/1 - \eta$ is the work done by the external agent in withdrawing the plate. This ensures conservation of energy.

3.201 Initially, capacitance of the system = $C\varepsilon$.

So, initial energy of the system $U_i = \frac{1}{2}(C\varepsilon)V^2$

and final energy of the capacitor $U_f = \frac{1}{2}CV^2$

Hence capacitance energy increment

$$\Delta U = \frac{1}{2}CV^2 - \frac{1}{2}(C\varepsilon)V^2 = -\frac{1}{2}CV^2(\varepsilon - 1) = -0.5\,\text{mJ}$$

From energy conservation

$$\Delta U = A_{\text{cell}} + A_{\text{agent}}$$

as there is no heat being released.

But, $$A_{\text{cell}} = (C_f - C_i)V^2 = (C - C\varepsilon)V^2$$

$$A_{\text{agent}} = \Delta U - A_{\text{cell}}$$

Hence, $$= \frac{1}{2}C(\varepsilon - 1)V^2 = 0.5\,\text{mJ}$$

3.202 If C_0 is the initial capacitance of the condenser before water rises in it then

$$U_i = \frac{1}{2}C_0V^2, \text{ where } C_0 = \frac{\varepsilon_0 2l\pi R}{d}$$

(R is the mean radius and l is the length of the capacitor plates.)

Suppose the liquid rises to a height h in it. Then capacitance of the condenser is

$$C = \frac{\varepsilon\varepsilon_0 h 2\pi R}{d} + \frac{\varepsilon(l-h)2\pi R}{d} = \frac{\varepsilon_0 2\pi R}{d}(l + (\varepsilon - 1)h)$$

and energy of the capacitor and the liquid (including both gravitational and electrostatic contributions) is

$$\frac{1}{2}\frac{\varepsilon_0 2\pi R}{d}(l+(\varepsilon-1)h)V^2+\rho g(2\pi R\, hd)\frac{h}{2}$$

If the capacitor were not connected to a battery this energy would have to be minimized. But the capacitor is connected to the battery and, in effect the potential energy of the whole system has to be minimized. Suppose we increase h by δh. Then the energy of the capacitor and the liquid increases by

$$\delta h\left(\frac{\varepsilon_0 2\pi R}{2d}(\varepsilon-1)V^2+\rho g(2\pi Rd)h\right)$$

and that of the cell diminishes by the quantity A_{cell} which is the product of charge flown and V

$$\delta h\frac{\varepsilon_0(2\pi R)}{d}(\varepsilon-1)V^2$$

In equilibrium, the two must balance, so

$$\rho g d h=\frac{\varepsilon_0(\varepsilon-1)V^2}{2d}$$

Hence,

$$h=\frac{\varepsilon_0(\varepsilon-1)V^2}{2\rho g d^2}$$

3.203 (a) Let us imagine a thin spherical layer with inner and outer radii r and $r+dr$, respectively. Lines of current at all the points of this layer are perpendicular to it and therefore such a layer can be treated as a spherical conductor of thickness, dr and cross sectional area $4\pi r^2$. Now, we know that resistance,

$$dR=\rho\frac{dr}{S(r)}=\rho\frac{dr}{4\pi r^2} \tag{1}$$

Integrating Eq. (1) between the limits,

$$\int_0^R dR=\int_a^b\rho\frac{dr}{4\pi r^2} \quad \text{or} \quad R=\frac{\rho}{4\pi}\left[\frac{1}{a}-\frac{1}{b}\right] \tag{2}$$

Capacitance of the network,

$$C=\frac{4\pi\varepsilon_0\varepsilon}{\left[\frac{1}{a}-\frac{1}{b}\right]} \tag{3}$$

and

$$q=C\varphi\left[\begin{array}{c}\text{where } q \text{ is the charge}\\ \text{at any arbitrary moment}\end{array}\right] \tag{4}$$

also,

$$\varphi=\left(\frac{-dq}{dt}\right)R, \text{ as capacitor is discharging.} \tag{5}$$

From Eqs. (2), (3), (4) and (5), we get

$$q=\frac{4\pi\varepsilon_0\varepsilon}{\left[\frac{1}{a}-\frac{1}{b}\right]}\frac{\left[-\frac{dq}{dt}\right]\rho\left[\frac{1}{a}-\frac{1}{b}\right]}{4\pi} \quad \text{or} \quad -\frac{dq}{q}=\frac{dt}{\rho\varepsilon\varepsilon_0}$$

Integrating

$$\int_{q_0}^{q}-\frac{dq}{q}=\frac{1}{\rho\varepsilon_0\varepsilon}\int_0^t dt=\frac{dt}{\rho\varepsilon\varepsilon_0}$$

Hence,

$$q=q_0 e^{-t/r\varepsilon_0\varepsilon}$$

(b) From energy conservation, heat generated during spreading of the charge

$$H = U_i - U_f \text{ (because } A_{\text{cell}} = 0)$$

$$= \frac{1}{2}\frac{q_0^2}{4\pi\varepsilon_0\varepsilon}\left[\frac{1}{a}-\frac{1}{b}\right] - 0 = \frac{q_0^2}{8\pi\varepsilon_0\varepsilon}\frac{b-a}{ab}$$

3.204 (a) Let, at any moment of time, charge on the plates be $(q_0 - q)$ then, current through the resistor

$$I = -\frac{d(q_0 - q)}{dt}$$

because the capacitor is discharging.

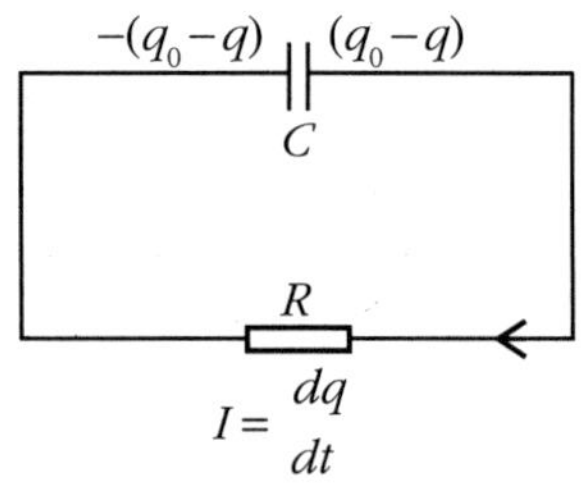

or $$I = \frac{dq}{dt}$$

Now, applying loop rule in the circuit

$$IR - \frac{q_0 - q}{C} = 0$$

or $$\frac{dq}{dt}R = \frac{q_0 - q}{C}$$

or $$\frac{dq}{q_0 - q} = \frac{1}{RC}dt$$

At $t = 0$, $q = 0$ and at $t = \tau$, $q = q$.

So, $$\ln\frac{q_0 - q}{q_0} = \frac{-\tau}{RC}$$

Thus, $$q = q_0(1 - e^{-\tau/RC}) = 0.18 \text{ mC}$$

(b) Amount of heat generated = decrease in capacitance energy

$$= \frac{1}{2}\frac{q_0^2}{C} - \frac{1}{2}\frac{[q_0 - q_0(1 - e^{-\tau/RC})]^2}{C}$$

$$= \frac{1}{2}\frac{q_0^2}{C}[1 - e^{-\tau/RC}] = 82 \text{ mJ}$$

3.205 Let at any moment of time, charge flown be q, then current $I = dq/dt$.

Applying loop rule in the circuit and using $\Delta\varphi = 0$, we get

$$\frac{dq}{dt}R - \frac{(CV_0 - q)}{C} + \frac{q}{C} = 0$$

or $$\frac{dq}{CV_0 - 2q} = \frac{1}{RC}dt$$

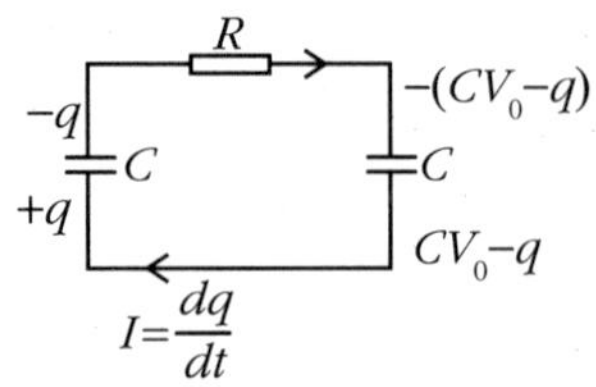

So, $$\ln\frac{CV_0 - 2q}{CV_0} = -2\frac{t}{RC} \text{ (for } 0 \le t \le t)$$

or $$q = \frac{CV_0}{2}(1 - e^{-2t/RC})$$

Hence, $$I = \frac{dq}{dt} = \frac{CV_0}{2}\frac{2}{RC}e^{-2t/RC} = \frac{V_0}{R}e^{-2t/RC}$$

Now, heat liberated will be given by

$$Q = \int_0^\infty I^2 R dt = \frac{V_0^2}{R^2} R \int_0^\infty e^{-4t/RC} dt = \frac{1}{4} C V_0^2$$

3.206 In a rotating frame, to first order in ω, the main effect is a Coriolis force given by $2m(\mathbf{v} \times \omega)$. This unbalanced force will cause electrons to react by setting up a magnetic field $\mathbf{B}$ so that the magnetic force $e(\mathbf{v} \times \mathbf{B})$ balances the Coriolis force.

Thus, $$-\frac{e}{2m}\mathbf{B} = \omega \text{ or } \mathbf{B} = -\frac{2m}{e}\boldsymbol{\omega}$$

The flux associated with this is

$$\Phi = N\pi r^2 B = N\pi r^2 \frac{2m}{e}\omega$$

where $N = l/2\pi r$ is the number of turns of the ring. If ω changes (and there is time for the electron to rearrange) then, B also changes and so does Φ. An e.m.f. will be induced and a current will flow, which is given by

$$I = N\pi r^2 \frac{2m}{e}\dot{\omega}/R$$

The total charge flowing through the ballastic galvanometer, as the ring is stopped, is

$$q = N\pi r^2 \frac{2m}{e}\omega/R$$

So, $$\frac{e}{m} = \frac{2N\pi r^2 \omega}{qR} = \frac{l\omega r}{qR}$$

$$= 1.8 \times 10^{11} \text{ C/kg (on substituting values)}$$

3.207 Let n_0 be the total number of electrons then, total momentum of electrons,

$$p = n_0 m_e v_d \quad \text{(where } v_d \text{ is drift velocity)} \tag{1}$$

Now, $$I = \rho S_x v_d = \frac{n_0 e}{V} S_x v_d = \frac{ne}{l} v_d \tag{2}$$

Here S_x = cross sectional area, ρ = electron charge density, V = volume of sample.

From Eqs. (1) and (2)

$$p = \frac{m_e}{e} Il = 0.40\,\mu\text{Ns}$$

3.208 By definition $nev_d = j$, where v_d is the drift velocity, n is number density of electrons.

Then $$\tau = \frac{l}{v_d} = \frac{nel}{j}$$

So distance actually travelled

$$S = \langle v \rangle \tau = \frac{nel\langle v \rangle}{j} = 10^7 \text{ m}$$

($\langle v \rangle$ = mean velocity of thermal motion of an electron.)

3.209 (a) Let n be the volume density of electrons, then from $I = \rho S v_d$, we get

$$I = neS|\langle \mathbf{v} \rangle| = neS\frac{l}{t}$$

So, $$t = \frac{neSl}{I} = 3\,\text{Ms}$$

(b) Sum of electric forces is given by

$$|(nv)\,e\mathbf{E}| = |n\;Sle\;\rho\mathbf{j}| \quad \text{(where } \rho \text{ is resistivity of the material)}$$

$$= nSle\rho\frac{I}{S} = nel\,\rho I = 1.0\,\text{MN}$$

3.210 From Gauss theorem, field strength at a surface of a cylindrical shape equals, $\lambda/2\pi\varepsilon_0 r$, where λ is the linear charge density.

Now, $$eV = \frac{1}{2}m_e v^2 \text{ or } v = \sqrt{\frac{2eV}{m_e}} \tag{1}$$

Also, $$dq = \lambda dx \text{ so, } \frac{dq}{dt} = \lambda\frac{dx}{dt}$$

or $$I = \lambda v \quad \text{or} \quad \lambda = \frac{I}{v} = \frac{I}{\sqrt{\frac{2eV}{m_e}}} \text{(using 1)}$$

Hence, $$E = \frac{I}{2\pi\varepsilon_0 r}\sqrt{\frac{m_e}{2eV}} = 32\text{ V/m}$$

For the point inside the solid charged cylinder, applying Gauss' theorem

$$2\pi rhE = \pi r^2 h\frac{q}{\varepsilon_0 \pi R^2 l}$$

or $$E = \frac{q/l}{2\pi\varepsilon_0 R^2}r = \frac{\lambda r}{2\pi\varepsilon_0 R^2}$$

So, from $$E = -\frac{d\varphi}{dr}$$

we get $$\int_{\varphi_1}^{\varphi_2} -d\varphi = \int_0^R \frac{\lambda}{2\pi\varepsilon_0 R^2} r\,dr$$

or $$\varphi_1 - \varphi_2 = \frac{\lambda}{2\pi\varepsilon_0 R^2}\left[\frac{r^2}{2}\right]_0^R = \frac{\lambda}{4\pi\varepsilon_0}$$

Hence, $$\varphi_1 - \varphi_2 = \frac{I}{4\pi\varepsilon_0}\sqrt{\frac{m_e}{2eV}} = 0.80\,\text{V}$$

3.211 (a) Between the plates

$$\varphi = ax^{4/3}$$

or $$\frac{\partial\varphi}{\partial x} = a\times\frac{4}{3}x^{1/3}$$

$$\frac{\partial^2\varphi}{\partial x^2} = \frac{4}{9}ax^{-2/3} = \frac{-\rho}{\varepsilon_0}$$

or $$\rho = -\frac{4\varepsilon_0 a}{9}x^{-2/3}$$

(b) Let the charge on the electron be $-e$.

Then $$\frac{1}{2}mv^2 - e\varphi = \text{constant} = 0$$

(as the electron is initially emitted with negligible energy),

or $$v^2 = \frac{2e\varphi}{m} \text{ and } v = \sqrt{\frac{2e\varphi}{m}}$$

So, $$j = -\rho v = \frac{4\varepsilon_0 a}{9}\sqrt{\frac{2\varphi e}{m}}x^{-2/3} = \frac{4}{9}\varepsilon_0 a^{3/2}\sqrt{\frac{2e}{m}}$$

(j is measured from anode to cathode, hence the $-ve$ sign.)

3.212 we know that, $$E = \frac{V}{d}$$

So by the definition of the mobility

$$v^+ = u_0^+\frac{v}{d},\quad v^- = u_0^-\frac{v}{d}$$

and $$j = (n_+u_0^+ + n_-u_0^-)\frac{eV}{d}$$

The negative ions move towards the anode and the positive ions towards the cathode and the total current is the sum of the currents due to them.

On the other hand, in equilibrium

$$n_+ = n_-$$

So, $$n_+ = n_- = \frac{I}{S}\Big/(u_0^+ + u_0^-)\frac{eV}{d}$$

$$= \frac{Id}{eVS(u_0^+ + u_0^-)} = 2.3\times10^8 \text{ cm}^{-3}$$

3.213 Velocity = mobility × field

or $$v = u\frac{v_0}{l}\sin\omega t \text{ (which is positive for } 0 \le \omega t \le \pi)$$

So, the maximum displacement in one direction is

$$x_{max} = \int_0^{\pi} u\frac{V_0}{l}\sin\omega t dt = \frac{2uV_0}{l\omega}$$

At $\omega = \omega_0, x_{max} = l$ so, $\dfrac{2uV_0}{l\omega_0} = l$

Thus, $$u = \frac{\omega_0 l^2}{2V_0}$$

3.214 When the current is saturated, all the ions produced reach the plate.

Then, $$\dot{n}_i = \frac{I_{sat}}{eV} = 6\times10^9 \text{ cm}^{-3}\text{ s}^{-1}$$

(Both positive ions and negative ions are counted here.)

The equation of balance is

$$\frac{dn}{dt} = \dot{n}_i - rn^2$$

The first term on the right is the production rate and the second term is the recombination rate which by the usual statistical arguments is proportional to n^2 (no. of positive ions × no. of $-ve$ ions). In equilibrium

$$\frac{dn}{dt} = 0$$

So,
$$n_{eq} = \sqrt{\frac{\dot{n}_i}{r}} = 6 \times 10^7 \text{ cm}^{-3}$$

3.215 Initially,
$$n = n_0 = \sqrt{\dot{n}_i / r}$$

Since we can assume that long exposure to the ionizer has caused equilibrium to be set up. After the ionizer is switched off,

$$\frac{dn}{dt} = -rn^2$$

or
$$rdt = -\frac{dn}{n^2} \quad \text{or} \quad rt = \frac{1}{n} + \text{constant}$$

But ,
$$n = n_0 \text{ at } t = 0 \text{ so, } rt = \frac{1}{n} - \frac{1}{n_0}$$

The concentration will decrease by a factor η when

$$rt_0 = \frac{1}{n_0/\eta} - \frac{1}{n_0} = \frac{\eta - 1}{n_0}$$

or
$$t_0 = \frac{\eta - 1}{\sqrt{r\dot{n}_i}} = 13 \text{ ms}$$

3.216 Ions produced will cause charge to decay. Clearly,

$$\eta CV = \text{decrease of charge} = \dot{n}_i\, eAdt = \frac{\varepsilon_0 A}{d} V\eta$$

or
$$t = \frac{\varepsilon_0 V \eta}{\dot{n}_i e d^2} = 4.6 \text{ days}$$

Note that n_i here, is the number of ion pairs produced.

3.217 If ν = number of electrons moving to the anode at distance x, then

$$\frac{dv}{dx} = \alpha v \quad \text{or} \quad v = v_0 e^{\alpha x}$$

Assuming saturation,

$$I = ev_0 e^{\alpha d}$$

3.218 Since the electrons are produced uniformly through the volume, the total current attaining saturation is clearly,

$$I = \int_0^d e(\dot{n}_i\, A dx)\, e^{\alpha x} = e\dot{n}_i A\left(\frac{e^{\alpha d} - 1}{\alpha}\right)$$

Thus,
$$j = \frac{I}{A} = e\dot{n}_i\left(\frac{e^{\alpha d} - 1}{\alpha}\right)$$

3.5 Constant Magnetic Field. Magnetics

3.219 (a) From Biot-Savart's law

$$d\mathbf{B} = \frac{\mu_0}{4\pi} I \frac{d\mathbf{l} \times \mathbf{r}}{r^3}$$

So,
$$d\mathbf{B} = \frac{\mu_0}{4\pi} I \frac{(Rd\theta)R}{R^3} \text{ (as } d\mathbf{l} \perp \mathbf{r})$$

From symmetry

$$B = \int dB = \int_0^{2\pi} \frac{\mu_0}{4\pi} \frac{I}{R} d\theta = \frac{\mu_0}{2} \frac{I}{R} = 6.3\ \mu T$$

(b) From Biot-Savart's law

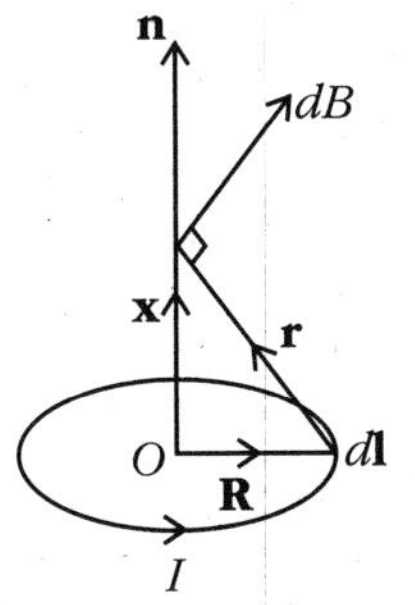

$$\mathbf{B} = \frac{\mu_0}{4\pi} I \int \frac{d\mathbf{l} \times \mathbf{r}}{r^3} \text{ (here } \mathbf{r} = \mathbf{x} - \mathbf{R})$$

So,
$$\mathbf{B} = \frac{\mu_0}{4\pi} \frac{I}{r^3} \left[\oint d\mathbf{l} \times \mathbf{x} - \oint \mathbf{dl} \times \mathbf{R} \right]$$

Since $\mathbf{x}$ is a constant vector and $|\mathbf{R}|$ is also constant, so,

$$\oint d\mathbf{l} \times \mathbf{x} = \left(\oint d\mathbf{l}\right) \times \mathbf{x} = 0 \left(\text{because } \oint d\mathbf{l} = 0\right)$$

and
$$-\oint d\mathbf{l} \times \mathbf{R} = \oint Rdl\, \mathbf{n} = \mathbf{n}R \oint dl = 2\pi R^2 \mathbf{n}$$

Here $\mathbf{n}$ is a unit vector perpendicular to the plane containing the current loop (see figure) and in the direction of $\mathbf{x}$.

Thus, we get
$$\mathbf{B} = \frac{\mu_0}{4\pi} \frac{2\pi R^2 I}{(x^2 + R^2)^{3/2}} \mathbf{n}$$
$$= 2.3\ \mu\text{T}$$

3.220 Since $\angle AOB = 2\pi/n$, OC or perpendicular distance of any segment from centre equals $R \cos \pi/n$. Now magnetic induction at O, due to the right current carrying element AB

$$= \frac{\mu_0}{4\pi} \frac{I}{R \cos \pi/n} 2 \sin \frac{\pi}{n}$$

From Biot–Savart's law, the magnetic field at O due to any section such as AB is perpendicular to the plane of the figure and has the magnitude

$$B = \int \frac{\mu_0}{4\pi} I \frac{dx}{r^2} \cos\theta = \int_{-\frac{\pi}{n}}^{\frac{\pi}{n}} \frac{\mu_0 I}{4\pi} \frac{R \cos \pi/n \sec^2 \theta d\theta}{R^2 \cos 2\pi/n \sec^2 \theta} \cos\theta = \frac{\mu_0 I}{4\pi} \frac{1}{R \cos \pi/n} 2 \sin \frac{\pi}{n}$$

As there are n number of sides and magnetic induction vectors, due to each side at O, are equal in magnitude and direction. So

$$B_0 = \frac{\mu_0}{4\pi} \frac{nI}{R \cos \frac{\pi}{n}} 2 \sin \frac{\pi}{n} \cdot n$$
$$= \frac{\mu_0}{2\pi} \frac{nI}{R} \tan \frac{\pi}{n}$$

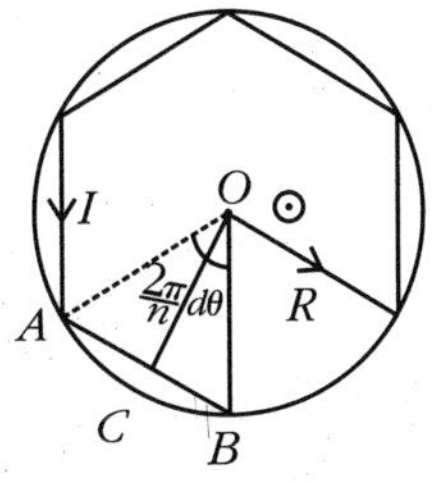

For $n \to \infty$
$$B_0 = \frac{\mu_0}{2}\frac{I}{R}\lim_{n\to\infty}\left(\frac{\tan\frac{\pi}{n}}{\pi/n}\right)$$
$$= \frac{\mu_0}{2}\frac{I}{R}$$

3.221 We know that magnetic induction due to a straight current carrying wire at any point, at a perpendicular distance from it is given by

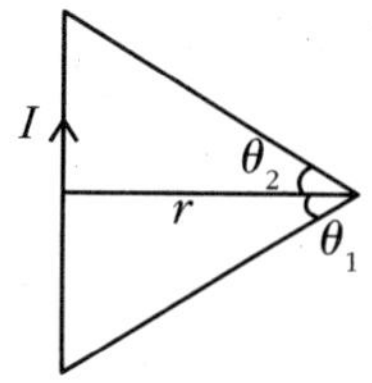

$$B = \frac{\mu_0}{4\pi}\frac{I}{r}(\sin\theta_1 + \sin\theta_2) \text{ (as shown in figure)}$$

Here, $$B_1 = B_3 = \frac{\mu_0}{4\pi}\frac{I}{(d/2)\sin\frac{\varphi}{2}}\left\{\cos\frac{\varphi}{2} + \cos\frac{\varphi}{2}\right\}$$

and $$B_2 = B_4 = \frac{\mu_0}{4\pi}\frac{I}{(d/2)\cos\frac{\varphi}{2}}\left(\sin\frac{\varphi}{2} + \sin\frac{\varphi}{2}\right)$$

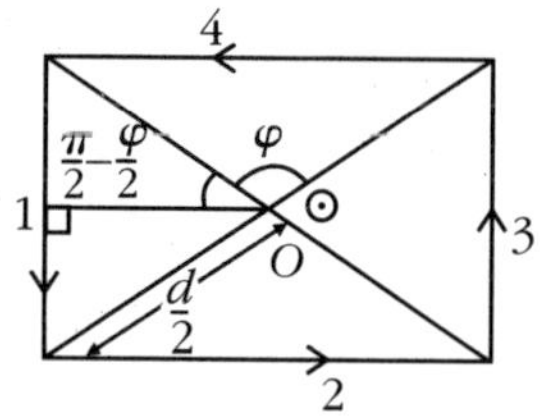

Hence, the magnitude of total magnetic induction at O,
$$B_0 = B_1 + B_2 + B_3 + B_4$$
$$= \frac{\mu_0}{4\pi}\frac{4I}{d/2}\left[\frac{\cos\frac{\varphi}{2}}{\sin\frac{\varphi}{2}} + \frac{\sin\frac{\varphi}{2}}{\cos\frac{\varphi}{2}}\right]$$
$$= \frac{4\mu_0 I}{\pi d\sin\varphi} = 0.10 \text{ mT}$$

3.222 Magnetic induction due to the arc segment at O is given by

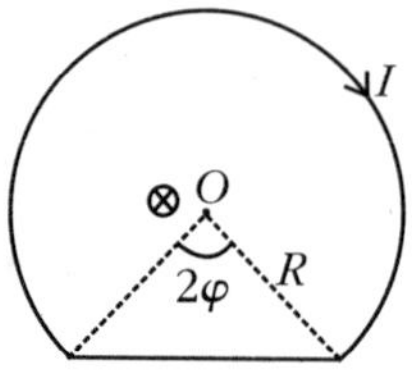

$$B_{\text{arc}} = \frac{\mu_0}{4\pi}\frac{I}{R}(2\pi - 2\varphi)$$

and magnetic induction due to the line segment at O is given by
$$B_{\text{line}} = \frac{\mu_0}{4\pi}\frac{I}{R\cos\varphi}[2\sin\varphi]$$

So, total magnetic induction
$$B_0 = B_{\text{arc}} + B_{\text{line}} = \frac{\mu_0}{2\pi}\frac{I}{R}[\pi - \varphi + \tan\varphi] = 28\ \mu\text{T}$$

3.223 (a) From the Biot-Savart law
$$d\mathbf{B} = \frac{\mu_0}{4\pi}I\frac{(d\mathbf{I}\times\mathbf{r})}{r^3}$$

So, magnetic field induction due to the arc of radius a at O is given by
$$B_a = \frac{\mu_0}{4\pi}\frac{I}{a}(2\pi - \varphi) \text{ (directed inward)}$$

and magnetic field induction due to the arc of radius b at O is given by
$$B_b = \frac{\mu_0}{4\pi}\frac{I}{a}\varphi \text{ (directed inward)}$$

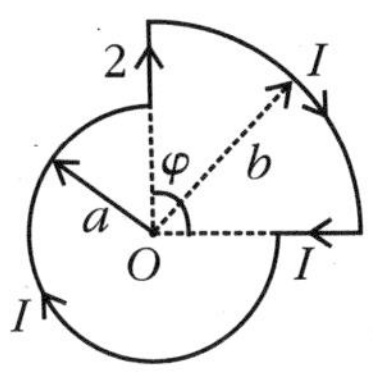

Due to each of straight current segments 1 and 2, magnetic induction is zero, because $d\mathbf{l} \parallel \mathbf{r}$.

Hence,
$$B_0 = \frac{\mu_0}{4\pi} I \left[\frac{2\pi - \varphi}{a} + \frac{\varphi}{b}\right]$$

(b) Due to each of straight currents directed towards (away) centre, magnetic induction vector at the centre is zero. From the symmetry, direction of magnetic induction vector due to rest of two straight wires and the arc current, all are directed normally into the plane of the figure. Magnetic induction due to each of straight current is

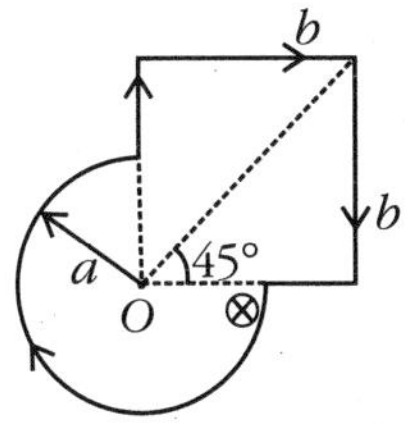

$$\frac{\mu_0}{4\pi}\frac{I}{b}\sin 45°$$

and due to arc current is

$$\frac{\mu_0}{4\pi}\frac{I}{a}\left(\frac{3\pi}{2}\right)$$

Hence, the total magnetic induction at O is given by

$$B_0 = \frac{\mu_0}{4\pi}\frac{I}{a}\left(\frac{3\pi}{2}\right) + 2\times\frac{\mu_0}{4\pi}\frac{I}{b}\sin 45° = \frac{\mu_0}{4\pi}\frac{I}{n}\left[\frac{3\pi}{2a} + \frac{\sqrt{2}}{b}\right]$$

3.224 The thin walled tube with a longitudinal slit can be considered equivalent to a full tube and a strip carrying the same current density in the opposite direction. Inside the tube, the former does not contribute so the total magnetic field is simply due to the strip. It is

$$B = \frac{\mu_0}{2\pi}\frac{(I/2\pi R)h}{r} = \frac{\mu_0}{4\pi^2}\frac{Ih}{Rr}$$

where r is the distance of the field point from the strip.

3.225 First of all let us find out the direction of vector **B** at point O. For this purpose, we divide the entire conductor into elementary fragments with current dI. It is obvious that the sum of any two symmetric fragments gives a resultant along **B** as shown in the figure and consequently, vector **B** will also be directed as shown.

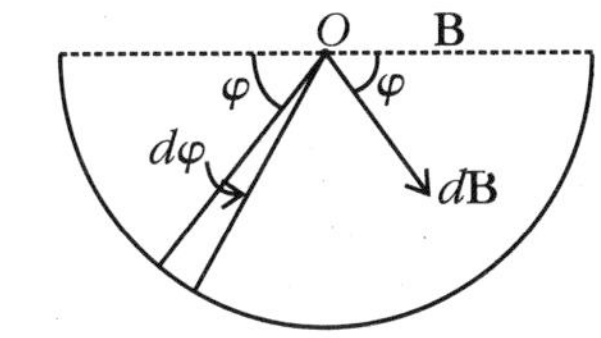

So,
$$|\mathbf{B}| = \int dB \sin\varphi$$
$$= \int \frac{\mu_0}{2\pi R} dI \sin\varphi$$
$$= \int_0^{\pi} \frac{\mu_0}{2\pi^2 R} I \sin\varphi\, d\varphi \left(as\ dI = \frac{I}{\pi} d\varphi\right)$$

Hence,
$$B = \frac{\mu_0 I}{\pi^2 R}$$

3.226 From symmetry, for Fig. (a)

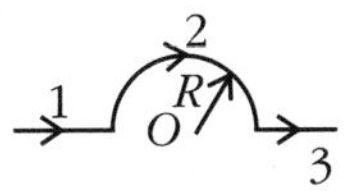

(a)

$$B_0 = B_1 + B_2 + B_3$$
$$= 0 + \frac{\mu_0}{4\pi}\frac{I}{R}\pi + 0 = \frac{\mu_0}{4}\frac{I}{R}$$

From symmetry, for Fig. (b)

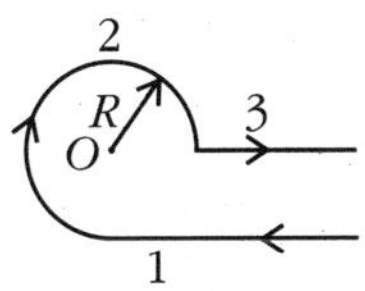

(b)

$$B_0 = B_1 + B_2 + B_3$$
$$= \frac{\mu_0}{4\pi}\frac{I}{R} + \frac{\mu_0}{2\pi}\frac{I}{R}\frac{3\pi}{2} + 0$$
$$= \frac{\mu_0}{4\pi}\frac{I}{R}\left[1 + \frac{3\pi}{2}\right]$$

From symmetry, for Fig. (c)

$$B_0 = B_1 + B_2 + B_3$$
$$= \frac{\mu_0}{4\pi}\frac{I}{R} + \frac{\mu_0}{4\pi}\frac{I}{R}\pi + \frac{\mu_0}{4\pi}\frac{I}{R}$$
$$= \frac{\mu_0}{4\pi}\frac{I}{R}(2+\pi)$$

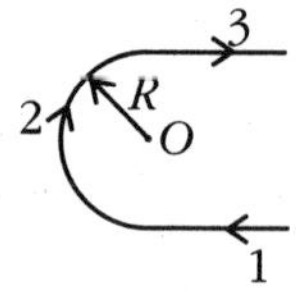

(c)

3.227 $$\mathbf{B}_0 = \mathbf{B}_1 + \mathbf{B}_2$$

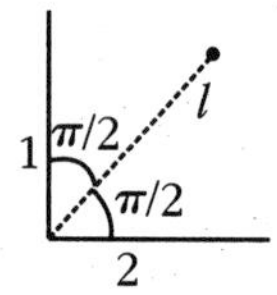

or $$|\mathbf{B}_0| = \frac{\mu_0 I}{4\pi l}\sqrt{2} = 2.0\ \mu\text{T} \text{ (using Problem 3.221)}$$

3.228 (a) $$\mathbf{B}_0 = \mathbf{B}_1 + \mathbf{B}_2 + \mathbf{B}_3$$
$$= \frac{\mu_0}{4\pi}\frac{I}{R}(-\mathbf{k}) + \frac{\mu_0}{4\pi}\frac{I}{R}\pi(-\mathbf{i}) + \frac{\mu_0}{4\pi}\frac{I}{R}(-\mathbf{k})$$
$$= -\frac{\mu_0}{4\pi}\frac{I}{R}[2\mathbf{k} + \pi\mathbf{i}]$$

So, $$|\mathbf{B}_0| = \frac{\mu_0}{4\pi}\frac{I}{R}\sqrt{\pi^2 + 4} = 0.030\ \mu\text{T}$$

(b) $$\mathbf{B}_0 = \mathbf{B}_1 + \mathbf{B}_2 + \mathbf{B}_3$$
$$= \frac{\mu_0}{4\pi}\frac{I}{R}(-\mathbf{k}) + \frac{\mu_0}{4\pi}\frac{I}{R}\pi(-\mathbf{i}) + \frac{\mu_0}{4\pi}\frac{I}{R}(-\mathbf{i})$$
$$= -\frac{\mu_0}{4\pi}\frac{I}{R}[\mathbf{k} + (\pi+1)\mathbf{i}]$$

So, $$|\mathbf{B}_0| = \frac{\mu_0}{4\pi}\frac{I}{R}\sqrt{1 + (\pi+1)^2} = 0.34\ \mu\text{T}$$

(c) Here, using the law of parallel resistances

$$I_1 + I_2 = I \quad \text{and} \quad \frac{I_1}{I_2} = \frac{1}{3}$$

So, $$\frac{I_1 + I_2}{I_2} = \frac{4}{3}$$

Hence, $$I_2 = \frac{3}{4}I \quad \text{and} \quad I_1 = \frac{1}{4}I$$

Thus, $$\mathbf{B}_0 = \frac{\mu_0}{4\pi}\frac{I}{R}(-\mathbf{k}) + \frac{\mu_0}{4\pi}\frac{I}{R}(-\mathbf{j}) + \left[\frac{\mu_0}{4\pi}\left(\frac{3\pi}{2}\right)\frac{I_1}{R}(-\mathbf{i}) + \frac{\mu_0}{4\pi}\frac{(\pi/2)I_2}{R}\mathbf{i}\right]$$
$$= \frac{-\mu_0}{4\pi}\frac{I}{R}(\mathbf{j}+\mathbf{k}) + 0$$

Thus, $$|\mathbf{B}_0| = \frac{\mu_0}{4\pi}\frac{\sqrt{2}I}{R} = 0.11\,\mu\text{T}$$

3.229 (a) We apply circulation theorem as shown in the figure. The current is vertically upwards in the plane and the magnetic field is horizontal and parallel to the plane.

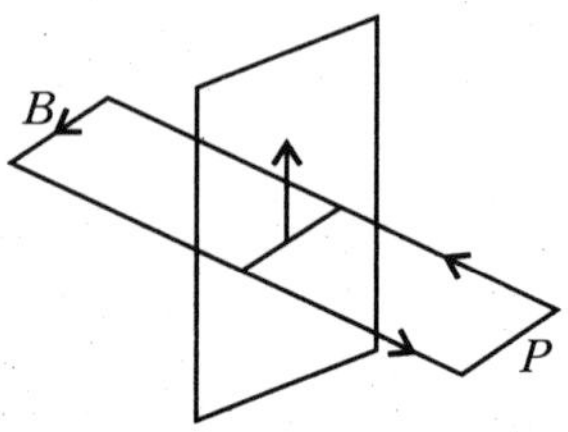

Then, $$\oint \mathbf{B}\cdot d\mathbf{l} = 2Bl = \mu_0 Il \quad \text{or} \quad B = \frac{\mu_0 I}{2}$$

(b) Each plane contributes $\mu_0 I/2$ between the planes and outside the plane that is cancelled out.

Thus,
$$B = \begin{cases} \mu_0 I \text{ between the plane} \\ 0 \text{ outside} \end{cases}$$

3.230 We assume that the current flows perpendicular to the plane of the paper, then by circulation theorem, inside the plate

$$2Bdl = \mu_0(2x\,dl)j$$

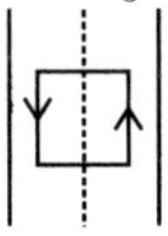

or
$$B = \mu_0 xj,\ |x| \le d$$

Outside the plate,
$$2Bdl = \mu_0(2d\,dl)j$$

or
$$B = \mu_0\,dj\,|x| \ge d$$

3.231 It is easy to convince oneself that in both the regions 1 and 2, there can only be a circular magnetic field (i.e., the component B_φ). Any radial field in region 1 or any B_z away from the current plane will imply a violation of Gauss' law of magnetostatics. B_φ must obviously be symmetrical about the straight wire. Then in 1,

$$B_\varphi 2\pi r = \mu_0 I$$

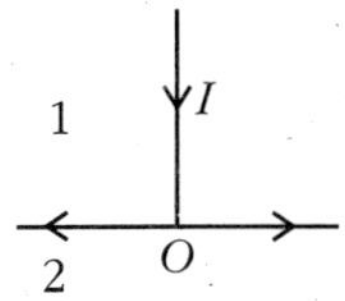

or
$$B_\varphi = \frac{\mu_0}{2\pi}\frac{I}{r}$$

In 2,
$$B_\varphi \cdot 2\pi r = 0 \quad \text{or} \quad B_\varphi = 0$$

3.232 Along the axis of loop,

$$B = \frac{\mu_0 IR^2}{2(R^2 + x^2)^{3/2}} = B_x \text{ (along the axis)}$$

Thus,
$$\int \mathbf{B}\cdot d\mathbf{r} = \int_{-\infty}^{\infty} B_x dx = \frac{\mu_0 IR^2}{2}\int_{-\infty}^{\infty}\frac{dx}{(R^2+x^2)^{3/2}}$$

$$= \frac{\mu_0 IR^2}{2}\int_{-\pi/2}^{\pi/2}\frac{R\sec^2\theta\,d\theta}{R^3\sec^3\theta} \text{ (on putting } x = R\tan\theta)$$

$$= \mu_0 I\frac{1}{2}\int_{-\pi/2}^{\pi/2}\cos\theta\,d\theta = \mu_0 I$$

The physical interpretation of this result is that $\int_{-\infty}^{\infty} B_x dx$ can be thought of as the circulation of B over a closed loop by imagining that the two ends of the axis are connected by a line at infinity (e.g., a semicircle of infinite radius).

3.233 By circulation theorem, inside the conductor

$$B_\varphi 2\pi r = \mu_0 j_z \pi r^2 \quad \text{or} \quad B_\varphi = \mu_0 j_z r/2$$

i.e.,
$$\mathbf{B} = \frac{1}{2}\mu_0 \mathbf{j}\times\mathbf{r}$$

Similarly, outside the conductor

$$B_\varphi 2\pi r = \mu_0 j_z \pi R^2 \quad \text{or} \quad B_\varphi = \frac{1}{2}\mu_0 j_z \frac{R^2}{r}$$

So,
$$\mathbf{B} = \frac{1}{2}\mu_0(\mathbf{j}\times\mathbf{r})\frac{R^2}{r^2}$$

2.234 The given current, which will be assumed uniform, arises due to a negative current flowing in the cavity superimposed on the true current everywhere including the cavity. Then from the previous problem, by superposition,

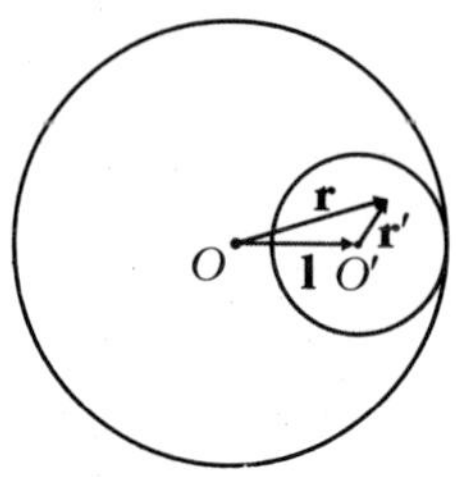

$$\mathbf{B} = \frac{\mu_0}{2}[\mathbf{j}\times\mathbf{r}] - \frac{\mu_0}{2}[\mathbf{j}\times\mathbf{r}'] = \frac{\mu_0}{2}[\mathbf{j}\times(\mathbf{r}-\mathbf{r}')]$$

$$\mathbf{B} = \frac{\mu_0}{2}(\mathbf{j}\times\mathbf{l})$$

If $\mathbf{l}$ vanishes so that the cavity is concentric with the conductor, there is no magnetic field in the cavity.

3.235 By circulation theorem

$$B_\varphi \cdot 2\pi r = \mu_0 \int_0^r j(r')\times 2\pi r'\, dr'$$

Using $B_{\varphi'} = br^\alpha$, inside the stream,

$$br^{\alpha+1} = \mu_0 \int_0^r j(r')r'dr'$$

So by differentiation,

$$(\alpha+1)br^\alpha = \mu_0 j(r) r$$

Hence,

$$j(r) = \frac{b(\alpha+1)}{\mu_0} r^{\alpha-1}$$

3.236 On the surface of the solenoid there is a surface current density, given by

$$\mathbf{j}_s = nI\mathbf{e}_\varphi$$

Then,

$$\mathbf{B} = -\frac{\mu_0}{4\pi} nI \int R\, d\varphi\, dz \frac{\mathbf{e}_\varphi\times\mathbf{r}_0}{r_0^3}$$

where $\mathbf{r}_0$ is the vector from the current element to the field point, which is the centre of the solenoid,

Now,

$$-\mathbf{e}_\varphi\times\mathbf{r}_0 = R\mathbf{e}_z$$

$$r_0 = (z^2+R^2)^{1/2}$$

Thus,

$$B = B_z = \frac{\mu_0 nI}{4\pi}\times 2\pi R^2 \int_{-l/2}^{l/2} \frac{dz}{(R^2+z^2)^{3/2}}$$

$$= \frac{1}{2}\mu_0 nI \int_{-\tan^{-1}\frac{l}{2R}}^{+\tan^{-1}\frac{l}{2R}} \cos\alpha\, d\alpha \text{ (on putting } z = R\tan\alpha)$$

$$= \mu_0 nI \sin\alpha = \mu_0 nI \frac{l/2}{\sqrt{(l/2)^2+R^2}}$$

$$= \mu_0 nI \Big/ \sqrt{1+\left(\frac{2R}{l}\right)^2}$$

3.237 (a) We proceed exactly as in the previous problem. Then the magnetic induction on the axis at a distance x from one end is clearly,

$$B = \frac{\mu_0 nI}{4\pi}\times 2\pi R^2 \int_0^\infty \frac{dz}{[R^2+(z-x)^2]^{3/2}} = \frac{1}{2}\mu_0 nIR^2 \int_x^\infty \frac{dz}{(z^2+R^2)^{3/2}}$$

$$= \frac{1}{2}\mu_0 nI \int_{\tan^{-1}\frac{x}{R}}^{\pi/2} \cos\theta\, d\theta = \frac{1}{2}\mu_0 nI\left(1-\frac{x}{\sqrt{x^2+R^2}}\right)$$

$x > 0$ means that the field point is outside the solenoid. B then decreases with (x). $x < 0$ means that the point gets more and more inside the solenoid. B then increases with (x) and eventually becomes constant, equal to $\mu_0 nI$. The $B - x$ graph is as given in the answer sheet.

(b) We have,

$$\frac{B_0 - \delta B}{B_0} = \frac{1}{2}\left[1 - \frac{x_0}{\sqrt{R^2 + x_0^{\,2}}}\right] = 1 - \eta$$

or
$$-\frac{x_0}{\sqrt{R^2 + x_0^{\,2}}} = 1 - 2\eta$$

Since η is small ($\approx 1\%$), x_0 must be negative.

Thus,
$$x_0 = -|x_0|$$

and
$$-\frac{|x_0|}{\sqrt{R^2 + |x_0|^2}} = 1 - 2\eta$$

$$|x_0|^2 = (1 - 4\eta + 4\eta^2)(R^2 + |x_0|^2)$$

$$0 = (1 - 2\eta)^2 R^2 - 4\eta(1 - \eta)|x_0|^2$$

or
$$|x_0| = \frac{(1 - 2\eta)R}{2\sqrt{\eta(1 - \eta)}}$$

$$\approx 5\,\mathrm{R}$$

3.238 If the strip is tightly wound, it must have a pitch of h. This means that the current will flow obliquely, partly along $\mathbf{e}_\varphi$ and partly along $\mathbf{e}_z$. Obviously, the surface current density is

$$\mathbf{j}_s = \frac{I}{b}\left[\sqrt{1 - (b/2\pi R)^2}\,\mathbf{e}_\varphi + \frac{b}{2\pi R}\mathbf{e}_z\right]$$

By comparision with the case of a solenoid and a hollow straight conductor, we see that field inside the coil $(r > R)$ is

$$= \mu_0 \frac{I}{b}\sqrt{1 - (b/2\pi R)^2}$$

$$(\text{Cf. } B = \mu_0 nI)$$

For the field outside the coil $(r < R)$, only the other term contributes, so

$$B_\varphi \times 2\pi r = \mu_0 \frac{I}{b} \times \frac{b}{2\pi R} \times 2\pi R$$

or
$$B_\varphi = \frac{\mu_0}{4\pi} \cdot \frac{2I}{r}$$

$$= 0.3\,\mathrm{mT}$$

Note: *Surface current density is defined as current flowing normally across a unit length over a surface.*

3.239 Suppose a is the radius of cross-section of the core. The winding has a pitch $2\pi R/N$, so the surface current density is

$$\mathbf{j}_s \cong \frac{I}{2\pi R/N}\mathbf{e}_1 + \frac{I}{2\pi a}\mathbf{e}_2$$

where $\mathbf{e}_1$ is a unit vector along the cross section of the core and $\mathbf{e}_2$ is a unit vector along its length.

The magnetic field inside the cross section of the core is due to first term above, and is given by

$$B_\varphi \cdot 2\pi R = \mu_0 NI$$

(NI is total current due to the above surface current (first term).)

Thus,
$$B_\varphi = \mu_0\, NI/2\pi R.$$

The magnetic field at the centre of the core can be obtained from the basic formula:

$$d\mathbf{B} = \frac{\mu_0}{4\pi}\frac{\mathbf{j}_s \times \mathbf{r}_0}{r_0^3}dS$$

and is due to the second term.

So, $$\mathbf{B} = B_z\mathbf{e}_z = \mathbf{e}_z\frac{\mu_0}{4\pi}\frac{I}{2\pi a}\int\frac{1}{R^3}Rd\varphi \times 2\pi a$$

or $$B_z = \frac{\mu_0 I}{2R}$$

The ratio of the two magnetic fields, $\eta = \dfrac{N}{\pi} = 8 \times 10^2$

3.240 We need the flux through the shaded area.

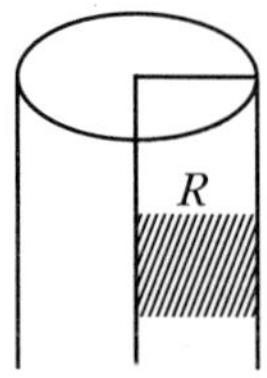

Now by Ampere's theorem,

$$B_\varphi 2\pi r = \mu_0\frac{I}{\pi R^2}\cdot \pi r^2$$

or $$B_\varphi = \frac{\mu_0}{2\pi}I\frac{r}{R^2}$$

The flux through the shaded region is

$$\Phi_1 = \int_0^R 1\cdot drB_\varphi(r)$$

$$= \int_0^R dr\frac{\mu_0}{2\pi}I\frac{r}{R^2} = \frac{\mu_0}{4\pi}I$$

$$= 1.0\ \mu\text{Wb/m}$$

3.241 Using solution of Problem 3.237, the magnetic field is given by,

$$B = \frac{1}{2}\mu_0 nI\left(1 - \frac{x}{\sqrt{x^2+R^2}}\right)$$

At the end, $$B = \frac{1}{2}\mu_0 nI = \frac{1}{2}B_0 \quad (\text{where } B_0 = \mu_0 nI)$$

is the field deep inside the solenoid. Thus,

$$\Phi = \frac{1}{2}\mu_0\, nIs = \frac{\Phi_0}{2} \text{ where } \Phi_0 = \mu_0 nIs$$

is the flux of the vector B through the cross-section deep inside the solenoid.

3.242 $$B_\varphi 2\pi r = \mu_0 NI$$

or $$B_\varphi = \frac{\mu_0 NI}{2\pi r}$$

Then, $$\Phi = \int_b^a B_\varphi b dr \ (\text{for } a \le r \le b)$$

$$= \frac{\mu_0}{4\pi}2NIb\ln\eta \ \left(\text{where } \eta = \frac{b}{a}\right)$$

$$= 8\ \mu\text{Wb}$$

3.243 Magnetic moment of a current loop is given by $p_m = nIS$, where n is the number of turns and S is the cross sectional area. In our problem,

$$n = 1,\ S = \pi R^2 \quad \text{and} \quad B = \frac{\mu_0}{2}\frac{I}{R}$$

So,
$$p_m = \frac{2BR}{\mu_0}\pi R^2 = \frac{2\pi BR^3}{\mu_0} = 30 \text{ mA m}^2$$

3.244 Take an element of length $r\,d\theta$ containing $\frac{N}{\pi r}\cdot r d\theta$ turns. Its magnetic moment is

$$\frac{N}{\pi}d\theta\cdot\frac{\pi}{4}d^2I$$

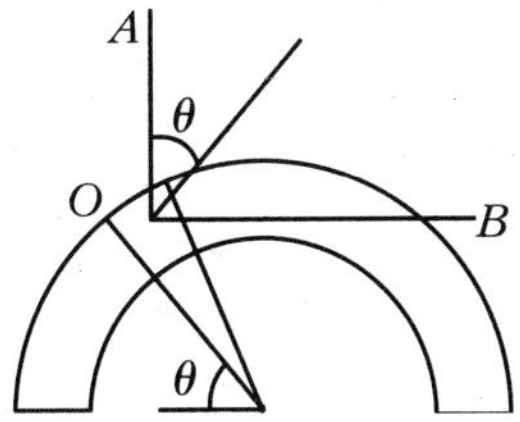

normal to the plane of cross section. We resolve it along OA and OB. The moment along OA integrates to

$$p_m = \int_0^{\pi}\frac{N}{4}d^2I\cos\theta\,d\theta = 0$$

while that along OB gives

$$p_m = \int_0^{\pi}\frac{Nd^2I}{4}\sin\theta\,d\theta = \frac{1}{2}Nd^2I = 0.5 \text{ Am}^2$$

3.245 (a) From Biot-Savart's law, the magnetic induction due to a circular current carrying wire loop at its centre is given by

$$B_r = \frac{\mu_0}{2r}I$$

The plane spiral is made up of concentric circular loops, having different radii, varying from a to b. Therefore, the total magnetic induction at the centre,

$$B_0 = \int\frac{\mu_0}{2r}IdN \tag{1}$$

where $\frac{\mu_0}{2r}I$ is the contribution of one turn of radius r and dN is the number of turns in the interval $(r, r + dr)$ i.e.,

$$dN = \frac{N}{b-a}dr$$

Substituting in Eq. (1) and integrating the result over r between a and b, we obtain

$$B_0 = \int_a^b\frac{\mu_0 I}{2r}\frac{N}{(b-a)}dr = \frac{\mu_0 IN}{2(b-a)}\ln\frac{b}{a} = 7\ \mu\text{T}$$

(b) The magnetic moment of a turn of radius r is $p_m = I\pi r^2$ and of all turns,

$$p = \int p_m dN = \int_a^b I\pi r^2\frac{N}{b-a}dr = \frac{\pi IN(b^3-a^3)}{3(b-a)}$$
$$= 15 \text{ mA m}^2$$

3.246 (a) Let us take a ring element of radius r and thickness dr. Then charge on the ring element

$$dq = \sigma\,2\pi rdr$$

and current due to this element

$$dI = \frac{(\sigma 2\pi rdr)\omega}{2\pi} = \sigma\,\omega r\,dr$$

So, magnetic induction at the centre, due to this element

$$dB = \frac{\mu_0}{2}\frac{dI}{r}$$

and hence, from symmetry

$$B = \int dB = \int_0^R \frac{\mu_0 \sigma \omega r dr}{r} = \frac{\mu_0}{2}\sigma\omega R$$

(b) Magnetic moment of the element,

$$dp_m = (dI)\pi r^2 = \sigma\omega dr \pi r^2 = \sigma\pi\omega r^3\, dr$$

Hence, the sought magnetic moment

$$p_m = \int dp_m = \int_0^R \sigma\pi\omega r^3 dr = \sigma\omega\pi\frac{R^4}{4}$$

3.247 Since only the outer surface of the sphere is charged, consider the element as a ring, as shown in the figure.

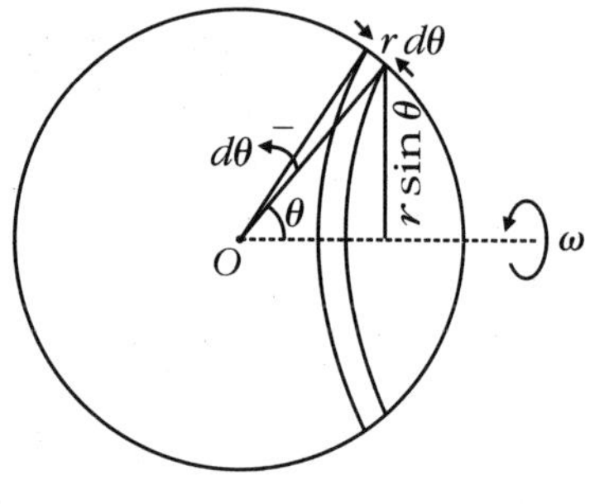

The equivalent current due to the ring element,

$$dI = \frac{\omega}{2\pi}(2\pi r \sin\theta r d\,\theta)\sigma \qquad (1)$$

and magnetic induction due to this loop element at the centre of the sphere O is given by

$$dB = \frac{\mu_0}{4\pi} dI \frac{2\pi r \sin\theta r \sin\theta}{r^3} = \frac{\mu_0}{4\pi} dI \frac{\sin^2\theta}{r}$$

Hence, total magnetic induction due to the sphere at centre O is given by

$$B = \int dB = \int_0^{\pi/2} \frac{\mu_0}{4\pi}\frac{\omega}{2\pi}\frac{2\pi r^2 \sin\theta d\theta \sin^2\theta\sigma}{r} \qquad \text{(using Eq. 1)}$$

Hence,

$$B = \int_0^{\pi/2} \frac{\mu_0\sigma\omega r}{4\pi}\sin^3\theta\, d\theta = \frac{2}{3}\mu_0\,\sigma\omega r = 29\ \text{pT}$$

3.248 The magnetic moment must clearly be along the axis of rotation. Consider a volume element dV. It contains a charge $3q/4\pi R^3 dV$. The rotation of the sphere causes this charge to revolve around the axis and constitute a current, given by

$$\frac{3q}{4\pi R^3}dV \times \frac{\omega}{2\pi}$$

Its magnetic moment will be

$$\frac{3q}{4\pi R^3}dV \times \frac{\omega}{2\pi} \times \pi r^2 \sin^2\theta$$

So the total magnetic moment is

$$p_m = \int_0^R\int_0^\pi \frac{3q}{2R^3} r^2 \sin\theta d\theta \times \frac{\omega r^2 \sin^2\theta}{2} dr$$

$$= \frac{3q}{2R^3}\times\frac{\omega}{2}\times\frac{R^5}{5}\times\frac{4}{3} = \frac{1}{5}qR^2\omega$$

The mechanical moment is

$$M = \frac{2}{5}mR^2\,\omega \quad \text{so,} \quad \frac{p_m}{M} = \frac{q}{2m}$$

3.249 Because of polarization, a space charge is present within the cylinder. Its density is

$$\rho_p = -\text{div}\,\mathbf{P} = -2\alpha$$

Since the cylinder as a whole is neutral, a surface charge density σ_p, must be present on the surface of the cylinder also. This has the magnitude (algebraically)

$$\sigma_p \times 2\pi R = 2\alpha\pi R^2 \quad \text{or} \quad \sigma_p = \alpha R$$

When the cylinder rotates, currents are set up which give rise to magnetic fields. The contribution of ρ_p and σ_p can be calculated separately and then added.

For the surface charge the current is (for a particular element),

$$\sigma R \times 2\pi R\, dx \times \frac{\omega}{2\pi} = \alpha R^2 \omega dx$$

Its contribution to the magnetic field at the centre is

$$\frac{\mu_0 R^2(\alpha R^2 \omega dx)}{2(x^2 + R^2)^{3/2}}$$

and the total magnetic field is

$$B_s = \int_{\infty}^{-\infty} \frac{\mu_0 R^2(\alpha R^2\, \omega\, dx)}{2(x^2 + R^2)^{3/2}} = \frac{\mu_0 \alpha R^4 \omega}{2} \int_{-\infty}^{\infty} \frac{dx}{(x^2 + R^2)^{3/2}}$$

$$= \frac{\mu_0 \alpha R^4 \omega}{2} \times \frac{2}{R^2} = \mu_0 \alpha R^2 \omega$$

For the volume charge density, consider a circle of radius r, radial thickness dr and length dx.

The current is

$$-2\alpha \times 2\pi r\, dr\, dx \times \frac{\omega}{2\pi} = -2\,\alpha r dr\, \omega\, dx$$

The total magnetic field due to the volume charge distribution is

$$B_v = -\int_0^R dr \int_{-\infty}^{\infty} dx 2\pi r \omega \frac{\mu_0 r^2}{2(x^2 + r^2)^{3/2}} = -\int_0^R \alpha\mu_0\, \omega r^3 dr \int_{-\infty}^{\infty} dx(x^2 + r^2)^{3/2}$$

$$= -\int_0^R \alpha\mu_0\, \omega r \times 2 = -\mu_0\, \alpha\, \omega\, R^2 \quad \text{so,} \quad B = B_s + B_v = 0$$

3.250 Force of magnetic interaction, $\mathbf{F}_{\text{mag}} = e(\mathbf{v} \times \mathbf{B})$

where, $$\mathbf{B} = \frac{\mu_0}{4\pi} \frac{e(\mathbf{v} \times \mathbf{r})}{r^3}$$

So, $$\mathbf{F}_{\text{mag}} = \frac{\mu_0}{4\pi} \frac{e^2}{r^3}[\mathbf{v} \times (\mathbf{v} \times \mathbf{r})]$$

$$= \frac{\mu_0 e^2}{4\pi r^3}[(\mathbf{v} \cdot \mathbf{r}) \times \mathbf{v} - (\mathbf{v} \cdot \mathbf{v}) \times \mathbf{r}] = \frac{\mu_0}{4\pi} \frac{e^2}{r^3}(-v^2 \mathbf{r})$$

and $$\mathbf{F}_{\text{ele}} = e\mathbf{E} = e \frac{1}{4\pi\varepsilon_0} \frac{e\mathbf{r}}{|\mathbf{r}|^3}$$

Hence, $$\frac{|\mathbf{F}_{\text{mag}}|}{|\mathbf{F}_{\text{ele}}|} = -v^2 \mu_0 \varepsilon_0 = \left(\frac{v}{c}\right)^2 = 1.00 \times 10^{-6}$$

3.251 (a) The magnetic field at O is only due to the curved path, as for the line element, $d\mathbf{l} \uparrow\uparrow \mathbf{r}$.

Hence, $$\mathbf{B} = \frac{\mu_0 I}{4\pi R} \pi(-\mathbf{k}) = \frac{\mu_0 I}{4R}(-\mathbf{k})$$

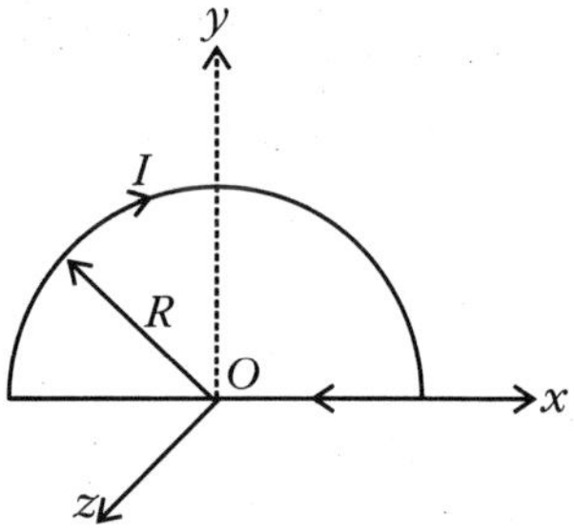

Thus, $$\mathbf{F}_u = IB(-\mathbf{j}) = \frac{\mu_0 I^2}{4R}(-\mathbf{j})$$

So, $$F_u = \frac{\pi_0 I^2}{4R} = 0.20 \text{ mN/m}$$

(b) In this part, magnetic induction **B** at O will be effective only due to the two semi infinite segments of wire. Hence,

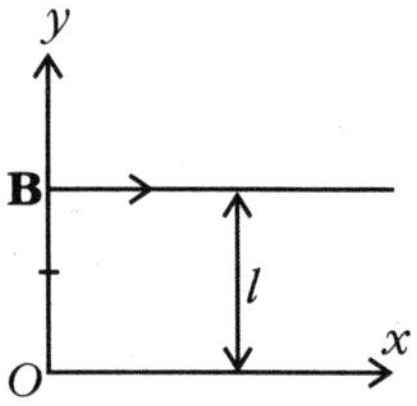

$$\mathbf{B} = 2\cdot\frac{\mu_0 I}{4\pi\left(\frac{l}{2}\right)}\sin\frac{\pi}{2}(-\mathbf{k})$$

$$= \frac{\mu_0 I}{\pi l}(-\mathbf{k})$$

Thus force per unit length,

$$F_u = -\frac{\mu_0 I^2}{\pi l}(-\mathbf{i})$$
$$= 0.13 \text{ mN/m}$$

3.252 Each element of length $dl = R d\alpha$ experiences a force $BI\,R d\alpha$, where $d\alpha$ is the angle subtended by the element at centre. This causes a tension T in the wire.

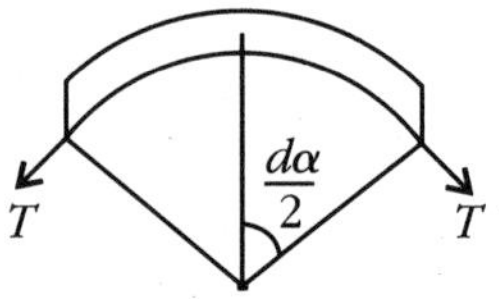

For equilibrium, ampere force $BI\,R d\alpha$ must be balanced by the radial component of tensile forces, which is

$$2T\sin(d\alpha/2) = T d\alpha$$

So, $$T d\alpha = BI\,R d\alpha,$$

Hence, $$T = BIR$$

Thus, the wire experiences a stress

$$\frac{BIR}{\pi d^2/4}$$

This must equal the breaking stress σ_m for rupture. Thus,

$$B_{\max} = \frac{\pi d^2 \sigma_m}{4IR}$$
$$= 8 \text{ kT}$$

3.253 The Ampere forces on the sides OP and $O'P'$ at their mid points are directed along the same line, in opposite directions and have equal values, hence the net force as well as the net torque of these forces about the axis OO' is zero. The Ampere force on the segment PP' and the corresponding moment of this force about the axis OO' is effective and deflecting in nature.

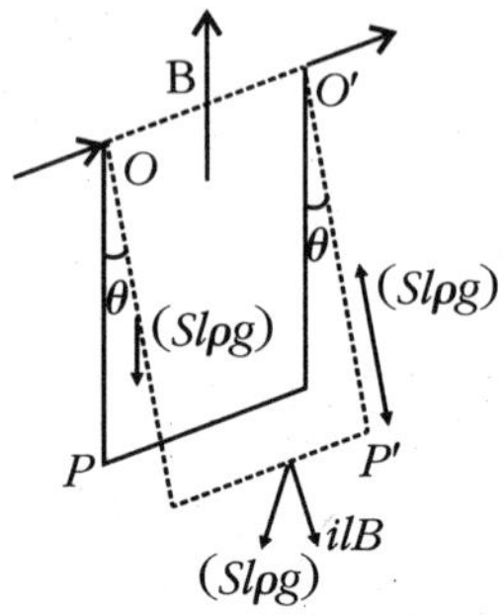

The weight of each segment develops restoring torque. Let the length of each side be l and ρ be the density of the material then, mass of each segment $m = Sl\rho$. The Ampere force on the segment PP' is IlB, from the formula of Ampere force on a straight wire in a uniform magnetic field $\mathbf{F} = I(\mathbf{l}\times\mathbf{B})$.

In equilibrium the deflecting torque must be equal to the restoring torque. So,

$$Fl\cos\varphi = 2mg(l/2)\sin\varphi + mgl\sin\varphi$$
$$I\,lB(l\cos\varphi) = 2S\rho g l^2\sin\varphi$$

Hence, $$B = \frac{2S\rho g}{I}\tan\varphi = 10\,\text{mT}$$

3.254 We know that the torque acting on a magnetic dipole

$$\mathbf{N} = \mathbf{p}_m \times \mathbf{B}$$

But, $\mathbf{p}_m = IS\mathbf{n}$, where $\mathbf{n}$ is the normal on the plane of the loop and is directed in the direction of advancement of a right handed screw, if we rotate the screw in the sense of current in the loop.

On passing current through the coil, the torque acting on the magnetic dipole is counterbalanced by the moment of additional weight about O.

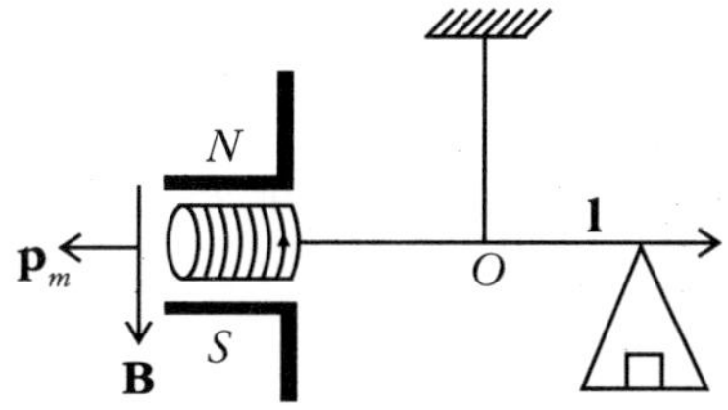

Hence, the direction of current in the loop must be in the direction shown in the figure.

$$\mathbf{p}_m \times \mathbf{B} = -\mathbf{1} \times \Delta m\mathbf{g}$$

or $$NI\,SB = \Delta\,mgl$$

So, $$B = \frac{\Delta mgl}{NIS} = 0.4\text{ T (on substituting values).}$$

3.255 (a) As is clear from the condition, Ampere's forces on the sides 2 and 4 are equal in magnitude but opposite in direction. Hence, the net effective force on the frame is the resultant of the forces experienced by sides 1 and 3.

Now, the Ampere force on side 1,

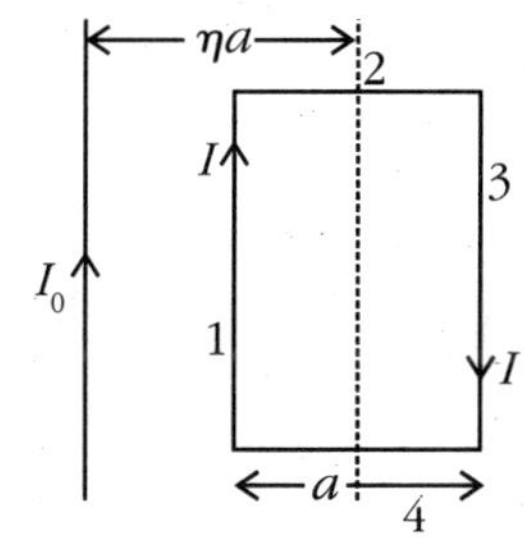

$$F_1 = \frac{\mu_0}{2\pi}\frac{II_0}{\left(\eta - \frac{1}{2}\right)}$$

and that on side 3,

$$F_3 = \frac{\mu_0}{2\pi}\frac{I_0 I}{\left(\eta + \frac{1}{2}\right)}$$

So, the resultant force on the frame $= F_1 - F_3$ (as they are opposite in nature).

$$= \frac{2\mu_0 II_0}{\pi(4\eta^2 - 1)} = 0.40\,\mu N$$

(b) Work done in turning the frame through some angle,

$$A = \int I d\Phi = I(\Phi_f - \Phi_i)$$

where Φ_f is the flux through the frame in final position and Φ_i in the initial position.

Here, $$|\Phi_f| = |\Phi_i| = \Phi \quad \text{and} \quad \Phi_i = -\Phi_f$$

So, $$\Delta\Phi = 2\,\Phi \quad \text{and} \quad A = I\,2\,\Phi$$

Hence, $$A = 2I\int \mathbf{B}\cdot d\mathbf{S}$$

$$= 2I\int_{a\left(\eta-\frac{1}{2}\right)}^{a\left(\eta+\frac{1}{2}\right)} \frac{\mu_0}{2\pi}\frac{I_0 a}{r}dr = \frac{\mu_0 II_0 a}{\pi}\ln\left(\frac{2\eta+1}{2\eta-1}\right)$$

$$= 0.10\,\mu\text{J}$$

3.256 There are excess surface charges on each wire (irrespective of whether the current is flowing through them or not). Hence, in addition to the magnetic force $\mathbf{F}_{mag}$, we must take into account the electric force $\mathbf{F}_{ele}$. Suppose that an excess charge λ corresponds to a unit length of the wire, then electric force exerted per unit length of the wire by other wire can be found with the help of Gauss' theorem.

$$F_{ele} = \lambda E = \lambda \frac{1}{4\pi\varepsilon_0}\frac{2\lambda}{l} = \frac{2\lambda^2}{4\pi\varepsilon_0 l} \quad (1)$$

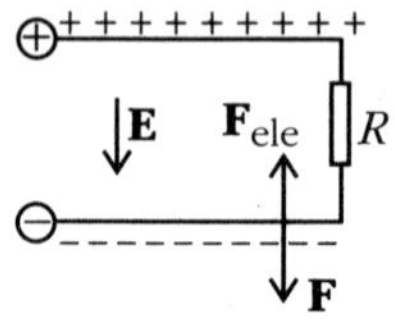

where l is the distance between the axes of the wires. The magnetic force acting per unit length of the wire can be found with the help of the theorem on circulation of vector **B**

$$F_{mag} = \frac{\mu_0}{4\pi}\frac{2I^2}{l} \quad (2)$$

where I is the current in the wire.

Now, from the relation, $\lambda = C\varphi$, where C is the capacitance of the wires per unit length, as given in solution Problem 3.108 and $\varphi = IR$. Substituting we get,

$$\lambda = \frac{\pi\varepsilon_0}{\ln\eta} IR \quad \text{or} \quad \frac{I}{\lambda} = \frac{\ln\eta}{\pi\varepsilon_0 R} \quad (3)$$

Dividing Eq. (2) by (1) and then substituting the value of $\frac{I}{\lambda}$ from Eq. (3), we get,

$$\frac{F_m}{F_e} = \frac{\mu_0}{\varepsilon_0}\frac{(\ln\eta)^2}{\pi^2 R^2}$$

The resultant force of interaction vanishes when this ratio equals unity. This is possible when $R = R_0$, where

$$R_0 = \sqrt{\frac{\mu_0}{\varepsilon_0}}\frac{\ln\eta}{\pi} = 0.36\ \text{k}\Omega$$

3.257 Using solution of Problem 3.225, the magnetic field due to the conductor with semicircular cross-section is given by,

$$B = \frac{\mu_0 I}{\pi^2 R}$$

Then

$$\frac{\partial F}{\partial l} = BI = \frac{\mu_0 I^2}{\pi^2 R}$$

3.258 We know that Ampere's force per unit length on a wire element in magnetic field is given by

$$d\mathbf{F}_n = I(\mathbf{n}\times\mathbf{B}) \quad (1)$$

where **n** is the unit vector along the direction of current.

Now, let us take an element of the conductor I_2, as shown in the figure. This wire element is in the magnetic field, produced by the current I_1, which is directed normally into the sheet of the paper and its magnitude is given by

$$|\mathbf{B}| = \frac{\mu_0 I_1}{2\pi r} \quad (2)$$

From Eqs. (1) and (2)

$$d\mathbf{F}_n = \frac{I_2}{b} dr(\mathbf{n}\times\mathbf{B}) \quad \text{(because the current through the element} = I_2/b\ dr\text{)}$$

So,

$$d\mathbf{F}_n = \frac{\mu_0}{2\pi}\frac{I_1 I_2}{b}\frac{dr}{r} \quad \text{(towards left, as } \mathbf{n}\perp\mathbf{B}\text{)}$$

Hence the magnetic force on the conductor:

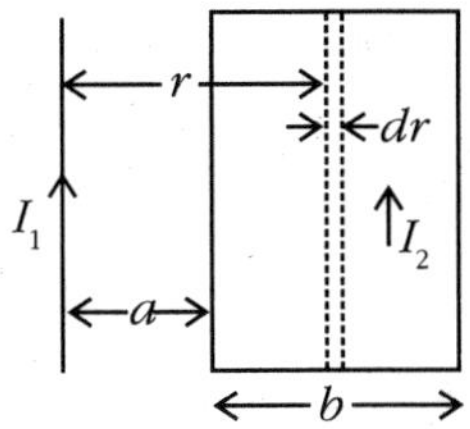

$$\mathbf{F}_n = \frac{\mu_0}{2\pi}\frac{I_1 I_2}{b}\int_a^{a+b}\frac{dr}{r} \text{ (towards left)}$$

$$= \frac{\mu_0}{2\pi}\frac{I_1 I_2}{b}\ln\frac{a+b}{a} \text{ (towards left)}$$

Then according to the Newton's third law, the magnitude of sought magnetic interaction force

$$= \frac{\mu_0 I_1 I_2}{2\pi b}\ln\frac{a+b}{a}$$

3.259 By the circulation theorem $B = \mu_0 I$, where I = current per unit length flowing along the plane perpendicular to the paper. Currents flow in the opposite sense in the two planes and produce the given field B by superposition. The field due to one of the plates is just 1/2 B. The force on the plate is 1/2 $B \times I \times$ Length $\times$ breadth $= B^2/2\mu_0$ per unit area.

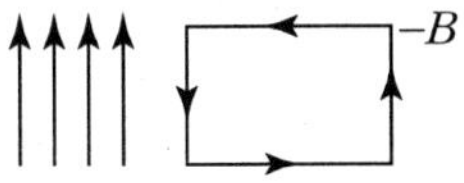

(Recall the formula $F = BIl$ on a straight wire.)

3.260 In Fig. (a), the external field must be $B_1 + B_2/2$, which when superposed with the internal field $B_1 - B_2/2$ (of opposite sign on the two sides of the plate), must give the actual field.

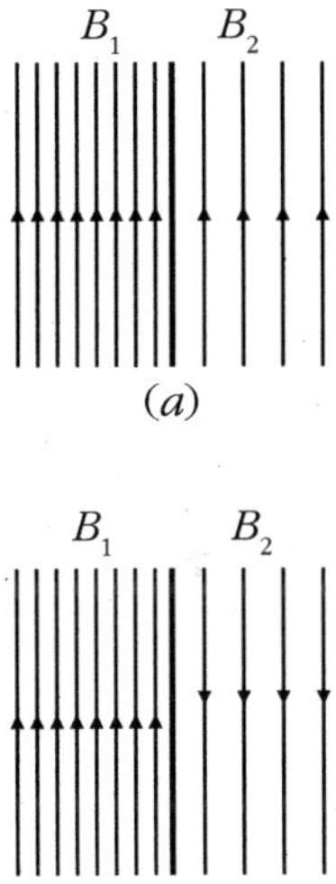

(a)
(b)

Now $$\frac{B_1 - B_2}{2} = \frac{1}{2}\mu_0 I$$

or $$I = \frac{B_1 - B_2}{\mu_0}$$

Thus, $$F = \frac{B_1^2 - B_2^2}{2\mu_0}$$

In Fig. (b), the external field must be $B_1 - B_2/2$ upward with an internal field, $B_1 + B_2/2$, upward on the left and downward on the right. Thus,

$$I = \frac{B_1 + B_2}{\mu_0} \quad \text{and} \quad F = \frac{B_1^2 - B_2^2}{2\mu_0}$$

In Fig. (c), the boundary condition from Gauss' theorem is, $B_1 \cos\theta_1 = B_2 \cos\theta_2$. Also $(B_1 \sin\theta_1 + B_2 \sin\theta_2) = \mu_0 I$, where I = current per unit length. The external field parallel to the plate must be

$$\frac{B_1 \sin\theta - B_2 \sin\theta_2}{2}$$

The perpendicular component $B_1 \cos\theta_1$, does not matter since the corresponding force is tangential.

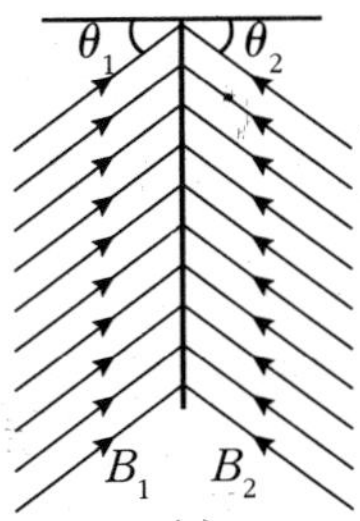

(c)

Thus, $$F = \frac{B_1^2 \sin^2\theta_1 - B_2^2 \sin^2\theta_2}{2\mu_0} \text{ per unit area}$$

$$= \frac{B_1^2 - B_2^2}{2\mu_0} \text{ per unit area}$$

The direction of the current in the plane conductor is perpendicular to the paper and beyond the drawing.

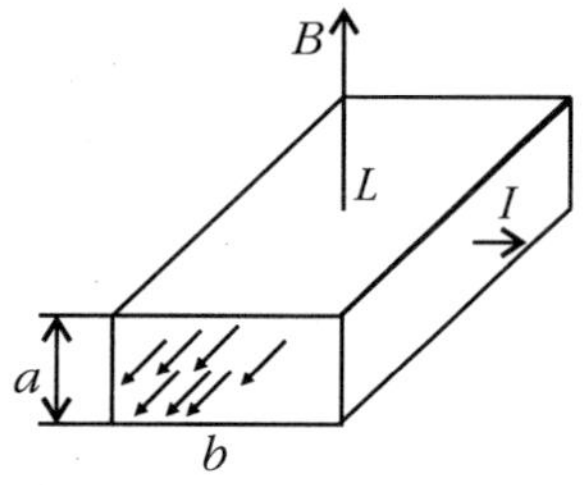

3.261 The current density is I/aL, where L is the length of the section. The difference in pressure produced must be

$$\Delta p = \frac{I}{aL} \times B \times (abL)/ab = \frac{IB}{a} = 0.5 \text{ kPa}$$

3.262 Let t = thickness of the wall of the cylinder. Then,

$$J = I/2\,\pi Rt \text{ (along } z \text{ axis)}$$

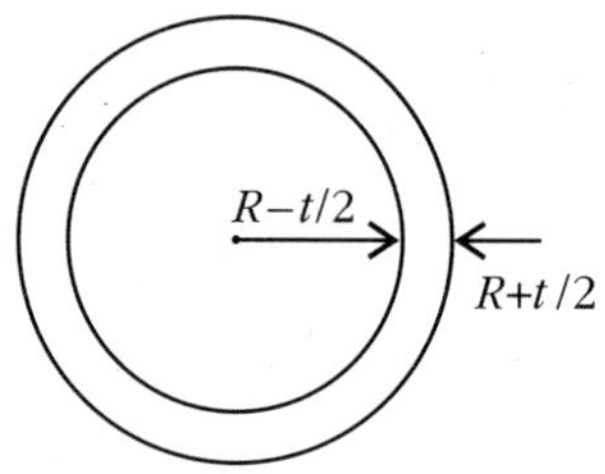

The magnetic field due to this at a distance r such that $r\,(R - t/2 < r < R + t/2)$, is given by

$$B_\varphi(2\pi r) = \mu_0 \frac{I}{2\pi Rt} \pi\{r^2 - (R - t/2)^2\}$$

$$B_\varphi = \frac{\mu_0 I}{4\pi Rrt}\{r^2 - (R - t/2)^2\}$$

Now, $$\mathbf{F} = \int \mathbf{J} \times \mathbf{B}\, dV$$

and $$p = \frac{F_r}{2\pi RL}$$

$$= \frac{1}{2\pi RL} \int_{R-\frac{t}{2}}^{R+\frac{t}{2}} \frac{\mu_0 I^2}{8\pi^2 R^2 t^2 r} \left\{ r^2 - \left(R - \frac{t}{2}\right)^2 \right\} \times 2\pi rL dr$$

$$= \frac{\mu_0 I^2}{8\pi^2 R^3 t^2} \int_{R-\frac{t}{2}}^{R+\frac{t}{2}} \left\{ r^2 - \left(R - \frac{t}{2}\right)^2 \right\} dr$$

$$= \frac{\mu_0 I^2}{8\pi^2 R^3 t^2} \left[\frac{\left(R + \frac{t}{2}\right)^3 - \left(R - \frac{t}{2}\right)^3}{3} - \left(R - \frac{t}{2}\right)^2 t \right]$$

$$= \frac{\mu_0 I^2}{8\pi^2 R^3 t}[Rt + 0(t^2)] \approx \frac{\mu_0 I^2}{8\pi^2 R^2}$$

3.263 When self-forces are involved, a typical factor of 1/2 comes into play. For example, the force on a current carrying straight wire in a magnetic induction B is BIl. If the magnetic induction B is due to the current itself then the force can be written as

$$F = \int_0^I B(I')\, dI' l$$

If $B\,(I') \propto I'$, then this becomes, $F = \frac{1}{2} B(I)\, Il$.

In the present case, $B\,(I) = \mu_0 nI$ and this acts on nI ampere turns per unit length, so, pressure is given by

$$p = \frac{F}{\text{Area}} = \frac{1}{2}\mu_0 n \frac{I \times nI \times 1 \times l}{1 \times l} = \frac{1}{2}\mu_0 n^2 I^2$$

3.264 The magnetic induction B in the solenoid is given by $B = \mu_0 nI$. The force on an element $dl = Rd\alpha$ of the current carrying conductor is

$$dF = \frac{1}{2}\mu_0 nII\, Rd\alpha = \frac{1}{2}\mu_0\, nI^2 Rd\alpha$$

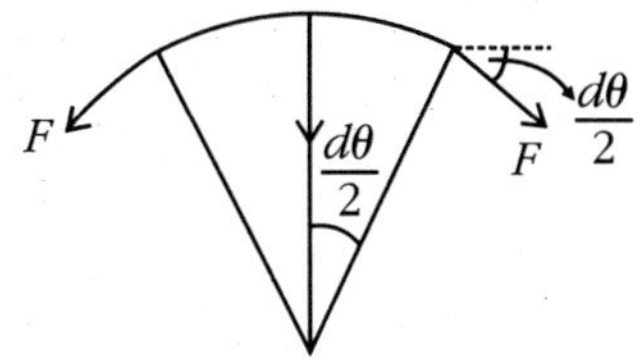

This is radially outwards. The factor 1/2 is explained above. For equilibrium, it should be balanced by the radial component of the longitudinal tensile forces, $T d\alpha$.

Hence, $$T = \frac{1}{2}\mu_0 n I^2 R$$

This equals $F_{\lim}$ when, $I = I_m = \sqrt{\frac{2F_{\lim}}{\mu_0 n R}}$

Note that $F_{\lim}$, here, is actually a force and not a stress.

3.265 Resistance of the liquid between the plates $= \frac{\rho d}{S}$

Voltage between the plates $= Ed = vBd$

Current through plates$= \frac{vBd}{R + \rho d/S}$

Power generated in the external resistance R, is given by

$$P = \frac{v^2B^2d^2R}{\left(R + \frac{\rho d}{S}\right)^2} = \frac{v^2B^2d^2}{\left(\sqrt{R} + \frac{\rho d}{S\sqrt{R}}\right)^2} = \frac{v^2B^2d^2}{\left[\left\{R^{1/4} - \left(\frac{\rho d}{S\sqrt{R}}\right)^{1/2}\right\}^2 + 2\sqrt{\frac{\rho d}{S}}\right]^2}$$

This is maximum when

$$R = \frac{\rho D}{S} \quad \text{and} \quad P_{\max} = \frac{v^2B^2Sd}{4\rho}$$

3.266 The electrons in the conductor are drifting with a speed of

$$v_d = \frac{J}{ne} = \frac{I}{\pi R^2 ne}$$

where e = magnitude of the charge on the electron, n = concentration of the conduction electrons.

The magnetic field inside the conductor due to this current is given by

$$B_\varphi(2\pi r) = \pi r^2 \frac{I}{\pi R^2}\mu_0 \quad \text{or} \quad B_\varphi = \frac{\mu_0}{2\pi}\frac{Ir}{R^2}$$

A radial electric field vB_φ must come into being in equilibrium. Its potential difference is

$$\begin{aligned}\Delta\varphi &= \int_0^R \frac{I}{\pi R^2 ne}\frac{\mu_0}{2\pi}\frac{Ir}{R^2}dr \\ &= \frac{I}{\pi R^2 ne}\left(\frac{\mu_0 I}{2\pi R^2}\right)\frac{1}{2}R^2 \\ &= \frac{\mu_0 I^2}{4\pi^2 R^2 ne} \\ &= 2 \text{ pV}\end{aligned}$$

3.267 Here, $v_d = \frac{E}{B}$ and $j = nev_d$

So, $$n = \frac{jB}{eE} = \frac{200 \times 10^4 \text{ A/m}^2 \times 1\text{T}}{1.6 \times 10^{-19}\text{ C} \times 5 \times 10^{-4} \text{ V/m}}$$

$$= 2.5 \times 10^{28} \text{ per m}^3 = 2.5 \times 10^{22} \text{ cm}^{-3}$$

Atomic weight of Na being 23 and its density ≈1, molar volume is 23 cm³. Thus, number of atoms per unit volume is

$$\approx \frac{6 \times 10^{23}}{23} = 2.6 \times 10^{22} \text{ cm}^{-3}$$

Thus, there is almost one conduction electron per atom.

3.268 By definition, mobility $= \frac{\text{Drift velocity}}{\text{Electric field component causing this drift}}$

or $$\mu = \frac{v_d}{E_L}$$

On the other hand, $E_T = v_d B = \frac{E_L}{\eta}$ (as given)

So, $$\mu = \frac{1}{\eta B} = 3.2 \times 10^{-3} \text{ m}^2/(\text{V} \cdot \text{s})$$

3.269 Due to the straight conductor, $B_\varphi = \frac{\mu_0 I}{2\pi r}$

We use the formula, $$\mathbf{F} = (\mathbf{p}_m \cdot \nabla)\mathbf{B}$$

(a) The vector $\mathbf{p}_m$ is parallel to the straight conductor, so

$$\mathbf{F} = p_m \frac{\partial}{\partial Z}\mathbf{B} = 0$$

because neither the direction nor the magnitude of $\mathbf{B}$ depends on z.

(b) The vector $\mathbf{p}_m$ is oriented along the radius vector $\mathbf{r}$, so

$$\mathbf{F} = p_m \frac{\partial}{\partial r}\mathbf{B}$$

The direction of $\mathbf{B}$ at $r + dr$ is parallel to the direction at r. Thus only the φ component of $\mathbf{F}$ will survive.

$$F_\varphi = p_m \frac{\partial}{\partial r}\frac{\mu_0 I}{2\pi r} = -\frac{\mu_0 I p_m}{2\pi r^2}$$

(c) The vector $\mathbf{p}_m$ coincides in direction with the magnetic field, produced by the conductor carrying current I.

$$\mathbf{F} = p_m \frac{\partial}{r\partial\varphi}\frac{\mu_0 I}{2\pi}\mathbf{e}_\varphi = \frac{\mu_0 I p_m}{2\pi r^2}\frac{\partial \mathbf{e}_\varphi}{\partial \varphi}$$

So, $$\mathbf{F} = -\frac{\mu_0 I p_m}{2\pi r^2}\mathbf{e}_r, \quad \text{as,} \quad \frac{\partial \mathbf{e}_\varphi}{\partial \varphi} = -\mathbf{e}_r$$

3.270 We have, $$F_x = p_m \frac{\partial B_x}{\partial x}$$

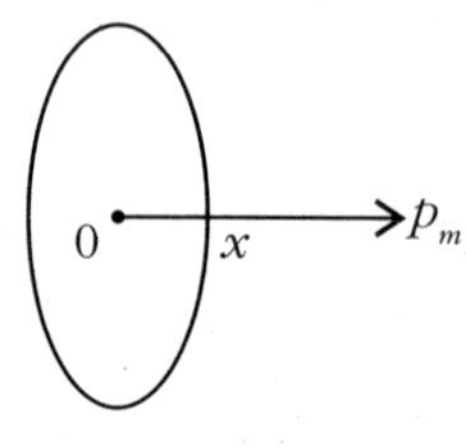

But, $$B_x = \frac{\mu_0 I}{4\pi}\int\frac{R\,dl}{(x^2+R^2)^{3/2}} = \frac{\mu_0 IR^2}{2(x^2+R^2)^{3/2}}$$

So, $$F = \frac{\mu_0}{4\pi}\frac{I\cdot 2\pi R^2}{(x^2+R^2)^{5/2}}\frac{3}{2}\cdot 2x\cdot p_m = \frac{\mu_0}{4\pi}\frac{6\pi R^2 Ip_m x}{(x^2+R^2)^{5/2}}$$

3.271 $$F = p_{2m}\frac{\partial}{\partial l}\left|\frac{(\mu_0 3\mathbf{p}_{1m}\cdot\mathbf{r})\,\mathbf{r}-\mathbf{p}_{1m}r^2}{4\pi}\frac{}{r^5}\right|, \text{ where } r = l$$

$$= p_{2m}\frac{\partial}{\partial l}\left[\frac{\mu_0}{2\pi}\frac{p_{1m}}{l^3}\right] = \frac{-3}{2}\frac{\mu_0 p_{1m}p_{2m}}{\pi l^4} = 9 \text{ nN}$$

3.272 From solution of Problem 3.270, for $x >> R$,

$$B_x = \frac{\mu_0 I' R^2}{2x^3}$$

or $$I' \approx \frac{2B_x x^3}{\mu_0 R^2} = \frac{2\times 3\times 10^{-5}T\times(10^{-1}\text{ m})^3}{1.26\times 10^{-6}\times(10^{-2}\text{ m})^4} \approx 0.5 \text{ kA}$$

3.273 $$B_n' = B\cos\alpha,$$

$$H_t' = \frac{1}{\mu_0}B\sin\alpha,$$

$$B_t' = \mu B\sin\alpha$$

So, $$B' = B\sqrt{\mu^2\sin^2\alpha+\cos^2\alpha}$$

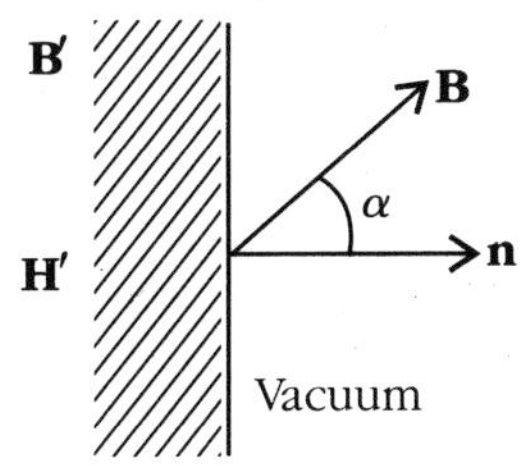

3.274 (a) $\oint \mathbf{H}\cdot d\mathbf{S} = \oint\left(\frac{\mathbf{B}}{\mu_0}-\mathbf{J}\right)\cdot d\mathbf{S} = -\oint \mathbf{J}\cdot d\mathbf{S}$, since $\oint \mathbf{B}\cdot d\mathbf{S} = 0$.

Now $\mathbf{J}$ is non-vanishing only in the bottom half of the sphere.

Here, $B_n' = B\cos\theta$, $H_t' = \frac{1}{\mu_0}B\sin\theta$, $B_t' = \mu B\sin\theta$, $H_n' = \frac{B}{\mu\mu_0}\cos\theta$

$$J_n = \frac{B\cos\theta}{\mu_0}\left(1-\frac{1}{\mu}\right) \quad\text{and}\quad J_t = \frac{\mu-1}{\mu_0}B\sin\theta$$

Only J_n contributes to the surface integral, so

$$-\oint \mathbf{J}\cdot d\mathbf{S} = \oint_{\text{lower}} J_n dS = \frac{\pi R^2 B\cos\theta}{\mu_0}\left(1-\frac{1}{\mu}\right)$$

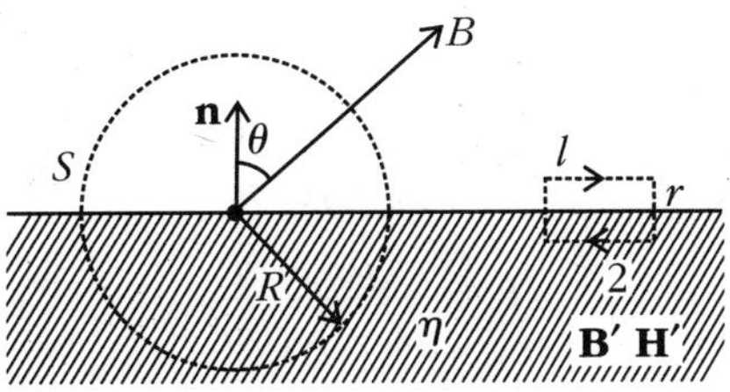

(b) $$\oint \mathbf{B}\cdot d\mathbf{r} = (B_t - B_t')l = (\mathbf{1}-\mu)/Bl\sin\theta$$

3.275 Inside the cylindrical wire, there is an external current of density $I/\pi R^2$. This gives a magnetic field H_φ with

$$H_\varphi 2\pi r = I\frac{r^2}{R^2} \quad \text{or} \quad H_\varphi = \frac{Ir}{2\pi R^2}$$

From this
$$B_\varphi = \frac{\mu\mu_0 Ir}{2\pi R^2} \quad \text{and} \quad J_\varphi = \frac{\mu - 1}{2\pi}\frac{Ir}{R^2} = \frac{\chi Ir}{2\pi R^2}$$

Hence, total volume of molecular current is

$$\oint_{r=R} \mathbf{J}_\varphi \cdot d\mathbf{r} = \int \frac{\chi I}{2\pi R} dl = \chi I$$

The surface current is obtained by using the equivalence of the surface current density to $\mathbf{J} \times \mathbf{n}$. This gives rise to a surface current density in the z-direction of

$$\frac{-\chi I}{2\pi R}$$

The total molecular surface current is

$$I_s' = -\frac{\chi I}{2\pi R}(2\pi R) = -\chi I$$

The two currents have opposite signs.

3.276 We can obtain the form of the curves required here, by qualitative arguments.

From $$\oint H \cdot dl = I$$

we get, $$H\,(x >> 0) = H\,(x << 0) = nI$$

Then $$B\,(x >> 0) = \mu\mu_0 nI$$
$$B\,(x << 0) = \mu_0 nI$$

Also,
$$B\,(x < 0) = \mu_0 H\,(x < 0)$$
$$J\,(x < 0) = 0$$

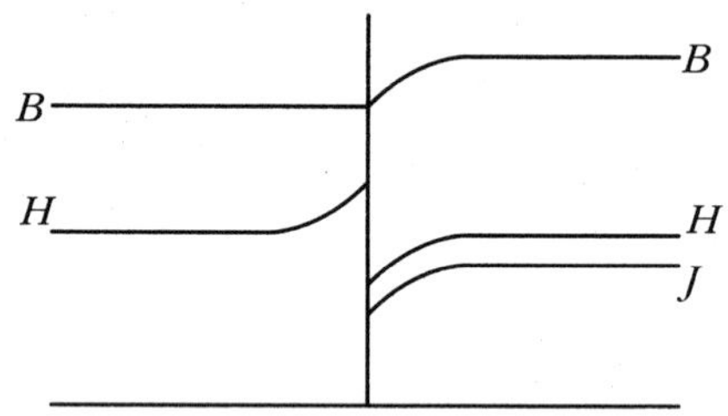

B is continuous at $x = 0$, H is not. These give the required curves as shown in the answer sheet.

3.277 The lines of B as well as H field are circles around the wire. Thus,

$$H_1\pi r + H_2\pi r = I \quad \text{or} \quad H_1 + H_2 = \frac{I}{\pi r}$$

Also $$\mu_0\mu_1 H_1 = \mu_2 H_2 \mu_0 = B_1 = B_2 = B$$

Thus,
$$H_1 = \frac{\mu_1}{\mu_1 + \mu_2}\frac{I}{\pi r}$$
$$H_2 = \frac{\mu_1}{\mu_1 + \mu_2}\frac{I}{\pi r}$$

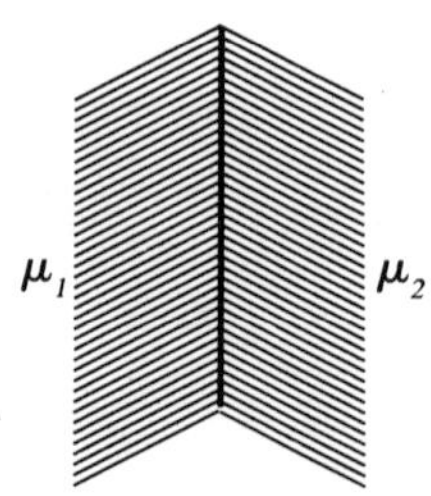

and $$B = \mu_0 \frac{\mu_1\mu_2}{\mu_1 + \mu_2}\frac{I}{\pi r}$$

3.278 The medium I is vacuum and contains a circular current carrying coil with current I. The medium II is magnetic with permeability μ. The boundary is the plane $z = 0$ and the coil is in the plane $z = l$. To find the magnetic induction, we note that the effect of the magnetic medium can be written as due to an image coil in II as far as the medium I is concerned. On the other hand, the induction in **I** can be written as due to the coil in I, carrying a different current. It is sufficient to consider the far away fields and ensure that the boundary conditions are satisfied there. Now for actual coil in medium I,

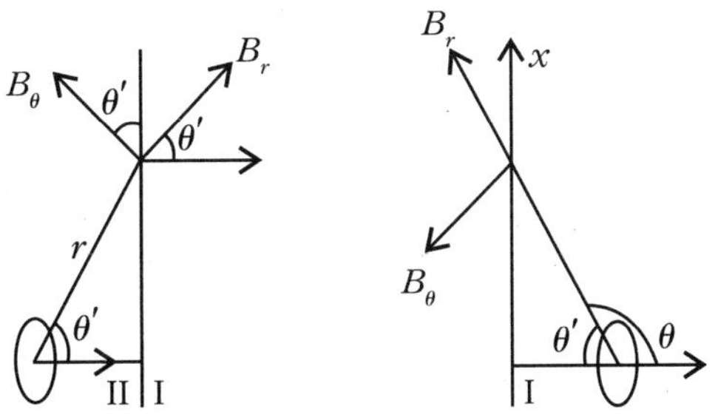

$$B_r = -\frac{2p_m \cos\theta'}{r^3}\cdot\left(\frac{\mu_0}{4\pi}\right) \quad \text{and} \quad B_\theta = \frac{p_m \sin\theta'}{r^3}\left(\frac{\mu_0}{4\pi}\right)$$

So, $$B_z = \frac{\mu_0 p_m}{4\pi}(2\cos^2\theta' - \sin^2\theta') \quad \text{and} \quad B_x = \frac{\mu_0 p_m}{4\pi}(-3\sin\theta'\cos\theta')$$

where, $p_m = I(\pi a^2)$, a = radius of the coil.

Similarly due to the image coil,

$$B_z = \frac{\mu_0 P'_m}{4\pi}(2\cos^2\theta' - \sin^2\theta'), B_x = \frac{\mu_0 P'_m}{4\pi}(3\sin\theta'\cos\theta'), P'_m = I'(\pi a^2)$$

As far as the medium II is concerned, we write similarly

$$B_z = \frac{\mu_0 P''_m}{4\pi}(2\cos^2\theta' - \sin^2\theta'), B_x = \frac{\mu_0 P''m}{4\pi}(-3\sin\theta'\cos\theta'), P''_m = I''(\pi a^2)$$

The boundary conditions are

$$p_m + p'_m = p''_m \text{ (from } B_{1n} = B_{2n})$$

$$-p_m + p'_m = -\frac{1}{\mu}p''_m \text{ (from } H_{1t}=H_{2t})$$

Thus, $$I'' = \frac{2\mu}{\mu+1}I \quad \text{and} \quad I' = \frac{\mu-1}{\mu+1}I$$

In the limit when the coil is on the boundary, the magnetic field everywhere can be obtained by taking the current to be

$$\frac{2\mu}{\mu+1}I$$

Thus, $$\mathbf{B} = \frac{2\mu}{\mu+1}\mathbf{B}_0$$

3.279 We use the fact that within an isolated uniformly magnetized ball, $\mathbf{H}' = -\mathbf{J}/3$, $\mathbf{B}' = 2\mu_0\mathbf{J}/3$, where $\mathbf{J}$ is the magnetization vector. Then, in a uniform magnetic field with induction $\mathbf{B}_0$, we have by superposition,

$$\mathbf{B}_{in} = \mathbf{B}_0 + \frac{2\mu_0\mathbf{J}}{3}, \mathbf{H}_{in} = \frac{\mathbf{B}_0}{\mu_0} - \frac{\mathbf{J}}{3}$$

or

$$\mathbf{B}_{in} + 2\mu_0\mathbf{H}_{in} = 3\mathbf{B}_0$$

Also,

$$\mathbf{B}_{in} = \mu\mu_0\mathbf{H}_{in}$$

Thus,

$$\mathbf{H}_{in} = \frac{3\mathbf{B}_0}{\mu_0(\mu+2)} \quad \text{and} \quad \mathbf{B}_{in} = \frac{3\mu\mathbf{B}_0}{\mu+2}$$

3.280 The coercive force H_c is just the magnetic field within the cylinder. This is by circulation theorem,

$$H_C = \frac{NI}{l} = 6 \text{ kA/m}$$

(from $\oint \mathbf{H}\cdot d\mathbf{r} = I$, total current, considering a regular contour).

3.281 We use,

$$\oint \mathbf{H}\cdot dl = 0$$

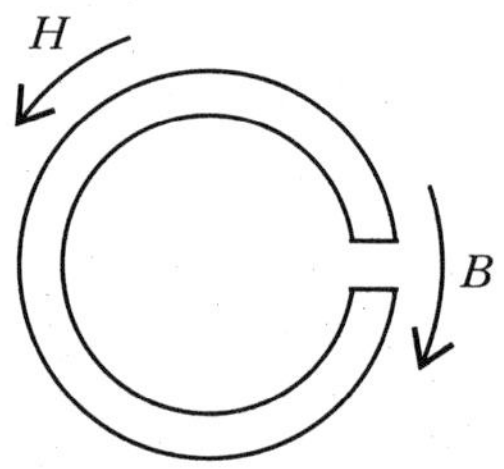

Neglecting the fringing of the lines of force, we write this as

$$H(\pi d - b) + \frac{B}{\mu_0}b = 0$$

or

$$H \approx \frac{-Bb}{\mu_0 \pi d} = 101 \text{ A/m}$$

The sense of H is opposite to B.

3.282 Here, $\oint \mathbf{H}\cdot d\mathbf{l} = NI$ or $H(2\pi R) + \frac{Bb}{\mu_0} = NI$ so, $H = \frac{NI\mu_0 - Bb}{2\pi R\mu_0}$

Hence,

$$\mu = \frac{B}{\mu_0 H} = \frac{2\pi RB}{\mu_0 NI - Bb} = 3700$$

3.283 One has to draw the graph of $\mu = \frac{B}{\mu_0 H}$ versus H from the given graph. The $\mu - H$ graph starts out horizontally, and then rises steeply at about $H = 0.04$ kA/m before falling again. It is easy to check that $\mu_{max} \approx 10{,}000$.

3.284 From the theorem on circulation of vector **H**

$$H\pi d + \frac{Bb}{\mu_0} = NI \quad \text{or} \quad B = \frac{\mu_0 NI}{b} - \frac{\mu_0 \pi d}{b}H = (1.51 - 0.987)H$$

where B is in Tesla and H in kA/m. Besides, B and H are interrelated as in the Fig. 3.76 of the problem book. Thus we have to solve for **B**, **H** graphically by simultaneously drawing the two curves (the hysteresis curve and the straight line) and find the point of intersection. It is at $H \approx 0.26$ kA/m, $B = 1.25$ T.

Then,
$$\mu = \frac{B}{\mu_0 H} \approx 4000$$

3.285 From the formula,

$$\mathbf{F} = (\mathbf{p}_m \cdot \nabla)\mathbf{B} \rightarrow \mathbf{F} = \int (\mathbf{J} \cdot \nabla)\mathbf{B}\, dV$$

Thus,
$$\mathbf{F} = \frac{\chi}{\mu\mu_0} \int (\mathbf{B} \cdot \nabla)\mathbf{B}\, dV$$

Since $\mathbf{B}$ is predominantly along the x-axis,

So,
$$F_x = \frac{\chi}{\mu\mu_0} \int B_x \frac{\partial B_x}{\partial x} S dx = \frac{\chi S}{2\mu\mu_0} \int_{x=0}^{x=L} dB_x^2 = -\frac{\chi SB^2}{2\mu\mu_0} = \frac{\chi SB^2}{2\mu_0}$$

3.286 The force in question is

$$\mathbf{F} = (\mathbf{p}_m \cdot \nabla)\mathbf{B} = \frac{\chi BV}{\mu\mu_0} \frac{d\mathbf{B}}{dx}$$

since B is essentially in the x-direction,

so,
$$F_x \approx \frac{\chi V}{2\mu_0} \frac{dB^2}{dx} = \frac{\chi B_0^2 V}{2\mu_0} \frac{d}{dx}(e^{-2ax^2}) = -4ax\, e^{-2ax^2} \frac{\chi B_0^2}{2\mu_0} V$$

This is maximum when its derivative vanishes i.e.,

$$16a^2x^2 - 4a = 0 \quad \text{or} \quad x_m = \frac{1}{\sqrt{4a}}$$

The maximum force is

$$F_{\max} = 4a \frac{1}{\sqrt{4a}} e^{-1/2} \frac{\chi B_0^2 V}{2\mu_0} = \frac{\chi B_0^2 V}{\mu_0} \sqrt{\frac{a}{e}}$$

So,
$$\chi = \frac{\mu_0 F_{\max} \sqrt{\dfrac{e}{a}}}{VB_0^2} = 3.6 \times 10^{-4}$$

3.287 The force is given by

$$F_x = (\mathbf{p}_m \cdot \nabla)B_x = \frac{\chi BV}{\mu\mu_0} \frac{dB}{dx} \approx \frac{\chi V}{2\mu_0} \frac{dB^2}{dx}$$

This force is attractive and an equal force must be applied for balance. The work done by applied forces is

$$A = \int_{x=0}^{x=L} -F_x dx = \frac{\chi V}{2\mu_0} (-B^2)_{x=0}^{x=L} \approx \frac{\chi VB^2}{2\mu_0}$$

3.6 Electromagnetic Induction. Maxwell's Equations

3.288 Taking the direction of vector **B** normally inward and direction on normal **n** to the plane pointing outward. Here,

$$d\Phi = \mathbf{B}\cdot d\mathbf{s} = -2Bx\,dy$$

Since
$$y = ax^2 \quad \text{or} \quad x = \sqrt{\frac{y}{a}}$$

So,
$$d\Phi = -2B\sqrt{\frac{y}{a}}\,dy$$

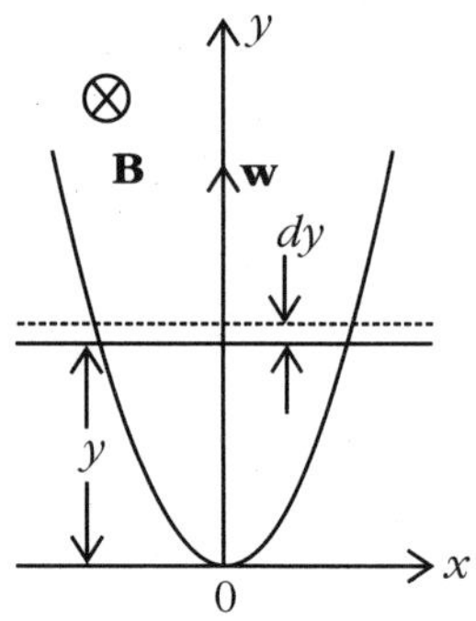

From Faraday's law of electromagnetic induction,

$$\xi_{\text{in}} = -\frac{d\Phi}{dt}$$

Hence,
$$\xi_{\text{in}} = 2B\sqrt{\frac{y}{a}}\frac{dy}{dt}$$
$$= By\sqrt{\frac{8w}{a}} \quad \left(\text{using}\,\frac{dy}{dt} = \sqrt{2wy}\right)$$

From Lenz's law, and right handed screw rule, the induced current and hence the induced e.m.f. in the loop is anticlockwise.

3.289 Let us assume **B** is directed into the plane of the loop. Then the motional e.m.f.

$$\xi_{\text{in}} = \int(\mathbf{v}\times\mathbf{B})\cdot d\mathbf{l} = vBl$$

and is directed in the sense of ($\mathbf{v}\times\mathbf{B}$).

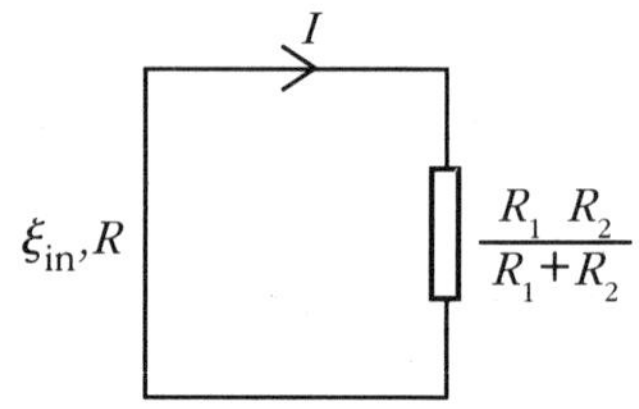

So,
$$I = \frac{\xi_{\text{in}}}{R + \dfrac{R_1R_2}{R_1+R_2}} = \frac{Bvl}{R+R_u}\left(\text{where } R_u = \frac{R_1R_2}{R_1+R_2}\right)$$

as R_1 and R_2 are in parallel connections.

3.290 (a) As the metal disk rotates, any free electron also rotates with it with the same angular velocity ω, and that is why an electron must have an acceleration of $\omega^2 r$ directed towards the disk's center, where r is the separation of electron from the center of the disk. We know from Newton's second law that if a particle has some acceleration then there must be a net effective force on it in the direction of acceleration. We also know that a charged particle can be influenced by two fields-electric and magnetic. In our problem, magnetic field is absent hence there must be an electric field near an electron and is directed opposite to the acceleration of the electron.

If E is the electric field strength at a distance r from the center of the disk, we have from Newton's second law

$$F_n = mw_n$$
$$eE = mr\omega^2$$

or
$$E = \frac{m\omega^2 r}{e}$$

and the potential difference is given by

$$\varphi_{\text{cen}} - \varphi_{\text{rim}} = \int_0^a \mathbf{E}\cdot d\mathbf{r} = \int_0^a \frac{m\omega^2 r}{e}dr \quad \text{as} \quad \mathbf{E}\uparrow\downarrow d\mathbf{r}$$

Thus, $$\varphi_{cen} - \varphi_{rim} = \Delta\varphi = \frac{m\omega^2}{e}\frac{a^2}{2} = 3.0 \text{ nV}$$

(b) When field **B** is present, by definition of motional e.m.f.

$$\varphi_1 - \varphi_2 = \int_1^2 -(\mathbf{v}\times\mathbf{B})\cdot d\mathbf{r}$$

Hence the sought potential difference,

$$\varphi_{cen} - \varphi_{rim} = \int_0^a -vBdr = \int_0^a -\omega rBdr \text{ (as } v = \omega r)$$

Thus, $$\varphi_{rim} - \varphi_{cen} = \varphi = \frac{1}{2}\omega Ba^2 = 20 \text{ mV}$$

In general $\omega < eB/m$, so we can neglect the effect diskussed in part (a) here.

3.291 By definition,

$$\mathbf{E} = -(\mathbf{v}\times\mathbf{B})$$

So, $$\int_A^C \mathbf{E}\cdot d\mathbf{r} = \int_A^C -(\mathbf{v}\times\mathbf{B})\cdot d\mathbf{r} = \int_0^d -vBdr$$

But, $v = \omega r$, where r is the perpendicular distance of the point from A.

Hence, $$\int_A^C \mathbf{E}\cdot d\mathbf{r} = \int_0^d -\omega Brdr = -\frac{1}{2}\omega Bd^2 = -10 \text{ mV}$$

This result can be generalized to a wire AC of arbitrary planar shape. We have

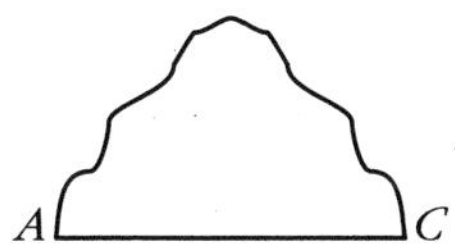

$$\int_A^C \mathbf{E}\cdot d\mathbf{r} = -\int_A^C (\mathbf{v}\times\mathbf{B})\cdot dr$$
$$= -\int_A^C ((\boldsymbol{\omega}\times\mathbf{r})\times\mathbf{B})\cdot d\mathbf{r}$$
$$= -\int_A^C (\mathbf{B}\cdot\mathbf{r}\boldsymbol{\omega} - \mathbf{B}\cdot\boldsymbol{\omega}\mathbf{r})\cdot d\mathbf{r}$$
$$= -\frac{1}{2}B\omega d^2$$
$$= -10 \text{ mV}$$

d being AC and $\boldsymbol{r}$ being measured from A.

3.292 Magnitude of flux at any moment of time,

$$|\Phi(t)| = \mathbf{B}\cdot\mathbf{S} = B\left(\frac{1}{2}R^2\varphi\right)$$

where φ is the sector angle, enclosed by the field.

Now, magnitude of induced e.m.f. is given by,

$$\xi_{in} = \left|\frac{d\Phi_i}{dt}\right| = \left|\frac{BR^2}{2}\frac{d\varphi}{dt}\right| = \frac{BR^2}{2}\omega$$

where ω is the angular velocity of the disk. But as it starts rotating from rest at $t = 0$ with an angular acceleration β, its angular velocity $\omega(t) = \beta t$. So,

$$\xi_{in} = \frac{BR^2}{2}\beta t$$

According to Lenz's law the first half cycle current in the loop is in anticlockwise direction, and in subsequent half cycle it is in clockwise direction.

Thus, $$\xi_{in} = (-1)^n \frac{BR^2}{2}\beta t \quad \text{(where } n \text{ is number of half revolutions)}$$

The plot $\xi_m(t)$, where $t_n = \sqrt{2\pi n/\beta}$, is as shown in the answer sheet.

3.293 Field, due to the current carrying wire in the region, right to it, is directed into the plane of the paper and its magnitude is given by

$$B = \frac{\mu_0}{2\pi r}\frac{I}{r}$$

where r is the perpendicular distance from the wire.

As B is same along the length of the rod thus, motional e.m.f.

$$\xi_{in} = \left|\int_1^2 (\mathbf{v}\times\mathbf{B})\cdot d\mathbf{l}\right| = vBl$$

and it is directed in the sense of ($\mathbf{v} \times \mathbf{B}$).

So, current (induced) in the loop

$$I_{in} = \frac{\xi_{in}}{R} = \frac{1}{2}\frac{\mu_0 lvI}{\pi Rr}$$

3.294 Field due to the current carrying wire at a perpendicular distance x from it is given by

$$B(x) = \frac{\mu_0}{2\pi}\frac{I}{x}$$

There will be no induced e.m.f. in the segments (2) and (4) as, $\mathbf{v} \uparrow\uparrow d\mathbf{l}$. The magnitude of e.m.f. induced in 1 and 3, will be

$$\xi_1 = vB(a+x)a = v\left(\frac{\mu_0}{4\pi}\frac{2I}{a+x}\right)a = \frac{\mu_0 avI}{2\pi(a+x)}$$

$$\xi_2 = vB(x)a = v\left(\frac{\mu_0}{4\pi}\frac{2I}{x}\right)a = \frac{\mu_0 av}{2\pi}\frac{I}{x}$$

respectively, and their sense will be in the direction of ($\mathbf{v} \times \mathbf{B}$).

So, e.m.f. induced in the network

$$= \xi_1 - \xi_2 \text{ (as } \xi_1 > \xi_2)$$

$$= \frac{av\mu_0 I}{2\pi}\left[\frac{1}{x} - \frac{1}{a+x}\right] = \frac{va^2\mu_0 I}{2\pi x(a+x)}$$

3.295 As the rod rotates, motional e.m.f. is induced in it

$$\xi = \int_0^a (\omega r) B\, dr = \frac{1}{2} a^2 B\omega$$

The net current in the conductor is then

$$\frac{\xi(t) - \frac{1}{2}a^2 B\omega}{R}$$

A magnetic force will then act on the conductor of magnitude BI per unit length. Its direction will be normal to B and the rod and its torque will be

$$\int_0^a \left(\frac{\xi(t) - \frac{1}{2}a^2 B\omega}{R} \right) (dx) Bx$$

Obviously both magnetic and mechanical torque acting on the C.M. of the rod must be equal but opposite in sense. Then for equilibrium at constant ω,

$$\frac{\xi(t) - \frac{1}{2}a^2 B\omega}{R} \frac{Ba^2}{2} = \frac{1}{2} mga \sin \omega t$$

or
$$\xi(t) = \frac{1}{2}a^2 B\omega + \frac{mg R}{aB} \sin \omega t = \frac{1}{2ab}(a^3 B^2 \omega + 2mg R \sin \omega t)$$

(The answer given in the book is incorrect dimensionally.)

3.296 The rod behaves like a cell of e.m.f. $= vBl$, where v is the velocity of the rod at an arbitrary instant of time. From Lenz's law, current in the loop is in clockwise direction.

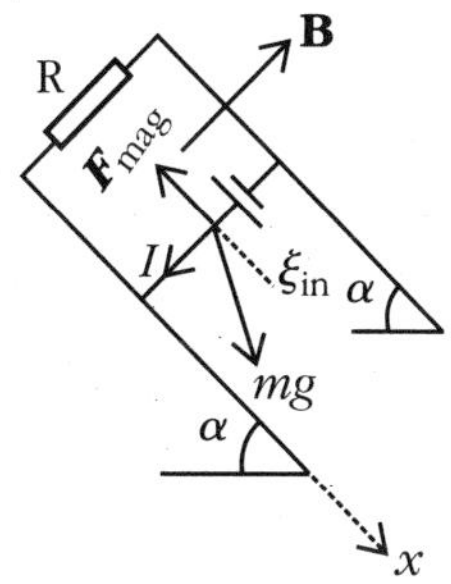

From Newton's second law for the rod, $F_x = mw_x$

or $\qquad mg \sin \alpha - IlB = mw$

For steady state, acceleration of the rod must be equal to zero.

Hence, $\qquad mg \sin \alpha = IlB \qquad (1)$

But, $\qquad I = \frac{\xi_{in}}{R} = \frac{vBl}{R} \qquad (2)$

From Eqs. (1) and (2)

$$v = \frac{mg \sin \alpha R}{B^2 l^2}$$

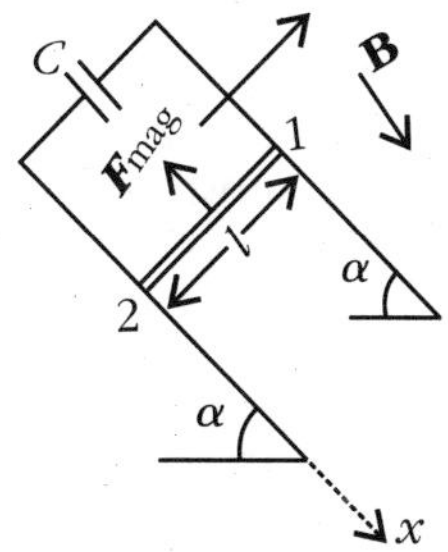

3.297 From Lenz's law, the current through the copper bar is directed from 1 to 2 or in other words, the induced current in the circuit is in clockwise direction.

Potential difference across the capacitor plates

$$\frac{q}{C} = \xi_{in} \quad \text{or} \quad q = C\xi_{in}$$

Hence, the induced current in the loop

$$I = \frac{dq}{dt} = C\frac{d\xi_{\text{in}}}{dt}$$

But the variation of magnetic flux through the loop is caused by the movement of the bar.

So, the induced e.m.f. $\qquad \xi_{\text{in}} = Blv$

On differentiation $\qquad \frac{d\xi_{\text{in}}}{dt} = Bl\frac{dv}{dt} = Blw$

Hence, $\qquad I = C\frac{d\xi}{dt} = CBlw$

Now, the forces acting on the bars are the weight and the Ampere's force, where $F_{\text{amp}} = IlB(CB\,lw)$ $B = Cl^2B^2w$.

From Newton's second law, for the rod,

$$F_x = mw_x$$

or $\qquad mg\sin\alpha - Cl^2B^2w = mw$

Hence, $\qquad w = \frac{mg\sin\alpha}{Cl^2B^2 + m} = \frac{g\sin\alpha}{1+\frac{l^2B^2C}{m}}$

3.298 Flux of **B**, at an arbitrary moment of time t is given by

$$\Phi_t = \mathbf{B}\cdot\mathbf{S} = B\frac{\pi a^2}{2}\cos\omega t$$

From Faraday's law, induced e.m.f.

$$\xi_{\text{in}} = -\frac{d\Phi}{dt}$$

$$= -\frac{d\left(B\pi\frac{a^2}{2}\cos\omega t\right)}{dt} = \frac{B\pi a^2\omega}{2}\sin\omega t$$

and induced current

$$I_{\text{in}} = \frac{\xi_{\text{in}}}{R} = \frac{B\pi a^2}{2R}\omega\sin\omega t$$

Now, thermal power generated in the circuit at the moment at time t

$$P(t) = \xi_{in}\times I_{in} = \left(\frac{B\pi a^2\omega}{2}\right)^2\frac{1}{R}\sin^2\omega t$$

and mean thermal power generated is given by

$$\langle P\rangle = \frac{\left[\frac{B\pi a^2\omega}{2}\right]^2\frac{1}{2}\int_0^T\sin^2\omega t\,dt}{\int_0^T dt} = \frac{1}{2R}\left(\frac{B\pi a^2\omega}{2}\right)^2$$

Note: *The calculation of ξ_{in}, which can also be checked by using motional e.m.f. is correct even though the conductor is not a closed semicircle, because the flux linked to the rectangular part containing the resistance R is not changing. The answer given in the book is off by a factor* 1/4.

3.299 The flux through the coil changes sign. Initially it is BS per turn. Finally it is $-BS$ per turn. Now if flux is Φ at an intermediate state, then the current at that moment will be

$$I = \frac{-N\dfrac{d\Phi}{dt}}{R}$$

So, charge that flows during a sudden turning of the coil is

$$q = \int I\,dt = -\frac{N}{R}[\Phi - (-\Phi)]$$
$$= 2N\frac{BS}{R}$$

Hence, $$B = \frac{1}{2}\frac{qR}{NS} = 0.5\text{ T} \quad \text{(on substituting values)}$$

3.300 According to Ohm's law and Faraday's law of induction, the current I_0 appearing in the frame, during its rotation, is determined by the formula

$$I_0 = -\frac{d\Phi}{dt} = -L\frac{dI_0}{dt}$$

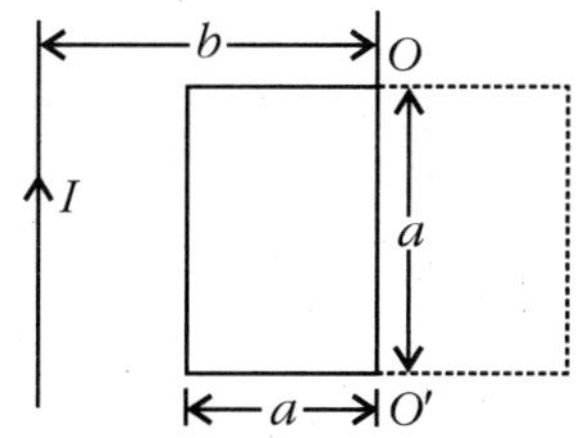

Hence, the required amount of electricity (charge) is

$$q = \int I_0 dt$$
$$= -\frac{1}{R}\int(d\Phi + L dI_0)$$
$$= -\frac{1}{R}(\Delta\Phi + L\Delta I_0)$$

Since the frame has been stopped after rotation, the current in it vanishes, and hence $\Delta I_0 = 0$. We have to find the increment of the flux $\Delta\Phi$ through the frame ($\Delta\Phi = \Phi_2 - \Phi_1$).

Let us choose the normal **n** to the plane of the frame, for instance, so that in the final position, **n** is directed behind the plane of the figure (along **B**).

Then it can be easily seen that in the final position, $\Phi_2 > 0$, while in the initial position, $\Phi_1 < 0$ (the normal is opposite to **B**), and $\Delta\varphi$ turns out to be simply equal to the flux through the surface bounded by the final and initial positions of the frame:

$$\Delta\Phi = \Phi_2 + |\Phi_1| = \int_{b-a}^{b+a} B\,a dr$$

where B is a function of r, whose form can be easily found with the help of the theorem of circulation. Finally omitting the minus sign, we obtain

$$q = \frac{\Delta\Phi}{R} = \frac{\mu_0 a I}{2\pi R}\ln\frac{b+a}{b-a}$$

3.301 As **B**, due to the straight current carrying wire, varies along the rod (connector) and enters linearly so to make the calculations simple, **B** is made constant by taking its average value in the range $[a, b]$.

$$<B> = \frac{\int_a^b B\,dr}{\int_a^b dr} = \frac{\int_a^b \frac{\mu_0}{2\pi}\frac{I_0}{r}\,dr}{\int_a^b dr}$$

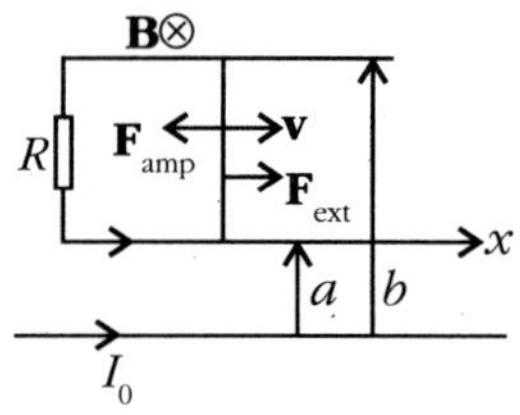

or
$$<B> = \frac{\mu_0}{2\pi}\frac{I_0}{(b-a)}\ln\frac{b}{a}$$

(a) Flux of **B** changes through the loop due to movement of the connector. According to Lenz's law, the current in the loop will be anticlockwise. The magnitude of motional e.m.f.

$$\xi_{in} = v<B>(b-a)$$
$$= \frac{\mu_0}{2\pi}\frac{I_0}{(b-a)}\left(\ln\frac{b}{a}\right)(b-a)\frac{dx}{dt} = \frac{\mu_0}{2\pi}I_0\ln\frac{b}{a}v$$

So, induced current

$$I_{in} = \frac{\xi_{in}}{R} = \frac{\mu_0}{2\pi}\frac{I_0 v}{R}\ln\frac{b}{a}$$

(b) The force required to maintain the constant velocity of the connector must be the magnitude equal to that of Ampere's force acting on the connector, but in opposite direction.

So,
$$F_{ext} = I_{in}l<B> = \left(\frac{\mu_0}{2\pi}\frac{I_0}{R}\ln\frac{b}{a}\right)(b-a)\left(\frac{\mu_0}{2\pi}\frac{I_0}{(b-a)}\ln\frac{b}{a}\right)$$
$$= \frac{v}{R}\left(\frac{\mu_0}{2\pi}I_0\ln\frac{b}{a}\right)^2$$

and will be directed as shown in the figure.

3.302 (a) The flux through the loop changes due to the movement of the rod AB. According to Lenz's law, current should be anticlockwise, since we have assumed that **B** is directed into the plane of the loop. The motional e.m.f. $\xi_{in}(t) = Blv$ and induced current $I_{in} = vBl/R$.

From Newton's law in projection form,

$$F_x = mw_x$$

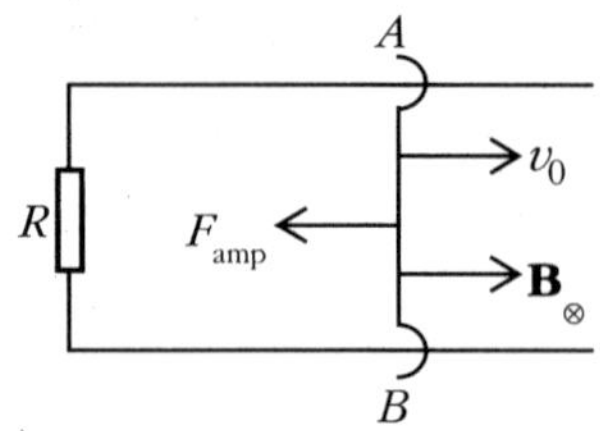

So,
$$-F_{amp} = m\frac{v\,dv}{dx}$$

But
$$F_{amp} = I_{in}lB = \frac{vB^2l^2}{R}$$

So,
$$-\frac{vB^2l^2}{R} = mv\frac{dv}{dx}$$

or
$$\int_0^x dx = -\frac{mR}{B^2l^2}\int_{v_0}^{0} dv \quad \text{or} \quad x = \frac{mRv_0}{B^2l^2}$$

(b) Sum of the works done by Ampere's force and induced e.m.f. is zero, so from energy conservation

$$\Delta E + Q = 0 \quad \text{(where } Q \text{ is heat liberated)}$$

or

$$\left[0 - \frac{1}{2}mv_0^2\right] + Q = 0$$

So,

$$Q = \frac{1}{2}mv_0^2$$

3.303 Using the calculation done in the previous problem, Ampere's force on the connector,

$$\mathbf{F}_{amp} = \frac{vB^2l^2}{R} \text{(directed towards left)}$$

Now from Newton's second law,

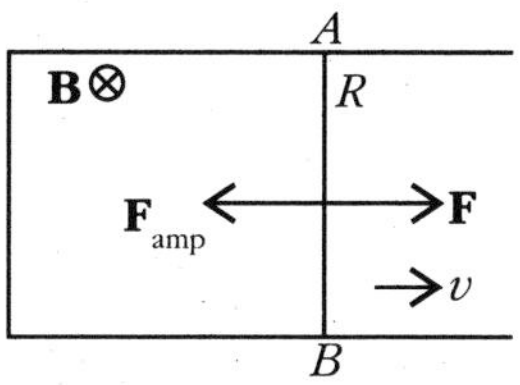

$$F - F_{amp} = m\frac{dv}{dt}$$

So,

$$F = \frac{vB^2l^2}{R} + m\frac{dv}{dt}$$

or

$$t = m\int_0^v \frac{dv}{F - \dfrac{vB^2l^2}{R}}$$

or

$$\frac{t}{m} = -\frac{R}{B^2l^2}\ln\left(\frac{F - \dfrac{vB^2l^2}{R}}{F}\right)$$

Thus,

$$v = (1 - e^{-tB^2l^2/Rm})\frac{RF}{B^2l^2}$$

3.304 According to Lenz's law, the sense of induced e.m.f. is such that it opposes the cause of change of flux. In our problem, magnetic field is directed away from the reader and is diminishing. So,

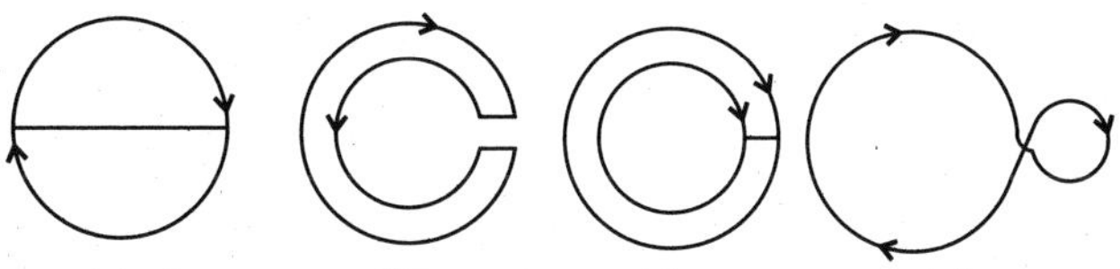

(a) (b) (c) (d)

in Fig. (a), in the round conductor, it is clockwise and there is no current in the connector.

in Fig. (b), in the outside conductor, it is clockwise.

in Fig. (c), in both the conductors, it is clockwise; and there is no current in the connector to obey the charge conservation.

in Fig. (d), in the left side of the figure, it is clockwise.

3.305 The loops are connected in such a way that if the current is clockwise in one, it is anticlockwise in the other. Hence the e.m.f. in loop b opposes the e.m.f. in loop a.

In the loop with side equal to a, e.m.f. is $\dfrac{d}{dt}(a^2B) = a^2\dfrac{d}{dt}(B_0 \sin \omega t)$

Similarly, e.m.f. in loop with side equal to $b = b^2 B_0 \omega \cos \omega t$.

Hence, net e.m.f. in the circuit = $(a^2 - b^2)B_0\omega \cos \omega t$, as both the e.m.f's are in opposite sense, and resistance of the circuit = $4(a + b)\rho$.

Therefore, the amplitude of the current

$$= \frac{(a^2 - b^2)B_0\omega}{4(a+b)\rho} = 0.5 \text{ A}$$

3.306 The flat shape is made up of concentric loops, having different radii, varying from O to a. Let us consider an elementary loop of radius r, then, e.m.f. induced due to this loop is

$$\frac{-d(\mathbf{B}\cdot\mathbf{S})}{dt} = \pi r^2 B_0 \omega \cos \omega t$$

and the total induced e.m.f.

$$\xi_{in} = \int_0^a (\pi r^2 B_0 \omega \cos \omega t)\, dN \qquad (1)$$

where $\pi r^2 \omega \cos \omega t$ is the contribution of one turn of radius r and dN is the number of turns in the interval $(r, r + dr)$.

So, $$dN = \left(\frac{N}{a}\right) dr \qquad (2)$$

From Eqs. (1) and (2),

$$\xi = \int_0^a -(\pi r^2 B_0 \omega \cos \omega t)\frac{N}{a}\, dr = \frac{\pi B_0 \omega \cos \omega t N a^2}{3}$$

Maximum value of e.m.f. amplitude

$$\xi_{max} = \frac{1}{3}\pi B_0\, \omega N a^2$$

3.307 The flux through the loop changes due to the variation in **B** with time and also due to the movement of the connector.

So, $$\xi_{in} = \left|\frac{d(\mathbf{B}\cdot s)}{dt}\right| = \left|\frac{d(Bs)}{dt}\right| \text{ (as } s \text{ and } \mathbf{B} \text{ are collinear)}$$

But, B, after t seconds of beginning of motion = Bt, and s becomes = $1/2\, l\, wt^2$, as connector starts moving from rest with a constant acceleration w.

So, $$\xi_{in} = \frac{3}{2} Bl\, wt^2$$

$$= 12 \text{ mV (on substituting values)}$$

3.308 We use $$B = \mu_0 n I$$

Then, from the law of electromagnetic induction

$$\oint \mathbf{E}\cdot d\mathbf{l} = -\frac{d\Phi}{dt}$$

So, for $r < a$

$$E_\varphi 2\pi r = -\pi r^2 \mu_0 n\dot{I} \quad \text{or} \quad E_\varphi = -\frac{1}{2}\mu_0\, nr\dot{I} \text{ (where } \dot{I} = dI/dt)$$

For $r > a$,

$$E_\varphi 2\pi r = -\pi a^2 \mu_0 n\dot{I} \quad \text{or} \quad E_\varphi = -\mu_0\, n\dot{I}a^2/2r$$

The meaning of minus sign can be deduced from Lenz's law.

3.309 The e.m.f. induced in the turn is $\mu_0 n\dot{I}\ \pi\, d^2/4$. The resistance is $\pi\, d/S\,\rho$. So, the current is $\mu_0 n\dot{I}Sd/4\rho = 2\,\text{mA}$, where ρ is the resistivity of copper.

3.310 The changing magnetic field will induce an e.m.f. in the ring, which is obviously equal in the two parts by symmetry (the e.m.f. induced by electromagnetic induction does not depend on resistance). The current that will flow due to this will be different in the two parts momentarily. This will cause an accummulation of charge, leading to the setting up of a static electric field E which has opposite sign in the two parts. Thus,

$$\frac{\xi}{2} - \pi aE = rI \quad \text{and} \quad \frac{\xi}{2} + \pi aE = \eta rI$$

where ξ is the total induced e.m.f. From this,

$$\xi = (\eta + 1)rI$$

and

$$E = \frac{1}{2\pi a}(\eta - 1)rI = \frac{1}{2\pi a}\frac{\eta - 1}{\eta + 1}\xi$$

But by Faraday's law,

$$\xi = \pi a^2 b$$

So,

$$E = \frac{1}{2}ab\frac{\eta - 1}{\eta + 1}$$

3.311 Consider the rotating frame with an instantaneous angular velocity $\omega(t)$. In this frame, a Coriolis force, 2 $mv' \times \boldsymbol{\omega}(t)$ acts, which must be balanced by the magnetic force $e\mathbf{v} \times \mathbf{B}(t)$.

Thus,

$$\boldsymbol{\omega}(t) = \frac{e}{2m}\mathbf{B}(t)$$

It is assumed that ω is small and varies slowly, so ω^2 and ω can be neglected.

3.312 The solenoid has an inductance,

$$L = \mu_0 n^2 \pi b^2 l$$

where n = number of turns of the solenoid per unit length. When the solenoid is connected to the source, an e.m.f. is set up, which, because of the inductance and resistance, rises slowly according to the equation

$$RI + L\dot{I} = V$$

This has the well-known solution

$$I = \frac{V}{R}(1 - e^{-tR/L})$$

Corresponding to this current, an e.m.f. is induced in the ring. Its magnetic field $B = \mu_0 nI$, in the solenoid, produces a force per unit length given by

$$\frac{dF}{dl} = BI = \frac{\mu_0^2 n^2 \pi a^2 II}{r}$$
$$= \frac{\mu_0^2 \pi a^2 V^2}{r}\left(\frac{n^2}{RL}\right) e^{-tR/L}(1 - e^{-tR/L}),$$

acting on each segment of the ring. This force is zero initially and zero for a large t. Its maximum value is for some finite t. The maximum value of

$$e^{-tR/L}(1 - e^{-tR/L}) = \frac{1}{4} - \left(\frac{1}{2} - e^{-tR/L}\right)^2 = \frac{1}{4}$$

So,

$$\frac{dF_{\max}}{dl} = \frac{\mu_0^2 \pi a^2 V^2}{r}\frac{n^2}{4RL} = \frac{\mu_0 a^2 V^2}{4rRlb^2}$$

3.313 The amount of heat generated in the loop during a small time interval dt

$$dQ = \xi^2/R\,dt, \quad \text{but,} \quad \xi = -\frac{d\Phi}{dt} = 2at - a\tau$$

So,

$$dQ = \frac{(2at - a\tau)^2}{R}dt$$

and hence, the amount of heat generated in the loop during the time interval 0 to τ

$$Q = \int_0^\tau \frac{(2at - a\tau)^2}{R}dt = \frac{1}{3}\frac{a^2\tau^3}{R}$$

3.314 Take an elementary ring of radius r and width dr. The e.m.f. induced in this elementary ring is $\pi r^2 \beta$. Now the conductance of this ring is

$$d\left(\frac{1}{R}\right) = \frac{bdr}{\rho 2\pi r} \quad \text{so} \quad dI = \frac{brdr}{2\rho}\beta$$

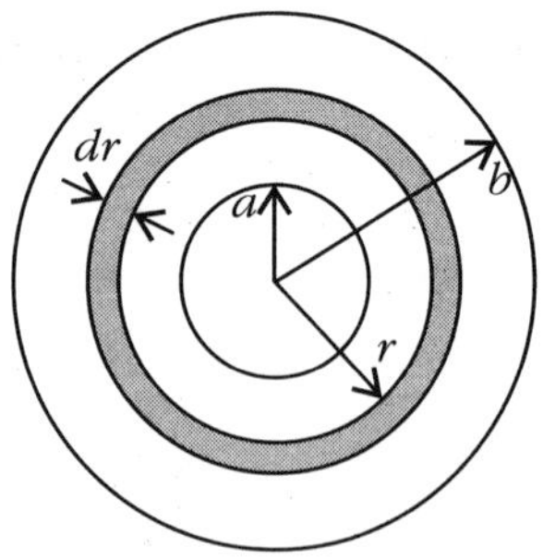

Integrating we get the total current as

$$I = \int_0^b \frac{brdr}{2\rho}\beta = \frac{b\beta(b^2 - a^2)}{4\rho}$$

3.315 Given $L = \mu_0 n^2 V = \mu_0 n^2 l_0 \pi R^2$, where R is the radius of the solenoid.

Thus,

$$n = \sqrt{\frac{L}{\mu_0 l_0 \pi}}\frac{1}{R}$$

So, length of the wire required is

$$l = nl_0 2\pi R = \sqrt{\frac{4\pi L l_0}{\mu_0}} = 0.10 \text{ km}$$

3.316 From the previous problem, we know that, of the wire needed

$$l' = \sqrt{\frac{Ll4\pi}{\mu_0}} \quad \text{(where } l = \text{length of solenoid)}$$

Now,
$$R = \frac{\rho_0 l'}{S}$$

where S = area of cross section of the wire, also $m = \rho S l'$.

Thus,
$$l' = \frac{RS}{\rho_0} = \frac{Rm}{\rho\rho_0 l'} \quad \text{or} \quad l' = \sqrt{\frac{Rm}{\rho\rho_0}}$$

where ρ_0 = resistivity of copper and ρ = density of copper.

Equating,
$$\frac{Rm}{\rho\rho_0} = \frac{Ll}{\mu_0/4\pi}$$

or
$$L = \frac{\mu_0}{4\pi}\frac{mR}{\rho\rho_0 l}$$

3.317 The current at time t is given by

$$I(t) = \frac{V}{R}(1 - e^{-tR/L})$$

The steady state value is

$$I_0 = \frac{V}{R}$$

and
$$\frac{I(t)}{I_0} = \eta = 1 - e^{-tR/L} \quad \text{or} \quad e^{-tR/L} = 1 - \eta$$

or
$$t_0\frac{R}{L} = \ln\frac{1}{1-\eta} \quad \text{or} \quad t_0 = \frac{L}{R}\ln\frac{1}{1-\eta} = 1.49 \text{ s}$$

3.318 The time constant τ is given by

$$\tau = \frac{L}{R} = \frac{L}{\rho_0 \dfrac{l_0}{S}}$$

where ρ_0 = resistivity, l_0 = length of the winding wire, S = cross section of the wire.

But
$$m = l\rho_0 S$$

So eliminating S,
$$\tau = \frac{L}{\dfrac{\rho_0 l_0}{m/\rho l_0}} = \frac{mL}{\rho\rho_0 l_0^2}$$

From solution of Problem 3.315, $l_0 = \sqrt{\dfrac{4\pi l L}{\mu_0}}$

(Note the interchange of l and l_0 because of difference in notation here.)

Thus,
$$\tau = \frac{mL}{\rho\rho_0 \dfrac{4\pi}{\mu_0} Ll} = \mu_0 4\pi \frac{m}{\rho\rho_0 l} = 0.7 \text{ ms}$$

3.319 Between the cables, where $a < r < b$, the magnetic field **H** satisfies

$$H_\varphi 2\pi r = I \quad \text{or} \quad H_\varphi = \frac{I}{2\pi r}$$

So,
$$B_\varphi = \frac{\mu\mu_0 I}{2\pi r}$$

The associated flux per unit length is

$$\Phi = \int_{r=a}^{r=b} \frac{\mu\mu_0 I}{2\pi r} \times 1 \times dr = \frac{\mu\mu_0 I}{2\pi} \ln\frac{b}{a}$$

Hence, the inductance per unit length is

$$L_1 = \frac{\Phi}{I} = \frac{\mu\mu_0}{2\pi} \ln\eta \left(\text{where } \eta = \frac{b}{a}\right)$$

We get
$$L_1 = 0.26 \frac{\mu\text{H}}{\text{m}}$$

3.320 Within the solenoid

$$H_\varphi \cdot 2\pi r = NI \quad \text{or} \quad H_\varphi = \frac{NI}{2\pi r}, B_\varphi = \mu\mu_0 \frac{NI}{2\pi r}$$

and the flux is
$$\Phi = N\Phi_1 = N\frac{\mu\mu_0}{2\pi} NI \int_b^{a+b} \frac{adr}{r}$$

Finally,
$$L = \frac{\mu\mu_0}{2\pi} N^2 a \ln\left(1 + \frac{a}{b}\right)$$

3.321 Neglecting end effects, the magnetic field B between the plates, which is mainly parallel to the plates, is

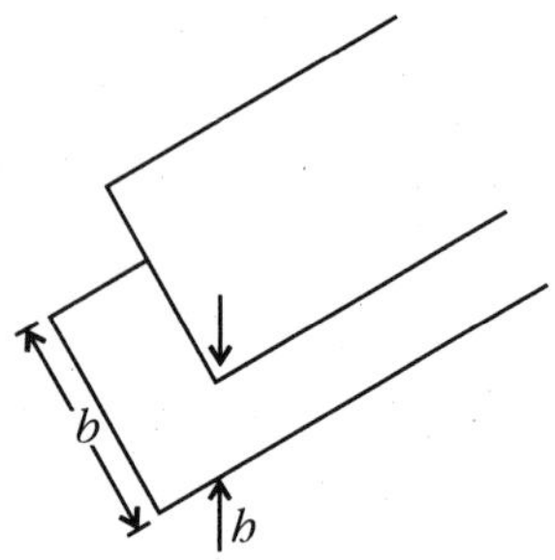

$$B = \mu_0 \frac{I}{b}$$

Thus, the associated flux per unit length of the plates is

$$\Phi = \mu_0 \frac{I}{b} \times h \times 1 = \left(\mu_0 \frac{h}{b}\right) \times I$$

So, L_1 = inductance per unit length $= \mu_0 \dfrac{h}{b} = 25$ nH/m.

3.322 For a single current carrying wire,

$$B_\varphi = \frac{\mu_0 I}{2\pi r} \quad (\text{for } r > a)$$

For the double line cable with current flowing in opposite direction in the two conductors $B_\varphi \approx \mu_0 I/\pi r$ between the cables, by superposition. The associated flux is

$$\Phi = \int_a^{d-a} \frac{\mu_0 I}{\pi} \frac{dr \times 1}{r} \approx \frac{\mu_0 I}{\pi} \ln\frac{d}{a} = \frac{\mu_0}{\pi} \ln\eta \times I \text{ per unit length}$$

Hence,
$$L_1 = \frac{\mu_0}{\pi}\ln\eta$$

is the inductance per unit length.

3.323 In a superconductor there is no resistance. Hence,

$$L\frac{dI}{dt} = +\frac{d\Phi}{dt}$$

So integrating,
$$I = \frac{\Delta\Phi}{L} = \frac{\pi a^2 B}{L}$$

because
$$\Delta\Phi = \Phi_f - \Phi_i$$
$$\Phi_f = \pi a^2 B \text{ as } \Phi_i = 0$$

Also, the work done is

$$A = \int \xi I dt = \int I dt \frac{d\Phi}{dt} = \frac{1}{2}LI^2 = \frac{1}{2}\frac{\pi^2 a^4 B^2}{L}$$

3.324 In a solenoid, the inductance

$$L = \mu\mu_0 n^2 V = \mu\mu_0 \frac{N S}{l}$$

where S = area of cross-section of the solenoid, l = its length, $V = Sl$, $N = nl$ = total number of turns.

When the length of the solenoid is increased, for example, by pulling it, its inductance will decrease. If the current remains unchanged, the flux, linked to the solenoid, will also decrease. An induced e.m.f. will then come into play, which by Lenz's law will try to oppose the decrease of flux, for example, by increasing the current. In the superconducting state, the flux will not change and so, I/l = constant.

Hence,
$$\frac{I}{l} = \frac{I_0}{l_0} \quad \text{or} \quad I = I_0\frac{l}{l_0} = I_0(1+\eta) = 2\text{ A} \text{ (on substituting values)}$$

3.325 The flux linked to the ring cannot change on transition to the super-conduction state, for reasons similar to those given above. Thus, a current I must be induced in the ring, where

$$I = \frac{\Phi}{L} = \frac{\pi a^2 B}{\mu_0 a\left(\ln\frac{8a}{b} - 2\right)} = \frac{\pi a B}{\mu_0\left(\ln\frac{8a}{b} - 2\right)} = 50 \text{ A (on substituting values)}$$

3.326 We write the equation of the circuit as

$$RI + \frac{L}{\eta}\frac{dI}{dt} = \xi$$

for $t \geq 0$. The current at $t = 0$ just after inductance is changed is $I = \eta\xi/R$, so that the flux through the inductance is unchanged.

We look for a solution of the above equation in the form

$$I = A + Be^{-t/C}$$

Substituting $$C = \frac{L}{\eta R},\ B = \eta - 1,\ A = \frac{\xi}{R}$$

We get, $$I = \frac{\xi}{R}(1 + (\eta - 1)e^{-\eta Rt/L})$$

3.327 In the left loop, $\xi - I_0 R = (I_0 - I)R$

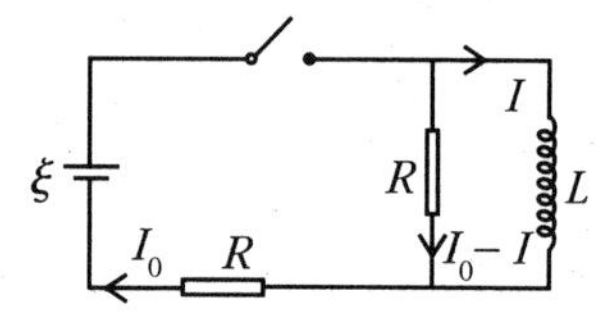

or $$\xi + IR = 2I_0 R$$

So, $$\xi - I_0 R = \xi - \frac{(\xi + IR)}{2} = \frac{\xi - IR}{2}$$

But in the bigger loop

$$\xi - I_0 R = L\frac{dI}{dt}$$

or $$\frac{\xi - IR}{2} = L\frac{dI}{dt}$$

So, $$\xi - IR = L\frac{dI}{dt}$$

3.328 The equations are

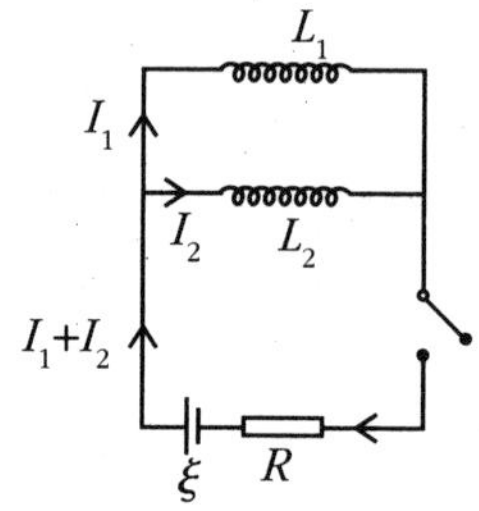

$$L_1\frac{dI_1}{dt} = L_2\frac{dI_2}{dt} = \xi - R(I_1 + I_2)$$

Then, $$\frac{d}{dt}(L_1 I_1 - L_2 I_2) = 0$$

or $$L_1 I_1 - L_2 I_2 = \text{constant}$$

But initially at $t = 0$, $I_1 = I_2 = 0$.

So, constant must be zero and at all times. Therefore,

$$L_1 I_1 = L_2 I_2$$

In the final steady state, current must obviously be $I_1 + I_2 = \xi/R$. Thus in steady state,

$$I_1 = \frac{\xi L_2}{R(L_1 + L_2)} \quad \text{and} \quad I_2 = \frac{\xi L_1}{R(L_1 + L_2)}$$

3.329 Here,

$$B = \frac{\mu_0 I}{2\pi r} \quad \text{(at a distance } r \text{ from the wire)}$$

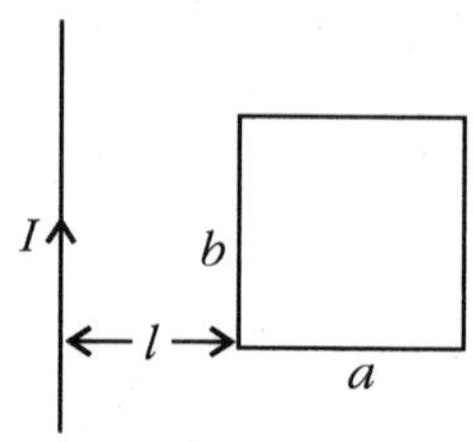

The flux through the frame is,

$$\Phi_{12} = \int_l^{a+l} \frac{\mu_0 I}{2\pi r} b\,dr = \frac{\mu_0 b}{2\pi} I \ln\left(1 + \frac{a}{l}\right)$$

Thus, $$L_{12} = \frac{\Phi_{12}}{I} = \frac{\mu_0 b}{2\pi}\ln\left(1 + \frac{a}{l}\right)$$

3.330 Here, $$B = \frac{\mu_0 I}{2\pi r} \quad \text{and} \quad \Phi = \mu_0 \mu \frac{l}{2\pi} \int_a^b \frac{h\,dr}{r} N$$

Thus, $$L_{12} = \frac{\mu \mu_0 h N}{2\pi} \ln \frac{b}{a}$$

3.331 The direct calculation of the flux Φ_2 is a rather complicated problem, since the configuration of the field itself is complicated. However, the application of the reciprocity theorem simplifies the solution of the problem. Indeed, let the same current I flow through loop 2. Then the magnetic flux created by this current through loop 1 can be easily found.

As the loop is very small, so, induction value of B at the center is for the whole loop. Magnetic induction at the center of the loop is

$$B = \frac{\mu_0 I}{2b}$$

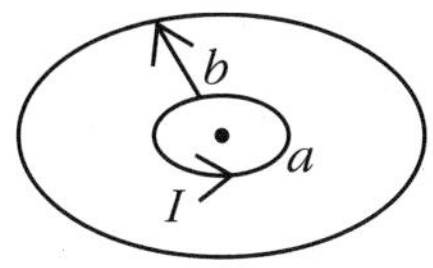

So, flux through loop 1 is

$$\Phi_{12} = \pi a^2 \frac{\mu_0 I}{2b}$$

(a) So, $$L_{12} = \frac{\Phi_{21}}{I} = \frac{1}{2} \frac{\mu_0 \pi a^2}{b}$$

(b) From reciprocity theorem,

$$\Phi_{12} = \Phi_{21}, \Phi_{21} = \frac{\mu_0 \pi a^2 I}{2b}$$

3.332 In the process of removal of the magnet, the total magnetic flux through the coil was changing, which resulted in the emergence of induced current defined by the equation:

$$RI = -N\frac{d\Phi}{dt} - L\frac{dI}{dt}$$

We multiply both sides of this equation by dt and take into account that $I\,dt = dq$. This gives

$$Rdq = -Nd\Phi - LdI$$

Having integrated this expression, we obtain $R_q = -N\Delta\Phi - L\,\Delta I$.

Considering now that $\Delta I = 0$ (the current was equal to zero at the beginning as well as at the end of the process), we obtain

$$q = N\Delta\Phi/R = \Phi R, \qquad (1)$$

where Φ is the magnetic flux through the coil at the beginning of the process (we omitted the minus sign since it is immaterial).

Thus, the problem is reduced to the calculation of the flux Φ through the coil, which can be easily obtained by using the reciprocity theorem. We mentally replace the magnet by a small current loop creating in the surrounding space the same magnetic field as the magnet. If the area of the loop is S and the current is I, their product will give the magnetic moment p_m of the magnet: $p_m = IS$. According to the reciprocity theorem, $L_{12}I = L_{21}I$ and the problem is reduced to determining the magnetic flux through the area S of the loop, which creates the same current I, but flowing the coil. Assuming the field to be uniform within the loop, we obtain

$$\Phi = BS = \mu_0 NIS/2a \qquad (2)$$

It remains for us to substitute (2) into (1) and recall that $IS = p_m$.

Then $q = \mu_0 N p_m/2aR$, and $p_m = 2aRq/\mu_0 N$.

3.333 If a current I flows in one of the coils, the magnetic field at the center of the other coil is

$$B = \frac{\mu_0 a^2 I}{2(l^2 + a^2)^{3/2}} = \frac{\mu_0 a^2 I}{2l^3}, \text{ as } l >> a$$

The flux associated with the second coil is then approximately $\mu_0 \pi\, a^4 I/2l^3$.

Hence,
$$L_{12} = \frac{\mu_0 \pi a^4}{2l^3}$$

3.334 When the current in one of the loops is $I_1 = \alpha t$, an e.m.f. $L_{12}\, dI_1/dt = L_{12}\, \alpha$, is induced in the other loop. Then, if the current in the other loop is I_2, we must have,

$$L_2 \frac{dI_2}{dt} + RI_2 = L_{12}\alpha$$

This equation has the solution,

$$I_2 = \frac{L_{12}\alpha}{R}(1 - e^{-tR/L_2})$$

which is the required current.

3.335 Initially, after a steady current is set up, the current is flowing as shown in the figure. In steady conditions, $I_{20} = \xi/R$ and $I_{10} = \xi/R_0$.

The current in the inductance cannot change suddenly. We then have the equation,

$$L\frac{dI_2}{dt} + (R + R_0)I_2 = 0$$

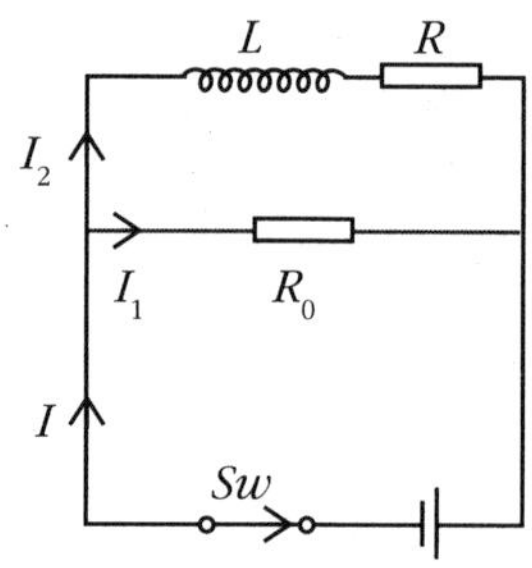

This equation has the solution

$$I_2 = I_{20} e^{-t(R+R_0)/L}$$

The heat dissipated in the coil is

$$Q = \int_0^\infty I_2^2\, R dt = I_{20}^2 R \int_0^\infty e^{-2t(R+R_0/L)} dt$$

$$= RI_{20}^2 \times \frac{L}{2(R + R_0)} = \frac{L\xi^2}{2R(R + R_0)} = 3\mu\text{J}$$

3.336 To find the magnetic field energy we recall that the flux varies linearly with current. Thus, when the flux is Φ for current I, we can write $\phi = AI$. The total energy enclosed in the field, when the current is I, is

$$W = \int \xi I dt = \int N \frac{d\Phi}{dt} I dt$$

$$= \int N d\Phi I = \int_0^I NA\, I dI = \frac{1}{2} NAI^2 = \frac{1}{2} N\Phi I$$

The characteristic factor 1/2 appears in this way.

3.337 We apply circulation theorem,

$$H \cdot 2\pi b = NI \quad \text{or} \quad H = \frac{NI}{2\pi b}$$

Thus the total energy,

$$W = \frac{1}{2} BH \cdot 2\pi b \cdot \pi a^2 = \pi^2 a^2 bBH$$

Given N, I, b, we know H, and can find out B from the $B - H$ curve. Then W can be calculated = 2.0 J.

3.338 From

$$\oint \mathbf{H} \cdot d\mathbf{r} = NI$$

$$H \cdot \pi d + \frac{B}{\mu_0} \cdot b \approx NI, (d >> b)$$

Also,

$$B = \mu\mu_0 H$$

Thus,

$$H = \frac{NI}{\pi d + \mu d}$$

Since B is continuous across the gap, B is given by,

$$B = \mu\mu_0 \frac{NI}{\pi d + \mu b}$$

both in the magnetic field and the gap.

(a) $$\frac{W_{\text{gap}}}{W_{\text{mag}}} = \frac{\frac{B^2}{2\mu_0} \times S \times b}{\frac{B^2}{2\mu\mu_0} \times S \times \pi d} = \frac{\mu b}{\pi d} = 3.0$$

(b) The flux is $N \int \mathbf{B} \cdot d\mathbf{S} = N\mu\mu_0 \frac{NI}{\pi d + \mu b} \cdot S = \mu_0 \frac{SN^2 I}{b + \frac{\pi d}{\mu}}$

So,

$$L \approx \frac{\mu_0 SN^2}{b + \frac{\pi d}{\mu}}$$

Total energy

$$= \frac{B^2}{2\mu_0}\left(\frac{\pi d}{\mu} + b\right)S = \frac{1}{2}\frac{\mu_0 N^2 S}{b + \frac{\pi d}{\mu}} \cdot I^2 = \frac{1}{2}LI^2 = 0.15\text{H}$$

The L found in one way is similar with the one found in the other way. Note that in calculating the flux, we do not consider the field in the gap since it is not linked to the winding. But the total energy includes that of the gap.

3.339 When the cylinder with a linear charge density λ rotates with a circular frequency ω, a surface current density (charge/length × time) of $I = \lambda\omega/2\pi$ is set up.

The direction of the surface current is normal to the plane of paper at Q and the contribution of this current to the magnetic field at P is

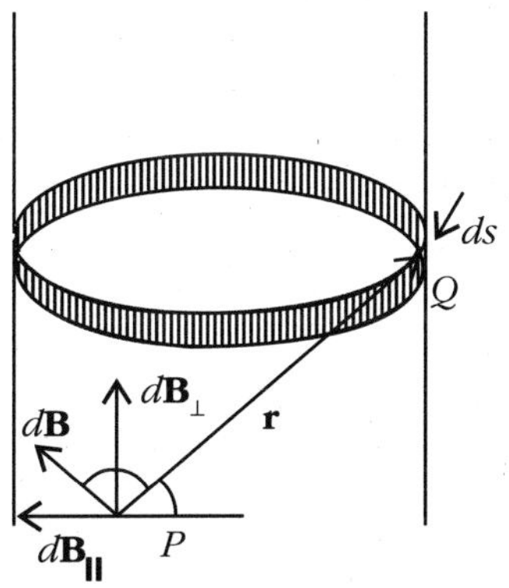

$$d\mathbf{B} = \frac{\mu_0}{4\pi} \frac{I(\mathbf{e} \times \mathbf{r})}{r^3} dS \quad \text{(where } \mathbf{e} \text{ is the direction of the current)}$$

In magnitude, $|\mathbf{e} \times \mathbf{r}| = r$, since $\mathbf{e}$ is normal to $\mathbf{r}$ and the direction of $d\mathbf{B}$ is as shown in the figure.

Its component, $d\mathbf{B}_{||}$, cancels out by cylindrical symmetry. The component that remains is

$$|\mathbf{B}_{\perp}| = \frac{\mu_0}{4\pi}\int \frac{IdS}{r^2}\cos\theta = \frac{\mu_0 I}{4\pi}\int d\Omega = \mu_0 I$$

where we have used,

$$\frac{dS\cos\theta}{r^2} = d\Omega \quad \text{and} \quad \int d\Omega = 4\pi$$

which is the total solid angle around any point.

The magnetic field vanishes outside the cylinder by similar argument.

The total energy per unit length of the cylinder is

$$W_1 = \frac{1}{2\mu_0}\mu_0^2\left(\frac{\lambda\omega}{2\pi}\right)^2 \times \pi a^2 = \frac{\mu_0}{8\pi}a^2\lambda^2\omega^2$$

3.340 $$w_{ele} = \frac{1}{2}\varepsilon_0 E^2 \text{ (for the electric field)}$$

$$w_{mag} = \frac{1}{2\mu_0}B^2 \text{ (for the magnetic field)}$$

Thus, $$\frac{1}{2\mu_0}B^2 = \frac{1}{2}\varepsilon_0 E^2$$

or $$E = \frac{B}{\sqrt{\varepsilon_0\mu_0}} = 3\times 10^8 \text{ V/m}$$

3.341 The electric field at P is

$$E_P = \frac{ql}{4\pi\varepsilon_0(a^2+l^2)^{3/2}}$$

To get the magnetic field, note that the rotating ring constitutes a current $I = q\,\omega/2\pi$, and the corresponding magnetic field at P is

$$B_P = \frac{\mu_0 a^2 I}{2(a^2+l^2)^{3/2}}$$

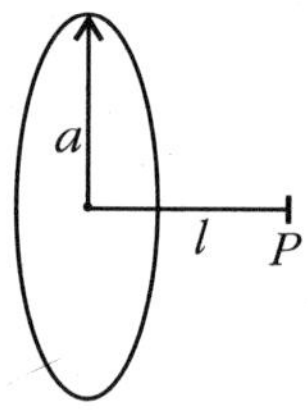

Thus, $$\frac{w_{ele}}{w_{mag}} = \frac{\varepsilon_0\mu_0 E^2}{B^2} = \varepsilon_0\mu_0\left(\frac{ql\times 2}{4\pi\varepsilon_0\mu_0 a^2 I}\right)^2$$

$$= \frac{1}{\varepsilon_0\mu_0}\left(\frac{l}{a^2\omega}\right)^2$$

or $$\frac{w_M}{w_E} = \frac{\varepsilon_0\mu_0\omega^2 a^4}{l^2}$$

$$= 1.1\times 10^{-15}$$

3.342 The total energy of the magnetic field is

$$\frac{1}{2}\int(\mathbf{B}\cdot\mathbf{H})dV = \frac{1}{2}\int\mathbf{B}\cdot\left(\frac{\mathbf{B}}{\mu_0} - \mathbf{J}\right)dV$$

$$= \frac{1}{2\mu_0}\int\mathbf{B}\cdot\mathbf{B}dV - \frac{1}{2}\int\mathbf{J}\cdot\mathbf{B}dV$$

The second term can be interpreted as the energy of magnetization, and has the density

$$-\frac{1}{2}\mathbf{J}\cdot\mathbf{B}.$$

3.343 (a) In series, the current I flows through both the coils, and the total e.m.f. induced when the current changes is

$$-2L\frac{dI}{dt} = -L'\frac{dI}{dt}$$

or $$L' = 2L$$

(b) In parallel, the current flowing through either coil is, $\frac{I}{2}$ and the e.m.f. induced is

$$-L\left(\frac{1}{2}\frac{dI}{dt}\right)$$

Since,
$$-L\left(\frac{1}{2}\frac{dI}{dt}\right) = -L'\frac{dI}{dt}$$

We get,
$$L' = \frac{1}{2}L$$

3.344 We use
$$L_1 = \mu_0 n_1^2 V, L_2 = \mu_0 n_2^2 V$$

So,
$$L_{12} = \mu_0 n_1 n_2 V = \sqrt{L_1 I_2}$$

3.345 The interaction energy is

$$\frac{1}{2\mu_0}\int |\mathbf{B}_1 + \mathbf{B}_2|^2 dV - \frac{1}{2\mu_0}\int |\mathbf{B}_1|^2 dV - \frac{1}{2\mu_0}\int |\mathbf{B}_2|^2 dV$$
$$= \frac{1}{\mu_0}\int \mathbf{B}_1 \cdot \mathbf{B}_2 dV$$

Here, if $\mathbf{B_1}$ is the magnetic field produced by the first of the current carrying loops, and $\mathbf{B_2}$ that of the second one, then the magnetic field due to both the loops will be $\mathbf{B_1} + \mathbf{B_2}$.

3.346 We can think of the smaller coil as constituting a magnet of dipole moment,

$$p_m = \pi a^2 I_1$$

Its direction is normal to the loop and it makes an angle θ with the direction of the magnetic field, due to the bigger loop. This magnetic field is

$$B_2 = \frac{\mu_0 I_2}{2b}$$

The interaction energy has the magnitude

$$|W| = \frac{\mu_0 I_1 I_2}{2b}\pi a^2 \cos\theta$$

Its sign depends on the sense of the currents.

3.347 (a) There is a radial, outward conduction current. Let Q be the instantaneous charge on the inner sphere, then,

$$j \times 4\pi r^2 = -\frac{dQ}{dt} \quad \text{or} \quad \mathbf{j} = -\frac{1}{4\pi r^2}\frac{dQ}{dt}\mathbf{r}$$

On the other hand
$$\mathbf{j}_d = \frac{\partial \mathbf{D}}{\partial t} = \frac{d}{dt}\left(\frac{Q}{4\pi r^2}\mathbf{r}\right) = -\mathbf{j}$$

where D is electric displacement vector.

(b) At the given moment,

$$\mathbf{E} = \frac{q}{4\pi\varepsilon_0 \varepsilon r^2}\mathbf{r}$$

and by Ohm's law $$\mathbf{j} = \frac{\mathbf{E}}{\rho} = \frac{q}{4\pi\varepsilon_0\varepsilon\rho r^2}\mathbf{r}$$

Then $$\mathbf{j}_d = -\frac{q}{4\pi\varepsilon_0\varepsilon\rho r^2}\mathbf{r}$$

and $$\oint \mathbf{j}_d \cdot d\mathbf{S} = -\frac{q}{4\pi\varepsilon_0\varepsilon\rho}\int\frac{dS\cos\theta}{r^2} = \frac{q}{\varepsilon_0\varepsilon\rho}$$

The surface integral must be $-ve$ because $\mathbf{j}_d$, being opposite of $\mathbf{j}$, is inward.

3.348 Here also we see that on neglecting edge effects, $\mathbf{j}_d = -\mathbf{j}$. Thus Maxwell's equations reduce to, div $\mathbf{B} = 0$, Curl $\mathbf{H} = 0$, $\mathbf{B} = \mu\mathbf{H}$.

A general solution of this equation is $\mathbf{B}$ = constant = $\mathbf{B}_0$. Here, $\mathbf{B}_0$ can be thought of as an extraneous magnetic field. If it is zero, $\mathbf{B} = 0$.

3.349 Given $$I = I_m \sin\omega t$$

We see that $$j = \frac{I_m}{S}\sin\omega t = -j_d = -\frac{\partial D}{\partial t}$$

or $$D = \frac{I_m}{\omega S}\cos\omega t$$

So, amplitude of electric field

$$E_m = \frac{I_m}{\varepsilon_0\omega S} = 7 \text{ V/cm}$$

3.350 The electric field between the plates can be written as

$$E = Re\left(\frac{V_m}{d}e^{i\omega t}\right) \text{ instead of } \frac{V_m}{d}\cos\omega t$$

This gives rise to a conduction current,

$$j_c = \sigma E = Re\left(\frac{\sigma}{d}V_m e^{i\omega t}\right)$$

and a displacement current,

$$j_d = \frac{\partial D}{\partial t} = Re\left(\varepsilon_0\varepsilon i\omega\frac{V_m}{d}e^{i\omega t}\right)$$

The total current is

$$j_r = \frac{V_m}{d}\sqrt{\sigma^2 + (\varepsilon_0\varepsilon\,\omega)^2}\cos(\omega t + \alpha)$$

where, $$\tan\alpha = \frac{\sigma}{\varepsilon_0\varepsilon\omega}$$

The corresponding magnetic field is obtained by using circulation theorem,

$$H \cdot 2\pi r = \pi r^2 j_T$$

or $$H = H_m\cos(\omega t + \alpha)$$

where, $$H_m = \frac{rV_m}{2d}\sqrt{\sigma^2 + (\varepsilon_0 \varepsilon \omega)^2}$$

3.351 Inside the solenoid, there is a magnetic field, given by

$$B = \mu_0 n I_m \sin \omega t$$

Since this varies in time, there is an associated electric field. This is obtained by using,

$$\oint \mathbf{E} \cdot d\mathbf{r} = -\frac{d}{dt}\int_s \mathbf{B} \cdot d\mathbf{S}$$

For $r < R$, $$2\pi rE = -\dot{B} \cdot \pi r^2 \quad \text{or} \quad E = -\frac{\dot{B}r}{2}$$

For $r > R$, $$E = -\frac{\dot{B}R^2}{2r}$$

The associated displacement current density is

$$j_d = \varepsilon_0 \frac{\partial E}{\partial t} = \begin{bmatrix} -\varepsilon_0 \ddot{B} r/2 \\ -\varepsilon_0 \ddot{B} R^2/2r \end{bmatrix}$$

(The answer given in the book is dimensionally incorrect without the factor ε_0.)

3.352 In the non-relativistic limit,

$$\mathbf{E} = \frac{q}{4\pi\varepsilon_0 r^3}\mathbf{r}$$

(a) On a straight line coinciding with the charge path,

$$\mathbf{j}_d = \varepsilon_0 \frac{\partial \mathbf{E}}{dt} = \frac{q}{4\pi}\left[\frac{-\mathbf{v}}{r^3} - \frac{3\mathbf{r}\dot{r}}{r^4}\right]\left(\text{using } \frac{d\mathbf{r}}{dt} = -\mathbf{v}\right)$$

But in this case,

$$\dot{r} = -v \quad \text{and} \quad v\frac{\mathbf{r}}{r} = \mathbf{v} \quad \text{so,} \quad \mathbf{j}_d = \frac{2q\mathbf{v}}{4\pi r^3}$$

(b) In this case, $\dot{r} = 0$, as $\mathbf{r} \perp \mathbf{v}$.

Thus, $$\mathbf{j}_d = -\frac{q\mathbf{v}}{4\pi r^3}$$

3.353 We have, $$E_p = \frac{qx}{4\pi\varepsilon_0 (a^2 + x^2)^{3/2}}$$

then, $$j_d = \frac{\partial D}{\partial t} = \varepsilon_0 \frac{\partial E}{\partial t} = \frac{qv}{4\pi (a^2 + x^2)^{5/2}}(a^2 - 2x^2)$$

This is maximum, when $x = x_m = 0$, and minimum at some other value. The maximum displacement current density is

$$(j_d)_{\max} = \frac{qv}{4\pi a^3}$$

To check this, we calculate

$$\frac{\partial j_d}{\partial x} = \frac{qv}{4\pi}[(-4x(a^2 + x^2) - 5x(a^2 - 2x^2)]$$

This vanishes for $x = 0$ and for $x = \sqrt{\frac{3}{2}}a$. The latter is easily shown to be a smaller local minimum (negative maximum).

3.354 We use Maxwell's equations in the form

$$\oint \mathbf{B}\cdot d\mathbf{r} = \varepsilon_0\mu_0 \frac{\partial}{\partial t}\int \mathbf{E}\cdot d\mathbf{S}$$

when the conduction current vanishes at the site.

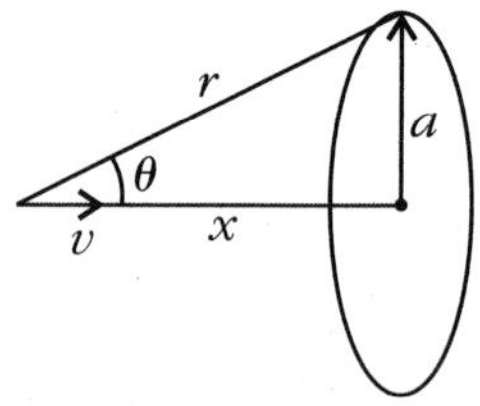

We know that,

$$\int \mathbf{E}\cdot d\mathbf{S} = \frac{q}{4\pi\varepsilon_0}\int \frac{d\mathbf{S}\cdot\mathbf{r}}{r^2}$$

$$= \frac{q}{4\pi\varepsilon_0}\int d\Omega = \frac{q}{4\pi\varepsilon_0}2\pi(1-\cos\theta)$$

where $2\pi(1-\cos\theta)$ is the solid angle, formed by the disk like surface, at the charge.

Thus, $$\oint \mathbf{B}\cdot d\mathbf{r} = 2\pi a B = \frac{1}{2}\mu_0 q\cdot\sin\theta\cdot\dot{\theta}$$

On the other hand, $$x = a\cot\theta$$

Differentiating and using $\frac{dx}{dt} = -v$

We get, $$v = a\,\mathrm{cosec}^2\theta\dot{\theta}$$

Thus, $$B = \frac{\mu_0 q v r\sin\theta}{4\pi r^3}$$

This can be written as

$$\mathbf{B} = \frac{\mu_0 q(\mathbf{v}\times\mathbf{r})}{4\pi r^3}$$

and $$\mathbf{H} = \frac{q}{4\pi}\frac{\mathbf{v}\times\mathbf{r}}{r^3}$$

(The sense has to be checked independently.)

3.355 (a) If $\mathbf{B} = \mathbf{B}(t)$, then,

$$\text{Curl }\mathbf{E} = \frac{-\partial\mathbf{B}}{\partial t} \neq 0$$

So, **E** cannot vanish.

(b) Here also, curl $\mathbf{E} \neq 0$, so **E** cannot be uniform.

(c) Suppose for instance,

$$\mathbf{E} = \mathbf{a}\,f(t)$$

where **a** is a spatially and temporally fixed vector. Then $-\partial\mathbf{B}/dt = \text{curl }\mathbf{E} = 0$. Generally speaking, this contradicts the other equation, curl $\mathbf{H} = \partial\mathbf{D}/\partial t \neq 0$ because in this case the left hand side is time independent but right-hand side is dependent on time. The only exception is when $f(t)$ is a linear function. Then the uniform field **E** can be time-dependent.

3.356 From the equation

$$\text{Curl }\mathbf{H} - \frac{\partial\mathbf{D}}{\partial t} = \mathbf{j}$$

We get on taking divergence of both sides,

$$-\frac{\partial}{\partial t}\text{div }\mathbf{D} = \text{div }\mathbf{j}$$

But div $\mathbf{D} = \rho$ and hence,

$$\text{div}\mathbf{j} + \frac{\partial\rho}{\partial t} = 0.$$

3.357 From the equation $$\nabla \times \mathbf{E} = -\frac{\partial \mathbf{B}}{\partial t}$$

On taking divergence of both sides,

$$0 = -\frac{\partial}{\partial t}\text{div}\,\mathbf{B}$$

This is compatible with div $\mathbf{B} = 0$.

3.358 A rotating magnetic field can be represented by

$$B_x = B\cos\omega t;\ B_y = B_0 \sin\omega t \text{ and } B_z = B_{zO}$$

Then, $$\text{Curl}\,\mathbf{E} = \frac{-\partial \mathbf{B}}{\partial t}$$

So, $$-(\text{Curl}\,\mathbf{E})_x = -\omega B_0 \sin\omega t = -\omega B_y$$

$$-(\text{Curl}\,\mathbf{E})_y = \omega B_0 \cos\omega t = \omega B_x \text{ and } -(\text{Curl}\,\mathbf{E})_z = 0$$

Hence, $$\text{Curl}\,\mathbf{E} = -\boldsymbol{\omega} \times \mathbf{B}$$

3.359 Consider a particle with charge e, moving with velocity $\mathbf{v}$, in frame K. It experiences a force $\mathbf{F} = e(\mathbf{v} \times \mathbf{B})$. In the frame K', moving with velocity $\mathbf{v}$, relative to K, the particle is at rest. This means that there must be an electric field $\mathbf{E}$ in K', so that the particle experiences a force given by

$$\mathbf{F}' = e\mathbf{E}' = \mathbf{F} = e\mathbf{v} \times \mathbf{B}$$

Thus, $$\mathbf{E}' = \mathbf{v} \times \mathbf{B}$$

3.360 Within the plate, there will appear a $(\mathbf{v} \times \mathbf{B})$ force, which will cause the charge inside the plate to drift, until a countervailing electric field is set up. This electric field is related to B, as $E = vB$, since v and B are mutually perpendicular, and E is perpendicular to both. The charge density $\pm\sigma$, on the force of the plate producing this electric field is given by

$$E = \frac{\sigma}{\varepsilon_0} \quad \text{or} \quad \sigma = \varepsilon_0 vB = 0.40 \text{ pC/m}^2$$

3.361 Choose $\boldsymbol{\omega} \uparrow\uparrow \mathbf{B}$ along the z-axis, and choose $\mathbf{r}$ as the cylindrical polar radius vector of a reference point (perpendicular distance from the axis). This point has the velocity

$$\mathbf{v} = \boldsymbol{\omega} \times \mathbf{r}$$

and experiences a $(\mathbf{v} \times \mathbf{B})$ force, which must be counterbalanced by an electric field

$$\mathbf{E} = -(\boldsymbol{\omega} \times \mathbf{r}) \times \mathbf{B} = -(\boldsymbol{\omega} \cdot \mathbf{B})\,\mathbf{r}$$

There must appear a space charge density

$$\rho = \varepsilon_0\, \text{div}\,\mathbf{E} = -2\varepsilon_0\, \boldsymbol{\omega} \cdot \mathbf{B} = -8 \text{ pC/m}^3$$

Since the cylinder as a whole is electrically neutral, the surface of the cylinder must acquire a positive charge of surface density,

$$\sigma = +\frac{2\varepsilon_0(\boldsymbol{\omega} \cdot \mathbf{B})\pi a^2}{2\pi a} = \varepsilon_0 a\, \boldsymbol{\omega} \cdot \mathbf{B} = +2 \text{ pC/m}^2$$

3.362 In the reference frame K' moving with the particle,

$$\mathbf{E}' \cong \mathbf{E} + \mathbf{v}_0 \times \mathbf{B} = \frac{q\mathbf{r}}{4\pi\varepsilon_0 r^3}$$

$$\mathbf{B}' \cong \mathbf{B} - \mathbf{v}_0 \times \frac{E}{c^2} = 0$$

Here, $\mathbf{v}_0$ = velocity of K' relative to the K frame, in which the particle has velocity $\mathbf{v}$. Clearly, $\mathbf{v}_0 = \mathbf{v}$. From the second equation,

$$\mathbf{B} \cong \frac{\mathbf{v} \times \mathbf{E}}{c^2} = \varepsilon_0 \mu_0 \times \frac{q}{4\pi\varepsilon_0} \frac{\mathbf{v} \times \mathbf{r}}{r^3} = \frac{\mu_0}{4\pi} \frac{q(\mathbf{v} \times \mathbf{r})}{r^3}$$

3.363 Suppose, there is only electric field $\mathbf{E}$ in K. Then in K', considering non-relativistic velocity $\mathbf{v}$,

$$\mathbf{E}' = \mathbf{E}, \mathbf{B}' = \frac{\mathbf{v} \times \mathbf{E}}{c^2}$$

So,

$$\mathbf{E}' \cdot \mathbf{B}' = 0$$

In the relativistic case,

$$\left.\begin{aligned} \mathbf{E}'_{\parallel} &= \mathbf{E}_{\parallel} \\ \mathbf{E}'_{\perp} &= \frac{\mathbf{E}_{\perp}}{\sqrt{1 - v^2/c^2}} \end{aligned}\right\} \begin{aligned} \mathbf{B}'_{\parallel} &= \mathbf{B}_{\parallel} = 0 \\ \mathbf{B}'_{\perp} &= \frac{-\mathbf{v} \times \mathbf{E}/c^2}{\sqrt{1 - v^2/c^2}} \end{aligned}$$

Now,

$$\mathbf{E}' \cdot \mathbf{B}' = \mathbf{E}_{\parallel} \cdot \mathbf{B}'_{\parallel} + \mathbf{E}'_{\perp} \quad \mathbf{B}'_{\perp} = 0, \text{ since}$$

$$\mathbf{E}'_{\perp} \cdot \mathbf{B}'_{\perp} = \frac{-\mathbf{E}_{\perp} \cdot (\mathbf{v} \times \mathbf{E})/(1 - v^2/c^2)}{-\mathbf{E}_{\perp} \cdot (\mathbf{v} \times \mathbf{E}_{\perp}) \Big/ \left(1 - \frac{v^2}{c^2}\right)} = 0$$

3.364 In K,

$$\mathbf{B} = b \frac{y\mathbf{i} - x\mathbf{j}}{x^2 + y^2} \quad (\text{where } b = \text{constant})$$

In K',

$$\mathbf{E}' = \mathbf{v} \times \mathbf{B} = bv \frac{y\mathbf{j} - x\mathbf{i}}{x^2 + y^2} = \frac{bv\mathbf{r}}{r^2}$$

The electric field is radial $(\mathbf{r} = x\mathbf{i} + y\mathbf{j})$.

3.365 In K,

$$\mathbf{E} = a\frac{\mathbf{r}}{r^2}, \mathbf{r} = (x\mathbf{i} + y\mathbf{j})$$

In K',

$$\mathbf{B}' = \frac{\mathbf{v} \times \mathbf{E}}{c^2} = \frac{a\mathbf{r} \times \mathbf{v}}{c^2 r^2}$$

The magnetic lines are circular.

3.366 In the non-relativistic limit, we neglect v^2/c^2 and write

$$\left.\begin{aligned} \mathbf{E}'_{\parallel} &= \mathbf{E}_{\parallel} \\ \mathbf{E}'_{\perp} &\cong \mathbf{E}_{\perp} + \mathbf{v} \times \mathbf{B} \end{aligned}\right\} \begin{aligned} \mathbf{B}'_{\parallel} &= \mathbf{B}_{\parallel} \\ \mathbf{B}'_{\perp} &\cong \mathbf{B}_{\perp} - \mathbf{v} \times \mathbf{E}/c^2 \end{aligned}$$

These two equations can be combined to give

$$\mathbf{E}' = \mathbf{E} + \mathbf{v} \times \mathbf{B} \text{ and } \mathbf{B}' = \mathbf{B} - \mathbf{v} \times \mathbf{E}/c^2$$

3.367 Choose $\mathbf{E}$ in the direction of the z-axis, $\mathbf{E} = (0,0,\mathrm{E})$. The frame K' is moving with velocity $\mathbf{v} = (v \sin\alpha,$ $0,\ v\cos\alpha)$, in the x–z plane. Then in the frame K',

$$\mathbf{E}'_{\parallel} = \mathbf{E}_{\parallel}, \mathbf{B}'_{\parallel} = 0$$

$$\mathbf{E}'_{\perp} = \frac{\mathbf{E}_{\perp}}{\sqrt{1 - v^2/c^2}} \quad \mathbf{B}'_{\perp} = \frac{-\mathbf{v} \times \mathbf{E}/c^2}{\sqrt{1 - v^2/c^2}}$$

The vector along $\mathbf{v}$ is $\mathbf{e} = (\sin\alpha, 0, \cos\alpha)$ and the perpendicular vector in the x–z plane is $\mathbf{f} = (-\cos\alpha, 0, \sin\alpha)$.

(a) Thus, using

$$\mathbf{E} = E\cos\alpha\,\mathbf{e} + E\sin\alpha\,\mathbf{f}$$

We get, $$E'_{\parallel} = E\cos\alpha \quad \text{and} \quad E'_{\perp} = \frac{E\sin\alpha}{\sqrt{1-v^2/c^2}}$$

So, $$E' = E\sqrt{\frac{1-\beta^2\cos^2\alpha}{1-\beta^2}} = 9 \text{ kV/m}$$

and $$\tan\alpha' = \frac{\tan\alpha}{\sqrt{1-v^2/c^2}} (\text{where } \beta = v/c)$$

(b) $$\mathbf{B}'_{\parallel} = 0, \mathbf{B}'_{\perp} = \frac{\mathbf{v}\times\mathbf{E}/c^2}{\sqrt{1-v^2/c^2}}$$

$$B' = \frac{\beta E\sin\alpha}{c\sqrt{1-\beta^2}} = 14 \ \mu\text{T}$$

3.368 Choose **B** in the z direction, and the velocity $\mathbf{v} = (v\sin\alpha, 0, v\cos\alpha)$ in the x–z plane, then in the K' frame

$$\mathbf{E}'_{\parallel} = \mathbf{E}_{\parallel} = 0 \qquad \mathbf{B}'_{\parallel} = \mathbf{B}_{\parallel}$$

$$\mathbf{E}'_{\perp} = \frac{\mathbf{v}\times\mathbf{B}}{\sqrt{1-v^2/c^2}} \qquad \mathbf{B}'_{\perp} = \frac{\mathbf{B}_{\perp}}{\sqrt{1-v^2/c^2}}$$

We find similarly, $$E' = \frac{c\beta B\sin\alpha}{\sqrt{1-\beta^2}} = 1.4 \text{ nV/m}$$

$$B' = B\sqrt{\frac{1-\beta^2\cos^2\alpha}{1-\beta^2}} = 0.9 \text{ T}$$

and $$\tan\alpha' = \frac{\tan\alpha}{\sqrt{1-\beta^2}}, \quad \alpha \approx 51^\circ$$

3.369 (a) We see that, $\mathbf{E}'\cdot\mathbf{B}' = \mathbf{E}'_{\parallel}\cdot\mathbf{B}'_{\parallel} + \mathbf{E}'_{\perp}\cdot\mathbf{B}'_{\perp}$

$$= \mathbf{E}_{\parallel}\cdot\mathbf{B}_{\parallel} + \frac{(\mathbf{E}_{\perp}+\mathbf{v}\times\mathbf{B})\cdot\left(\mathbf{B}_{\perp} - \dfrac{\mathbf{v}\times\mathbf{E}}{c^2}\right)}{1-\dfrac{v^2}{c^2}}$$

$$= \mathbf{E}_{\parallel}\cdot\mathbf{B}_{\parallel} + \frac{\mathbf{E}_{\perp}\cdot\mathbf{B}_{\perp} - (\mathbf{v}\times\mathbf{B})\cdot(\mathbf{v}\times\mathbf{E})/c^2}{1-v^2/c^2}$$

$$= \mathbf{E}_{\parallel}\cdot\mathbf{B}_{\parallel} + \frac{\mathbf{E}_{\perp}\mathbf{B}_{\perp} - (\mathbf{v}\times\mathbf{B}_{\perp})\cdot(\mathbf{v}\times\mathbf{E}_{\perp})/c^2}{1-\dfrac{v^2}{c^2}}$$

But, $$(\mathbf{A}\times\mathbf{B})\cdot(\mathbf{C}\times\mathbf{D}) = \mathbf{A}\cdot\mathbf{C} - \mathbf{B}\cdot\mathbf{D} - \mathbf{A}\cdot\mathbf{D}\mathbf{B}\cdot\mathbf{C}$$

So, $$\mathbf{E}'\cdot\mathbf{B}' = \mathbf{E}_{\parallel}\cdot\mathbf{B}_{\parallel} + \mathbf{E}_{\perp}\cdot\mathbf{B}_{\perp}\frac{\left(1-\dfrac{v^2}{c^2}\right)}{1-\dfrac{v^2}{c^2}} = \mathbf{E}\cdot\mathbf{B}$$

(b) We see that, $E'^2 - c^2B'^2$

$$= \left(E'^2_{\parallel} - c^2B'^2_{\parallel}\right) + \left(E'^2_{\perp} - c^2B'^2_{\perp}\right)$$

$$= E^2_{\parallel} - c^2B^2_{\parallel} + \frac{1}{1-\frac{v^2}{c^2}}\left[(\mathbf{E}_{\perp} + \mathbf{v}\times\mathbf{B})^2 - c^2\left(\mathbf{B}_{\perp} - \frac{\mathbf{v}\times\mathbf{E}}{c^2}\right)^2\right]$$

$$= E^2_{\parallel} - c^2B^2_{\parallel} + \frac{1}{1-\frac{v^2}{c^2}}\left[E^2_{\perp} - c^2B^2 + (\mathbf{v}\times\mathbf{B}_{\perp})^2 - \frac{1}{c^2}(\mathbf{v}\times\mathbf{E}_{\perp})^2\right]$$

$$= E^2_{\parallel} - c^2B^2_{\parallel} + \frac{1}{1-\frac{v^2}{c^2}}\left[E^2_{\perp} - c^2B^2_{\perp}\right]\left(1-\frac{v^2}{c^2}\right) = E^2 - c^2B^2$$

(Since, $(\mathbf{v}\times\mathbf{A}_{\perp})^2 = v^2A^2_{\perp}$.)

3.370 In this case, $\mathbf{E}\cdot\mathbf{B} = 0$, as the fields are mutually perpendicular. Also,

$$E^2 - c^2B^2 = -20\times10^8\left(\frac{\text{V}}{\text{m}}\right)^2$$

Thus, we can find a frame in which $E' = 0$, and

$$B' = \frac{1}{c}\sqrt{c^2B^2 - E^2} = B\sqrt{1-\frac{E^2}{c^2B^2}}$$

$$= 0.20\sqrt{1-\left(\frac{4\times10^4}{3\times10^8\times2\times10^{-4}}\right)^2} = 0.15 \text{ mT}$$

3.371 Suppose the charge q moves in the positive direction of the x-axis of the frame K. Let us go over to the moving frame K', at whose origin the charge is at rest. We take the x and x' axes of the two frames to be coincident, and the y-and y'-axes to be parallel.

In the K' frame,

$$\mathbf{E} = \frac{1}{4\pi\varepsilon_0}\frac{q\mathbf{r}'}{r'^3}$$

and this has the following components'

$$E'_x = \frac{1}{4\pi\varepsilon_0}\frac{qx'}{r'^3} \quad \text{and} \quad E'_y = \frac{1}{4\pi\varepsilon_0}\frac{qy'}{r'^3}$$

Now let us go back to the frame K. At the moment when the origins of the two frames coincide, we take $t = 0$. Then,

$$x = r\cos\theta = x'\sqrt{1-\frac{v^2}{c^2}} \quad \text{and} \quad y = r\sin\theta = y'$$

Also, $$E_x = E'_x, E_y = E'_y/\sqrt{1-v^2/c^2}$$

From these equations, $r'^2 = \dfrac{r^2(1-\beta^2\sin^2\theta)}{1-\beta^2}$

$$\mathbf{E}' = \frac{q}{4\pi\varepsilon_0}\frac{1}{r^3(1-\beta^2\sin^2\theta)^{3/2}}\left[(1-\beta^2)^{3/2}\left(x'\mathbf{i} + \frac{y'}{\sqrt{1-\beta^2}}\mathbf{j}\right)\right]$$

$$= \frac{q\mathbf{r}(1-\beta^2)}{4\pi\varepsilon_0 r^3(1-\beta^2\sin^2\theta)^{3/2}}$$

3.7 Motion of Charged Particles in Electric and Magnetic Fields

3.372 Let the electrons leave the negative plate of the capacitor at time $t = 0$

As, $$E_x = -\frac{d\varphi}{dx},\ E = \frac{\varphi}{l} = \frac{at}{l}$$

therefore, the acceleration of the electron is

$$w = \frac{eE}{m} = \frac{eat}{ml}$$

or $$\frac{dv}{dt} = \frac{eat}{ml}$$

or $$\int_0^v dv = \frac{ea}{ml}\int_0^t t\,dt$$

or $$v = \frac{1}{2}\frac{ea}{ml}t^2 \tag{1}$$

But from, $$s = \int v\,dt$$

$$l = \frac{1}{2}\frac{ea}{ml}\int_0^t t^2 dt$$

or $$l = \frac{eat^3}{6\,ml}$$

or $$t = \left(\frac{6\,ml^2}{ea}\right)^{1/3}$$

Putting the value of t in Eq. (1), we get

$$v = \frac{1}{2}\frac{ea}{ml}\left(\frac{6\,ml^2}{ea}\right)^{2/3} = \left(\frac{9}{2}\frac{ale}{m}\right)^{1/3} = 16\text{ km/s}$$

3.373 The electric field inside the capacitor varies with time as, $E = at$. Hence, electric force on the proton, $F = eat$ and subsequently, acceleration of the proton is

$$w = \frac{eat}{m}$$

Now, if t is the time elapsed during motion of the proton between the plates, then $t = l/v_{||}$, as no acceleration is effective in this direction. (Here $v_{||}$ is velocity along the length of the plate.)

From kinematics, $$\frac{dv_\perp}{dt} = w$$

So, $$\int_0^{v_\perp} dv_\perp = \int_0^t w\,dt$$

as initially, the component of velocity in the direction $\perp$ to plates was zero.

or
$$v_{\perp} = \int_0^t \frac{ea}{m} t\,dt = \frac{ea}{2m}\frac{l^2}{v_{\parallel}^2}$$

Now,
$$\tan\alpha = \frac{v_{\perp}}{v_{\parallel}} = \frac{eal^2}{2mv_{\parallel}^3}$$

$$= \frac{eal^2}{2m\left(\frac{2eV}{m}\right)^{\frac{3}{2}}}, \quad \text{as} \quad v_{\parallel} = \left(\frac{2eV}{m}\right)^{\frac{1}{2}} \text{ (from energy conservation)}$$

$$= \frac{al^2}{4}\sqrt{\frac{m}{2eV^3}}$$

3.374 The equation of motion is

$$\frac{dv}{dt} = v\frac{dv}{dx} = \frac{q}{m}(E_0 - ax)$$

Integrating,
$$\frac{1}{2}v^2 - \frac{q}{m}(E_0 x - \frac{1}{2}ax^2) = \text{constant}$$

But initially, $v = 0$ when $x = 0$, so "constant" = 0.

Thus,
$$v^2 = \frac{2q}{m}\left(E_0 x - \frac{1}{2}ax^2\right)$$

Thus $v = 0$, again for $x = x_m = \frac{2E_0}{a}$

The corresponding acceleration is

$$\left(\frac{dv}{dt}\right)_{x_m} = \frac{q}{m}(E_0 - 2E_0) = -\frac{qE_0}{m}$$

3.375 From the law of relativistic conservation of energy

$$\frac{m_0 c^2}{\sqrt{1-(v^2/c^2)}} - e\,Ex = m_0 c^2$$

as the electron is at rest ($v = 0$ for $x = 0$) initially.

Thus clearly,
$$T = eEx$$

On the other hand, $\sqrt{1-(v^2/c^2)} = \dfrac{m_0 c^2}{m_0 c^2 + eEx}$

or
$$\frac{v}{c} = \frac{\sqrt{(m_0 c^2 + eEx)^2 - m_0^2 c^4}}{m_0 c^2 + eEx}$$

or
$$ct = \int c\,dt = \int \frac{(m_0 c^2 + eEx)\,dx}{\sqrt{(m_0 c^2 + eEx)^2 - m_0^2 c^4}}$$

$$= \frac{1}{2\,eE}\int \frac{dy}{\sqrt{y - m_0^2 c^4}} = \frac{1}{eE}\sqrt{(m_0 c^2 + eEx)^2 - m_0^2 c^4} + \text{constant}$$

The "constant" = 0, at $t = 0$, for $x = 0$.

So, $$ct = \frac{1}{eE}\sqrt{(m_0c^2 + eEx)^2 - m_0^2c^4}$$

Finally, using $T = eEx$, we get

$$ce \cdot Et_0 = \sqrt{T(T + 2m_0c^2)}$$

or $$t_0 = \frac{\sqrt{T(T + 2m_0c^2)}}{eEc} = 3.0 \text{ ns}$$

3.376 As before, $T = eEx$. Now in linear motion,

$$\frac{d}{dt}\frac{m_0v}{\sqrt{1 - v^2/c^2}} = \frac{m_0w}{\sqrt{1 - v^2/c^2}} + \frac{m_0w}{(1 - v^2/c^2)^{3/2}}\frac{v}{c^2}w$$

$$= \frac{m_0}{(1 - v^2/c^2)^{3/2}}w = \frac{(T + m_0c^2)^3}{m_0^2c^6}w = eE$$

So, $$w = \frac{eEm_0^2c^6}{(T + m_0c^2)^3} = \frac{eE}{m_0}\left(1 + \frac{T}{m_0c^2}\right)^{-3}$$

3.377 The equations are

$$\frac{d}{dt}\left(\frac{m_0v_x}{\sqrt{1 - (v^2/c^2)}}\right) = 0 \quad \text{and} \quad \frac{d}{dt}\left(\frac{m_0v_y}{\sqrt{1 - v^2/c^2}}\right) = eE$$

Hence, $$\frac{v_x}{\sqrt{1 - v^2/c^2}} = \text{constant} = \frac{v_0}{\sqrt{1 - (v_0^2/c^2)}}$$

Also, by energy conservation

$$\frac{m_0c^2}{\sqrt{1 - (v^2/c^2)}} = \frac{m_0c^2}{\sqrt{1 - (v_0^2/c^2)}} + eEy$$

Dividing, $$v_x = \frac{v_0\varepsilon_0}{\varepsilon_0 + eEy}, \quad \varepsilon_0 = \frac{m_0c^2}{\sqrt{1 - (v_0^2/c^2)}}$$

Also, $$\frac{m_0}{\sqrt{1 - (v^2/c^2)}} = \frac{\varepsilon_0 + eEy}{c^2}$$

Thus, $(\varepsilon_0 + eEy)v_y = c^2eEt + \text{constant}$

This "constant" = 0 as $v_y = 0$ at $t = 0$.

Integrating again,

$$\varepsilon_0 y + \frac{1}{2}eEy^2 = \frac{1}{2}c^2Et^2 + \text{constant}$$

Again, this "constant" = 0, as $y = 0$, at $t = 0$.

Thus, $$(ce\,Et)^2 = (eyE)^2 + 2\varepsilon_0 eEy + \varepsilon_0^2 - \varepsilon_0^2$$

or $$ceEt = \sqrt{(\varepsilon_0 + eEy)^2 - \varepsilon_0^2}$$

or $$\varepsilon_0 + eEy = \sqrt{\varepsilon_0^2 + c^2e^2E^2t^2}$$

Hence, $$v_x = \frac{v_0 \mathcal{E}_0}{\sqrt{\mathcal{E}_0^2 + c^2e^2E^2t^2}} \quad \text{also,} \quad v_y = \frac{c^2eEt}{\sqrt{\mathcal{E}_0^2 + c^2e^2E^2t^2}}$$

and $$\tan\theta = \frac{v_y}{v_x} = \frac{eEt}{m_0 v_0}\sqrt{1-(v_0^2/c^2)}$$

3.378 From the figure,

$$\sin\alpha = \frac{d}{R} = \frac{dqB}{mv}$$

as radius of the arc $R = mv/qB$, where v is the velocity of the particle when it enters into the field. From initial condition of the problem

$$qV = \frac{1}{2}mv^2 \quad \text{or} \quad v = \sqrt{\frac{2qV}{m}}$$

Hence, $$\sin\alpha = -\frac{dqB}{m\sqrt{\frac{2qV}{m}}} = dB\sqrt{\frac{q}{2\,mV}}$$

and $$\alpha = \sin^{-1}\left(dB\sqrt{\frac{q}{2\,mV}}\right) = 30° \text{ (on substituting values)}$$

3.379 (a) For motion along circle, the magnetic force acted on the particle will provide the centripetal force necessary for its circular motion,

i.e., $$\frac{mv^2}{r} = evB \quad \text{or} \quad v = \frac{eBr}{m} = 100 \text{ km/s}$$

and the period of revolution, $T = \dfrac{2\pi}{\omega} = \dfrac{2\pi r}{v} = \dfrac{2\pi m}{eB} = 6.5\ \mu\text{s}$

(b) Generally, $\dfrac{d\mathbf{p}}{dt} = \mathbf{F}$

But, $$\frac{d\mathbf{p}}{dt} = \frac{d}{dt}\frac{m_0\mathbf{v}}{\sqrt{1-(v^2/c^2)}} = \frac{m_0\dot{\mathbf{v}}}{\sqrt{1-(v^2/c^2)}} + \frac{m_0}{(1-(v^2/c^2))^{3/2}}\frac{\mathbf{v}(\mathbf{v}\cdot\dot{\mathbf{v}})}{c^2}$$

For transverse motion, $\mathbf{v}\cdot\dot{\mathbf{v}} = 0$ so,

$$\frac{d\mathbf{p}}{dt} = \frac{m_0\dot{\mathbf{v}}}{\sqrt{1-(v^2/c^2)}} = \frac{m_0}{\sqrt{1-(v^2/c^2)}}\frac{v^2}{r}$$

Thus, $$\frac{m_0v^2}{r\sqrt{1-v^2/c^2}} = Bev \quad \text{or} \quad \frac{v/c}{\sqrt{1-(v^2/c^2)}} = \frac{Ber}{m_0c}$$

or $$\frac{v}{c} = \frac{Ber}{\sqrt{B^2e^2r^2 + m_0^2c^2}}$$

Finally, $$T = \frac{2\pi r}{v} = \frac{2\pi m_0}{eB\sqrt{1-v^2/c^2}} = \frac{2\pi}{cBe}\sqrt{B^2e^2r^2 + m_0^2c^2} = 4.1 \text{ ns}$$

3.380 (a) As before, $$p = B\,qr$$

(b) $$T = \sqrt{c^2p^2 + m_0^2c^4} = \sqrt{c^2B^2q^2r^2 + m_0^2c^4}$$

(c) $$w = \frac{v^2}{r} = -\frac{c^2}{r[1 + (m_0c/Bqr)^2]}$$

using the result for v from the previous problem.

3.381 From solution of Problem 3.279,

$$T = \frac{2\pi\varepsilon}{c^2eB}(\text{relativistic}), T_0 = \frac{2\pi m_0c^2}{c^2eB} \text{ (non-relativistic)}$$

Here, $$m_0c^2/\sqrt{1 - v^2/c^2} = \varepsilon$$

Thus, $$\delta T = \frac{2\pi T}{c^2\,eB}, (T = \text{K.E.})$$

Now, $$\frac{\delta T}{T_0} = \eta = \frac{T}{m_0c^2} \quad \text{so,} \quad T = \eta\, m_0c^2$$

3.382 The given potential difference is not large enough to cause significant deviations from the non-relativistic formula. So,

$$T = eV = \frac{1}{2}mv^2$$

Thus, $$v = \sqrt{\frac{2eV}{m}}$$

So, $$v_{\parallel} = \sqrt{\frac{2eV}{m}}\cos\alpha \quad v_{\perp} = \sqrt{\frac{2eV}{m}}\sin\alpha$$

Now, $$\frac{mv_{\perp}^2}{r} = Bev_{\perp} \quad \text{or} \quad r = \frac{mv_{\perp}}{Be}$$

and $$T = \frac{2\pi r}{v_{\perp}} = \frac{2\pi m}{Be}$$

Pitch $$p = v_{\parallel}T = \frac{2\pi m}{Be}\sqrt{\frac{2eV}{m}}\cos\alpha = 2\pi\sqrt{\frac{2mV}{eB^2}}\cos\alpha = 2.0 \text{ cm}$$

3.383 The charged particles will traverse a helical trajectory and will be focused on the axis after traversing a number of turns. Thus,

$$\frac{1}{v_0} \approx n\cdot\frac{2\pi m}{qB_1} = (n+1)\frac{2\pi m}{qB_2}$$

So, $$\frac{n}{B_1} = \frac{n+1}{B_2} = \frac{1}{B_2 - B_1}$$

Hence, $$\frac{1}{v_0} = \frac{2\pi m}{q(B_2 - B_1)}$$

or
$$\frac{l^2}{2qV/m} = \frac{(2\pi)^2}{(B_2 - B_1)^2} \times \frac{1}{(q/m)^2}$$

or
$$\frac{q}{m} = \frac{8\pi^2 V}{l^2(B_2 - B_1)^2}$$

3.384 Let us take the point A as origin O and the axis of solenoid as z-axis. At an arbitrary moment of time let us resolve the velocity of electron into its two rectangular components, $\mathbf{v}_{||}$ along the axis and $\mathbf{v}_\perp$ perpendicular to the axis of solenoid. We know the magnetic force does no work, so the kinetic energy as well as the speed of the electron $|\mathbf{v}_\perp|$ will remain constant in the x-y plane. Thus $\mathbf{v}_\perp$ can change only its direction as shown in the problem figure, $\mathbf{v}_{||}$ will remain constant as it is parallel to $\mathbf{B}$. Thus at time $= t$,

$$v_x = v_\perp \cos\omega t = v \sin\alpha \cos\omega t$$

$$v_y = v_\perp \sin\omega t = v \sin\alpha \sin\omega t$$

and
$$v_z = v\cos\alpha \text{ (where } \omega = eB/m)$$

As at $t = 0$, we have $x = y = z = 0$, so the equation of motion of the electron is

$$z = v\cos\alpha t$$

$$x = \frac{v\sin\alpha}{\omega}\sin\omega t$$

$$y = \frac{v\sin\alpha}{\omega}(\cos\omega t - 1)$$

the equation of the helix.

On the screen $z = l$, so $t = \dfrac{l}{v\cos\alpha}$

Then,
$$r^2 = x^2 + y^2 = \frac{2v^2\sin^2\alpha}{\omega^2}\left(1 - \cos\frac{\omega l}{v\cos\alpha}\right)$$

$$r = \frac{2v\sin\alpha}{\omega}\left|\sin\frac{\omega l}{2v\cos\alpha}\right| = 2\frac{mv}{eB}\sin\alpha\left|\sin\frac{leB}{2mv\cos\alpha}\right|$$

3.385 Choose the wire along the z-axis, and the initial direction of the electron, along the x-axis. Then the magnetic field in the x-z plane is along the y-axis and outside the wire it is

$$B = B_y = \frac{\mu_0 I}{2\pi x} \quad (\text{as } B_x = B_z = 0, \text{ if } y = 0)$$

The motion must be confined to the x-z plane. Then the equations of motion are

$$\frac{d}{dt}mv_x = -ev_x B_y$$

$$\frac{d(mv_z)}{dt} = +ev_x B_y$$

Multiplying the first equation by v_x and the second by v_z and then adding, we get

$$v_x\frac{dv_x}{dt} + v_z\frac{dv_z}{dt} = 0$$

or $$v_x^2 + v_z^2 = v_0^2, \text{ say} \quad \text{or} \quad v_z = \sqrt{v_0^2 - v_x^2}$$

Then, $$v_x \frac{dv_x}{dx} = -\frac{e}{m}\sqrt{v_0^2 - v_x^2}\,\frac{\mu_0 I}{2\pi x}$$

or $$\frac{v_x dv_x}{\sqrt{v_0^2 - v_x^2}} = \frac{\mu_0 Ie}{2\pi m}\frac{dx}{x}$$

Integrating, $$\sqrt{v_0^2 - v_x^2} = \frac{\mu_0 Ie}{2\pi m}\ln\frac{x}{a}$$

on using, $v_x = v_0$, if $x = a$ (i.e., initially).

Now, $$v_x = 0, \quad \text{when} \quad x = x_m$$

So, $$x_m = ae^{v_0/b}, \quad \text{where} \quad b = \frac{\mu_0 Ie}{2\pi m}$$

3.386 Inside the capacitor, the electric field follows the $1/r$ law, and so the potential can be written as

$$\varphi = \frac{V \ln r/a}{\ln b/a} \quad \text{and} \quad E = \frac{-V}{\ln b/a}\frac{1}{r},$$

Here r is the distance from the axis of the capacitor.

Also, $$\frac{mv^2}{r} = \frac{qV}{\ln b/a}\frac{1}{r}$$

or $$mv^2 = \frac{qV}{\ln b/a}$$

On the other hand, $mv = qBr$ in the magnetic field.

Thus, $$v = \frac{V}{Br \ln b/a} \quad \text{and} \quad \frac{q}{m} = \frac{v}{Br} = \frac{V}{B^2 r^2 \ln(b/a)}$$

3.387 (a) The equations of motion are

$$m\frac{dv_x}{dt} = -qBv_z, \quad m\frac{dv_y}{dt} = qE \quad \text{and} \quad m\frac{dv_z}{dt} = qv_x B$$

These equations can be solved easily.

First, $$v_y = \frac{qE}{m}t, \; y = \frac{qE}{2m}t^2$$

Then, $$v_x^2 + v_z^2 = \text{constant} = v_0^2 \text{ (as before)}$$

In fact, $v_x = v_0 \cos\omega t$ and $v_z = v_0 \sin\omega t$ as one can check.

Integrating again and using $x = z = 0$, at $t = 0$, we get

$$x = \frac{v_0}{\omega}\sin\omega t, \quad z = \frac{v_0}{\omega}(1 - \cos\omega t)$$

Thus, $$x = z = 0 \quad \text{for} \quad t = t_n = n\frac{2\pi}{\omega}$$

At that instant, $$y_n = \frac{qE}{2m} \times \frac{2\pi}{qB/m} \times n^2 \times \frac{2\pi}{qB/m} = \frac{2\pi^2 mEn^2}{qB^2}$$

(b) Also, $\tan \alpha_n = \frac{v_x}{v_y}$ ($v_z = 0$ at this moment)

$$= \frac{mv_0}{qEt_n} = \frac{mv_0}{qE} \times \frac{qB}{m} \times \frac{1}{2\pi n} = \frac{Bv_0}{2\pi En}$$

3.388 The equation of the trajectory is

$$x = \frac{v_0}{\omega}\sin \omega t,\ z = \frac{v_0}{\omega}(1 - \cos \omega t),\ y = \frac{qE}{2m}t^2 \text{ (as in problem 3.384)}$$

Now on the screen $x = l$, so

$$\sin \omega t = \frac{\omega l}{v_0} \quad \text{or} \quad \omega t = \sin^{-1}\frac{\omega l}{v_0}$$

At that moment,

$$y = \frac{qE}{2m\omega^2}\left(\sin^{-1}\frac{\omega l}{v_0}\right)^2$$

So,

$$\frac{\omega l}{v_0} = \sin\sqrt{\frac{2m\omega^2 y}{qE}} = \sin\sqrt{\frac{2qB^2 y}{Em}}$$

and

$$z = \frac{v_0}{\omega} 2\sin^2\frac{\omega t}{2} = l\tan\frac{\omega t}{2}$$

$$= l\tan\frac{1}{2}\left[\sin^{-1}\frac{\omega l}{v_0}\right] = l\tan\sqrt{\frac{qB^2 y}{2mE}}$$

For small z,

$$\frac{qB^2 y}{2mE} = \left(\tan^{-1}\frac{z}{l}\right)^2 \approx \frac{z^2}{l^2}$$

or

$$y = \frac{2mE}{qB^2l^2}\cdot z^2$$

This is the equation of a parabola.

3.389 In crossed field, $eE = evB$ so $v = \frac{E}{B}$

Then force exerted on the plate, $F = \frac{I}{e} \times m\frac{E}{\mathbf{B}} = \frac{mIE}{eB} = 20\ \mu\text{N}$

3.390 When the electric field is switched off, the path followed by the particle will be helical and pitch.

$$\Delta l = v_{||}T$$

(where $v_{||}$ is the velocity of the particle, parallel to **B** and T the time period of revolution).

$$= v\cos(90 - \varphi)T = v\sin\varphi\ T$$

$$= v\sin\varphi\frac{2\pi m}{qB} \quad \left(\text{as } T = \frac{2\pi}{qB}\right) \tag{1}$$

Now, when both the fields were present, $qE = qvB\sin(90 - \varphi)$, as no net force was effective on the system.

So,

$$v = \frac{E}{B\cos\varphi} \tag{2}$$

From Eqs. (1) and (2),

$$\Delta l = \frac{E}{B}\frac{2\pi m}{qB}\tan\varphi = 6\ \text{cm}$$

3.391 When there is no deviation,

$$-q\mathbf{E} = q\,(\mathbf{v} \times \mathbf{B})$$

or in scalar from, $$E = vB(\text{as } \mathbf{v} \perp \mathbf{B}) \quad \text{or} \quad v = \frac{E}{B} \tag{1}$$

Now, when the magnetic field is switched on, let the deviation in the field be x. Then,

$$x = \frac{1}{2}\left(\frac{qvB}{m}\right)t^2$$

where t is the time required to pass through this region.

Also, $$t = \frac{a}{v}$$

Thus, $$x = \frac{1}{2}\left(\frac{qvB}{m}\right)\left(\frac{a}{v}\right)^2 = \frac{1}{2}\frac{q}{m}\frac{a^2B^2}{E} \tag{2}$$

For the region where the field is absent, velocity in the upward direction

$$= \left(\frac{qvB}{m}\right)t = \frac{q}{m}aB \tag{3}$$

Now, $$\Delta x - x = \frac{qaB}{m}t'$$

$$= \frac{qa}{m}\frac{B^2b}{E}, \text{ where } t' = \frac{b}{v} = \frac{bB}{E} \tag{4}$$

From Eqs. (2) and (4), $$\Delta x - \frac{1}{2}\frac{q}{m}\frac{a^2B^2}{E} = \frac{q}{m}\frac{aB^2b}{E}$$

or $$\frac{q}{m} = \frac{2E\Delta x}{aB^2(a+2b)}$$

3.392 (a) The equation of motion is

$$m\frac{d^2\mathbf{r}}{dt^2} = q(\mathbf{E} + \mathbf{v} \times \mathbf{B})$$

Now, $$\mathbf{v} \times \mathbf{B} = \begin{vmatrix} \mathbf{i} & \mathbf{j} & \mathbf{k} \\ \dot{x} & \dot{y} & \dot{z} \\ 0 & 0 & B \end{vmatrix} = \mathbf{i}B\dot{y} - \mathbf{j}B\dot{x}$$

So, the equation becomes,

$$\frac{dv_x}{dt} = \frac{qBv_y}{m}; \quad \frac{dv_y}{dt} = \frac{qE}{m} - \frac{qB}{m}v_x \quad \text{and} \quad \frac{dv_z}{dt} = 0$$

(Here, $v_x = \dot{x}$, $v_y = \dot{y}$, $v_z = \dot{z}$).

The last equation is easy to integrate as v_z = constant = 0, since v_z is zero initially. Thus integrating again, z = constant = 0, and motion is confined to the x–y plane.

Method 1:

Let us put $$\xi = v_x + iv_y$$

or
$$\begin{aligned}\dot{\xi} &= \frac{dv_x}{dt} + i\frac{dv_y}{dt}\\ &= \frac{qB}{m}v_y + i\left(\frac{qE}{m} - \frac{qB}{m}v_x\right)\\ &= \omega v_y + i\left(\frac{\omega E}{B} - \omega v_x\right) \left(\text{where } \omega = \frac{qB}{m}\right)\\ &= \omega v_y + i\omega\frac{E}{B} - i\omega v_x\\ &= i\omega\frac{E}{B} - i\omega(v_x + iv_y) \text{ (because } i^2 = -1)\end{aligned}$$

This equation after being multiplied by $e^{i\omega t}$ can be rewritten as
$$\frac{d}{dt}(\xi e^{i\omega t}) = i\,\omega e^{i\omega t}\,\frac{E}{B}$$

and integrated to give,
$$\xi = \frac{E}{B} + Ce^{-i\omega t - i\alpha}$$

where C and α are two real constants. Taking real and imaginary parts,
$$v_x = \frac{E}{B} + C\cos(\omega t + \alpha) \quad \text{and} \quad v_y = -C\sin(\omega t + \alpha)$$

Since $v_y = 0$, when $t = 0$, we can take $\alpha = 0$, then $v_x = 0$ at $t = 0$ gives,
$$C = -\frac{E}{B}$$

and we get,
$$v_x = \frac{E}{B}(1 - \cos\omega t) \quad \text{and} \quad v_y = \frac{E}{B}\sin\omega t$$

Integrating again and using $x = y = 0$, at $t = 0$, we get
$$x(t) = \frac{E}{B}\left(t - \frac{\sin\omega t}{\omega}\right), y(t) = \frac{E}{\omega B}(1 - \cos\omega t)$$

This is the equation of a cycloid.

Method 2:

We have,
$$\dot{v}_x = \frac{qB}{m}v_y \quad \text{and} \quad \dot{v}_y = \frac{qE}{m} - \frac{qB}{m}v_x$$

or
$$\dot{v}_x = \omega v_y \quad \text{and} \quad \dot{v}_y = \omega\left(\frac{E}{B} - v_x\right)$$

After differentiating $\dot{v}_y$ with respect to time, we get
$$\ddot{v}_y = -\omega\dot{v}_x = -\omega^2 v_y$$

At $t = 0$, $v_y = 0$, so its solution becomes

$v_y = v_{ym}\sin\omega t$ (where v_{ym} is velocity amplitude)

At $t = 0$, $$\dot{v}_y = \frac{qE}{m}$$

So, $$\omega v_{ym} = \frac{\omega E}{B} \Rightarrow v_{ym} = \frac{E}{B}$$

Hence, $$v_y = \frac{E}{B}\sin \omega t$$

because at $t = 0$, $x = y = 0$.

Now differentiating $\dot{v}_x = \omega v_y$ with respect to time, we get

$$\ddot{v}_x = \omega \dot{v}_y = \omega^2[E/B - v_x]$$

or $$\ddot{v}_x + \omega^2 v_x = \omega^2 E \qquad (1)$$

This equation is the form $\ddot{x} + \omega^2 x = A$

(where A is any constant).

or $$\ddot{x} + \omega^2\left(x - \frac{A}{\omega^2}\right) = 0$$

Putting $$x' = x - \frac{A}{\omega^2}$$

and $\ddot{x}' = \ddot{x}$ in the solutions

We get $$x' = a\cos(\omega t + \delta)$$

or $$x - \frac{A}{\omega^2} = a\cos(\omega t + \delta)$$

or $$x = a\cos(\omega t + \delta) + \frac{A}{\omega^2}$$

Thus the solution of Eq. (1) is

$$v_x = v_{xm}\cos(\omega t + \delta) + \frac{E}{B}$$

On differentiation

$$\dot{v}_x = -\omega v_{xm}\sin(\omega t + \delta)$$

At $t = 0$, $\dot{v}_x = 0$, $\delta = 0$, hence

$$v_x = v_{xm}\cos \omega t + \frac{E}{B}$$

Again at $t = 0$,

$$v_x = 0, \quad \text{so,} \quad v_{xm} = -\frac{E}{B}$$

or $$v_x = \frac{E}{B}[1 - \cos \omega t]$$

Alternate:

Suppose we have two inertial systems of reference: the system K and the system K' moving relative to the first system at a velocity $\mathbf{v}_0$. We know the magnitudes of the fields **E** and **B** at a certain point in space and time in the system K. If $v_0 << c$ the fields $\mathbf{E}'$ and $\mathbf{B}'$ at the same point in space and time in the system $\mathbf{E}'$ are given by

$$\mathbf{E} = \mathbf{E} + (v_0 \times \mathbf{B}),\ \mathbf{B}' = \mathbf{B} - (v_0 \times \mathbf{E})/c^2$$

The particle moves under the action of the Lorentz force. It can be easily seen that the particle always remains in the plane xy. Its motion can be described most easily in a certain system K' where only the magnetic field is present. Let us find this reference system. It follows from transformation $\mathbf{E}' = \mathbf{E} + (\mathbf{v}_0 \times \mathbf{B})$, $\mathbf{E}' = 0$ in a reference system that moves with a velocity $\mathbf{v}_0$ satisfying the relation $\mathbf{E} = -(\mathbf{v}_0 \times \mathbf{B})$. It is more convenient to choose the system K' whose velocity v_0 is directed towards the positive values on the x-axis, since in such system the particle will move perpendicular to vector $\mathbf{B}'$ and its motion will be the simplest.

Thus in the system K' that moves to the right at a velocity $v_0 = E/B$, the field $\mathbf{E}' = 0$ and only the field $\mathbf{B}'$ is observed. In accordance with transformation $\mathbf{B}' = \mathbf{B} - (v_0 \times \mathbf{E})/c^2$ and figure, we have

$$\mathbf{B}' = \mathbf{B} - (v_0 \times \mathbf{E})/c^2 = \mathbf{B}(1 - v_0^2/c^2)$$

Since for a non-relativistic particle $v_0 << c$, we can assume that $\mathbf{B}' = \mathbf{B}$.

In the system K', the particle will move only in the magnetic field, it's velocity being perpendicular to this field. The equation of motion for this particle in the system K' will have the form

$$\frac{mv_0^2}{R} = qv_0 B$$

This equation is written for the instant $t = 0$, where the particle moved in the system K' as is shown in Fig. (a). Since the Lorentz force $\mathbf{F}$ is always perpendicular to the velocity of the particle, $v_0 =$ constant, and it follows from above equation that in the system K' the particle will move in a circle of radius

$$R = \frac{mv_0}{qB}$$

Thus the particle moves uniformly with the velocity v_0 in a circle in the system K', which, in turn, moves uniformly to the right with the same velocity $v_0 = E/B$. This motion can be compared with the motion of the point q at the rim of a wheel Fig. (b) rolling with the angular velocity $\omega = v_0/R = qB/m$.

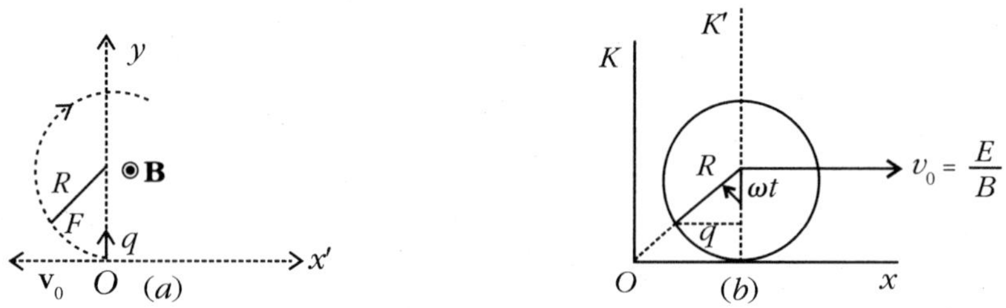

Fig. (b) readily shows the coordinates of the particle q at the instant t are given by

$$x = v_0 t - R\sin\omega t = a(\omega t - \sin\omega t)$$

$$y = R - R\cos\omega t = a(1 - \cos\omega t)$$

where $$a = \frac{mE}{qB^2} \quad \text{and} \quad w = \frac{qB}{m}$$

(b) The velocity is zero, when $\omega t = 2n\pi$. We see that

$$v^2 = v_x^2 + v_y^2 = \left(\frac{E}{B}\right)^2 (2 - 2\cos\omega t)$$

or $$v = \frac{ds}{dt} = \frac{2E}{B}\left|\sin\frac{\omega t}{2}\right|$$

The quantity inside the modulus is positive for $0 < \omega t < 2\pi$. Thus we can drop the modulus and write for the distance traversed between two successive zeroes of velocity as

$$S = \frac{4E}{\omega B}\left(1 - \cos\frac{\omega t}{2}\right)$$

Putting $\omega t = 2\pi$, we get

$$S = \frac{8E}{\omega B} = \frac{8mE}{qB^2}$$

(c) The drift velocity is in the x-direction and has the magnitude

$$<v_x> = \left\langle \frac{E}{B}(1-\cos\omega t)\right\rangle = \frac{E}{B}$$

3.393 When a current I flows along the axis, a magnetic field $B_\varphi = \frac{\mu_0 I}{2\pi\rho}$ is set up, where $\rho^2 = x^2 + y^2$. In terms of components,

$$B_x = \frac{\mu_0 I y}{2\pi\rho^2}, B_y = \frac{\mu_0 I x}{2\pi\rho^2} \quad \text{and} \quad B_z = 0$$

Suppose a potential difference V is set up between the inner cathode and the outer anode. This means a potential function of the form

$$\varphi = V\frac{\ln\rho/b}{\ln a/b} \quad (a > \rho > b)$$

as one can check by solving Laplace equation.

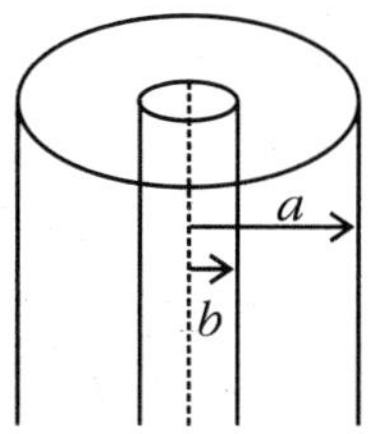

The electric field corresponding to this is

$$E_x = -\frac{Vx}{\rho^2\ln a/b}, \; E_y = -\frac{Vy}{\rho^2\ln a/b}, \; E_z = 0$$

The equations of motion are

$$\frac{d}{dt}mv_x = +\frac{|e|Vx}{\rho^2\ln a/b} + \frac{|e|\mu_0 I}{2\pi\rho^2}x\dot{z}$$

$$\frac{d}{dt}mv_y = +\frac{|e|Vy}{\rho^2\ln a/b} + \frac{|e|\mu_0 I}{2\pi\rho^2}y\dot{z}$$

and

$$\frac{d}{dt}mv_z = -|e|\frac{\mu_0 I}{2\pi\rho^2}(x\dot{x} + y\dot{y}) = |e|\frac{\mu_0 I}{2\pi}\frac{d}{dt}\ln\rho$$

$(-|e|)$ is the charge on the electron.

Integrating the last equations, we get

$$mv_z = -|e|\frac{\mu_0 I}{2\pi}\ln\frac{\rho}{a} = m\dot{z}$$

Since $v_z = 0$ when $\rho = a$, we now substitute this $\dot{z}$ in the other two equations to get

$$\frac{d}{dt}\left(\frac{1}{2}mv_x^2 + \frac{1}{2}mv_y^2\right)$$

$$= \left[\frac{|e|V}{\ln a/b} - \frac{|e|^2}{m}\left(\frac{\mu_0 I}{2\pi}\right)^2 \ln\frac{\rho}{b}\right]\cdot\frac{x\dot{x}+y\dot{y}}{\rho^2}$$

$$= \left[\frac{|e|V}{\ln a/b} - \frac{|e|^2}{m}\left(\frac{\mu_0 I}{2\pi}\right)^2 \ln\frac{\rho}{b}\right]\cdot\frac{1}{2\rho^2}\frac{d}{dt}\rho^2$$

$$= \left[\frac{|e|V}{\ln a/b} - \frac{|e|^2}{m}\left(\frac{\mu_0 I}{2\pi}\right)^2 \ln\frac{\rho}{a}\right]\frac{d}{dt}\ln\frac{\rho}{b}$$

Integrating and using $v^2 = 0$, at $\rho = b$, we get,

$$\frac{1}{2}mv^2 = \left[\frac{|e|V}{\ln a/b}\ln\frac{\rho}{a} - \frac{1}{2m}|e|^2\left(\frac{\mu_0 I}{2\pi}\right)^2\left(\ln\frac{\rho}{b}\right)\right]$$

The RHS must be positive, for all $a > \rho > b$. The condition for this is

$$V \geq \frac{1}{2}\frac{|e|}{m}\left(\frac{\mu_0 I}{2\pi}\right)^2 \ln\frac{a}{b}$$

3.394 This differs from the previous problem in $(a \leftrightarrow b)$ and the magnetic field is along the z-direction. Thus $B_x = B_y = 0,\ B_z = B$.

Assuming as usual the charge of the electron to be $-|e|$, we write the equation of motion as

$$\frac{d}{dt}mv_x = \frac{|e|V_x}{\rho^2 \ln b/a} - |e|B\dot{y},\quad \frac{d}{dt}mv_y = \frac{|e|V_y}{\rho^2 \ln b/a} + |e|B\dot{x}$$

and $$\frac{d}{dt}mv_z = 0 => z = 0$$

The motion is confined to the plane $z = 0$. Eliminating B from the first two equations,

we get $$\frac{d}{dt}\left(\frac{1}{2}mv^2\right) = \frac{|e|V}{\ln b/a}\frac{x\dot{x} + y\dot{y}}{\rho^2}$$

or $$\frac{1}{2}mv^2 = |e|V\frac{\ln \rho/a}{\ln b/a}$$

So, as expected, since magnetic forces do not work

$$v = \sqrt{\frac{2|e|V}{m}} \quad (\text{at } \rho = b)$$

On the other hand, eliminating V, we also get

$$\frac{d}{dt}m(xv_y - yv_x) = |e|B(x\dot{x} + y\dot{y})$$

i.e., $$(xv_y - yv_x) = \frac{|e|B}{2m}\rho^2 + \text{constant}$$

The constant is easily evaluated, since v is zero at $\rho = a$. Thus,

$$(xv_y - yv_x) = \frac{|e|B}{2m}(\rho^2 - a^2) > 0$$

At $\rho = b$, $(xv_y - yv_x) \le vb$. Thus,

$$vb \ge \frac{|e|B}{2m}(b^2 - a^2)$$

or $$B \le \frac{2mb}{b^2 - a^2}\sqrt{\frac{2|e|V}{m}} \times \frac{1}{|e|}$$

or $$B \le \frac{2b}{b^2 - a^2}\sqrt{\frac{2mV}{|e|}}$$

3.395 The equations are as in soultion of Problem 3.392.

$$\frac{dv_x}{dt} = \frac{qB}{m}v_y,\ \frac{dv_y}{dt} = \frac{qE_m}{m}\cos\omega t - \frac{qB}{m}v_x \quad \text{and} \quad \frac{dv_z}{dt} = 0$$

with $$\omega = \frac{qB}{m},\ \xi = v_x + iv_y$$

We get, $$\frac{d\xi}{dt} = i\frac{E_m}{B}\omega\cos\omega t - i\omega\xi$$

Or multiplying by e^{iwt}, we get

$$\frac{d}{dt}(\xi e^{i\omega t}) = i\frac{E_m}{2B}\omega(e^{2i\omega t} + 1)$$

Integrating, $$\xi e^{i\omega t} = \frac{E_m}{4B}e^{2i\omega t} + \frac{E_m}{2B}i\omega t$$

or $$\xi = \frac{E_m}{4B}(e^{i\omega t} + 2i\,\omega t e^{i\omega t}) + Ce^{i\omega t}$$

Since $$\xi = 0 \quad \text{at} \quad t = 0,\; C = -\frac{E_m}{4B}$$

Thus, $$\xi = i\frac{E_m}{2B}\sin\omega t + i\frac{E_m}{2B}\omega t\, e^{i\omega t}$$

or $$v_x = \frac{E_m}{2B}\omega t \sin\omega t \quad \text{and} \quad v_y = \frac{E_m}{2B}\sin\omega t + \frac{E_m}{2B}\omega t\cos\omega t$$

Integrating again,

$$x = \frac{a}{2\omega^2}(\sin\omega t - \omega t\cos\omega t),\; y = \frac{a}{2\omega}t\sin\omega t$$

where $a = qE_m/m$, and we have used $x = y = 0$, at $t = 0$.

The trajectory is an unwinding spiral.

3.396 We know that for a charged particle (proton) in a magnetic field,

$$\frac{mv^2}{r} = Bev \quad \text{or} \quad mv = Ber$$

But, $$\omega = \frac{eB}{m}$$

Thus, $$E = \frac{1}{2}mv^2 = \frac{1}{2}m\omega^2 r^2$$

So, $$\Delta E = m\omega^2 r\Delta r = 4\pi\nu^2\, mr\,\Delta r$$

On the other hand $\Delta E = 2eV$, where V is the effective acceleration voltage across the dees, there being two crossings per revolution.

So, $$V \geq 2\pi^2\nu^2 mr\,\Delta r/e$$

$$= 0.10 \text{ MV}$$

3.397 (a) From $$\frac{mv^2}{r} = Bev \quad \text{or} \quad mv = Ber$$

and $$T = \frac{(Ber)^2}{2m} = \frac{1}{2}mv^2 = 12 \text{ MeV}$$

(b) From $$\frac{2\pi}{\omega} = \frac{2\pi r}{v}$$

We get, $$f_{\min} = \frac{v}{2\pi r} = \frac{1}{\pi r}\sqrt{\frac{T}{2m}} = 20 \text{ MHz}$$

3.398 (a) The total time of acceleration is

$$t = \frac{1}{2\nu}\cdot n$$

where n is the number of passages of the dees.

But, $$T = neV = \frac{B^2e^2r^2}{2m}$$

or $$n = \frac{B^2er^2}{2mV}$$

So, $$t = \frac{\pi}{eB/m} \times \frac{B^2er^2}{2mV} = \frac{\pi Br^2}{2V} = \frac{\pi^2 m\nu r^2}{eV} = 17\ \mu\text{s}$$

(b) The distance covered is

$$s = \sum \nu_n \frac{1}{2\nu}$$

But, $$v_n = \sqrt{\frac{2eV}{m}}\sqrt{n}$$

So, $$s = \sqrt{\frac{eV}{2m\nu^2}}\sum\sqrt{n} = \sqrt{\frac{eV}{2m\nu^2}}\int\sqrt{n}\,dn = \sqrt{\frac{eV}{2m\nu^2}}\frac{2}{3}n^{3/2}$$

But, $$n = \frac{B^2e^2r^2}{2eVm} = \frac{2\pi^2 m\nu^2 r^2}{eV}$$

Thus, $$s \approx \frac{4\pi^3\nu^2 mr^2}{3eV} = 0.74 \text{ km}$$

3.399 In the nth orbit,

$$\frac{2\pi r_n}{v_n} = nT_0 = \frac{n}{\nu}$$

We ignore the rest mass of the electron and write $v_n \approx c$. Also $W \approx cp = cBer_n$.

Thus, $$\frac{2\pi W}{Bec^2} = \frac{n}{\nu}$$

or $$n = \frac{2\pi W\nu}{Bec^2} = 9$$

3.400 The basic condition is the relativistic equation

$$\frac{mv^2}{r} = Bqv \quad \text{or} \quad mv = \frac{m_0 v}{\sqrt{1-v^2/c^2}} = Bqr$$

Using, $$\omega = \frac{Bq}{m}$$

We get, $$\omega = \frac{\omega_0}{\sqrt{1+\dfrac{\omega_0^2 r^2}{c^2}}} \quad \text{and} \quad \omega_0 = \frac{Bq}{m_0}r$$

The time of acceleration is

$$t = \sum_{n=1}^{N}\frac{1}{2\nu_n} = \sum_{n=1}^{N}\frac{\pi}{\omega_n} = \sum_{n}^{N}\frac{\pi W_n}{qBc^2}$$

N is the number of crossings of either Dee.

But,

$$W_n = m_0c^2 + \frac{n\Delta W}{2}$$

there being two crossings of the Dees per revolution.

So, $$t = \sum\frac{\pi m_0 c^2}{qBc^2} + \sum\frac{\pi\Delta W_n}{2qBc^2}$$

$$= N\frac{\pi}{\omega_0} + \frac{N(N+1)}{4}\frac{\pi\Delta W}{qBc^2} \cong N^2\frac{\pi\Delta W}{4qBc^2} \quad (N >> 1)$$

Also, $$r = r_N \frac{v_N}{\omega_N} \approx \frac{c}{\pi}\frac{\partial t}{\partial N} \cong \frac{\Delta W}{2qBc} N$$

Hence finally, $$\omega = \frac{\omega_0}{\sqrt{1 + \frac{q^2B^2}{m_0^2c^2} \times \frac{\Delta W^2}{4q^2B^2c^2} N^2}}$$

$$= \frac{\omega_0}{\sqrt{1 + \frac{(\Delta W)^2}{4m_0^2c^4} \times \frac{4qBc^2}{\pi \Delta W} t}} = \frac{\omega_0}{\sqrt{1 + at}}$$

$$a = \frac{qB\Delta W}{\pi m_0^2 c^2}$$

3.401 When the magnetic field is set up in the solenoid, an electric field is induced in it, which will accelerate the charged particle. If B is the rate at which the magnetic field is increasing, then

$$\pi r^2 \dot{B} = 2\pi r E \quad \text{or} \quad E = \frac{1}{2} r\dot{B}$$

Thus, $$m\frac{dv}{dt} = \frac{1}{2} r\dot{B}q \quad \text{or} \quad v = \frac{qBr}{2m}$$

After the field is set up, the particle will execute a circular motion of radius ρ, where

$$mv = Bq\rho \quad \text{or} \quad \rho = \frac{1}{2} r$$

3.402 The increment in energy per revolution is $e\Phi$, so the number of revolutions is

$$N = \frac{W}{e\Phi} = 5 \times 10^6 \text{ revolutions}$$

The distance traversed is $s = 2\pi r N = 8 \times 10^3$ km.

3.403 We know that,

$$\frac{dp}{dt} \aleph eE \quad \frac{e}{2\pi r}\frac{d\Phi}{dt} \quad \frac{e}{2\pi r}\frac{d}{dt}\int_0^r \pi r' B r' dr'$$

On the other hand, $$p = B(r)er \quad (\text{where } r = \text{constant})$$

So, $$\frac{dp}{dt} = er\frac{d}{dt}B(r) = er\,\dot{B}(r)$$

Hence, $$er\,\dot{B}(r) = \frac{e}{2\pi r}\pi r^2 \frac{d}{dt}\langle B\rangle$$

So, $$\dot{B}(r) = \frac{1}{2}\frac{d}{dt}\langle B\rangle$$

This equation is most easily satisfied by taking

$$B(r_0) = \frac{1}{2}\langle B\rangle$$

3.404 The condition, $B(r_0) = \frac{1}{2}\langle B\rangle = \frac{1}{2}\int_0^{r_0} B \cdot 2\pi r \frac{dr}{\pi r_0^2}$

or $$B(r_0) = \frac{1}{r_0^2}\int_0^{r_0} Brdr$$

This gives r_0.

In the present case,

$$B_0 - ar_0^2 = \frac{1}{r_0^2}\int_0^{r_0}(B - ar^2)rdr$$

$$= \frac{1}{2}\left(B_0 - \frac{1}{2}ar_0^2\right)$$

or

$$\frac{3}{4}ar_0^2 = \frac{1}{2}B_0$$

or

$$r_0 = \sqrt{\frac{2B_0}{3a}}$$

3.405 The induced electric field (or eddy current field) is given by

$$E(r) = \frac{1}{2\pi r}\frac{d}{dt}\int_0^r 2\pi r'(r')\,B(r')\,dr'$$

Hence,

$$\frac{dE}{dr} = -\frac{1}{2\pi r^2}\frac{d}{dt}\int_0^r 2\pi r'\,B(r')\,dr' + \frac{dB(r)}{dt}$$

$$= -\frac{1}{2}\frac{d}{dt}<B> + \frac{dB(r)}{dt}$$

This vanishes for $r = r_0$ by the betatron condition, where r_0 is the radius of the equilibrium orbit.

Hence,

$$\frac{dE}{dr} = \dot{B}(r_0) - \frac{1}{2}\dot{B} = 0$$

3.406 From the betatron condition,

$$\frac{1}{2}\frac{d}{dt}<B> = \frac{dB}{dt}(r_0) = \frac{B}{\Delta t}$$

Thus,

$$\frac{d}{dt}<B> = \frac{2B}{\Delta t}$$

and

$$\frac{d\Phi}{dt} = \pi r^2\frac{d<B>}{dt} = \frac{2\pi r^2 B}{\Delta t}$$

So, energy increment per revolution is

$$e\frac{d\Phi}{dt} = \frac{2\pi r^2 eB}{\Delta t} = 0.10\text{ keV}$$

3.407 (a) Even in the relativistic case, we know that $p = Ber$.

Thus, $W = \sqrt{c^2p^2 + m_0^2c^4} - m_0c^2 = m_0c^2(\sqrt{1 + (Ber/m_0c)^2} - 1)$

(b) The distance traversed is

$$2\pi r\frac{W}{e\Phi} = 2\pi r\frac{W}{2\pi r^2 eB/\Delta t} = \frac{W\Delta t}{Ber}$$

on using the result of the previous problem.